# STRUCTURAL HEALTH MONITORING
# OF AEROSPACE COMPOSITES

# STRUCTURAL HEALTH MONITORING OF AEROSPACE COMPOSITES

Victor Giurgiutiu

AMSTERDAM • BOSTON • HEIDELBERG • LONDON
NEW YORK • OXFORD • PARIS • SAN DIEGO
SAN FRANCISCO • SINGAPORE • SYDNEY • TOKYO

Academic Press is an imprint of Elsevier

Academic Press is an imprint of Elsevier
32 Jamestown Road, London NW1 7BY, UK
525 B Street, Suite 1800, San Diego, CA 92101-4495, USA
225 Wyman Street, Waltham, MA 02451, USA
The Boulevard, Langford Lane, Kidlington, Oxford OX5 1GB, UK

**British Library Cataloguing-in-Publication Data**
A catalogue record for this book is available from the British Library.

**Library of Congress Cataloging-in-Publication Data**
A catalog record for this book is available from the Library of Congress.

ISBN: 978-0-12-409605-9

For Information on all Academic Press publications
visit our website at http://store.elsevier.com/

Working together
to grow libraries in
developing countries

www.elsevier.com • www.bookaid.org

*Publisher*: Joe Hayton
*Acquisition Editor*: Carrie Bolger
*Editorial Project Manager*: Charlotte Kent
*Production Project Manager*: Nicky Carter
*Designer*: Greg Harris

Typeset by MPS Limited, Chennai, India
www.adi-mps.com

Printed and bound in the USA

# Dedication

To my Loving and Understanding Family

# Contents

## 6. Piezoelectric Wafer Active Sensors

## 11. Summary and Conclusions

C H A P T E R

# 1

# Introduction

## 1.1 PREAMBLE

The concept of *composites* has attracted the interest of both the engineers and the business professionals. To engineers, composites are the opportunity to create *designer materials* with palettes of properties that cannot be found in existing mineral materials. To the business professional, composites offer unprecedented business growth especially in areas where

*Structural Health Monitoring of Aerospace Composites*
DOI: http://dx.doi.org/10.1016/B978-0-12-409605-9.00001-5

unprecedented material properties are in high demand. Not surprisingly, the aerospace market is one of the largest and arguably the most important to the composites industry. Commercial aircraft, military aircraft, helicopters, business jets, general aviation aircraft, and spacecraft all make substantial use of high-performance composites. The aerospace usage of high-performance composites has experienced a continuously growing over several decades (Figure 1).

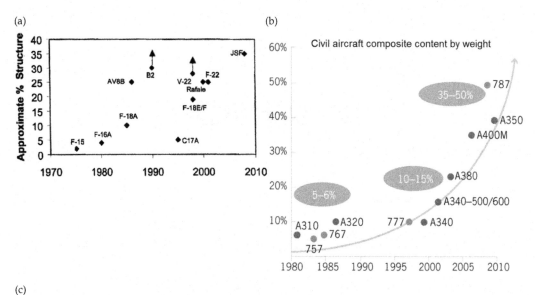

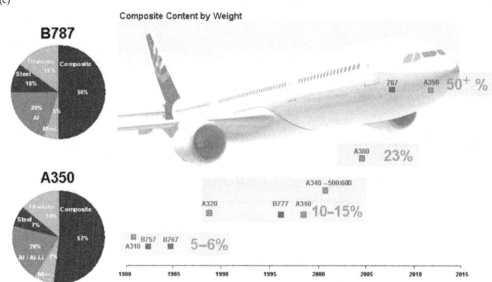

**FIGURE 1** Increase of the weight content of composites in aircraft structures over a 30-year time span: (a) trends in military aircraft composite usage [1]; (b) trends in civil aircraft composite usage [2]; (c) breakdown of weight content by material types in Boeing 787 and Airbus A350 XWB [3].

Composites have good tensile strength and resistance to compression, making them suitable for use in aircraft manufacture. The tensile strength of the material comes from its fibrous nature. When a tensile force is applied, the fibers within the composite line up with the direction of the applied force, giving its tensile strength. The good resistance to compression can be attributed to the adhesive and stiffness properties of the matrix which must maintain the fibers as straight columns and prevent them from buckling.

## 1.2  WHY AEROSPACE COMPOSITES?

The primary needs for all the advanced composites used in aerospace applications remain the same, i.e., lighter weight, higher operating temperatures, greater stiffness, higher reliability, and increased affordability. Some other special needs can be also achieved only with composites, like good radio-frequency compatibility of fiberglass radomes and low-observability airframes for stealth aircraft.

High-performance composites were developed because no single homogeneous structural material could be found that had all of the desired attributes for a given application. Fiber-reinforced composites were developed in response to demands of the aerospace community, which is under constant pressure for materials development in order to achieve improved performance. Aluminum alloys, which provide high strength and fairly high stiffness at low weight, have provided good performance and have been the main materials used in aircraft structures over many years. However, both corrosion and fatigue in aluminum alloys have produced problems that have been very costly to remedy. Fiber-reinforced composites have been developed and widely applied in aerospace applications to satisfy requirements for enhanced performance and reduced maintenance costs.

## 1.3  WHAT ARE AEROSPACE COMPOSITES?

Aerospace composites are a class of engineered materials with a very demanding palette of properties. High strength combined with low weight and also high stiffness are common themes in the aerospace composites world. Nowadays, engineers and scientists thrive to augment these high-performance mechanical properties with other properties such as electric and thermal conductivity, shape change, self-repair capabilities, etc.

### 1.3.1  Definition of Aerospace Composites

From a pure lexical point of view, "composites" seem to have a variety of definitions and there is no completely universal accepted one. One school prefers the word *composite* to include only those materials consisting of a strong structural reinforcement encapsulated in a binding matrix, while the purists believe that the word *composite* should include everything except homogeneous or single-phase materials. In a generic sense, a composite material can be defined as a macroscopic combination of two or more distinct materials, having a recognizable interface between them. One material acts as a supporting matrix, while another material builds on this base scaffolding and reinforces the entire material. Thus, the aerospace definition of composite materials can be restricted to include only those engineered materials that contain a reinforcement (such as fibers or particles) supported by a matrix material.

Fiber-reinforced composites, which dominate the aerospace applications, contain reinforcements having lengths much greater than their thickness or diameter. Most continuous-fiber (or continuous-filament) composites, in fact, contain fibers that are comparable in length to the overall dimensions of the composite part. Composite laminates are obtained through the superposition of several relatively thin layers having two of their dimensions much larger than their third.

High-performance composites are composites that have superior performance compared to conventional structural materials such as steel, aluminum, and titanium. Polymer matrix composites have gained the upper hand in airframe applications, whereas metal matrix composites, ceramic matrix composites, and carbon matrix composites are being considered for more demanding aerospace applications such as aero-engines, landing gear, reentry nose cones, etc. However, there are significant dissimilarities between polymer-matrix composites and those made with metal, ceramic, and carbon matrices. Our emphasis in this book will be on polymer matrix composites for airframe applications.

Polymer matrix composites provide a synergistic combination of high-performance fibers and moldable polymeric matrices. The fiber provides the high strength and modulus while the polymeric matrix spreads the load as well as offers resistance to weathering and corrosion. Composite tensile strength is almost directly proportional to the basic fiber strength, whereas other properties depend on the matrix—fiber interaction. Fiber-reinforced composites are ideally suited to anisotropic loading situations where weight is critical. The high strengths and moduli of these composites can be tailored to the high load direction(s), with little material wasted on needless reinforcement.

## 1.3.2 High-Performance Fibers for Aerospace Composites Applications

Fiber composites offer many superior properties. Almost all high-strength/high-stiffness materials fail because of the propagation of flaws. A fiber of such a material is inherently stronger than the bulk form because the size of a flaw is limited by the small diameter of the fiber. In addition, if equal volumes of fibrous and bulk material are compared, it is found that even if a flaw does produce failure in a fiber, it will not propagate to fail the entire assemblage of fibers, as would happen in the bulk material. Furthermore, preferred orientation may be used to increase the lengthwise modulus, and perhaps strength, well above isotropic values. When this material is also lightweight, there is a tremendous potential advantage in strength-to-weight and/or stiffness-to-weight ratios over conventional materials.

Glass fibers were the first to be considered for high-performance applications because of their high strength when drawn in very thin filaments. Considering that bulk glass is quite brittle, the surprising high strength of these ultra-thin glass fibers gave impetus to this line of research. Subsequently, a variety of other high-performance fibers have been developed: S-glass fibers (which are even stronger that ordinary E glass), aramid (Kevlar) fibers, boron fibers, Spectra fibers, etc.

The fiber that has eventually attained widespread usage in aerospace composites has been the carbon fiber (a.k.a. graphite fiber) that is used in carbon-fiber reinforced polymer (CFRP) composites. High-strength, high-modulus carbon fibers are about 5–6 $\mu m$ in diameter and consist of small crystallites of "turbostratic" graphite, one of the allotropic forms of carbon. Two major carbon-fiber fabrication processes have been developed, one based on polyacrylonitrile (PAN), the other based on pitch. Refinements in carbon-fiber fabrication technology have led to considerable improvements in tensile strength ($\sim$4.5 GPa) and in strain to fracture (more than 2%) for PAN-based fibers. These can now be supplied in three basic forms, high modulus ($\sim$380 GPa), intermediate modulus ($\sim$290 GPa), and high strength (with a modulus of $\sim$230 GPa and tensile strength of 4.5 GPa). The tensile stress–strain response is elastic up to failure, and a large amount of energy is released when the fibers break in a brittle manner. The selection of the appropriate fiber depends very much on the application. For military aircraft, both high modulus and high strength are desirable. Satellite applications, in contrast, benefit from the use of high-modulus fibers that improve stiffness and stability of reflector dishes, antennas, and their supporting structures.

### 1.3.3  High-Performance Matrices for Aerospace Composites Applications

The desirable properties of the reinforcing fibers can be converted to practical application when the fibers are embedded in a matrix that binds them together, transfers load to and between the fibers, and protects them from environments and handling. The polymeric matrices considered for composite applications include both thermosetting polymers (epoxy, polyester, phenolic, polyimide resins) and thermoplastic polymers (polypropylene, Nylon 6.6, polymethylmethacrylate a.k.a. PMMA, polyetheretherketone a.k.a. PEEK). In current aerospace composites, the epoxy thermosetting resin has achieved widespread utilization; however, efforts are under way toward the introduction of thermoplastic polymers which may present considerable manufacturing advantages.

The polymeric matrix of aerospace composites performs a number of functions such as (i) stabilizing the fiber in compression (providing lateral support); (ii) conveying the fiber properties into the laminate; (iii) minimizing damage due to impact by exhibiting plastic deformation and providing out-of-plane properties to the laminate. Matrix-dominated composite properties (interlaminar strength, compressive strength) are reduced when polymer matrix is exposed to higher temperatures or to the inevitable absorption of environmental moisture.

### 1.3.4  Advantages of Composites in Aerospace Usage

The primary advantage of using composite materials in aerospace applications is the weight reduction: weight savings in the range of 20–50% are often quoted. Unitization is another advantage: it is easy to assemble complex components as unitized composite parts using automated layup machinery and rotational molding processes. For example, the single-barrel fuselage concept used in Boeing 787 Dreamliner is a monocoque ("single-shell") molded structure that delivers higher strength at much lower weight.

Aerodynamic benefits can be achieved with composites that were impossible with metals. The majority of aircraft control-lift surfaces have a single degree of curvature due to limitation of metal fabrication techniques. But further improvements in aerodynamic efficiency can be obtained by adopting a double-curvature design, e.g., variable-camber twisted wings. Composites and modern molding tools allow the shape to be tailored to meet the required performance targets at various points in the flying envelope.

The tailoring of mechanical properties along preferential stress directions is an extraordinary design advantage offered by aerospace composites that cannot be duplicated in isotropic metallic airframes. Aerospace composites can be tailored by "layup" design, with tapering thicknesses as needed to maintain optimal strength-to-weight ratio (Figure 2a). In addition, local reinforcing layup can be placed at required orientation at design hot spots.

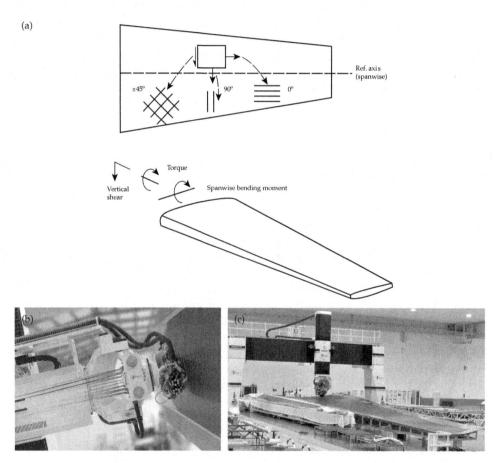

**FIGURE 2**   Unique advantages of using composites in aerospace structures: (a) the concept of strength and stiffness tailoring along major loading directions in an aircraft wing actual wing buildup processes [1]; (b) automated fiber placement (AFP) [4]; (c) advanced tape laying (ATL) [5].

A further advantage of using composites in airplane design is the ability to tailor the aeroelastic behavior to further extend the flying envelope. This tailoring can involve adopting specialized laminate configurations that allow the cross-coupling of flexure and torsion such that wing twist can result from bending and vice versa. Modern analysis techniques allow this process of aeroelastic tailoring, along with strength and dynamic stiffness (flutter) requirements to be performed automatically with a minimum of post-analysis testing and verification

Thermal stability of composites is another advantage that is especially relevant in CFRP composites. The basic carbon fiber has a small negative coefficient of thermal expansion (CTE) which, when combined with the positive CTE of the resin yields the temperature stability of the CFRP composite. This means that CFRP composites do not expand or contract excessively with rapid change in the environmental temperature (as, for example, during the climb from a 90°F runway to $-67$°F at 35,000 ft altitude in a matter of minutes).

Another major advantage of using high-performance composites in aerospace application is that the problems of combined fatigue/corrosion that appear in conventional airframes are virtually eliminated. High-performance polymeric composites do not corrode and the fatigue life of fibrous materials is much higher than that of bulk materials. Nonetheless, environmental effects will eventually affect the matrix polymeric material and some form of fatigue (though different from that of metals) will develop in the composite. However, fracture of composite materials seldom occurs catastrophically without warning as it does in some metallic alloys. In composites, fatigue and fracture is a progressive phenomenon with substantial damage (and the accompanying loss of stiffness) being widely dispersed throughout the material before final failure takes place.

## 1.3.5 Fabrication of Aerospace Composites

Most carbon-fiber composites used in safety-critical primary structures are fabricated by placing uncured layer upon layer of unidirectional plies to achieve the design stacking sequence and orientation requirements. A number of techniques have been developed for the accurate placement of the composite layers in or over a mold, ranging from labor-intensive hand layup techniques to those requiring high capital investment such as automatic fiber placement (AFP, Figure 2b) and in advanced tape laying (ATL, Figure 2c) equipment. Large cylindrical and conical shapes can be obtained through AFP or ATL fabrication over rotating molding mandrels. AFP and ATL machines operate under numerical control and significant effort is being directed laying complicated contoured surfaces.

After been laid up in the mold, the uncured composite is subjected to polymerization by exposure to temperature and pressure. This is usually done in an autoclave, a pressure vessel designed to contain a gas under pressures and fitted with a means of raising the internal temperature to that required to cure the resin. Vacuum bagging is also generally used to assist with removing trapped air and organic vapors from the composite. The process produces structures of low porosity, less than 1%, and high mechanical integrity. Large autoclaves have been installed in the aircraft industry capable of housing complete wing or tail sections.

Alternative lower-cost non-autoclave processing methods are also being investigated such as vacuum molding (VM), resin transfer molding (RTM), vacuum-assisted RTM (VARTM), and resin film infusion (RFI). The vacuum molding processes make use of atmospheric pressure to consolidate the material while curing, thereby obviating the need for an autoclave. The RTM process lays out the fiber reinforcement as a dry preform into a mold and then lets the polymeric resin infiltrate into the preform. The composite systems suitable for vacuum-only processing are cured at 60–120°C and then post-cured typically at 180°C to fully develop the resin properties. The RTM process is assisted by resin temperature fluidization, pumping pressure, and vacuum suction at specific mold vents.

## 1.4 EVOLUTION OF AEROSPACE COMPOSITES

Development of advanced composites for aerospace use has been both costly and potentially risky; therefore, initial development was done by the military where performance is the dominant factor. The Bell-Boeing V-22 Osprey military transport uses 50% composites, whereas Boeing's C-17 military transport has over 7300 kg of structural composites. Helicopter rotor blades and the space program were among the early adopters of composites technology (Figure 3).

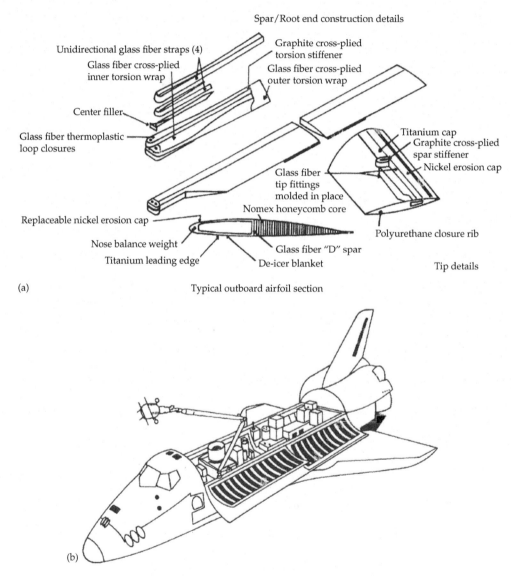

**FIGURE 3**  Early usage of composites in aerospace primary structures: (a) CH-46 helicopter main rotor blade; (b) composite bay-bay doors on the Space Shuttle [6].

As service experience with the use of advanced composites has accumulated, they have started to penetrate into the civilian aerospace usage. Composites have flown on commercial aircraft safety-critical primary structures for more than 30 years, but only recently have they conquered the fuselage, wing-box, and wings. This evolutionary process has recently culminated with the introduction of "all-composite" airliners, the Boeing 787 Dreamliner and the Airbus A350 XWB, which have more than 80% by volume composites in their construction.

Early composite designs were replicas of the corresponding metallic parts and the resulting high production costs jeopardized their initial acceptance. Expensive raw materials ("exotic" fibers and specialty resins) as well as labor-intensive hand layup techniques contributed to these high initial costs. The production cost was further increased by the machining and drilling difficulties since these new fibrous materials behaved radically different than metals under these circumstances. Since this cost is in direct relation to the number of assembled parts, design and manufacturing solutions were sought to reduce the part count and the number of associated fasteners. Automated layup methods, integrally stiffened structures, co-cured or co-bonded of substructures, and the use of honeycomb sandwich solutions have decreased the part count by order of magnitudes while revealing the manufacturing advantages of using composites instead of conventional metals.

## 1.4.1 Early Advances

World War II promoted a need for materials with improved structural properties. In response, fiber-reinforced composites were developed. By the end of the war, fiberglass-reinforced plastics had been used successfully in filament-wound rocket motors and in various other structural applications. These materials were put into broader use in the 1950s, and initially seemed to be the only viable approach available for the elimination the problems of corrosion and crack formation observed in high-performance metallic structures.

## 1.4.2 Composite Growth in the 1960s and 1970s

Although developments in metallic materials have led to some solutions to the crack and corrosion problems, fiber-reinforced composites continued to offer other benefits to designers and manufacturers. The 1960s and 1970s have experience a flurry of research into the development of a variety of advanced fiber for high-performance composites such as boron, S-glass, Spectra fiber, and Kevlar fibers. But the fiber that had eventually captured the market was the carbon fiber (a.k.a. graphite fiber) because of its excellent strength and modulus weight ratios and relative manufacturing ease. However, early industrial implementation of carbon-fiber development was not without surprises as, for example, their unique impact behavior, discovered by Rolls Royce in the 1960s when the innovative RB211 jet engine with carbon-fiber compressor blades failed catastrophically due to bird strikes.

In large commercial aircraft, composites have found application because of the weight considerations that were highlighted by the energy crisis of the 1970s. Spurred by these events, the use of composites in the aerospace industry has increased dramatically since the 1970s. Traditional materials for aircraft construction include aluminum, steel, and titanium. The primary benefits that composite components can offer are reduced weight and assembly simplification. The performance advantages associated with reducing the weight of aircraft structural elements has been the major impetus for military aviation composites development. Although commercial carriers have increasingly been concerned with fuel economy, the potential for reduced production and maintenance costs has proven to be a major factor in the push toward composites. Composites are also being used increasingly as replacements for metal parts and for composite patch repairs on older aircraft.

### 1.4.3 Composites Growth Since the 1980s

Since 1980s, the use of high-performance polymer-matrix fiber composites in aircraft structures has grown steadily, although not as dramatically as initially predicted. This is despite the significant weight-saving and other advantages that advanced composites could provide. One reason for the slower-than-anticipated advancement might be that the aircraft components made of aerospace composites have a higher cost than similar structures made from aerospace metals. Other factors include the high cost of certification of new components and their relatively low resistance to mechanical damage, low through-thickness strength, and (compared with titanium alloys) temperature limitations. Thus, metals have continued to be favored for many airframe applications. CFRP composites have eventually emerged as the most favored advanced composite for aerospace applications. Although the raw material costs of this and similar composites are still relatively high, their advantages over metals in both strength-to-weight ratio, tailored design, and unitized manufacturability are increasingly recognized. Nonetheless, competition remains intense with continuing developments in structural metals such as aluminum alloys: improved toughness and corrosion resistance; new lightweight alloys (such as aluminum lithium); low-cost aerospace-grade castings; mechanical alloying leading to high-temperature alloys; and superplastic forming. For titanium, powder preforms, casting, and superplastic-forming/diffusion bonding are to be mentioned. Advanced joining techniques such as laser and friction stir welding, automated riveting techniques, and high-speed (numerically controlled) machining also make metallic structures more affordable. And the use of hybrid metal−composite combinations (such as the GLARE[1] material used on Airbus A380) which seems to have the best of both worlds also gains popularity with certain designers.

## 1.5 TODAY'S AEROSPACE COMPOSITES

Though the growth has not been as fast as initially predicted, the penetration of high-performance composites into the civilian aerospace has been steady on a continuous upward trend. The drivers for lightweight aircraft structures have continued to push engineers and scientists in looking for unprecedented structural solutions and materials. These major drivers for lightweight structures have been nicely summarized in the 2001 study of the Advisory Council of Aeronautical Research in Europe (ACARE) which identified the aeronautical research needs to be achieved by 2020 [7]. The ACARE goals include: (i) noise reduction to one-half of current average levels; (ii) elimination of noise nuisance outside the airport boundary by quieter aircraft; (iii) a 50% reduction in $CO_2$ emissions per passenger-kilometer (which means a 50% cut in fuel consumption in the new aircraft of

---

[1]GLARE is a proprietary glass-reinforced fiber metal laminate material.

2020); and (iv) an 80% reduction in nitrogen oxide ($NO_X$) emissions. A more detailed vision of the aerospace goals in the 2050 time frame is given in the report "Flightpath 2050: Europe's Vision for Aviation" [8]. Similar requirements have been put forward in the United States and elsewhere. As a result, the civilian aerospace industry is now producing large almost all-composite passenger aircraft like the Boeing 787 Dreamliner and Airbus A350 XWB airliners (Figure 4). These unprecedented engineering achievements have over 80% composites by volume.

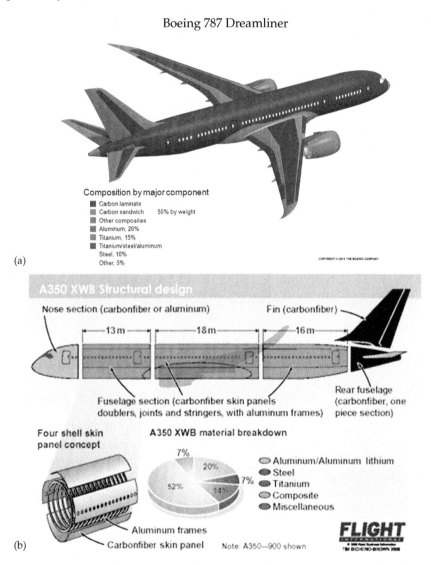

**FIGURE 4**  Composite content of all-composite airliners: (a) Boeing 787 Dreamliner has ∼80% by volume (∼50% by weight) composites [9]; (b) Airbus A350 XWB has ∼83% by volume (∼52% by weight) composites [10]. The lower by-weight ratio is due to the fact that other materials are much heavier than composites.

The main features of the Boeing 787 and Airbus A350 XWB are briefly discussed in the following sections.

### 1.5.1 Boeing 787 Dreamliner

The Boeing 787 Dreamliner (Figure 5) is a family of long-range, midsize wide-body, twin-engine jet airliners that can seat 242–335 passengers in a typical three-class seating configuration. This aircraft, the world's first major commercial airliner to use composite materials as the primary material in its airframe, is Boeing's most fuel-efficient airliner [11]. The Boeing 787 maiden flight took place on December 15, 2009, and completed flight testing in mid-2011. Final Federal Aviation Administration (FAA) and European Aviation Safety Agency (EASA) type certification was received in August 2011 and the first 787-8 model was delivered to All Nippon Airways in September 2011.

**FIGURE 5**   Boeing 787 Dreamliner [12].

The Boeing 787 aircraft is 80% composite by volume. By weight, the material contents is 50% composite, 20% aluminum, 15% titanium, 10% steel, and 5% other [11]. Aluminum is used for the wing and tail leading edges; titanium is used mainly on engines and fasteners, with steel used in various areas.

Each Boeing 787 aircraft contains approximately 32,000 kg of CFRP composites, made with 23 tons of carbon fiber [11]. Composites are used on fuselage, wings, tail, doors, and interior. Boeing 787 fuselage sections are laid up on huge rotating mandrels (Figure 6a). AFP and ATL robotic heads robotically layers of carbon-fiber epoxy resin prepreg to contoured surfaces. Reinforcing fibers are oriented in specific directions to deliver maximum strength along maximum load paths. The fuselage sections are cured in huge autoclaves. The resulting monocoque shell has internal longitudinal stiffeners already built in (Figure 6b,c). This highly integrated structure requires orders of magnitude less fasteners than the conventional built-up airframes. Similar composite manufacturing techniques are applied to the wings.

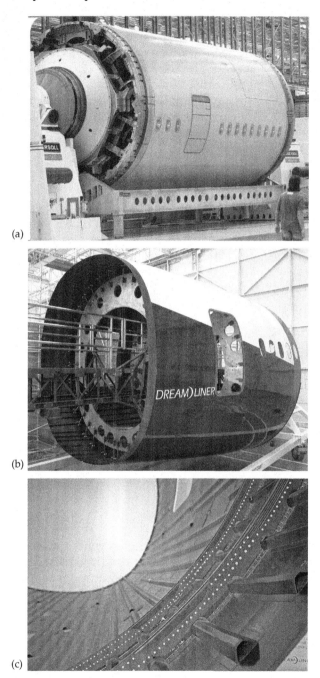

(a)

(b)

(c)

**FIGURE 6**    Composite fuselage of the Boeing 787 Dreamliner: (a) the fuselage barrel is a continuous construction build on a rotating mandrel through automated tape laying [13]; (b) the resulting monocoque shell has internal longitudinal stiffeners already built in [12]; (c) the highly integrated internal structure of the fuselage requires orders of magnitude less fasteners than the conventional built-up airframes [11].

Boeing 787 has composite wings with raked wingtips where the tip of the wing has a higher degree of sweep than the rest of the wing. This aerodynamic design feature improves fuel efficiency and climb performance while shortening takeoff length. It does this in much the same way that winglets do, by increasing the effective aspect ratio of the wing and interrupting harmful wingtip vortices thus decreasing the amount of lift-induced drag experienced by the aircraft. This capability of applying various camber shapes along the wing span as well as a double-curvature configuration is particular to composite wings and cannot be efficiently achieved in metallic wings.

## 1.5.2 Airbus A350 XWB

The Airbus A350 XWB (Figure 7) is a family of long-range, midsize wide-body twin-engine jet airliners that can seat 250–350 passengers in a typical three-class seating configuration. The Airbus A350 XWB maiden flight took place on June 14, 2013. The Airbus A350 XWB received EASA type certification in September 2014 and FAA certification in November 2014. The first Airbus A350 XWB was delivered to Qatar Airways in December 2014 with the first commercial flight in January 2015 [14].

**FIGURE 7**    Airbus A350 XWB [15].

The Airbus A350 XWB airframe includes a range of advanced materials: composites in the fuselage, wings and tail; aluminum–lithium alloys in floor beams, frames, ribs and landing gear bays; titanium alloys in main landing gear supports, engine pylons, and some attachments. The Airbus A350 XWB fuselage section has a four-panel construction such that the major fuselage sections are created by the assembly of four large panels which are joined with longitudinal riveted joints (Figure 8). The fuselage composite panels are mounted on composite fuselage frames. Airbus designers see in this approach a better management of construction tolerances when the jetliner's composite fuselage sections come together on the final assembly. Another perceived benefit of the four-panel concept might be the improved reparability in operational service, as an individual panel can be replaced in the event of significant damage—avoiding major repair work that could require extensive composite patching.

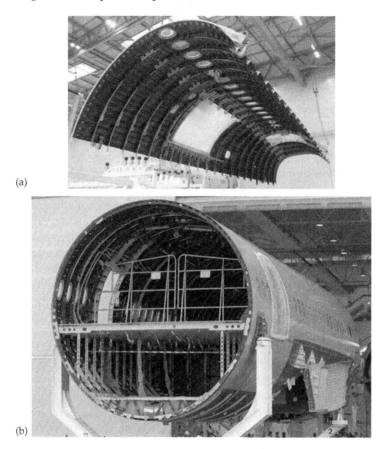

(a)

(b)

FIGURE 8    Airbus A350 XWB four-panel concept: (a) one of the four panels [16]; (b) fuselage assembled from four panels [15].

The Airbus A350 XWB has composite wings with a blended tip winglets thus departing significantly from Airbus's traditional wingtip fences. The wings curve upward over the final 4.4 m in a "sabre-like" shape. This capability of applying various camber shapes along the wing span as well as a double-curvature configuration is particular to composite wings and cannot be efficiently achieved in metallic wings.

## 1.6  CHALLENGES FOR AEROSPACE COMPOSITES

Though greatly popular and very attractive for development, the aerospace composites activity is not without challenges. Some of these challenges could be grouped in safety concerns, not surprisingly since the commercial use of composites in flight-critical primary structures is still at the beginning. Other challenges are related to future developments, where composites are expected to deliver the "unobtainium" material that would make our engineering dreams come true. Both of these challenges are briefly discussed in the following section.

### 1.6.1 Concerns About the Aerospace Use of Composites

Several concerns have been voiced about the aerospace use of composites. One issue that has been raised concerns barely visible damage (BVD), i.e., damage of the composite material that cannot be detected by preflight visual inspection (a routine procedure that identifies dents and other damages on current metallic aircraft). In fact, composite materials may suffer internal damage due to a low-velocity impact (e.g., a tool drop during routine maintenance) without any obvious changes to its surface.

Another often voiced concern is about the fact that the polymeric matrix constituent of the composite materials may collect moisture and change its properties over time. Moisture may also accumulate in matrix microcrack and minor delaminations between the layers of the composite laminate. As the aircraft goes at altitude and temperature drops below freezing, this trapped water would expand and promote further microcracking. Over several flight cycles, the freezing and unfreezing phenomenon will make cracks to expand and eventually cause delamination.

The aircraft designers are well aware of these issues and all necessary measures are being taken to maintain the aircraft safety and integrity. These measures have included extensive testing under accelerated climatic and environmental conditions to ensure that the composite will maintain its integrity over the whole design life of the aircraft. In some cases, these measures may also have included excessive design factors such that considerations other than pure operational stress and strain have been dominant in sizing some composite aircraft parts.

Recent technology has provided a variety of reinforcing fibers and matrices that can be combined to form composites having a wide range of very exceptional properties. In many instances, the sheer number of available material combinations can make selection of materials for evaluation a difficult and almost overwhelming task. In addition, once a material is selected, the choice of an optimal fabrication process can be very complex.

### 1.6.2 The November 2001 Accident of AA Flight 587

The Nov. 2001 Accident of AA flight 587 is one of the worst aviation accidents on US soil resulting in the death of all 260 people aboard the aircraft and five people on the ground [17]. On November 12, 2001, the Airbus A300-600 of American Airlines flight 587 crashed in Queens, New York City, shortly after takeoff [18]. The aircraft vertical stabilizer (tail fin) detached from the aircraft causing the aircraft to crash. The A300-600 vertical stabilizer is connected to the fuselage with six attaching points (Figure 9). Each point has two sets of attachment lugs, one made of composite material, another of aluminum, all connected by a titanium bolt; damage analysis showed that the bolts and aluminum lugs were intact, but not the composite lugs [18]. This, coupled with two events earlier in the life of the aircraft, namely delamination in part of the vertical stabilizer prior to its delivery from Airbus's Toulouse factory and an encounter with heavy turbulence in 1994, caused investigators to examine the use of composites [18].

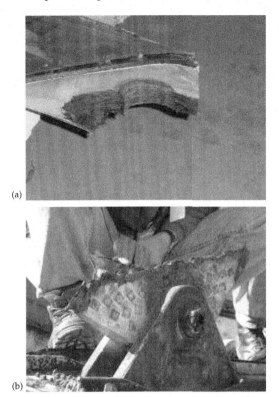

**FIGURE 9**   Composite vertical stabilizer lug (tail fin) broken during the AA flight 587 of Airbus A300-600 in Queens, New York City, November 12, 2001: (a) vertical stabilizer (tail fin) attachment point; (b) close-up of center vertical stabilizer attachment clevis at crash site [19].

The possibility that the composite materials might not be as strong as previously supposed was a cause of concern because they are used in other areas of the plane, including the engine mounting and the wings. Tests carried out on the vertical stabilizers from the accident aircraft, and from another similar aircraft, found that the strength of the composite material had not been compromised, and the National Transportation Safety Board (NTSB) concluded that the material had failed because it had been stressed beyond its design limit, despite 10 previous recorded incidents where A300 tail fins had been stressed beyond their design limitation in which none resulted in the separation of the vertical stabilizer in-flight [18,20].

## 1.6.3 Fatigue Behavior of Composite Materials

Whereas aerospace metals, such as aluminum, have a well-known fatigue behavior, the composites fatigue life is much more complicated and less understood. The aerospace metallic materials have been extensively studied and their fatigue behavior is well understood by now. The situation is drastically different in the case of composites. The fatigue life of a metallic aircraft part can be directly deduced from two basic ingredients: (i) knowledge of the aluminum fatigue data and (ii) knowledge of cyclic stress distribution. In the case of metallic materials, the ingredient (i) i.e., the material fatigue data, is well known and easily accessible. In the case of composites, the ingredient (i), i.e., the composite fatigue data, is far from being universally understood. In fact, the fatigue behavior of a composite material depends not only on that of its constituent fibers and matrix, but also on the layup sequence and hence it may vary from part to part. This observation explains in simple terms why the aerospace composites fatigue still remains a very fruitful research topic.

Nonetheless, the basic fatigue superiority of composites over the metals exists due to the fact that a fibrous material is less susceptible to catastrophic failure than a conventional metal. So far, the aircraft designers have relied on extensive certification tests and procedures to ensure that the composite materials used in their designs have an adequate fatigue behavior such that the aircraft safety is always ensured. However, these certification procedures are lengthy and expensive; for that reason, the introduction of other composite solutions is currently somehow retarded and certain conservatism exists with the tendency of using only the solutions that have been already certified and approved. This situation will persist until a better way of designing and in-service monitoring of aerospace composites is implemented.

## 1.6.4 The Future of Composites in Aerospace

When it comes to aerospace, composite materials are here to stay. With ever increasing fuel costs and environmental regulations, commercial flying remains under sustained pressure to improve performance, and weight reduction is a key factor for achieving this goal. With their excellent strength-to-weight ratio, advanced composites are an obvious choice. Beyond the day-to-day operating costs, the aircraft maintenance programs are a heavy burden on the airline budgets. Aircraft maintenance can be simplified by the reduction of component count and elimination of corrosion issues. Again, composites are the obvious choice. The competitive nature of the aircraft construction business ensures that any opportunity to reduce operating costs is explored and exploited wherever possible. Competition also exists in the military, with continuous pressure to increase payload and range, flight performance characteristics, and "survivability," in both airplanes and missiles.

Composite technology continues to advance, and the advent of new fiber and matrix types as well as new manufacturing techniques is certain to accelerate and extend composite usage. Several research and development areas are of special interest to both scientists and engineers.

One technological shortcoming that requires a solution is the elimination of mechanical fasteners from the composite assemblies. At present, even the Boeing 787 and the Airbus A350 XWB still use thousands of mechanical fasteners during assembly. Why not use adhesive bonding? Because, in order to ensure safety, current certification requirements mandate that proof must be made that each and every *adhesively bonded* joint will not separate and cause structural failure should it reach its critical design load. Using mechanical fasteners is still the easiest and least expensive way to meet certification requirements. However, the full realization of cost and weight savings through composite materials will only be attained if our scientific understanding and technical trust in bonded joints reaches the point where certification can be attained without additional fasteners.

One of the most exciting upcoming opportunities for aerospace composites is in the commercial space flight arena. For example, the Virgin Galactic LLC air-launch space travel concepts consider all-composite solutions consisting of a space vehicle (VSS Enterprise) being launched at altitude from a carrier aircraft (the White Knight) [21].

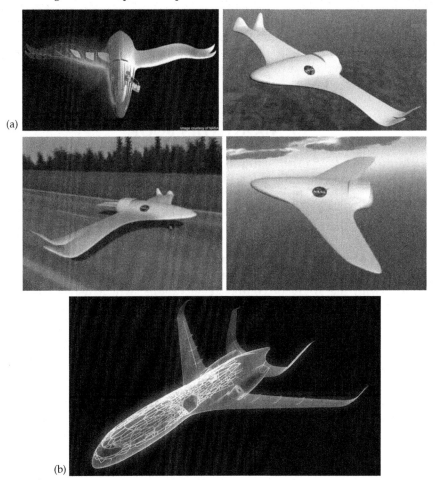

**FIGURE 10**   Composites-enabled future aircraft: (a) NASA morphing aircraft program aims at changing the aircraft shape as needed by various flight profiles [22,23]; (b) artist rendering of Airbus future composite aircraft concept [24].

Future composite aircraft are envisaged to be able to change their shape as required by the flight regime in which they operate. Figure 10a shows several artist renderings resulting from NASA morphing aircraft program [22,23]; one notices that the straight wide-span double wings are needed for short takeoff and landing morph into a single swept-wing required for high-speed flight, as well as the appearance of individual winglets as required for fast maneuvers. Thus, a morphing aircraft would be able to change its shape as needed by various flight profiles in which it has to operate.

Figure 10b presents a futuristic aircraft concept originating from one of the major aircraft manufacturers [24]. Besides special aerodynamic contours that are only possible through the use of composites, this aircraft concept also displays a network of sensors and interconnects that, similar to the animal nervous system, would be able to collect data about the aircraft state of health, relay it to a central unit, and advise appropriate corrective actions and/or future maintenance scheduling. Such an aircraft nervous is also depicted in Figure 10b, with the additional proviso that self-repairing aerospace composites are being considered in order to restore the composite aircraft to its full initial capability.

## 1.7  ABOUT THIS BOOK

This book addresses the field of structural health monitoring (SHM) and presents a review of the principal means and methods for SHM of aerospace composite structures. This very challenging issue is addressed in a step-by-step way such as the readers will become gradually aware of all the aspects of the problem as they progress through the book. The present introductory chapter has given an overview of why and how composites are used in the aerospace industry.

The next three chapters dealt with the analysis of composite behavior and response. Chapter 2 is dedicated to the discussion of fundamental aspect of composite materials. Chapter 3 deals with the study of composite vibrations, which may be used for SHM applications. Chapter 4 is dedicated to the study of wave propagation in thin-wall composite structures. Attention was focused on the propagation of ultrasonic guided waves in laminated composites which is an essential element of several SHM techniques discussed in subsequent chapters.

Chapter 5 presents a review of damage and failure in aerospace composites. The chapter starts with a discussion of basic failure mechanisms and then advances through the treatment of various lamination and loading options. The tension damage and failure of a unidirectional composite is discussed first. Then, tension damage and failure in a cross-ply composite laminate is considered. Linear and nonlinear aspects are discussed starting with the ply discount method and continuing with the discussion of matrix cracking, interfacial stresses, local interface failure, delamination, etc. The characteristic damage state (CDS) concept is introduced and the evolutionary decrease of stiffness with damage accumulation is discussed. The next section of Chapter 5 deals with the analysis of fatigue damage and the long-term behavior of aerospace composites. This is followed by the discussion of compression fatigue and failure which is fundamentally different from the tension behavior. The presentation of other composite damage types such as fastener hole damage, impact damage, and damage specific to sandwich composites and adhesive joints is done next. Chapter 5 ends with a discussion of what could and/or or should be detected by a permanently installed SHM system and what should be expected to be detected by the nondestructive inspection (NDI), nondestructive testing (NDT), and non-destructive evaluation (NDE) processes during composite fabrication and during scheduled maintenance events.

Chapters 6–8 treat the subject of sensors that could be used for SHM of aerospace composites. Chapter 6 deals with piezoelectric sensors, in particular the surface-mounted piezoelectric wafer active sensors, a.k.a. PWAS. Chapter 7 covers fiber-optic sensors for SHM of aerospace composites. The chapter starts with a cursory review of the major fiber optics sensing principles, but attention is subsequently focused on fiber Bragg gratings (FBG) optical sensors, which have become the dominant fiber-optic sensing technology in SHM applications. The fabrication of FBG sensors, the conditioning equipment used with FBG sensors, and the FBG demodulation at ultrasonic frequencies, which is particularly important for SHM applications, are discussed. Chapter 7 also covers other optical sensor types with good potential for application in composite SHM system; of these, the distributed optical fiber sensing based on optical time domain reflectometry and Rayleigh backscatter are found of particular interest. Chapter 8 covers other sensors that could be used in aerospace composites SHM. Of particular interest are the conventional resistance strain gages and the electrical properties sensors.

Chapters 9 and 10 discuss methods for monitoring damage initiation and growth in aerospace composites. Chapter 9 focuses on the detection of impact and acoustic emission (AE) and on monitoring impact damage intensity and growth in aerospace composites. This area of research has received extensive attention because of the drastic possible effects of undetected BVD events. For impact monitoring, it was found that the passive sensing diagnostics (PSD) approach can tell if an impact has taken place, locate the impact position, and even estimate the impact intensity. Another aspect of the PSD methodology is that AE events could also be detected with the same sensor network installation. Simultaneous measurement of both AE and impact waves is also possible. The next part of Chapter 9 is dedicated to the discussion of active sensing diagnostics (ASD) in which the transducers installed on the composite structure are used to send interrogative wave signals that interact with the damaged area and produce scatter waves that are picked up by same and/or other transducers. The ASD methodology, which has similarities to acousto-ultrasonics, has been used with piezo transmitters and piezo receivers, as well as with piezo transmitters and fiber-optic receivers. Other methods for impact detection discussed in Chapter 9 include direct methods for impact damage detection, strain mapping method for damage detection, vibration SHM methods, frequency transfer methods, and local area sensing with the electromechanical impedance spectroscopy method. Also presented in Chapter 9 are electrical and electromagnetic methods used for impact damage detection in aerospace composites. The detection of delaminations with the electrical resistance method in CFRP composites was found to have received extensive attention in the SHM community. This approach is specific to CFRP composites because their carbon fibers have electrical conductivity which is imparted to the overall composite through the fact that individual fibers embedded in the polymeric matrix make occasional contact when bunched together in the composite system. This conductivity is changed by impact damage and delaminations.

Chapter 10 covers the monitoring of fatigue damage of aerospace composites that may appear during normal in-service operation of the composite aircraft or spacecraft. Such damage may be due to operational loads and environmental factors and may result in a gradual degradation of composite properties rather than the sudden changes that may appear due to accidental events, like the impact damage discussed in the previous chapter. Chapter 10 starts with a discussion of passive SHM methods, e.g., monitoring strain, acoustic emission, and operational loads that were used in damage monitoring. The next major section of Chapter 10 dealt with monitoring the actual fatigue damage induced in the composite by repeated application of service loads. Various passive SHM and active SHM methods for fatigue monitoring are discussed including fiber-optic measurements, pitch-catch piezo measurements, and electrochemical impedance spectroscopy. The electrical resistance method is used to monitor in service degradation and fatigue of CFRP composites. Early tests have shown that as the CFRP material is fatigue loaded, its electrical resistance changes thus acting as a built-in indicator of microcracks, delamination, and other fatigue damage taking place in the CFRP composite. The last section of Chapter 10 covers a variety of methods used in the monitoring of disbonds and delamination in composite patch repairs, composite adhesive joints, in nonconductive GFRP composites, etc. Guided-wave measurements as well as dielectric measurements are presented and briefly discussed.

A major conclusion of Chapter 10 is that monitoring of aerospace composites damage that may appear during normal operation is possible, but this damage type has not received as much attention as the monitoring of impact events and of resulting damage as discussed in Chapter 9. However, this situation may change in the future as major airlines are acquiring and entering into service new composite-intensive aircraft such as Boeing 787 Dreamliner and Airbus A350 XWB. Hence, the need for fundamental and applied research into developing SHM methods for monitoring the in-service degradation and fatigue of aerospace composites is likely to increase significantly. It is apparent that sustained research programs for developing such methods and technologies should be put into place such that the fruits of discovery are made available before dramatic events happen into practice.

Overall, it can be said that, whereas NDE technology is a rather mature technology, the SHM methodology and related technologies for aerospace composites have only just started to emerge. Considerable further research is needed to mature the development of aerospace composites SHM sensors and methods in order to achieve viable practical implementation of this promising new technology.

The introduction of SHM systems and SHM methodology should contribute to lower the maintenance costs of composite structures for which deterministic damage events, types, and limit sizes are difficult to predict. SHM would also facilitate the introduction of new composite materials and development of new composite structures by reducing the uncertainty component of the aircraft design cycle.

# References

[1] Baker, A.; Dutton, S.; Kelly, D. (2004) *Composite Materials for Aircraft Structures*, 2nd ed., AIAA Education Series, American Institute of Aeronautics and Astronautics, Reston, VA.

[2] Cookson, I. (2009) "Grant Thornton on US Aerospace Component M&A, 2008", *Defense Industry Daily*, March 15, 2009, http://www.defenseindustrydaily.com/Grant-ThorntononUS-Aerospace-Component-MA-2008-05334/ (accessed Nov. 2014) and http://www.grantthornton.com/staticfiles/GTCom/files/Industries/Consumer%20&%20industrial%20products/Publications/Aerospace%20components%20MA%20update%202009.pdf (accessed Dec. 2014).

[3] Anon. "Composites Penetration—Step Change Underway with Intermediate Modulus Carbon Fiber as the Standard", Hexcel Corporation, http://www.sec.gov/Archives/edgar/data/717605/000110465908021748/g97851bci012.jpg.

[4] Anon. (2014) "TORRESFIBERLAYUP—Automatic Fiber Placement Machine", M. Torres Group, http://www.mtorres.es/en/aeronautics/products/carbon-fiber/torresfiberlayup#sthash.qgLshFMe.dpuf, picture 4 (accessed Dec. 2014).

[5] TORRESLAYUP—Automatic Tape Layer Machine, M. Torres Group, http://www.mtorres.es/en/aeronautics/products/carbon-fiber/torreslayuppicture#10 (accessed Dec. 2014).

[6] Anon. (1989) "Composites II: Material Selection and Applications", Materials Engineering Institute course 31, ASM International, 9639 Kinsman Rd., Materials Park, OH.

[7] Anon. (2001) "European Aeronautics: A Vision for 2020—Report of a Group of Personalities", Advisor Council for Aviation Research and Innovation in Europe, http://www.acare4europe.org/documents/vision-2020 (accessed Dec. 2014).

[8] Anon. (2011) "Flightpath 2050: Europe's Vision for Aviation", Report of the High Level Group on Aviation Research, European Commission, 2011, http://ec.europa.eu/transport/modes/air/doc/flightpath2050.pdf (accessed Jan. 2015).

[9] Hale, J. (2006) "Boeing 787 from the Ground Up", *Boeing Aero Magazine*, Issue 24, 2006, Q-04, http://www.boeing.com/commercial/aeromagazine/articles/qtr_4_06/index.html (accessed Dec. 2014).

[10] Kinsley-Jones, M. (2006) "Airbus's A350 vision takes shape—Flight takes an in-depth look at the new twinjet", *Flight International* 12 Dec. 2006, http://www.flightglobal.com/news/articles/airbus39s-a350-vision-takes-shape-flight-takes-an-in-depth-look-at-the-new-211028/ (accessed Dec. 2014).

[11] Anon. (2014) "Boeing 787 Dreamliner", Wikipedia article http://en.wikipedia.org/wiki/Boeing_787_Dreamliner.

[12] Anon. (2014) "Boeing image gallery", The Boeing Company, http://www.boeing.com/boeing/companyoffices/gallery/images/commercial/787/index1.page.

[13] Norris, G.; Wagner, M. (2009) *Boeing 787 Dreamliner*, ISBN 0760328153, Zenith Press, www.zenithpress.com, MBI Pub. Co., Minneapolis, MN.

[14] Anon. (2014) "Airbus A350 XWB", Wikipedia article, https://en.wikipedia.org/wiki/Airbus_A350_XWB.

[15] Anon. (2014) "Airbus Photo gallery" http://www.airbus.com/galleries/photo-gallery/filter/a350-xwb-family/photo A350-MSN2-003.

[16] Anon. (2012) "A350 XWB Intelligent and aerodynamic airframe", Airbus International, http://www.a350xwb.com/advanced/fuselage/ (accessed Dec. 2014).

[17] Anon. (2004) "In-Flight Separation of Vertical Stabilizer American Airlines Flight 587 Airbus Industries A300-605R, N14053 Belle Harbor, New York November 12, 2001", Aircraft Accident Report NTSB/AAR-04-04, NTIS number PB2004-910404, Oct. 26, 2004.

[18] Anon. (2014) "American Airlines Flight 587", Wikipedia, http://en.wikipedia.org/wiki/American_Airlines_Flight_587.

[19] Anon. (2002) "NTSB schedules public investigative hearing on crash of American Airlines flight 587", for immediate release NTSB publication SB-02-31, file ntsb_020919.

[20] Anon. (2013) "S03E08 Plane Crash in Queens: American Airlines Flight 587", http://secondsfromdisaster.net/s03e08-plane-crash-in-queens-american-airlines-flight-587/.

[21] Anon. (2010) "Virgin Galactic's VSS Enterprise makes first manned glide flight", *High-Performance Composites*, Nov. 2010, page 11.

[22] Stories by Williams (2014) "The Future of Flight: Morphing Wings", http://storiesbywilliams.com/2014/02/01/the-future-of-flight-morphing-wings/, images 1 and 6 (accessed Dec. 2014).

[23] Anon. (2014) "NASA Morphing Aircraft Movie", https://www.youtube.com/watch?v = vR3T8mdpdTI (accessed Dec. 2014).

[24] Bowler, T. (2014) "Carbon fibre planes: lighter and stronger by design", BBC News Business, 27 Jan. 2014, http://www.bbc.com/news/business-25833264.

# 2.1 INTRODUCTION

Aerospace composite materials are made of high-strength fibers embedded in a polymeric matrix. Glass-fiber-reinforced polymer (GFRP), carbon-fiber-reinforced polymer (CFRP), and Kevlar-fiber-reinforced polymer (KFRP) are among the most common aerospace composite materials.

　Aerospace composite structures are obtained through the overlapping of several unidirectional layers with various angle orientations as required by the stacking sequence. Thus, we distinguish a stack of laminae (a.k.a. plies) bonded together to act as an integral structural element. Each ply (a.k.a. lamina) may have its own orientation $\theta$ with respect to a global system of axes $x - y$ (Figure 1). The information about the orientation of all the plies in the laminate is contained in the stacking sequence. For example, $[0/90/45/-45]_s$ signifies a laminate made of 0°, 90°, and ±45° plies placed in a sequence that is symmetric about the laminate mid-surface, i.e., 0°, 90°, +45°, −45°, −45°, +45°, 90°, 0°. This laminate has $N = 8$ plies and its stacking vector is

$$[\theta] = \begin{bmatrix} 0° & 90° & +45° & -45° & -45° & +45° & 90° & 0° \end{bmatrix}^t \tag{1}$$

The plies in the stacking sequence may be of same composite material (e.g., CFRP) or of different materials (e.g., some CFRP, some GFRP, others KFRP, etc.).

(a)

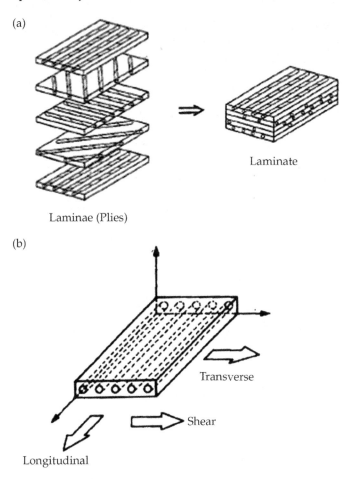

Laminate

Laminae (Plies)

(b)

Transverse

Shear

Longitudinal

**FIGURE 1**    Composite laminates: (a) layup made up of a stack of composite laminae (a.k.a. plies) with various orientations $\theta$; (b) longitudinal, transverse, and shear definitions in a lamina (ply) [1].

The question that composites lamination theory has to answer could be stated as follows: "Given a certain stacking sequence and a set of external loads, what is the structural response of the composite laminate?" In order to address this question, we need to analyze the mechanics of the composite laminate: first we would analyze the local mechanics of an individual layer (a.k.a. lamina) and then apply a stacking analysis (lamination theory) to determine the global properties of the laminated composite and its response under load.

## 2.2  ANISOTROPIC ELASTICITY

This section recalls some basic definitions and relations that are essential for the analysis of anisotropic elastic structures such as aerospace composites.

## 2.2.1 Basic Notations

$$\frac{\partial}{\partial x}() = (\cdot)' \quad \text{and} \quad \frac{\partial}{\partial t}() = (\dot{\cdot}) \tag{2}$$

$$\delta_{ij} = \begin{cases} 1 & \text{if } i = j \\ 0 & \text{otherwise} \end{cases} \quad \text{(Kronecker delta)} \tag{3}$$

$$()_{ii} = ()_{11} + ()_{22} + ()_{33} \quad \text{(Einstein implied summation)} \tag{4}$$

$$()_{i,j} = \frac{\partial ()_i}{\partial x_j} \quad \text{(differentiation shorthand)} \tag{5}$$

## 2.2.2 Stresses—The Stress Tensor

In $x_1 x_2 x_3$ notations, the stress tensor is defined as

$$\sigma_{ij} \quad\quad i,j = 1,2,3 \quad\quad \text{(stress tensor)} \tag{6}$$

where the first index indicates the surface on which the stress acts and the second index indicates the direction of the stress; thus, $\sigma_{ij}$ signifies the stress on the surface of normal $\vec{e}_i$ acting in the direction $\vec{e}_j$. The strain tensor is symmetric, i.e.,

$$\sigma_{ji} = \sigma_{ij} \quad\quad i,j = 1,2,3 \quad\quad \text{(symmetry of stress tensor in } x_1 x_2 x_3 \text{ notations)} \tag{7}$$

The stress tensor can be represented in an array form as

$$[\sigma_{ij}] = \begin{bmatrix} \sigma_{11} & \sigma_{12} & \sigma_{13} \\ \sigma_{13} & \sigma_{22} & \sigma_{23} \\ \sigma_{13} & \sigma_{23} & \sigma_{33} \end{bmatrix} \tag{8}$$

The array in Eq. (8) was written with the symmetry properties of Eq. (7) already included. In $xyz$ notations, Eq. (6) is written as

$$\sigma_{ij} \quad\quad i,j = x,y,z \tag{9}$$

Hence, Eq. (8) becomes

$$[\sigma_{ij}] = \begin{bmatrix} \sigma_{xx} & \sigma_{xy} & \sigma_{xz} \\ \sigma_{xy} & \sigma_{yy} & \sigma_{yz} \\ \sigma_{xz} & \sigma_{yz} & \sigma_{zz} \end{bmatrix} \tag{10}$$

The stress symmetry in $xyz$ notations is expressed as

$$\sigma_{yx} = \sigma_{xy} \quad\quad \sigma_{zy} = \sigma_{yz} \quad\quad \sigma_{zx} = \sigma_{xz} \quad\quad \text{(symmetry of stress tensor in } xyz \text{ notations)} \tag{11}$$

## 2.2.3 Strain–Displacement Relations—The Strain Tensor

In $x_1x_2x_3$ notations, the strain tensor is defined as

$$\varepsilon_{ij} = \tfrac{1}{2}(u_{i,j} + u_{j,i}) \quad i,j = 1,2,3 \tag{12}$$

where the differentiation shorthand Eq. (5) was used. In longhand, Eq. (12) is written as

$$\varepsilon_{ij} = \frac{1}{2}\left(\frac{\partial u_i}{\partial x_j} + \frac{\partial u_j}{\partial x_i}\right) \quad i,j = 1,2,3 \tag{13}$$

It is apparent that Eqs. (12), (13) are symmetric, i.e.,

$$\varepsilon_{ji} = \varepsilon_{ij}, \quad i,j = 1,2,3 \quad \text{(symmetry of strain tensor in } x_1x_2x_3 \text{ notations)} \tag{14}$$

The strain tensor can be represented in an array form as

$$\left[\varepsilon_{ij}\right] = \begin{bmatrix} \varepsilon_{11} & \varepsilon_{12} & \varepsilon_{13} \\ \varepsilon_{12} & \varepsilon_{22} & \varepsilon_{23} \\ \varepsilon_{13} & \varepsilon_{23} & \varepsilon_{33} \end{bmatrix} \tag{15}$$

In $xyz$ notations, Eq. (15) is written as

$$\left[\varepsilon_{ij}\right] = \begin{bmatrix} \varepsilon_{xx} & \varepsilon_{xy} & \varepsilon_{xz} \\ \varepsilon_{xy} & \varepsilon_{yy} & \varepsilon_{yz} \\ \varepsilon_{xz} & \varepsilon_{yz} & \varepsilon_{zz} \end{bmatrix} \tag{16}$$

which includes the symmetry relations in $xyz$ notations, i.e.,

$$\varepsilon_{yx} = \varepsilon_{xy}, \quad \varepsilon_{zy} = \varepsilon_{yz}, \quad \varepsilon_{xz} = \varepsilon_{zx} \quad \text{(symmetry of strain tensor in } xyz \text{ notations)} \tag{17}$$

The elements of Eq. (16) are obtained from the expansion of Eq. (12), i.e.,

$$\varepsilon_{xx} = \frac{\partial u_x}{\partial x} \qquad \varepsilon_{xy} = \frac{1}{2}\left(\frac{\partial u_x}{\partial y} + \frac{\partial u_y}{\partial x}\right)$$

$$\varepsilon_{yy} = \frac{\partial u_y}{\partial y} \qquad \varepsilon_{yz} = \frac{1}{2}\left(\frac{\partial u_y}{\partial z} + \frac{\partial u_z}{\partial y}\right) \tag{18}$$

$$\varepsilon_{zz} = \frac{\partial u_z}{\partial z} \qquad \varepsilon_{zx} = \frac{1}{2}\left(\frac{\partial u_z}{\partial x} + \frac{\partial u_x}{\partial z}\right)$$

## 2.2.4 Stress–Strain Relations

### 2.2.4.1 Stiffness Tensor; Compliance Tensor

The stress–strain displacement in tensor notations are written as, i.e.,

$$\sigma_{ij} = c_{ijkl}\varepsilon_{kl} \qquad i,j = 1,2,3 \tag{19}$$

where the Einstein implied summation rule of Eq. (4) applies. The term $c_{ijkl}$ is called the *stiffness tensor*, i.e.,

$$c_{ijkl} \qquad i,j,k,l = 1,2,3 \qquad \text{(stiffness tensor)} \qquad (20)$$

It can be shown that the stiffness tensor is symmetric, i.e.,

$$c_{ijkl} = c_{jikl} = c_{ijlk} = c_{klij} = c_{jilk} \qquad i,j,k,l = 1,2,3 \qquad (21)$$

The symmetry properties of the stiffness tensor Eq. (21) imply that of the 81 stiffness elements only 21 are independent. (For isotropic materials, these 21 independent stiffness constants reduce to only two that can be related to two independent elastic properties of the isotropic material, e.g., the moduli $E, G$ or the modulus $E$ and the Poisson constant $\nu$.)

The inverse of the stress–strain relation given by Eq. (19) is the strain–stress relation given by

$$\varepsilon_{ij} = s_{ijkl}\sigma_{kl} \qquad i,j = 1,2,3 \qquad (22)$$

where the term $s_{ijkl}$ is called the *compliance tensor*. The compliance tensor enjoys the same symmetry properties as the stiffness tensor, i.e.,

$$s_{ijkl} = s_{jikl} = s_{ijlk} = s_{klij} = s_{jilk} \qquad i,j,k,l = 1,2,3 \qquad (23)$$

### 2.2.4.2  From Tensor Notations to Voigt Matrix Notation

The stiffness tensor of Eq. (20) has four indices, i.e., it is a fourth order tensor and it cannot be represented in a plane array. To overcome this difficulty, Voigt matrix notation has been introduced to allow the stiffness tensor to be written as a matrix. This Voigt matrix notation consists of replacing $ij$ or $kl$ by $p$ or $q$, where $i,j,k,l = 1,2,3$ and $p,q = 1,2,3,4,5,6$ according to Table 1.

TABLE 1  Conversion from tensor to matrix indices for the Voigt notation

| $ij$ or $kl$ | $p$ or $q$ |
| --- | --- |
| 11 | 1 |
| 22 | 2 |
| 33 | 3 |
| 23 or 32 | 4 |
| 31 or 13 | 5 |
| 12 or 21 | 6 |

The transition from tensor notations to Voigt matrix notations is achieved as follows: Recall Eq. (19) giving the stress–strain displacement in tensor notations, i.e.,

$$\sigma_{ij} = c_{ijkl}\varepsilon_{kl} \qquad i,j = 1,2,3 \qquad (24)$$

Expand the implied summations in Eq. (24) to get

$$\sigma_{ij} = c_{ij11}\varepsilon_{11} + c_{ij12}\varepsilon_{12} + c_{ij13}\varepsilon_{13}$$
$$+ c_{ij21}\varepsilon_{21} + c_{ij22}\varepsilon_{22} + c_{ij23}\varepsilon_{23} \qquad i,j = 1,2,3 \qquad (25)$$
$$+ c_{ij31}\varepsilon_{31} + c_{ij32}\varepsilon_{32} + c_{ij33}\varepsilon_{33}$$

Recall Eq. (12) that expresses strains $\varepsilon_{kl}$ in terms of displacements, i.e.,

$$\varepsilon_{kl} = \tfrac{1}{2}(u_{k,l} + u_{l,k}) \qquad k,l = 1,2,3 \qquad (26)$$

Note the symmetry property of Eq. (26), i.e.,

$$\varepsilon_{lk} = \tfrac{1}{2}(u_{l,k} + u_{k,l}) = \tfrac{1}{2}(u_{k,l} + u_{l,k}) = \varepsilon_{kl} \quad k,l = 1,2,3 \qquad (27)$$

Recall Eq. (21) giving the symmetry properties of the stiffness sensor, i.e.,

$$c_{ijkl} = c_{jikl} = c_{ijlk} = c_{klij} = c_{jilk} \qquad (28)$$

Apply Eqs. (27), (28) into Eq. (25) to get

$$\sigma_{ij} = c_{ij11}\varepsilon_{11} + c_{ij12}\varepsilon_{12} + c_{ij31}\varepsilon_{31}$$
$$+ c_{ij12}\varepsilon_{12} + c_{ij22}\varepsilon_{22} + c_{ij23}\varepsilon_{23} \qquad i,j = 1,2,3 \qquad (29)$$
$$+ c_{ij31}\varepsilon_{31} + c_{ij23}\varepsilon_{23} + c_{ij33}\varepsilon_{33}$$

Collect on common terms and rearrange Eq. (29) as

$$\sigma_{ij} = c_{ij11}\varepsilon_{11} + c_{ij22}\varepsilon_{22} + c_{ij33}\varepsilon_{33} + 2c_{ij23}\varepsilon_{23} + 2c_{ij31}\varepsilon_{31} + 2c_{ij12}\varepsilon_{12} \quad i,j = 1,2,3 \qquad (30)$$

The order of indices used in writing Eq. (30) has followed the common practice associated with Voigt matrix notation. Now, let $i,j = 1,2,3$ and express Eq. (30) in matrix form, i.e.,

$$
\begin{Bmatrix} \sigma_{11} \\ \sigma_{22} \\ \sigma_{33} \\ \sigma_{23} \\ \sigma_{31} \\ \sigma_{12} \end{Bmatrix} =
\begin{bmatrix}
c_{1111} & c_{1122} & c_{1133} & c_{1123} & c_{1131} & c_{1112} \\
c_{2211} & c_{2222} & c_{2233} & c_{2223} & c_{2231} & c_{2212} \\
c_{3311} & c_{3322} & c_{3333} & c_{3323} & c_{3331} & c_{3312} \\
c_{2311} & c_{2322} & c_{2333} & c_{2323} & c_{2331} & c_{2312} \\
c_{3111} & c_{3122} & c_{3133} & c_{3123} & c_{3131} & c_{3112} \\
c_{1211} & c_{1222} & c_{1233} & c_{1223} & c_{1231} & c_{1212}
\end{bmatrix}
\begin{Bmatrix} \varepsilon_{11} \\ \varepsilon_{22} \\ \varepsilon_{33} \\ 2\varepsilon_{23} \\ 2\varepsilon_{31} \\ 2\varepsilon_{12} \end{Bmatrix}
\qquad (31)
$$

Recall Eq. (28) giving the symmetry properties of the stiffness tensor and write the matrix in Eq. (31) as a symmetric matrix, i.e.,

$$
\begin{Bmatrix} \sigma_{11} \\ \sigma_{22} \\ \sigma_{33} \\ \sigma_{23} \\ \sigma_{31} \\ \sigma_{12} \end{Bmatrix} =
\begin{bmatrix}
c_{1111} & c_{1122} & c_{1133} & c_{1123} & c_{1131} & c_{1112} \\
c_{1122} & c_{2222} & c_{2233} & c_{2223} & c_{2231} & c_{2212} \\
c_{1133} & c_{2233} & c_{3333} & c_{3323} & c_{3331} & c_{3312} \\
c_{1123} & c_{2223} & c_{3323} & c_{2323} & c_{2331} & c_{2312} \\
c_{1131} & c_{2231} & c_{3331} & c_{2331} & c_{3131} & c_{3112} \\
c_{1112} & c_{2212} & c_{3312} & c_{2312} & c_{3112} & c_{1212}
\end{bmatrix}
\begin{Bmatrix} \varepsilon_{11} \\ \varepsilon_{22} \\ \varepsilon_{33} \\ 2\varepsilon_{23} \\ 2\varepsilon_{31} \\ 2\varepsilon_{12} \end{Bmatrix}
\qquad (32)
$$

Use Voigt notations to rename the components of the stress and strain tensor by using a single index $p = 1, ..., 6$ and the indexing rule of Table 1, i.e.,

$$\sigma = \begin{Bmatrix} \sigma_1 \\ \sigma_2 \\ \sigma_3 \\ \sigma_4 \\ \sigma_5 \\ \sigma_6 \end{Bmatrix} = \begin{Bmatrix} \sigma_{11} \\ \sigma_{22} \\ \sigma_{33} \\ \sigma_{23} \\ \sigma_{31} \\ \sigma_{12} \end{Bmatrix} \quad \text{(stress matrix in Voigt notations)} \tag{33}$$

$$\varepsilon = \begin{Bmatrix} \varepsilon_1 \\ \varepsilon_2 \\ \varepsilon_3 \\ \varepsilon_4 \\ \varepsilon_5 \\ \varepsilon_6 \end{Bmatrix} = \begin{Bmatrix} \varepsilon_{11} \\ \varepsilon_{22} \\ \varepsilon_{33} \\ 2\varepsilon_{23} \\ 2\varepsilon_{31} \\ 2\varepsilon_{12} \end{Bmatrix} \quad \text{(strain matrix in Voigt notations)} \tag{34}$$

Using the indexing rule of Table 1, denote the elements of the $6 \times 6$ symmetric matrix of Eq. (32) as follows:

$$\begin{bmatrix} c_{1111} \rightarrow C_{11} & c_{1122} \rightarrow C_{12} & c_{1133} \rightarrow C_{13} & c_{1123} \rightarrow C_{14} & c_{1131} \rightarrow C_{15} & c_{1112} \rightarrow C_{16} \\ & c_{2222} \rightarrow C_{22} & c_{2233} \rightarrow C_{23} & c_{2223} \rightarrow C_{24} & c_{2231} \rightarrow C_{25} & c_{2212} \rightarrow C_{26} \\ & & c_{3333} \rightarrow C_{33} & c_{3323} \rightarrow C_{34} & c_{3331} \rightarrow C_{35} & c_{3312} \rightarrow C_{36} \\ & & & c_{2323} \rightarrow C_{44} & c_{2331} \rightarrow C_{45} & c_{2312} \rightarrow C_{46} \\ & sym. & & & c_{3131} \rightarrow C_{55} & c_{3112} \rightarrow C_{56} \\ & & & & & c_{1212} \rightarrow C_{66} \end{bmatrix} \tag{35}$$

### 2.2.4.3 Stiffness Matrix

The $6 \times 6$ matrix of Eq. (35) is known as the *stiffness matrix*, i.e.,

$$C = \begin{bmatrix} C_{11} & C_{12} & C_{13} & C_{14} & C_{15} & C_{16} \\ C_{12} & C_{22} & C_{23} & C_{24} & C_{25} & C_{26} \\ C_{13} & C_{23} & C_{33} & C_{34} & C_{35} & C_{36} \\ C_{14} & C_{24} & C_{34} & C_{44} & C_{45} & C_{46} \\ C_{15} & C_{25} & C_{35} & C_{45} & C_{55} & C_{56} \\ C_{16} & C_{26} & C_{36} & C_{46} & C_{56} & C_{66} \end{bmatrix} \quad \text{(stiffness matrix)} \tag{36}$$

Using Eqs. (33), (36) into Eq. (32) yields the compact stress–strain matrix relation, i.e.,

$$\sigma = C \varepsilon \quad \text{(compact stress–strain matrix relation)} \tag{37}$$

In expanded form, the stress–strain matrix relation of Eq. (37) can be written as

$$
\begin{Bmatrix} \sigma_{11} \\ \sigma_{22} \\ \sigma_{33} \\ \sigma_{23} \\ \sigma_{31} \\ \sigma_{12} \end{Bmatrix} = \begin{bmatrix} C_{11} & C_{12} & C_{13} & C_{14} & C_{15} & C_{16} \\ C_{12} & C_{22} & C_{23} & C_{24} & C_{25} & C_{26} \\ C_{13} & C_{23} & C_{33} & C_{34} & C_{35} & C_{36} \\ C_{14} & C_{24} & C_{34} & C_{44} & C_{45} & C_{46} \\ C_{15} & C_{25} & C_{35} & C_{45} & C_{55} & C_{56} \\ C_{16} & C_{26} & C_{36} & C_{46} & C_{56} & C_{66} \end{bmatrix} \begin{Bmatrix} \varepsilon_{11} \\ \varepsilon_{22} \\ \varepsilon_{33} \\ 2\varepsilon_{23} \\ 2\varepsilon_{31} \\ 2\varepsilon_{12} \end{Bmatrix} \quad \text{(matrix stress–strain relation)} \quad (38)
$$

### 2.2.4.4 Compliance Matrix

The stress–strain relation Eq. (37) can be inverted to give the strain–stress relation, i.e.,

$$\varepsilon = S\sigma \quad \text{(compact strain–stress matrix relation)} \quad (39)$$

where $S$ is the compliance matrix, i.e.,

$$
S = \begin{bmatrix} S_{11} & S_{12} & S_{13} & S_{14} & S_{15} & S_{16} \\ S_{12} & S_{22} & S_{23} & S_{24} & S_{25} & S_{26} \\ S_{13} & S_{23} & S_{33} & S_{34} & S_{35} & S_{36} \\ S_{14} & S_{24} & S_{34} & S_{44} & S_{45} & S_{46} \\ S_{15} & S_{25} & S_{35} & S_{45} & S_{55} & S_{56} \\ S_{16} & S_{26} & S_{36} & S_{46} & S_{56} & S_{66} \end{bmatrix} \quad \text{(compliance matrix)} \quad (40)
$$

The stiffness matrix $C$ and compliance matrix $S$ are mutually inverse, i.e.,

$$S = C^{-1} \quad \text{and} \quad C = S^{-1} \quad \text{(mutually-inverse stiffness and compliance matrices)} \quad (41)$$

Equation (41) permits the computation of stiffness matrix from a known compliance matrix, i.e.,

$$
\begin{bmatrix} C_{11} & C_{12} & C_{13} & C_{14} & C_{15} & C_{16} \\ C_{12} & C_{22} & C_{23} & C_{24} & C_{25} & C_{26} \\ C_{13} & C_{23} & C_{33} & C_{34} & C_{35} & C_{36} \\ C_{14} & C_{24} & C_{34} & C_{44} & C_{45} & C_{46} \\ C_{15} & C_{25} & C_{35} & C_{45} & C_{55} & C_{56} \\ C_{16} & C_{26} & C_{36} & C_{46} & C_{56} & C_{66} \end{bmatrix} = \begin{bmatrix} S_{11} & S_{12} & S_{13} & S_{14} & S_{15} & S_{16} \\ S_{12} & S_{22} & S_{23} & S_{24} & S_{25} & S_{26} \\ S_{13} & S_{23} & S_{33} & S_{34} & S_{35} & S_{36} \\ S_{14} & S_{24} & S_{34} & S_{44} & S_{45} & S_{46} \\ S_{15} & S_{25} & S_{35} & S_{45} & S_{55} & S_{56} \\ S_{16} & S_{26} & S_{36} & S_{46} & S_{56} & S_{66} \end{bmatrix}^{-1} \quad (42)
$$

In expanded form, Eqs. (39), (41) are written as

$$\begin{Bmatrix} \varepsilon_{11} \\ \varepsilon_{22} \\ \varepsilon_{33} \\ 2\varepsilon_{23} \\ 2\varepsilon_{31} \\ 2\varepsilon_{12} \end{Bmatrix} = \begin{bmatrix} S_{11} & S_{12} & S_{13} & S_{14} & S_{15} & S_{16} \\ S_{12} & S_{22} & S_{23} & S_{24} & S_{25} & S_{26} \\ S_{13} & S_{23} & S_{33} & S_{34} & S_{35} & S_{36} \\ S_{14} & S_{24} & S_{34} & S_{44} & S_{45} & S_{46} \\ S_{15} & S_{25} & S_{35} & S_{45} & S_{55} & S_{56} \\ S_{16} & S_{26} & S_{36} & S_{46} & S_{56} & S_{66} \end{bmatrix} \begin{Bmatrix} \sigma_{11} \\ \sigma_{22} \\ \sigma_{33} \\ \sigma_{23} \\ \sigma_{31} \\ \sigma_{12} \end{Bmatrix} \quad \text{(matrix strain–stress relation)} \quad (43)$$

### 2.2.4.5 Stress–Strain Relations for an Isotropic Material

In the spirit of completeness, the stress–strain matrix relation for an isotropic material is presented next; these relations are commonly known as Hooke's law. For an isotropic material, the stress–strain relations of Eq. (38) simplify to

$$\begin{Bmatrix} \sigma_{11} \\ \sigma_{22} \\ \sigma_{33} \\ \sigma_{23} \\ \sigma_{31} \\ \sigma_{12} \end{Bmatrix} = \begin{bmatrix} \lambda+2\mu & \lambda & \lambda & 0 & 0 & 0 \\ \lambda & \lambda+2\mu & \lambda & 0 & 0 & 0 \\ \lambda & \lambda & \lambda+2\mu & 0 & 0 & 0 \\ 0 & 0 & 0 & \mu & 0 & 0 \\ 0 & 0 & 0 & 0 & \mu & 0 \\ 0 & 0 & 0 & 0 & 0 & \mu \end{bmatrix} \begin{Bmatrix} \varepsilon_{11} \\ \varepsilon_{22} \\ \varepsilon_{33} \\ 2\varepsilon_{23} \\ 2\varepsilon_{31} \\ 2\varepsilon_{12} \end{Bmatrix} \quad \begin{pmatrix} \text{Hooke's Law:} \\ \text{isotropic stress–strain} \\ \text{relation} \end{pmatrix} \quad (44)$$

where $\lambda, \mu$ are the Lame constants given by

$$\lambda = \frac{\nu}{(1+\nu)(1-2\nu)}E$$

$$\mu = G = \frac{1}{2(1+\nu)}E \qquad \text{(Lame constants)} \qquad (45)$$

The corresponding strain–stress matrix relation for an isotropic material is

$$\begin{Bmatrix} \varepsilon_{11} \\ \varepsilon_{22} \\ \varepsilon_{33} \\ 2\varepsilon_{23} \\ 2\varepsilon_{31} \\ 2\varepsilon_{12} \end{Bmatrix} = \begin{bmatrix} 1/E & -\nu/E & -\nu/E & 0 & 0 & 0 \\ -\nu/E & 1/E & -\nu/E & 0 & 0 & 0 \\ -\nu/E & -\nu/E & 1/E & 0 & 0 & 0 \\ 0 & 0 & 0 & 1/G & 0 & 0 \\ 0 & 0 & 0 & 0 & 1/G & 0 \\ 0 & 0 & 0 & 0 & 0 & 1/G \end{bmatrix} \begin{Bmatrix} \sigma_{11} \\ \sigma_{22} \\ \sigma_{33} \\ \sigma_{23} \\ \sigma_{31} \\ \sigma_{12} \end{Bmatrix} \quad \begin{pmatrix} \text{isotropic} \\ \text{strain–stress} \\ \text{relation} \end{pmatrix} \quad (46)$$

where $G = E/2(1 + \nu)$. In view of Eqs. (44), (46), we identify the isotropic stiffness and compliance matrices as

$$
\mathbf{C}^{isotropic} = \begin{bmatrix} \lambda + 2\mu & \lambda & \lambda & 0 & 0 & 0 \\ \lambda & \lambda + 2\mu & \lambda & 0 & 0 & 0 \\ \lambda & \lambda & \lambda + 2\mu & 0 & 0 & 0 \\ 0 & 0 & 0 & \mu & 0 & 0 \\ 0 & 0 & 0 & 0 & \mu & 0 \\ 0 & 0 & 0 & 0 & 0 & \mu \end{bmatrix} \quad \text{(isotropic stiffness matrix)} \quad (47)
$$

$$
\mathbf{S}^{isotropic} = \begin{bmatrix} 1/E & -\nu/E & -\nu/E & 0 & 0 & 0 \\ -\nu/E & 1/E & -\nu/E & 0 & 0 & 0 \\ -\nu/E & -\nu/E & 1/E & 0 & 0 & 0 \\ 0 & 0 & 0 & 1/G & 0 & 0 \\ 0 & 0 & 0 & 0 & 1/G & 0 \\ 0 & 0 & 0 & 0 & 0 & 1/G \end{bmatrix} \quad \text{(isotropic compliance matrix)} \quad (48)
$$

The elements of the matrices in Eqs. (47), (48) are related through Eq. (45).

### 2.2.5  Equation of Motion in Terms of Stresses

The equation of motion in terms of stresses as obtained from the free-body analysis of an infinitesimal element $dx_1\, dx_2\, dx_3$ is

$$
\sigma_{ij,j} = \rho \ddot{u}_i \qquad i, j = 1, 2, 3 \tag{49}
$$

where the differentiation shorthand Eq. (5) was used. In longhand, Eq. (49) is written as

$$
\frac{\partial \sigma_{ij}}{\partial x_j} = \rho \ddot{u}_i \qquad i, j = 1, 2, 3 \tag{50}
$$

### 2.2.6  Equation of Motion in Terms of Displacements

Substitution of the stress–strain relation Eq. (24) into the right-hand side of the stress equation of motion Eq. (49) yields

$$
\sigma_{ij,j} = c_{ijkl} \varepsilon_{kl,j} \qquad i, j = 1, 2, 3 \tag{51}
$$

Recall the strain–displacement relations Eq. (12), i.e.,

$$
\varepsilon_{kl} = \tfrac{1}{2}(u_{k,l} + u_{l,k}) \qquad i, j = 1, 2, 3 \tag{52}
$$

where the subscripts $kl$ are used instead of $ij$ for convenience. Differentiation of Eq. (52) gives

$$
\varepsilon_{kl,j} = \tfrac{1}{2}(u_{k,lj} + u_{l,kj}) \qquad i, j = 1, 2, 3 \tag{53}
$$

Substitution of Eq. (53) into Eq. (51) yields

$$\sigma_{ij,j} = \tfrac{1}{2} c_{ijkl}(u_{k,lj} + u_{l,kj}) \qquad i,j = 1,2,3 \tag{54}$$

In virtue of its symmetry properties, Eq. (54) can be further simplified to give

$$\sigma_{ij,j} = c_{ijkl} u_{k,lj} \qquad i,j = 1,2,3 \tag{55}$$

*Proof of Eq. (55):* Recall the right-hand side of Eq. (54) and expand it as

$$\tfrac{1}{2} c_{ijkl}(u_{k,lj} + u_{l,kj}) = \tfrac{1}{2} c_{ijkl} u_{k,lj} + \tfrac{1}{2} c_{ijkl} u_{l,kj} \tag{56}$$

Consider the second term in Eq. (56) expansion and interchange $k$ with $l$ to write

$$\tfrac{1}{2} c_{ijkl} u_{l,kj} \underset{k \leftrightarrow l}{=} \tfrac{1}{2} c_{ijlk} u_{k,lj} \tag{57}$$

Recall Eq. (21) giving the symmetry properties of the stiffness matrix, i.e., $c_{ijlk} = c_{ijkl}$; hence, the second part of Eq. (57) can be written as

$$\tfrac{1}{2} c_{ijlk} u_{k,lj} = \tfrac{1}{2} c_{ijkl} u_{k,lj} \tag{58}$$

Combining Eqs. (57), (58), one gets

$$\tfrac{1}{2} c_{ijkl} u_{l,kj} = \tfrac{1}{2} c_{ijkl} u_{k,lj} \tag{59}$$

Substitution of Eq. (59) into Eq. (56) yields two identical terms in the right-hand side summation which can be summed up to cancel the 1/2 factor, i.e.,

$$\tfrac{1}{2} c_{ijkl}(u_{k,lj} + u_{l,kj}) = \tfrac{1}{2} c_{ijkl} u_{k,lj} + \tfrac{1}{2} c_{ijkl} u_{l,kj} = \tfrac{1}{2} c_{ijkl} u_{k,lj} + \tfrac{1}{2} c_{ijkl} u_{k,lj} = c_{ijkl} u_{k,lj} \tag{60}$$

Substitution of Eq. (60) into Eq. (54) yields Eq. (55). QED[1]

Now, substitution of Eq. (55) into Eq. (49) yields

$$c_{ijkl} u_{k,lj} = \rho \ddot{u}_i \quad i,j = 1,2,3 \quad \text{(tensor equation of motion in terms of displacements)} \tag{61}$$

Equation (61) is the tensor equation of motion in terms of displacements.

A matrix form of the equation of motion in terms of displacements can be obtained by expanding Eq. (61) and changing from tensor notations to Voigt notations using the rules of Table 1. However, the resulting expressions are rather long, as illustrated below for the first line in Eq. (61), i.e.,

$$C_{11}u_{1,11} + C_{12}u_{2,21} + C_{13}u_{3,31} + C_{16}(u_{1,21} + u_{2,11}) + C_{14}(u_{2,31} + u_{3,21}) + C_{15}(u_{1,31} + u_{3,11}) +$$

$$C_{16}u_{1,12} + C_{26}u_{2,22} + C_{36}u_{3,32} + C_{66}(u_{1,22} + u_{2,12}) + C_{46}(u_{2,32} + u_{3,22}) + C_{56}(u_{1,32} + u_{3,12}) +$$

$$C_{15}u_{1,13} + C_{25}u_{2,23} + C_{35}u_{3,33} + C_{56}(u_{1,23} + u_{2,13}) + C_{45}(u_{2,33} + u_{3,23}) + C_{55}(u_{1,33} + u_{3,13}) = \rho \ddot{u}_1 \tag{62}$$

---

[1] *quod erat demonstrandum*, http://en.wikipedia.org/wiki/Q.E.D, www.merriam-webster.com/dictionary/qed.

Equation (62) can be somehow simplified by reverting the order of differentiation in certain places and grouping by displacements, i.e.,

$$C_{11}u_{1,11} + C_{66}u_{1,22} + C_{55}u_{1,33} + 2C_{16}u_{1,12} + 2C_{56}u_{1,23} + C_{15}u_{1,31} +$$

$$C_{16}u_{2,11} + C_{26}u_{2,22} + C_{45}u_{2,33} + (C_{12} + C_{66})u_{2,12} + (C_{25} + C_{46})u_{2,23} + (C_{14} + C_{56})u_{2,31} + \qquad (63)$$

$$C_{15}u_{3,11} + C_{46}u_{3,22} + C_{35}u_{3,33} + (C_{14} + C_{56})u_{3,12} + (C_{36} + C_{45})u_{3,23} + (C_{13} + C_{55})u_{3,31} = \rho\ddot{u}_1$$

The longhand matrix notation of Eq. (61) is not recommended for use in the general case. However, a simplified version of Eq. (63) will be used in the case of monoclinic stiffness formulation, as shown later in Section 5.3.

## 2.3 UNIDIRECTIONAL COMPOSITE PROPERTIES

This section discusses the basic properties of a unidirectional lamina. These properties are usually determined experimentally. Micromechanics models will also be used to develop estimates of the composite elastic properties based on the elastic properties of their constituents.

### 2.3.1 Elastic Constants of a Unidirectional Composite

Aerospace composites are made of high-strength fibers embedded in a polymeric matrix. A unidirectional composite material consists of tightly packed high-stiffness fibers embedded in a rigid polymeric matrix. Such unidirectional composite material has orthotropic elastic properties: it is very stiff along the fibers and rather compliant across the fibers (Figure 2). They are also in-plane isotropic, i.e., properties in a plane transverse to the fibers do not depend on the direction in which they are measured. A unidirectional composite is defined by five independent elastic constants:

(1) $E_L$ or $E_1$, the longitudinal modulus measured along the fibers
(2) $E_T$ or $E_2$, the transverse modulus measured across the fibers
(3) $\nu_{LT}$ or $\nu_{12}$, the LT Poisson ratio measuring the transverse contraction when longitudinal stress is applied, i.e., $\nu_{LT} = -\varepsilon_T/\varepsilon_L$
(4) $G_{LT}$ or $G_{12}$, the shear modulus for LT shear in the plane of the lamina
(5) $G_{23}$, a.k.a. the interlaminar shear modulus [2], corresponds to shear taking place in a plane that is transverse to the fibers. The corresponding Poisson ratio, $\nu_{23}$, a.k.a. the interlaminar Poisson ratio, is related to $G_{23}$ through the formula $G_{23} = E_T/2(1 + \nu_{23})$; hence, only one of them can be independently defined.

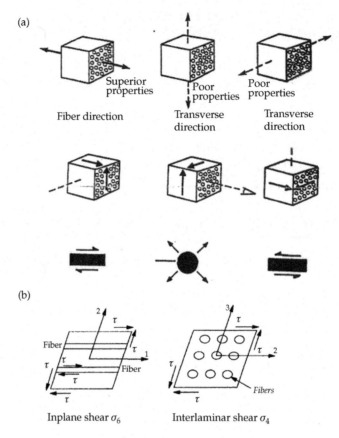

**FIGURE 2** (a) Overview of composite properties (stiff along fibers, compliant across fibers and in matrix shear) [3]; (b) definition of in-plane shear $\sigma_{LT}$ and interlaminar shear $\sigma_{23}$ [2].

## 2.3.2 Compliance Matrix of a Unidirectional Composite

The strain−stress matrix relation for a unidirectional composite is written as

$$
\begin{Bmatrix} \varepsilon_{11} \\ \varepsilon_{22} \\ \varepsilon_{33} \\ 2\varepsilon_{23} \\ 2\varepsilon_{31} \\ 2\varepsilon_{12} \end{Bmatrix} = \begin{bmatrix} 1/E_L & -\nu_{LT}/E_L & -\nu_{LT}/E_L & 0 & 0 & 0 \\ -\nu_{LT}/E_L & 1/E_T & -\nu_{23}/E_T & 0 & 0 & 0 \\ -\nu_{LT}/E_L & -\nu_{23}/E_T & 1/E_T & 0 & 0 & 0 \\ 0 & 0 & 0 & 1/G_{23} & 0 & 0 \\ 0 & 0 & 0 & 0 & 1/G_{LT} & 0 \\ 0 & 0 & 0 & 0 & 0 & 1/G_{LT} \end{bmatrix} \begin{Bmatrix} \sigma_{11} \\ \sigma_{22} \\ \sigma_{33} \\ \sigma_{23} \\ \sigma_{31} \\ \sigma_{12} \end{Bmatrix} \quad \begin{pmatrix} \text{strain−stress} \\ \text{matrix} \\ \text{relation} \end{pmatrix}
$$

(64)

The relation described by Eq. (64) represents an *orthotropic transversely isotropic* strain–stress relation. This relation is specific to unidirectional composites and is associated with their physical nature. The salient features of this relation are as follows:

- Direct stresses are decoupled from shear stress (orthotropic property)
- Shear stresses are decoupled from each other (orthotropic property)
- 12 and 13 matrix elements are the same (in-plane isotropic property)
- 55 and 66 matrix elements are the same (in-plane isotropic property)

From Eq. (64), we deduce that the compliance matrix $\mathbf{S}$ for a unidirectional composite is given by

$$\mathbf{S} = \begin{bmatrix} 1/E_L & -\nu_{LT}/E_L & -\nu_{LT}/E_L & 0 & 0 & 0 \\ -\nu_{LT}/E_L & 1/E_T & -\nu_{23}/E_T & 0 & 0 & 0 \\ -\nu_{LT}/E_L & -\nu_{23}/E_T & 1/E_T & 0 & 0 & 0 \\ 0 & 0 & 0 & 1/G_{23} & 0 & 0 \\ 0 & 0 & 0 & 0 & 1/G_{LT} & 0 \\ 0 & 0 & 0 & 0 & 0 & 1/G_{LT} \end{bmatrix} \begin{pmatrix} \text{compliance matrix} \\ \text{of} \\ \text{unidirectional} \\ \text{composite lamina} \end{pmatrix}$$

(65)

In generic notations, Eq. (65) can be written as

$$\mathbf{S} = \begin{bmatrix} S_{11} & S_{12} & S_{13} & 0 & 0 & 0 \\ S_{12} & S_{22} & S_{23} & 0 & 0 & 0 \\ S_{13} & S_{23} & S_{33} & 0 & 0 & 0 \\ 0 & 0 & 0 & S_{44} & 0 & 0 \\ 0 & 0 & 0 & 0 & S_{55} & 0 \\ 0 & 0 & 0 & 0 & 0 & S_{66} \end{bmatrix} \begin{pmatrix} \text{compliance matrix} \\ \text{of} \\ \text{unidirectional} \\ \text{composite lamina} \end{pmatrix}$$

(66)

where

$$S_{11} = \frac{1}{E_L} \qquad\qquad S_{23} = -\frac{\nu_{23}}{E_T}$$

$$S_{12} = S_{13} = -\frac{\nu_{LT}}{E_L} \qquad S_{44} = \frac{1}{G_{23}}$$

(67)

$$S_{22} = S_{33} = \frac{1}{E_T} \qquad S_{55} = S_{66} = \frac{1}{G_{LT}}$$

Equation (66) displays the same *orthotropic in-plane isotropic* features as Eq. (64), i.e.,

- Direct stresses are decoupled from shear stresses (orthotropic property)
- Shear stresses are decoupled from each other (orthotropic property)
- $S_{12}$ and $S_{13}$ matrix elements are the same (transversely isotropic property)
- $S_{55}$ and $S_{66}$ matrix elements are the same (transversely isotropic property)

The corresponding strain–stress matrix relation is

$$
\begin{Bmatrix} \varepsilon_{11} \\ \varepsilon_{22} \\ \varepsilon_{33} \\ 2\varepsilon_{23} \\ 2\varepsilon_{31} \\ 2\varepsilon_{12} \end{Bmatrix} = \begin{bmatrix} S_{11} & S_{12} & S_{13} & 0 & 0 & 0 \\ S_{12} & S_{22} & S_{23} & 0 & 0 & 0 \\ S_{13} & S_{23} & S_{33} & 0 & 0 & 0 \\ 0 & 0 & 0 & S_{44} & 0 & 0 \\ 0 & 0 & 0 & 0 & S_{55} & 0 \\ 0 & 0 & 0 & 0 & 0 & S_{66} \end{bmatrix} \begin{Bmatrix} \sigma_{11} \\ \sigma_{22} \\ \sigma_{33} \\ \sigma_{23} \\ \sigma_{31} \\ \sigma_{12} \end{Bmatrix} \quad \begin{pmatrix} \text{strain–stress} \\ \text{matrix relation} \\ \text{of} \\ \text{unidirectional} \\ \text{composite lamina} \end{pmatrix} \quad (68)
$$

## 2.3.3 Stiffness Matrix of a Unidirectional Composite

The stiffness matrix of a unidirectional composite $\mathbf{C}$ is obtained by inversion of the compliance matrix $\mathbf{S}$, i.e.,

$$
\mathbf{C} = \mathbf{S}^{-1} \quad \text{(stiffness matrix = inverse of compliance matrix)} \quad (69)
$$

The longhand expression of the stiffness matrix is

$$
\mathbf{C} = \begin{bmatrix} C_{11} & C_{12} & C_{13} & 0 & 0 & 0 \\ C_{12} & C_{22} & C_{23} & 0 & 0 & 0 \\ C_{13} & C_{23} & C_{33} & 0 & 0 & 0 \\ 0 & 0 & 0 & C_{44} & 0 & 0 \\ 0 & 0 & 0 & 0 & C_{55} & 0 \\ 0 & 0 & 0 & 0 & 0 & C_{66} \end{bmatrix} = \begin{bmatrix} S_{11} & S_{12} & S_{13} & 0 & 0 & 0 \\ S_{12} & S_{22} & S_{23} & 0 & 0 & 0 \\ S_{13} & S_{23} & S_{33} & 0 & 0 & 0 \\ 0 & 0 & 0 & S_{44} & 0 & 0 \\ 0 & 0 & 0 & 0 & S_{55} & 0 \\ 0 & 0 & 0 & 0 & 0 & S_{66} \end{bmatrix}^{-1} \quad (70)
$$

The corresponding stress–strain matrix relation is

$$
\begin{Bmatrix} \sigma_{11} \\ \sigma_{22} \\ \sigma_{33} \\ \sigma_{23} \\ \sigma_{31} \\ \sigma_{12} \end{Bmatrix} = \begin{bmatrix} C_{11} & C_{12} & C_{13} & 0 & 0 & 0 \\ C_{12} & C_{22} & C_{23} & 0 & 0 & 0 \\ C_{13} & C_{23} & C_{22} & 0 & 0 & 0 \\ 0 & 0 & 0 & C_{44} & 0 & 0 \\ 0 & 0 & 0 & 0 & C_{55} & 0 \\ 0 & 0 & 0 & 0 & 0 & C_{66} \end{bmatrix} \begin{Bmatrix} \varepsilon_{11} \\ \varepsilon_{22} \\ \varepsilon_{33} \\ 2\varepsilon_{23} \\ 2\varepsilon_{31} \\ 2\varepsilon_{12} \end{Bmatrix} \quad \text{(stress–strain matrix relation)} \quad (71)
$$

The stiffness matrix $\mathbf{C}$ of Eq. (70) has the same *orthotropic in-plane isotropic* features as the compliance matrix $\mathbf{S}$ of Eq. (66), i.e.,

- Direct stresses are decoupled from shear stresses (orthotropic property)
- Shear stresses are decoupled from each other (orthotropic property)
- $C_{12}$ and $C_{13}$ matrix elements are the same (in-plane isotropic property)
- $C_{55}$ and $C_{66}$ matrix elements are the same (in-plane isotropic property)

## 2.3.4 Estimation of Elastic Constants from the Constituent Properties

The elastic constants of a unidirectional composite described in Section 3.1 can be estimated from the properties of the composite constituents, i.e., from the elastic properties of the fiber and the matrix (Figure 3). The estimation is done through the rule of mixtures using the contribution ratios measured as volume fractions: $v_f$ for the fibers a $v_m$ for the matrix. In addition, since composite manufacturing cannot be perfect, a certain amount of voids and trapped gases are always present and represented by the void volume fraction $v_v$. The balance equation for the volume fractions is

$$v_f + v_m + v_v = 1 \quad \text{(volume fraction balance equation)} \tag{72}$$

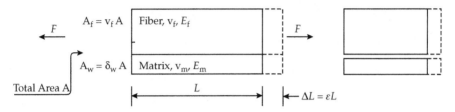

**FIGURE 3**   Schematic diagram for the estimation of the longitudinal modulus $E_L$.

### 2.3.4.1 Estimation of the Longitudinal Modulus $E_L$

Consider a unidirectional composite of length $L$ under longitudinal load $F$ applied over a total cross-sectional area $A$ as indicated in Figure 4. The corresponding proportional areas of the fiber and the matrix are

$$A_f = v_f A \quad \text{(area of the fibers)} \tag{73}$$

$$A_m = v_m A \quad \text{(area of the matrix)} \tag{74}$$

The fibers and matrix stretch together with the same strain $\varepsilon_L$. They act as parallel springs and their respective forces add whereas their stretch is the same. The stresses in the fibers and matrix are

$$\sigma_f = E_f \varepsilon_L \quad \text{(stress in the fibers)} \tag{75}$$

$$\sigma_m = E_m \varepsilon_L \quad \text{(stress in the matrix)} \tag{76}$$

where $E_f, E_m$ are the fiber and matrix elastic moduli. The total force is obtained by adding the forces from the fibers and the matrix, i.e.,

$$F = \sigma_f A_f + \sigma_m A_m = (\sigma_f v_f + \sigma_m v_m)A \quad \text{(total force)} \tag{77}$$

Substitution of Eqs. (75), (76) into Eq. (77) gives

$$F = (E_f v_f + E_m v_m)\varepsilon_L A \tag{78}$$

The same force $F$ is obtained by considering the effective modulus of the composite $E_L$ multiplied by the strain $\varepsilon_L$ and by the total area $A$, i.e.,

$$F = E_L \varepsilon_L A \tag{79}$$

Combining Eqs. (78), (79) yields the formula for calculating the longitudinal modulus $E_L$ as function of fiber and matrix moduli $E_f, E_m$ and volume fractions $v_f, v_m$, i.e.,

$$E_L = E_f v_f + E_m v_m \quad \text{(longitudinal modulus of the composite)} \tag{80}$$

### 2.3.4.2 Estimation of the Transverse Modulus $E_T$

Consider a unidirectional composite of width $l$ under transverse load $P$ applied over an area $A$ as indicated in Figure 4. The fibers and matrix are subject to the same transverse load $P$ but they stretch differently according to their different compliances. They act like series springs and their stretches add together while the force is the same in them both. The corresponding proportional transverse width of the fiber and the matrix is

$$l_f = v_f l \quad \text{(width of the fibers)} \tag{81}$$

$$l_m = v_m l \quad \text{(width of the matrix)} \tag{82}$$

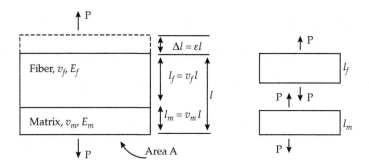

**FIGURE 4**   Schematic diagram for the estimation of the transverse modulus $E_T$.

Both the fibers and the matrix share the same transverse load $P$ applied over the same transverse area $A$. Hence, the stresses in the fiber and the matrix, $\sigma_f, \sigma_m$, are equal and are also equal to the overall stress $\sigma_T = P/A$, i.e.,

$$\sigma_f = \sigma_m = \frac{P}{A} = \sigma_T \quad \text{(transverse stresses in fiber and matrix)} \tag{83}$$

The transverse strains in the fiber and matrix are obtained from the stresses upon division by the respective elastic moduli, i.e.,

$$\varepsilon_f = \frac{\sigma_f}{E_f} = \frac{\sigma_T}{E_f} \quad \text{(transverse strain in fiber)} \tag{84}$$

$$\varepsilon_m = \frac{\sigma_m}{E_m} = \frac{\sigma_T}{E_m} \quad \text{(transverse strain in matrix)} \tag{85}$$

The total stretch $\Delta l$ is the sum of the individual stretches which are obtained by multiplication of the strains and respective lengths, i.e.,

$$\Delta l = \varepsilon_f l_f + \varepsilon_m l_m \quad \text{(total stretch)} \tag{86}$$

The overall strain $\varepsilon_T$ is obtained by dividing the total stretch $\Delta l$ by the overall length $l$, i.e.,

$$\varepsilon_T = \frac{\Delta l}{l} = \frac{\varepsilon_f l_f + \varepsilon_m l_m}{l} = \frac{\varepsilon_f v_f l + \varepsilon_m v_m l}{l} = \varepsilon_f v_f + \varepsilon_m v_m \quad \text{(overall strain)} \tag{87}$$

Substitution of Eqs. (84), (85) into Eq. (87) yields

$$\varepsilon_T = \varepsilon_f v_f + \varepsilon_m v_m = \frac{\sigma_T}{E_f} v_f + \frac{\sigma_T}{E_m} v_m \quad \text{(overall strain)} \tag{88}$$

But the overall strain $\varepsilon_T$ is the ratio between the overall stress $\sigma_T$ and the effective transverse modulus $E_T$, i.e.,

$$\varepsilon_T = \frac{\sigma_T}{E_T} \quad \text{(transverse strain in matrix)} \tag{89}$$

Combining Eqs. (88), (89) and simplifying through by $\sigma_T$ gives

$$\frac{1}{E_T} = \frac{1}{E_f} v_f + \frac{1}{E_m} v_m \tag{90}$$

Solution of Eq. (90) yields the formula for the transverse modulus $E_T$, i.e.,

$$E_T = \left( E_f^{-1} v_f + E_m^{-1} v_m \right)^{-1} \quad \text{(transverse modulus of the composite)} \tag{91}$$

Equations (90), (91) indicate that the transverse modulus is the inverse of the transverse compliance, which is obtained through the addition of the fiber and matrix compliances modulated by the respective volume fractions. Another way of expressing the formula given in Eq. (91) is [4,5]

$$E_T = \frac{E_f E_m}{E_f v_m + E_m v_f} \tag{92}$$

However, for numerical computation, the expression in Eq. (91) is preferred to that of Eq. (92) because it is less prone to human typing error.

The formulae in Eqs. (91), (92) combine the moduli of the fibers and the matrix like series springs, i.e., it adds their compliances after multiplying them by the appropriate volume fractions.

### 2.3.4.3 Estimation of Poisson Ratio $\nu_{LT}$

The Poisson ratio $\nu_{LT}$ is defined as

$$\nu_{LT} = -\frac{\varepsilon_T}{\varepsilon_L} \tag{93}$$

where the strains $\varepsilon_L$, $\varepsilon_T$ are measured while loading is applied in the L direction. To calculate an estimate for the Poisson ratio $\nu_{LT}$, consider the schematic in Figure 5.

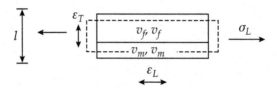

**FIGURE 5**    Schematic diagram for the estimation of the Poisson ratio $\nu_{LT}$.

The loading $\sigma_L$ in the L direction is assumed to produce the strain $\varepsilon_L$ which applies to both the fibers and the matrix, i.e.,

$$\begin{aligned} (\varepsilon_L)_f &= \varepsilon_L \\ (\varepsilon_L)_m &= \varepsilon_L \end{aligned} \quad \text{(same L strain in fibers and matrix)} \tag{94}$$

The corresponding transverse strains in the fibers and matrix are calculated using their respective Poisson ratios, i.e.,

$$\begin{aligned} (\varepsilon_T)_f &= -\nu_f\, \varepsilon_L \\ (\varepsilon_T)_m &= -\nu_m\, \varepsilon_L \end{aligned} \quad \text{(same L strain in fibers and matrix)} \tag{95}$$

The portions of the transverse width $l$ allocated to the fibers and the matrix are

$$l_f = v_f\, l \quad \text{(width allocated to the fibers)} \tag{96}$$

$$l_m = v_m\, l \quad \text{(width allocated to the matrix)} \tag{97}$$

Using Eqs. (95)–(97), calculate the total transverse elongation, i.e.,

$$\begin{aligned} \Delta l &= (\varepsilon_T)_f l_f + (\varepsilon_T)_m l_m \\ &= -\nu_f\, \varepsilon_L l_f - \nu_m\, \varepsilon_L l_m = -(\nu_f\, v_f + \nu_m\, v_m)\varepsilon_L l \end{aligned} \quad \text{(total transverse elongation)} \tag{98}$$

Note that the transverse elongation is negative, i.e., it is a shrinkage, as expected from the Poisson effect. The effective transverse strain is calculated by dividing the elongation by the nominal length, i.e.,

$$\varepsilon_T = \frac{\Delta l}{l} = \frac{-(\nu_f\, v_f + \nu_m\, v_m)\varepsilon_L \cancel{l}}{\cancel{l}} = -(\nu_f\, v_f + \nu_m\, v_m)\varepsilon_L \quad \text{(effective transverse strain)} \tag{99}$$

Recall the definition Eq. (93) to write

$$\nu_{LT} = -\frac{\varepsilon_T}{\varepsilon_L} = \nu_f\,v_f + \nu_m\,v_m \tag{100}$$

### 2.3.4.4 Estimation of the LT Shear Modulus $G_{LT}$

The estimation of the LT shear modulus $G_{LT}$, a.k.a. longitudinal shear modulus, is done under the assumption that the fibers and matrix undergo separate shear deformations and their shear deformations add up to create the shear deformation of the assemblage (Figure 6).

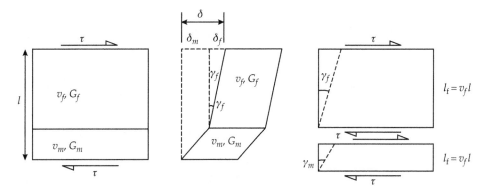

**FIGURE 6**   Schematic diagram for the estimation of the LT shear modulus $G_{LT}$.

The fibers and the matrix are subject to the same shear stress $\tau$, i.e.,

$$\tau_f = \tau_m = \tau \quad \text{(common shear stress in fibers and matrix)} \tag{101}$$

The corresponding shear strains are

$$\gamma_f = \frac{\tau_f}{G_f} = \frac{\tau}{G_f}$$
$$\gamma_m = \frac{\tau_m}{G_m} = \frac{\tau}{G_m} \quad \text{(shear strains in fibers and matrix)} \tag{102}$$

The corresponding sliding displacements in the fibers and matrix are

$$\delta_f = \gamma_f l_f = \gamma_f v_f l \quad \text{(sliding displacement of the fibers)} \tag{103}$$
$$\delta_m = \gamma_m l_m = \gamma_m v_m l \quad \text{(sliding displacement of the matrix)} \tag{104}$$

The total displacement $\delta$ is

$$\delta = \delta_f + \delta_m = \gamma_f v_f l + \gamma_m v_m l \quad \text{(overall shear strain)} \tag{105}$$

The overall shear strain $\gamma$ is obtained by dividing the total sliding displacement $\delta$ by the width $l$, i.e.,

$$\gamma = \frac{\delta}{l} = \gamma_f v_f + \gamma_m v_m \quad \text{(overall shear strain)} \tag{106}$$

By definition, the overall shear strain $\gamma$ is the ratio between the overall shear stress $\tau$ and the effective shear modulus $G_{LT}$ of the composite, i.e.,

$$\gamma = \frac{\tau}{G_{LT}} \tag{107}$$

Substitution of Eq. (102) into Eq. (106), followed by combination with Eq. (107) and division by $\tau$ gives

$$\frac{1}{G_{LT}} = \frac{1}{G_f} v_f + \frac{1}{G_m} v_m \tag{108}$$

Solution of Eq. (108) yields the formula for the LT shear modulus $G_{LT}$, i.e.,

$$G_{LT} = \left( G_f^{-1} v_f + G_m^{-1} v_m \right)^{-1} \quad \text{(in-plane shear modulus of the composite)} \tag{109}$$

Equations (108), (109) indicate that the LT shear modulus is the inverse of the LT shear compliance, which is obtained through the addition of the fiber and matrix compliances modulated by the respective volume fractions. Another way to express the formula given in Eq. (109) is [4,5]

$$G_{LT} = \frac{G_f G_m}{G_f v_m + G_m v_f} \tag{110}$$

However, for numerical computation, the expression in Eq. (91) is preferred to that of Eq. (92) because it is less prone to human typing error.

The formulae in Eqs. (109), (110) combine the shear moduli of the fibers and the matrix like series springs, i.e., it adds their compliances after multiplying them by the appropriate volume fractions.

An improved formula for the estimation of $G_{LT}$ is obtained through the elasticity solution of an cylindrical assemblage model (CAM) consisting of concentric fiber and matrix cylinders (Ref. [6] as cited in Ref. [2]) and zero void fraction ($v_f + v_m = 1$), i.e.,

$$G_{LT} = G_m \left[ \frac{1 + v_f + (1 - v_f)G_m/G_f}{1 - v_f + (1 + v_f)G_m/G_f} \right] \quad \text{(CAM in-plane shear modulus)} \tag{111}$$

### 2.3.4.5 Estimation of Transverse Shear Modulus $G_{23}$

A formula for the estimation of the transverse shear modulus $G_{23}$ was developed using the semiempirical stress-partitioning parameter (SPP) method (Ref. [7] cited in Ref. [2], p. 75), i.e.,

$$G_{23} = G_m \frac{v_f + \eta_{23} v_m}{\eta_{23} v_m + v_f G_m/G_f}$$

$$\eta_{23} = \frac{3 - 4\nu_m + G_m/G_f}{4(1 - \nu_m)} \quad \text{(transverse shear modulus)} \tag{112}$$

### 2.3.4.6 Matrix-Dominated Approximations

In practical applications, the matrix modulus is orders of magnitude smaller than the corresponding fiber modulus, i.e.,

$$E_m \ll E_f$$
$$G_m \ll E_f \tag{113}$$

Thus, the moduli estimation formulae of Eqs. (80), (91), (109), (111) simplify as follows:

$$E_L = E_f v_f \quad \text{(approx. longitudinal modulus of the composite)} \tag{114}$$

$$E_T = E_m / v_m \quad \text{(approx. transverse modulus of the composite)} \tag{115}$$

$$G_{LT} = G_m / v_m \quad \text{(in-plane shear modulus of the composite)} \tag{116}$$

$$G_{LT} = G_m \frac{1 + v_f}{1 - v_f} \quad \text{(approx. CAM shear modulus)} \tag{117}$$

Other micromechanics-based derivations of composite properties are available in Refs. [8–10].

# 2.4 PLANE-STRESS 2D ELASTIC PROPERTIES OF A COMPOSITE LAYER

This section presents the 2D elastic properties of a composite layer which is assumed to be in a state of plane stress. It will discuss the 2D stiffness matrix $\mathbf{Q}$ which is the plane-stress correspondent of the 3D stiffness matrix $\mathbf{C}$. The procedure for evaluating the rotated 2D stiffness matrix $\overline{\mathbf{Q}}$ will be established. The effect of the fiber orientation angle on the 2D stiffness matrix of a unidirectional composite layer will be studied.

In the analysis of thin-wall composite structures, it is common to assume that the thickness-wise stress $\sigma_{33}$ is zero everywhere throughout the thickness, i.e., $\sigma_{33} \equiv 0$. Therefore, a state of plane stress is assumed to exist, and the only stresses and strains of interest are $\sigma_{11}, \sigma_{22}, \sigma_{12}, \varepsilon_{11}, \varepsilon_{22}, \varepsilon_{12}$. For composite laminates made up through the stacking of several plies of various orientations, plane-stress conditions apply in each ply of the composite layup.

## 2.4.1 Plane-Stress 2D Compliance Matrix

Consider a thin unidirectional composite lamina in plane-stress condition $\sigma_{33} \equiv 0$; the stresses and strains of interest are contained in the lamina plane $O\,x_1 x_2$, i.e., $\sigma_{11}, \sigma_{22}, \sigma_{12}$ and $\varepsilon_{11}, \varepsilon_{22}, \varepsilon_{12}$. The stresses $\sigma_{13}, \sigma_{23}$ and the strains $\varepsilon_{33}, \varepsilon_{13}, \varepsilon_{23}$ are not necessarily zero, but they will simply not make the object of our attention which is focused on finding the relation between the stresses $\sigma_{11}$, $\sigma_{22}$, $\sigma_{12}$ and the corresponding strains $\varepsilon_{11}$, $\varepsilon_{22}$, $\varepsilon_{12}$. Recall Eq. (68) giving the complete strain−stress matrix relation, i.e.,

$$\begin{Bmatrix} \varepsilon_{11} \\ \varepsilon_{22} \\ \varepsilon_{33} \\ 2\varepsilon_{23} \\ 2\varepsilon_{31} \\ 2\varepsilon_{12} \end{Bmatrix} = \begin{bmatrix} S_{11} & S_{12} & S_{13} & 0 & 0 & 0 \\ S_{12} & S_{22} & S_{23} & 0 & 0 & 0 \\ S_{13} & S_{23} & S_{33} & 0 & 0 & 0 \\ 0 & 0 & 0 & S_{44} & 0 & 0 \\ 0 & 0 & 0 & 0 & S_{55} & 0 \\ 0 & 0 & 0 & 0 & 0 & S_{66} \end{bmatrix} \begin{Bmatrix} \sigma_{11} \\ \sigma_{22} \\ \sigma_{33} \\ \sigma_{23} \\ \sigma_{31} \\ \sigma_{12} \end{Bmatrix} \tag{118}$$

Retaining only the elements related to $\varepsilon_{11}$, $\varepsilon_{22}$, $\varepsilon_{12}$, $\sigma_{11}$, $\sigma_{22}$, $\sigma_{12}$, one gets the 2D plane-stress compliance relation

$$\begin{Bmatrix} \varepsilon_1 \\ \varepsilon_2 \\ 2\varepsilon_{12} \end{Bmatrix} = \begin{bmatrix} S_{11} & S_{12} & 0 \\ S_{12} & S_{22} & 0 \\ 0 & 0 & S_{66} \end{bmatrix} \begin{Bmatrix} \sigma_1 \\ \sigma_2 \\ \sigma_{12} \end{Bmatrix} \tag{119}$$

Associated with Eq. (119), define the 2D compliance matrix

$$\mathbf{S} = \begin{bmatrix} S_{11} & S_{12} & 0 \\ S_{12} & S_{22} & 0 \\ 0 & 0 & S_{66} \end{bmatrix} \quad \text{(2D compliance matrix)} \tag{120}$$

Using Eq. (64), one can write the 2D compliance matrix Eq. (120) explicitly in terms of the engineering constants, i.e.,

$$\mathbf{S} = \begin{bmatrix} 1/E_L & -\nu_{LT}/E_L & 0 \\ -\nu_{LT}/E_L & 1/E_T & 0 \\ 0 & 0 & 1/G_{LT} \end{bmatrix} \tag{121}$$

Equation (120) can be used to write Eq. (119) compactly as

$$\boldsymbol{\varepsilon} = \mathbf{S}\,\boldsymbol{\sigma} \quad \text{(strain−stress matrix relation)} \tag{122}$$

where $\boldsymbol{\varepsilon}$ and $\boldsymbol{\sigma}$ are the 2D strain and stress column matrices, i.e.,

$$\boldsymbol{\varepsilon} = \begin{Bmatrix} \varepsilon_1 \\ \varepsilon_2 \\ 2\varepsilon_{12} \end{Bmatrix} \qquad \boldsymbol{\sigma} = \begin{Bmatrix} \sigma_1 \\ \sigma_2 \\ \sigma_{12} \end{Bmatrix} \quad \text{(2D strain and stress column matrices)} \tag{123}$$

### 2.4.2 Plane-Stress 2D Stiffness Matrix

Upon inversion, Eq. (120) yields the plane-stress 2D stiffness matrix, i.e.,

$$\mathbf{Q} = \mathbf{S}^{-1} \quad \text{(2D stiffness matrix)} \tag{124}$$

In longhand, Eq. (124) is written as

$$\begin{bmatrix} Q_{11} & Q_{12} & 0 \\ Q_{12} & Q_{22} & 0 \\ 0 & 0 & Q_{66} \end{bmatrix} = \begin{bmatrix} S_{11} & S_{12} & 0 \\ S_{12} & S_{22} & 0 \\ 0 & 0 & S_{66} \end{bmatrix}^{-1} \quad \text{(2D stiffness matrix)} \tag{125}$$

The 2D stress–strain matrix relation is written using 2D stiffness matrix as

$$\sigma = Q\varepsilon \quad \text{(stress–strain matrix relation)} \tag{126}$$

In longhand, Eq. (126) is written as

$$\begin{Bmatrix} \sigma_1 \\ \sigma_2 \\ \sigma_{12} \end{Bmatrix} = \begin{bmatrix} Q_{11} & Q_{12} & 0 \\ Q_{12} & Q_{22} & 0 \\ 0 & 0 & Q_{66} \end{bmatrix} \begin{Bmatrix} \varepsilon_1 \\ \varepsilon_2 \\ 2\varepsilon_{12} \end{Bmatrix} \tag{127}$$

The 2D compliance matrix of Eq. (120) is directly related to the complete compliance matrix given by Eq. (66) and can be obtained from the latter by deletion of appropriate rows and columns. It can be said that the 2D compliance matrix of Eq. (120) is a submatrix of the complete compliance matrix of Eq. (66). In contrast, the 2D stiffness matrix of Eq. (125) is *not* a submatrix of the complete stiffness matrix of Eq. (70) and bears no direct relationship to it. For this reason, the 2D stiffness matrix is denoted differently than the complete stiffness matrix, i.e., by $Q$ instead of $C$. A closed-form expression of the 2D stiffness matrix can also be obtained explicitly [1], but the use of such closed-form expression is not recommended because manual coding could introduce errors which are hard to trace. For numerical work, Eq. (124), which uses the inversion of the compliance matrix, is always preferred.

### 2.4.3  Rotated 2D Stiffness Matrix

A laminated composite is made up of several layers of unidirectional plies (laminae) of different orientation. In order to be able to assemble the overall properties of the laminate from the properties of the individual laminae, we need to express all properties in the same coordinate system. Assume (Figure 7) that the local axes $x_1'x_2'$ are rotated through the angle $\theta$ with respect to the global axes $x_1x_2$ (The rotation takes place about the out-of-plane axis $x_3$, hence the local axis $x_3'$ coincides with the global axis $x_3$.)

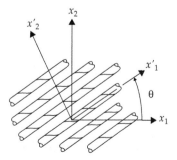

**FIGURE 7**   Definition of axes rotation by angle $\theta$; $x_1x_2$ are the global axes; $x_1'x_2'$ are the local axes aligned with the fiber direction (note that $x_1'$ is the L direction; $x_2'$ is the T direction).

Recall the matrix notations of Eq. (123), i.e.,

$$\boldsymbol{\sigma} = \left\{ \begin{array}{c} \sigma_1 \\ \sigma_2 \\ \sigma_{12} \end{array} \right\} \qquad \boldsymbol{\varepsilon} = \left\{ \begin{array}{c} \varepsilon_1 \\ \varepsilon_2 \\ 2\varepsilon_{12} \end{array} \right\} \qquad \text{(2D stress and strain column matrices)} \qquad (128)$$

The effect of the rotation $\theta$ on the stresses and strains is given by the "Mohr circle" rotation relations ([11], pp. 44, 71):

$$\left\{ \begin{array}{c} \sigma'_{11} \\ \sigma'_{22} \\ \sigma'_{12} \end{array} \right\} = \left[ \begin{array}{ccc} \cos^2\theta & \sin^2\theta & 2\sin\theta\cos\theta \\ \sin^2\theta & \cos^2\theta & -2\sin\theta\cos\theta \\ -\sin\theta\cos\theta & \sin\theta\cos\theta & \cos^2\theta - \sin^2\theta \end{array} \right] \left\{ \begin{array}{c} \sigma_{11} \\ \sigma_{22} \\ \sigma_{12} \end{array} \right\} \quad \left( \begin{array}{c} \text{rotation of} \\ \text{global stresses} \\ \text{into local stresses} \end{array} \right) \qquad (129)$$

$$\left\{ \begin{array}{c} \varepsilon'_{11} \\ \varepsilon'_{22} \\ \varepsilon'_{12} \end{array} \right\} = \left[ \begin{array}{ccc} \cos^2\theta & \sin^2\theta & 2\sin\theta\cos\theta \\ \sin^2\theta & \cos^2\theta & -2\sin\theta\cos\theta \\ -\sin\theta\cos\theta & \sin\theta\cos\theta & \cos^2\theta - \sin^2\theta \end{array} \right] \left\{ \begin{array}{c} \varepsilon_{11} \\ \varepsilon_{22} \\ \varepsilon_{12} \end{array} \right\} \quad \left( \begin{array}{c} \text{rotation of} \\ \text{global strains} \\ \text{into local strains} \end{array} \right) \qquad (130)$$

We wish to write Eqs. (129), (130) compactly using the matrix notation of Eq. (128) together with a $\theta$-dependent rotation matrix $\mathbf{T}$ defined as

$$\mathbf{T} = \left[ \begin{array}{ccc} \cos^2\theta & \sin^2\theta & 2\sin\theta\cos\theta \\ \sin^2\theta & \cos^2\theta & -2\sin\theta\cos\theta \\ -\sin\theta\cos\theta & \sin\theta\cos\theta & \cos^2\theta - \sin^2\theta \end{array} \right] \qquad \text{(rotation matrix)} \qquad (131)$$

Using Eqs. (128), (131), write the stress rotation Eq. (129) as

$$\boldsymbol{\sigma}' = \mathbf{T}\boldsymbol{\sigma} \qquad (132)$$

However, the strain rotation Eq. (130) cannot be as immediately in matrix notations as the stress rotation because the shear strain $\varepsilon_{12}$ of Eq. (130) appears with a factor of 2 in Eq. (128). To resolve this issue, introduce the $\mathbf{R}$ matrix [12] defined as

$$\mathbf{R} = \left[ \begin{array}{ccc} 1 & 0 & 0 \\ 0 & 1 & 0 \\ 0 & 0 & 2 \end{array} \right] \qquad \mathbf{R}^{-1} = \left[ \begin{array}{ccc} 1 & 0 & 0 \\ 0 & 1 & 0 \\ 0 & 0 & 1/2 \end{array} \right] \qquad (133)$$

Using Eq. (133), write

$$\left\{ \begin{array}{c} \varepsilon'_{11} \\ \varepsilon'_{22} \\ 2\varepsilon'_{12} \end{array} \right\} = \left[ \begin{array}{ccc} 1 & 0 & 0 \\ 0 & 1 & 0 \\ 0 & 0 & 2 \end{array} \right] \left\{ \begin{array}{c} \varepsilon'_{11} \\ \varepsilon'_{22} \\ \varepsilon'_{12} \end{array} \right\} \qquad (134)$$

$$\left\{ \begin{array}{c} \varepsilon_{11} \\ \varepsilon_{22} \\ \varepsilon_{12} \end{array} \right\} = \left[ \begin{array}{ccc} 1 & 0 & 0 \\ 0 & 1 & 0 \\ 0 & 0 & 1/2 \end{array} \right] \left\{ \begin{array}{c} \varepsilon_{11} \\ \varepsilon_{22} \\ 2\varepsilon_{12} \end{array} \right\} \qquad (135)$$

Substitution of Eq. (130) into Eq. (134) gives

$$
\begin{Bmatrix} \varepsilon'_{11} \\ \varepsilon'_{22} \\ 2\varepsilon'_{12} \end{Bmatrix} = \begin{bmatrix} 1 & 0 & 0 \\ 0 & 1 & 0 \\ 0 & 0 & 2 \end{bmatrix} \begin{Bmatrix} \varepsilon'_{11} \\ \varepsilon'_{22} \\ \varepsilon'_{12} \end{Bmatrix}
$$

$$
= \begin{bmatrix} 1 & 0 & 0 \\ 0 & 1 & 0 \\ 0 & 0 & 2 \end{bmatrix} \begin{bmatrix} \cos^2\theta & \sin^2\theta & 2\sin\theta\cos\theta \\ \sin^2\theta & \cos^2\theta & -2\sin\theta\cos\theta \\ -\sin\theta\cos\theta & \sin\theta\cos\theta & \cos^2\theta - \sin^2\theta \end{bmatrix} \begin{Bmatrix} \varepsilon_{11} \\ \varepsilon_{22} \\ \varepsilon_{12} \end{Bmatrix}
$$

(136)

Use of Eq. (135) into Eq. (136) yields

$$
\begin{Bmatrix} \varepsilon'_{11} \\ \varepsilon'_{22} \\ 2\varepsilon'_{12} \end{Bmatrix} = \begin{bmatrix} 1 & 0 & 0 \\ 0 & 1 & 0 \\ 0 & 0 & 2 \end{bmatrix} \begin{bmatrix} \cos^2\theta & \sin^2\theta & 2\sin\theta\cos\theta \\ \sin^2\theta & \cos^2\theta & -2\sin\theta\cos\theta \\ -\sin\theta\cos\theta & \sin\theta\cos\theta & \cos^2\theta - \sin^2\theta \end{bmatrix} \begin{bmatrix} 1 & 0 & 0 \\ 0 & 1 & 0 \\ 0 & 0 & 1/2 \end{bmatrix} \begin{Bmatrix} \varepsilon_{11} \\ \varepsilon_{22} \\ 2\varepsilon_{12} \end{Bmatrix}
$$

(137)

Utilizing the matrix notations of Eqs. (128), (131), (133), write Eq. (137) compactly as

$$
\boldsymbol{\varepsilon}' = \mathbf{R}\mathbf{T}\mathbf{R}^{-1}\boldsymbol{\varepsilon}
$$

(138)

Upon evaluation (see Section 4.5), one finds that $\mathbf{R}\mathbf{T}\mathbf{R}^{-1}$ is actually $\mathbf{T}^{-t}$, the transpose of the inverse of $\mathbf{T}$, i.e.,

$$
\mathbf{R}\mathbf{T}\mathbf{R}^{-1} = \mathbf{T}^{-t}
$$

(139)

Substitution of Eq. (139) into Eq. (138) yields

$$
\boldsymbol{\varepsilon}' = \mathbf{T}^{-t}\boldsymbol{\varepsilon}
$$

(140)

Our aim is to express the 2D stiffness matrix in global coordinates $\mathbf{Q}$ as function of the 2D stiffness matrix in local coordinates $\mathbf{Q}'$ and the rotation matrix $\mathbf{T}$ that depends on the rotation angle $\theta$. We know $\mathbf{Q}'$ from Eqs. (124), (125), which are related to the local properties through Eq. (121). Recall Eq. (126) and write the stress–strain relation in local coordinates, i.e.,

$$
\boldsymbol{\sigma}' = \mathbf{Q}'\boldsymbol{\varepsilon}' \quad \text{(stress – strain matrix relation in local coordinates)}
$$

(141)

where

$$
\mathbf{Q}' = \mathbf{S}'^{-1} = \begin{bmatrix} 1/E_L & -\nu_{LT}/E_L & 0 \\ -\nu_{LT}/E_L & 1/E_T & 0 \\ 0 & 0 & 1/G_{LT} \end{bmatrix}^{-1} \quad \text{(local stiffness matrix)}
$$

(142)

Use Eqs. (132), (140) into Eq. (141) to write

$$\mathbf{T}\sigma = \mathbf{Q}'\,\mathbf{T}^{-t}\varepsilon \tag{143}$$

Premultiplication of Eq. (143) by $\mathbf{T}^{-1}$ yields

$$\sigma = \mathbf{T}^{-1}\mathbf{Q}'\,\mathbf{T}^{-t}\varepsilon \tag{144}$$

Recall Eq. (126) which gives the stress−strain relation in global coordinates, i.e.,

$$\sigma = \mathbf{Q}\,\varepsilon \quad \text{(stress−strain matrix relation in global coordinates)} \tag{145}$$

Comparison of Eqs. (144), (145) yields the rotated stiffness matrix as

$$\mathbf{Q} = \mathbf{T}^{-1}\mathbf{Q}'\,\mathbf{T}^{-t} \tag{146}$$

Longhand expression of the rotated stiffness matrix is

$$\mathbf{Q} = \begin{bmatrix} Q_{11} & Q_{12} & Q_{16} \\ Q_{12} & Q_{22} & Q_{26} \\ Q_{16} & Q_{26} & Q_{66} \end{bmatrix} \tag{147}$$

The longhand expressions of the stress−strain matrix relation (145) is

$$\begin{Bmatrix} \sigma_{11} \\ \sigma_{22} \\ \sigma_{12} \end{Bmatrix} = \begin{bmatrix} Q_{11} & Q_{12} & Q_{16} \\ Q_{12} & Q_{22} & Q_{26} \\ Q_{16} & Q_{26} & Q_{66} \end{bmatrix} \begin{Bmatrix} \varepsilon_{11} \\ \varepsilon_{22} \\ 2\varepsilon_{12} \end{Bmatrix} \tag{148}$$

If the global coordinates are denoted $Oxyz$ instead of $Ox_1x_2x_3$, then Eq. (148) is written as ([4], p. 191)

$$\begin{Bmatrix} \sigma_{xx} \\ \sigma_{yy} \\ \sigma_{xy} \end{Bmatrix} = \begin{bmatrix} Q_{11} & Q_{12} & Q_{16} \\ Q_{12} & Q_{22} & Q_{26} \\ Q_{16} & Q_{26} & Q_{66} \end{bmatrix} \begin{Bmatrix} \varepsilon_{xx} \\ \varepsilon_{yy} \\ 2\varepsilon_{xy} \end{Bmatrix} \tag{149}$$

## 2.4.4 Rotated 2D Compliance Matrix

A similar argument can be employed to find the rotated 2D compliance matrix $\mathbf{S}$. Recall the strain−stress matrix relation Eq. (122) and write it in local coordinates, i.e.,

$$\varepsilon' = \mathbf{S}'\,\sigma' \quad \text{(strain − stress matrix relation in local coordinates)} \tag{150}$$

where

$$\mathbf{S}' = \begin{bmatrix} 1/E_L & -\nu_{LT}/E_L & 0 \\ -\nu_{LT}/E_L & 1/E_T & 0 \\ 0 & 0 & 1/G_{LT} \end{bmatrix} \quad \text{(local compliance matrix)} \tag{151}$$

Use Eqs. (132), (140) to express $\sigma'$, $\varepsilon'$ of Eq. (150) in terms of $\sigma$, $\varepsilon$ and get

$$\mathbf{T}^{-t}\varepsilon = \mathbf{S}'\,\mathbf{T}\sigma \qquad\qquad (152)$$

Multiply Eq. (152) by $\mathbf{T}^t$ and obtain

$$\varepsilon = \mathbf{T}^t\mathbf{S}'\,\mathbf{T}\sigma \qquad\qquad (153)$$

Recall Eq. (122) and write the strain−stress relation in global coordinates, i.e.,

$$\varepsilon = \mathbf{S}\,\sigma \quad \text{(strain−stress matrix relation in global coordinates)} \qquad\qquad (154)$$

Comparison of Eqs. (153), (154) yields the rotated compliance matrix $\mathbf{S}$, i.e.,

$$\mathbf{S} = \mathbf{T}^t\mathbf{S}'\,\mathbf{T} \qquad\qquad (155)$$

Note: Closed-form expressions for the elements of the rotated stiffness and compliance matrices, $\mathbf{Q}$ and $\mathbf{S}$, are possible and can be found in some textbooks. However, these closed-form expressions contain quite elaborate formulae; if manual coding of these formulae is attempted, then hard-to-trace errors may be inadvertently introduced. Hence, the use of such closed-form complicated formulae is not recommended.

## 2.4.5 Proof of $\mathbf{RTR}^{-1} = \mathbf{T}^{-t}$

In Section 4.3, Eq. (139), we cited without proof the formula $\mathbf{RTR}^{-1} = \mathbf{T}^{-t}$. Here we are going to prove it. It is convenient to define the following shorthand notations:

$$s = \sin\theta \quad c = \cos\theta \quad s^2 + c^2 = \sin^2\theta + \cos^2\theta = 1 \qquad\qquad (156)$$

Use Eq. (156) to write the rotation matrix $\mathbf{T}$ of Eq. (131) as

$$\mathbf{T} = \begin{bmatrix} c^2 & s^2 & 2sc \\ s^2 & c^2 & -2sc \\ -sc & sc & c^2 - s^2 \end{bmatrix} \quad \text{(2D rotation matrix in shorthand notations)} \qquad\qquad (157)$$

Calculate the inverse matrix $\mathbf{T}^{-1}$, i.e.,

$$\mathbf{T}^{-1} = \begin{bmatrix} c^2 & s^2 & -2sc \\ s^2 & c^2 & 2sc \\ sc & -sc & c^2 - s^2 \end{bmatrix} \quad \text{(inverse of the rotation matrix)} \qquad\qquad (158)$$

and verify that $T^{-1}T = I$, i.e.,

$$
T^{-1}T = \begin{bmatrix} c^2 & s^2 & -2sc \\ s^2 & c^2 & 2sc \\ sc & -sc & c^2 - s^2 \end{bmatrix} \begin{bmatrix} c^2 & s^2 & 2sc \\ s^2 & c^2 & -2sc \\ -sc & sc & c^2 - s^2 \end{bmatrix}
$$

$$
= \begin{bmatrix} c^4 + s^4 + 2s^2c^2 & c^2s^2 + s^2c^2 - 2s^2c^2 & \text{etc.} \\ & \cdots & & \cdots & & \cdots \\ & \cdots & & \cdots & & \cdots \end{bmatrix} = \begin{bmatrix} (c^2 + s^2)^2 & 2c^2s^2 - 2s^2c^2 & \text{etc.} \\ & \cdots & & \cdots & & \cdots \\ & \cdots & & \cdots & & \cdots \end{bmatrix} \tag{159}
$$

$$
= \begin{bmatrix} 1 & 0 & 0 \\ 0 & 1 & 0 \\ 0 & 0 & 1 \end{bmatrix} = I
$$

Calculate $T^{-t}$, the transpose of the inverse of the $T$ matrix, i.e.,

$$
T^{-t} = \begin{bmatrix} c^2 & s^2 & sc \\ s^2 & c^2 & -sc \\ -2sc & 2sc & c^2 - s^2 \end{bmatrix} \quad \text{(transposed of the inverse of } T\text{)} \tag{160}
$$

Calculate $RTR^{-1}$, i.e.,

$$
RTR^{-1} = \begin{bmatrix} 1 & 0 & 0 \\ 0 & 1 & 0 \\ 0 & 0 & 2 \end{bmatrix} \begin{bmatrix} c^2 & s^2 & 2sc \\ s^2 & c^2 & -2sc \\ -sc & sc & c^2 - s^2 \end{bmatrix} \begin{bmatrix} 1 & 0 & 0 \\ 0 & 1 & 0 \\ 0 & 0 & 1/2 \end{bmatrix}
$$

$$
= \begin{bmatrix} 1 & 0 & 0 \\ 0 & 1 & 0 \\ 0 & 0 & 2 \end{bmatrix} \begin{bmatrix} c^2 & s^2 & sc \\ s^2 & c^2 & -sc \\ -sc & sc & (1/2)(c^2 - s^2) \end{bmatrix} = \begin{bmatrix} c^2 & s^2 & sc \\ s^2 & c^2 & -sc \\ -2sc & 2sc & c^2 - s^2 \end{bmatrix} \tag{161}
$$

Comparison of Eqs. (160) and (161) reveals that

$$
RTR^{-1} = T^{-t} \quad \text{(QED)} \tag{162}
$$

## 2.5 FULLY 3D ELASTIC PROPERTIES OF A COMPOSITE LAYER

This section presents the 3D stiffness matrix of a composite layer. A procedure for evaluating the stiffness matrix from the orthotropic elastic properties will be given. The rotated 3D stiffness matrix will be introduced. The effect of the fiber orientation angle on the 3D stiffness matrix of a unidirectional composite layer will be discussed.

## 2.5.1 Orthotropic Stiffness Matrix

Consider a lamina of orthotropic composite material such as a unidirectional ply in laminated composite layup. Assume a local coordinate system with the $x_3'$ axis perpendicular to the lamina; the $Ox_1'x_2'$ plane is the plane of the lamina with the $x_1'$ axis along the fibers and the $x_2'$ axis perpendicular to the fibers. The coordinates $x_1', x_2', x_3'$ are the material coordinates (a.k.a. local coordinates). The stiffness matrix in material coordinates $\mathbf{C}'$ is given by Eq. (70), i.e.,

$$\mathbf{C}' = \begin{bmatrix} C_{11}' & C_{12}' & C_{13}' & 0 & 0 & 0 \\ C_{12}' & C_{22}' & C_{23}' & 0 & 0 & 0 \\ C_{13}' & C_{23}' & C_{33}' & 0 & 0 & 0 \\ 0 & 0 & 0 & C_{44}' & 0 & 0 \\ 0 & 0 & 0 & 0 & C_{55}' & 0 \\ 0 & 0 & 0 & 0 & 0 & C_{66}' \end{bmatrix} \tag{163}$$

As indicated by Eq. (70), the elements of $\mathbf{C}'$ are obtained through the inversion of the material compliance matrix $\mathbf{S}'$, which, in turn, can be expressed in terms of the engineering constants of the lamina $E_L, E_T, G_{LT}, \nu_{LT}, G_{23}, \nu_{23}$ through Eqs. (65) or (66) and (67), i.e.,

$$\mathbf{S} = \begin{bmatrix} S_{11} & S_{12} & S_{13} & 0 & 0 & 0 \\ S_{12} & S_{22} & S_{23} & 0 & 0 & 0 \\ S_{13} & S_{23} & S_{33} & 0 & 0 & 0 \\ 0 & 0 & 0 & S_{44} & 0 & 0 \\ 0 & 0 & 0 & 0 & S_{55} & 0 \\ 0 & 0 & 0 & 0 & 0 & S_{66} \end{bmatrix} \begin{pmatrix} \text{compliance matrix} \\ \text{in material coordinates} \end{pmatrix} \tag{164}$$

$$\mathbf{S} = \begin{bmatrix} 1/E_L & -\nu_{LT}/E_L & -\nu_{LT}/E_L & 0 & 0 & 0 \\ -\nu_{LT}/E_L & 1/E_T & -\nu_{23}/E_T & 0 & 0 & 0 \\ -\nu_{LT}/E_L & -\nu_{23}/E_T & 1/E_T & 0 & 0 & 0 \\ 0 & 0 & 0 & 1/G_{23} & 0 & 0 \\ 0 & 0 & 0 & 0 & 1/G_{LT} & 0 \\ 0 & 0 & 0 & 0 & 0 & 1/G_{LT} \end{bmatrix} \begin{pmatrix} \text{compliance} \\ \text{matrix} \\ \text{in} \\ \text{material} \\ \text{coordinates} \end{pmatrix} \tag{165}$$

Note that only five independent elastic constants are needed since $\nu_{23}$ and $G_{23}$ are related through the formula $G_{23} = E_T/2(1 + \nu_{23})$. Hence, recall Eq. (37) and write the stress–strain relation in local coordinates as

$$\sigma' = \mathbf{C}' \, \varepsilon' \quad \text{(stress–strain relation in local coordinates)} \tag{166}$$

## 2.5.2 Rotated Stiffness Matrix

When included in a composite laminate, the unidirectional lamina can be oriented at various angles. The stiffness matrix for a unidirectional layer matrix with arbitrary orientation can be obtained through the application of a rotation about the $x'_3$ axis. Recall from Section 4.3 the local coordinate system with the $x'_3$ axis perpendicular to the lamina, the $x'_1$ axis along the fibers, and the $x'_2$ axis contained in the plane of the lamina in a direction perpendicular to the fibers. Assume a global coordinate system $Ox_1x_2x_3$ which is obtained from the $Ox'_1x'_2x'_3$ system through a rotation about the $x'_3$ axis. This means that the axes $x_3$ and $x'_3$ coincide whereas the axes $x_1$ and $x'_1$ as well as $x_2$ and $x'_2$ make an angle $\theta$ to each other. The stiffness matrix in the local system $\mathbf{C}'$ is given by Eq. (163). The stiffness matrix in the global system $\mathbf{C}$ (i.e., the rotated stiffness matrix) is obtained from the local stiffness matrix $\mathbf{C}'$ through the application of rotation formulae as described next. The analysis of this 3D situation is sketched in Ref. [3], pp. 473–477 but not pursued; instead, the reader was sent to Ref. [13] for details. Reference [14] gives without proof analytical expressions of the elements of the 3D compliance matrix. For the sake of clarity, we will pursue in this section a complete 3D derivation and provide guidance for practical calculations. Recall Eqs. (33), (34) giving the 3D stress and strain column matrices, i.e.,

$$\sigma = \begin{Bmatrix} \sigma_{11} \\ \sigma_{22} \\ \sigma_{33} \\ \sigma_{23} \\ \sigma_{31} \\ \sigma_{12} \end{Bmatrix} \qquad \varepsilon = \begin{Bmatrix} \varepsilon_{11} \\ \varepsilon_{22} \\ \varepsilon_{33} \\ 2\varepsilon_{23} \\ 2\varepsilon_{31} \\ 2\varepsilon_{12} \end{Bmatrix} \qquad (167)$$

Recall Eq. (37) to write the stress–strain relation in global coordinates as

$$\sigma = \mathbf{C}\,\varepsilon \quad \text{(stress–strain relation in global coordinates)} \qquad (168)$$

To develop a relation between the global stiffness matrix and the local stiffness matrix, we need to find a rotation matrix $\mathbf{T}$ that permits the expression of the stress and strain matrices in local coordinates (material coordinates) $\sigma', \varepsilon'$ in terms of the stress and strain matrices in global coordinates $\sigma, \varepsilon$. Recall Eqs. (129), (130) for the rotation of 2D stresses and strains, $\sigma_{11}, \sigma_{22}, \sigma_{12}$ and $\varepsilon_{11}, \varepsilon_{22}, \varepsilon_{12}$, i.e.,

$$\begin{Bmatrix} \sigma'_{11} \\ \sigma'_{22} \\ \sigma'_{12} \end{Bmatrix} = \begin{bmatrix} \cos^2\theta & \sin^2\theta & 2\sin\theta\cos\theta \\ \sin^2\theta & \cos^2\theta & -2\sin\theta\cos\theta \\ -\sin\theta\cos\theta & \sin\theta\cos\theta & \cos^2\theta - \sin^2\theta \end{bmatrix} \begin{Bmatrix} \sigma_{11} \\ \sigma_{22} \\ \sigma_{12} \end{Bmatrix} \quad \text{(rotation of 2D stresses)} \qquad (169)$$

$$\begin{Bmatrix} \varepsilon'_{11} \\ \varepsilon'_{22} \\ \varepsilon'_{12} \end{Bmatrix} = \begin{bmatrix} \cos^2\theta & \sin^2\theta & 2\sin\theta\cos\theta \\ \sin^2\theta & \cos^2\theta & -2\sin\theta\cos\theta \\ -\sin\theta\cos\theta & \sin\theta\cos\theta & \cos^2\theta - \sin^2\theta \end{bmatrix} \begin{Bmatrix} \varepsilon_{11} \\ \varepsilon_{22} \\ \varepsilon_{12} \end{Bmatrix} \quad \text{(rotation of 2D strains)} \qquad (170)$$

We need to find similar relations for the other three stresses and strains, $\sigma_{33}, \sigma_{23}, \sigma_{31}$ and $\varepsilon_{33}, \varepsilon_{23}, \varepsilon_{31}$. The case of out-of-plane stress and strain $\sigma_{33}, \varepsilon_{33}$ is straightforward because the axis $x_3$ and $x_3'$ coincide and the transformation of stresses and strains is an identity, i.e.,

$$\sigma_{33}' = \sigma_{33} \quad \text{(rotation of out-of-plane stress)} \tag{171}$$

$$\varepsilon_{33}' = \varepsilon_{33} \quad \text{(rotation of out-of-plane strain)} \tag{172}$$

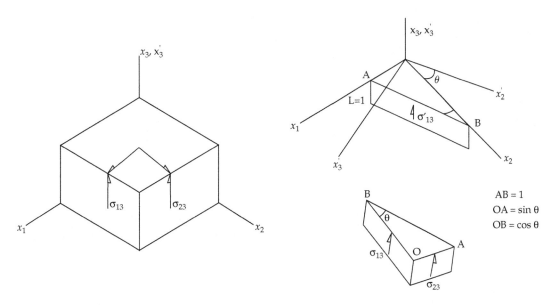

**FIGURE 8**   Rotated wedge element for the analysis of shear stresses $\sigma_{23}, \sigma_{31}$.

The case of the shear stresses and strains $\sigma_{23}, \sigma_{31}$ and $\varepsilon_{23}, \varepsilon_{31}$ is resolved by considering a small wedge element as shown in Figure 8. Vertical free-body analysis of the wedge element yields

$$\sigma_{13}' = \sigma_{13}\cos\theta + \sigma_{23}\sin\theta$$
$$\sigma_{23}' = \sigma_{13}\cos(\theta + \tfrac{\pi}{2}) + \sigma_{23}\sin(\theta + \tfrac{\pi}{2}) \tag{173}$$

Using the trigonometric relations $\sin(\theta + \tfrac{\pi}{2}) = \cos\theta$, $\cos(\theta + \tfrac{\pi}{2}) = -\sin\theta$, Eq. (173) becomes

$$\sigma_{13}' = \sigma_{13}\cos\theta + \sigma_{23}\sin\theta$$
$$\sigma_{23}' = -\sigma_{13}\sin\theta + \sigma_{23}\cos\theta \tag{174}$$

Equation (174) can be rearranged in the order of Eq. (167) and then expressed in matrix form as

$$\begin{Bmatrix} \sigma_{23}' \\ \sigma_{13}' \end{Bmatrix} = \begin{bmatrix} \cos\theta & -\sin\theta \\ \sin\theta & \cos\theta \end{bmatrix} \begin{Bmatrix} \sigma_{23} \\ \sigma_{13} \end{Bmatrix} \quad \begin{pmatrix} \text{rotation of} \\ \sigma_{23}, \sigma_{13} \text{ stresses} \end{pmatrix} \tag{175}$$

The rotation relation of Eq. (175) also applies to strains since the strain tensor and stress tensor behave similarly, i.e.,

$$\begin{Bmatrix} \varepsilon'_{23} \\ \varepsilon'_{13} \end{Bmatrix} = \begin{bmatrix} \cos\theta & -\sin\theta \\ \sin\theta & \cos\theta \end{bmatrix} \begin{Bmatrix} \varepsilon_{23} \\ \varepsilon_{13} \end{Bmatrix} \quad \begin{pmatrix} \text{rotation of} \\ \varepsilon_{23}, \varepsilon_{13} \text{ strains} \end{pmatrix} \tag{176}$$

Combining Eqs. (169), (171), (175) into a single matrix expression yields the 3D stress rotation relation

$$\begin{Bmatrix} \sigma'_{11} \\ \sigma'_{22} \\ \sigma'_{33} \\ \sigma'_{23} \\ \sigma'_{13} \\ \sigma'_{12} \end{Bmatrix} = \begin{bmatrix} \cos^2\theta & \sin^2\theta & 0 & 0 & 0 & 2\sin\theta\cos\theta \\ \sin^2\theta & \cos^2\theta & 0 & 0 & 0 & -2\sin\theta\cos\theta \\ 0 & 0 & 1 & 0 & 0 & 0 \\ 0 & 0 & 0 & \cos\theta & -\sin\theta & 0 \\ 0 & 0 & 0 & \sin\theta & \cos\theta & 0 \\ -\sin\theta\cos\theta & \sin\theta\cos\theta & 0 & 0 & 0 & \cos^2\theta - \sin^2\theta \end{bmatrix} \begin{Bmatrix} \sigma_{11} \\ \sigma_{22} \\ \sigma_{33} \\ \sigma_{23} \\ \sigma_{13} \\ \sigma_{12} \end{Bmatrix} \quad \begin{pmatrix} \text{3D stress} \\ \text{rotation} \end{pmatrix} \tag{177}$$

The expression in Eq. (177) agrees with Ref. [4], p. 477, Eq. (A.54) citing Ref. [13], p. 22, Eq. (102). Similarly, combining Eqs. (170), (172), (176) into a single matrix expression yields the 3D strain rotation relation

$$\begin{Bmatrix} \varepsilon'_{11} \\ \varepsilon'_{22} \\ \varepsilon'_{33} \\ \varepsilon'_{23} \\ \varepsilon'_{13} \\ \varepsilon'_{12} \end{Bmatrix} = \begin{bmatrix} \cos^2\theta & \sin^2\theta & 0 & 0 & 0 & 2\sin\theta\cos\theta \\ \sin^2\theta & \cos^2\theta & 0 & 0 & 0 & -2\sin\theta\cos\theta \\ 0 & 0 & 1 & 0 & 0 & 0 \\ 0 & 0 & 0 & \cos\theta & -\sin\theta & 0 \\ 0 & 0 & 0 & \sin\theta & \cos\theta & 0 \\ -\sin\theta\cos\theta & \sin\theta\cos\theta & 0 & 0 & 0 & \cos^2\theta - \sin^2\theta \end{bmatrix} \begin{Bmatrix} \varepsilon_{11} \\ \varepsilon_{22} \\ \varepsilon_{33} \\ \varepsilon_{23} \\ \varepsilon_{13} \\ \varepsilon_{12} \end{Bmatrix} \quad \begin{pmatrix} \text{3D strain} \\ \text{rotation} \end{pmatrix} \tag{178}$$

In view of Eqs. (177), (178), define the $\theta$-dependent 3D rotation matrix $\mathbf{T}$ as

$$\mathbf{T} = \begin{bmatrix} \cos^2\theta & \sin^2\theta & 0 & 0 & 0 & 2\sin\theta\cos\theta \\ \sin^2\theta & \cos^2\theta & 0 & 0 & 0 & -2\sin\theta\cos\theta \\ 0 & 0 & 1 & 0 & 0 & 0 \\ 0 & 0 & 0 & \cos\theta & -\sin\theta & 0 \\ 0 & 0 & 0 & \sin\theta & \cos\theta & 0 \\ -\sin\theta\cos\theta & \sin\theta\cos\theta & 0 & 0 & 0 & \cos^2\theta - \sin^2\theta \end{bmatrix} \quad \begin{pmatrix} \text{3D} \\ \text{rotation} \\ \text{matrix} \end{pmatrix} \tag{179}$$

The stress rotation Eq. (177) can be written directly in matrix notations using Eqs. (167), (179), i.e., as

$$\sigma' = \mathbf{T}\sigma \tag{180}$$

The strain rotation Eq. (178) cannot be written directly in matrix notations because the shear strains $\varepsilon_{23}, \varepsilon_{31}, \varepsilon_{12}$ appear with a factor of 2 in Eq. (167) but without such factor in Eq. (178). To resolve this issue, define the **R** matrix in 3D, i.e.,

$$\mathbf{R} = \begin{bmatrix} 1 & & & & & \\ & 1 & & & & \\ & & 1 & & & \\ & & & 2 & & \\ & & & & 2 & \\ & & & & & 2 \end{bmatrix} \qquad \mathbf{R}^{-1} = \begin{bmatrix} 1 & & & & & \\ & 1 & & & & \\ & & 1 & & & \\ & & & 1/2 & & \\ & & & & 1/2 & \\ & & & & & 1/2 \end{bmatrix} \tag{181}$$

Using Eq. (181), write

$$\begin{Bmatrix} \varepsilon'_{11} \\ \varepsilon'_{22} \\ \varepsilon'_{33} \\ 2\varepsilon'_{23} \\ 2\varepsilon'_{31} \\ 2\varepsilon'_{12} \end{Bmatrix} = \begin{bmatrix} 1 & & & & & \\ & 1 & & & & \\ & & 1 & & & \\ & & & 2 & & \\ & & & & 2 & \\ & & & & & 2 \end{bmatrix} \begin{Bmatrix} \varepsilon'_{11} \\ \varepsilon'_{22} \\ \varepsilon'_{33} \\ \varepsilon'_{23} \\ \varepsilon'_{13} \\ \varepsilon'_{12} \end{Bmatrix} \tag{182}$$

$$\begin{Bmatrix} \varepsilon_{11} \\ \varepsilon_{22} \\ \varepsilon_{33} \\ \varepsilon_{23} \\ \varepsilon_{31} \\ \varepsilon_{12} \end{Bmatrix} = \begin{bmatrix} 1 & & & & & \\ & 1 & & & & \\ & & 1 & & & \\ & & & 1/2 & & \\ & & & & 1/2 & \\ & & & & & 1/2 \end{bmatrix} \begin{Bmatrix} \varepsilon_{11} \\ \varepsilon_{22} \\ \varepsilon_{33} \\ 2\varepsilon_{23} \\ 2\varepsilon_{31} \\ 2\varepsilon_{12} \end{Bmatrix} \tag{183}$$

Substitution of Eq. (178) into Eq. (182) gives

$$\begin{Bmatrix} \varepsilon'_{11} \\ \varepsilon'_{22} \\ \varepsilon'_{33} \\ 2\varepsilon'_{23} \\ 2\varepsilon'_{31} \\ 2\varepsilon'_{12} \end{Bmatrix} = \begin{bmatrix} 1 & & & & & \\ & 1 & & & & \\ & & 1 & & & \\ & & & 2 & & \\ & & & & 2 & \\ & & & & & 2 \end{bmatrix} \begin{Bmatrix} \varepsilon'_{11} \\ \varepsilon'_{22} \\ \varepsilon'_{33} \\ \varepsilon'_{23} \\ \varepsilon'_{13} \\ \varepsilon'_{12} \end{Bmatrix} =$$

$$\begin{bmatrix} 1 & & & & & \\ & 1 & & & & \\ & & 1 & & & \\ & & & 2 & & \\ & & & & 2 & \\ & & & & & 2 \end{bmatrix} \begin{bmatrix} \cos^2\theta & \sin^2\theta & 0 & 0 & 0 & 2\sin\theta\cos\theta \\ \sin^2\theta & \cos^2\theta & 0 & 0 & 0 & -2\sin\theta\cos\theta \\ 0 & 0 & 1 & 0 & 0 & 0 \\ 0 & 0 & 0 & \cos\theta & -\sin\theta & 0 \\ 0 & 0 & 0 & \sin\theta & \cos\theta & 0 \\ -\sin\theta\cos\theta & \sin\theta\cos\theta & 0 & 0 & 0 & \cos^2\theta - \sin^2\theta \end{bmatrix} \begin{Bmatrix} \varepsilon_{11} \\ \varepsilon_{22} \\ \varepsilon_{33} \\ \varepsilon_{23} \\ \varepsilon_{13} \\ \varepsilon_{12} \end{Bmatrix} \tag{184}$$

Use of Eq. (183) into Eq. (184) yields

$$
\begin{Bmatrix} \varepsilon'_{11} \\ \varepsilon'_{22} \\ \varepsilon'_{33} \\ 2\varepsilon'_{23} \\ 2\varepsilon'_{31} \\ 2\varepsilon'_{12} \end{Bmatrix} =
\begin{bmatrix} 1 & & & & & \\ & 1 & & & & \\ & & 1 & & & \\ & & & 2 & & \\ & & & & 2 & \\ & & & & & 2 \end{bmatrix}
$$

$$
\times \begin{bmatrix}
\cos^2\theta & \sin^2\theta & 0 & 0 & 0 & 2\sin\theta\cos\theta \\
\sin^2\theta & \cos^2\theta & 0 & 0 & 0 & -2\sin\theta\cos\theta \\
0 & 0 & 1 & 0 & 0 & 0 \\
0 & 0 & 0 & \cos\theta & -\sin\theta & 0 \\
0 & 0 & 0 & \sin\theta & \cos\theta & 0 \\
-\sin\theta\cos\theta & \sin\theta\cos\theta & 0 & 0 & 0 & \cos^2\theta - \sin^2\theta
\end{bmatrix}
\tag{185}
$$

$$
\times \begin{bmatrix} 1 & & & & & \\ & 1 & & & & \\ & & 1 & & & \\ & & & 1/2 & & \\ & & & & 1/2 & \\ & & & & & 1/2 \end{bmatrix}
\begin{Bmatrix} \varepsilon_{11} \\ \varepsilon_{22} \\ \varepsilon_{33} \\ 2\varepsilon_{23} \\ 2\varepsilon_{31} \\ 2\varepsilon_{12} \end{Bmatrix}
$$

Use of Eqs. (179), (181) into Eq. (185) yields a compact matrix expression, i.e.,

$$
\varepsilon' = \mathbf{R}\mathbf{T}\mathbf{R}^{-1}\varepsilon
\tag{186}
$$

Upon evaluation, one finds that $\mathbf{R}\mathbf{T}\mathbf{R}^{-1}$ is actually $\mathbf{T}^{-t}$, the transpose of the inverse of $\mathbf{T}$ (see Section 5.6), i.e.,

$$
\mathbf{R}\mathbf{T}\mathbf{R}^{-1} = \mathbf{T}^{-t}
\tag{187}
$$

Substitution of Eq. (187) into Eq. (186) yields

$$
\varepsilon' = \mathbf{T}^{-t}\varepsilon
\tag{188}
$$

Recall Eq. (166) which gives the stress–strain relations in local coordinates, i.e.,

$$
\sigma' = \mathbf{C}'\,\varepsilon' \quad \text{(stress–strain matrix relation in local coordinates)}
\tag{189}
$$

Use Eqs. (180), (186) to express $\sigma'$, $\varepsilon'$ of Eq. (189) in terms of $\sigma$, $\varepsilon$ and get

$$
\mathbf{T}\sigma = \mathbf{C}'\,\mathbf{T}^{-t}\varepsilon
\tag{190}
$$

Premultiplication of Eq. (190) by $\mathbf{T}^{-1}$ yields

$$\boldsymbol{\sigma} = \mathbf{T}^{-1}\mathbf{C}'\,\mathbf{T}^{-t}\boldsymbol{\varepsilon} \tag{191}$$

Recall Eq. (168) expressing the stress−strain relation in global coordinates, i.e.,

$$\boldsymbol{\sigma} = \mathbf{C}\,\boldsymbol{\varepsilon} \quad \text{(stress−strain matrix relation in global coordinates)} \tag{192}$$

Comparison of Eqs. (191), (192) yields the rotated stiffness matrix as

$$\mathbf{C} = \mathbf{T}^{-1}\mathbf{C}'\,\mathbf{T}^{-t} \tag{193}$$

After performing the matrix operations, Eq. (193) yields the rotated stiffness matrix $\mathbf{C}$ in the form

$$\mathbf{C} = \begin{bmatrix} C_{11} & C_{12} & C_{13} & 0 & 0 & C_{16} \\ C_{12} & C_{22} & C_{23} & 0 & 0 & C_{26} \\ C_{13} & C_{23} & C_{33} & 0 & 0 & C_{36} \\ 0 & 0 & 0 & C_{44} & C_{45} & 0 \\ 0 & 0 & 0 & C_{45} & C_{55} & 0 \\ C_{16} & C_{26} & C_{36} & 0 & 0 & C_{66} \end{bmatrix} \tag{194}$$

where $C_{14} = C_{15} = C_{24} = C_{25} = C_{34} = C_{35} = C_{46} = C_{56} = 0$. The stress−strain relation of Eq. (38) is written as

$$\begin{Bmatrix} \sigma_{11} \\ \sigma_{22} \\ \sigma_{33} \\ \sigma_{23} \\ \sigma_{31} \\ \sigma_{12} \end{Bmatrix} = \begin{bmatrix} C_{11} & C_{12} & C_{13} & 0 & 0 & C_{16} \\ C_{12} & C_{22} & C_{23} & 0 & 0 & C_{26} \\ C_{13} & C_{23} & C_{33} & 0 & 0 & C_{36} \\ 0 & 0 & 0 & C_{44} & C_{45} & 0 \\ 0 & 0 & 0 & C_{45} & C_{55} & 0 \\ C_{16} & C_{26} & C_{36} & 0 & 0 & C_{66} \end{bmatrix} \begin{Bmatrix} \varepsilon_{11} \\ \varepsilon_{22} \\ \varepsilon_{33} \\ 2\varepsilon_{23} \\ 2\varepsilon_{31} \\ 2\varepsilon_{12} \end{Bmatrix} \tag{195}$$

A stiffness matrix of the form of Eq. (194) is also known as *monoclinic stiffness matrix* because it can be shown that the general stiffness matrix of Eq. (36) reduces to the form of Eq. (194) in the case of a material that shown monoclinic symmetry. In our case, the rotated composite lamina shows monoclinic symmetry with the symmetry plane being the $Ox_1x_2$ plane.

## 2.5.3 Equations of Motion for a Monoclinic Composite Layer

The monoclinic stiffness matrix of Eq. (194) has much fewer terms than the generally anisotropic stiffness matrix presented earlier in Eq. (36). As a consequence, the expanded form of the equation of motion in terms of displacements becomes more manageable. Recall that in the general case, the expanded form of the equation of motion in terms of displacements had 18 terms per degree of freedom (dof), as illustrated for the $\ddot{u}_1$-dof by

Eq. (63). However, in the case of monoclinic stiffness matrix, $C_{14} = C_{15} = C_{24} = C_{25} = C_{34} = C_{35} = C_{46} = C_{56} = 0$, and Eq. (63) simplifies, i.e.,

$$C_{11}u_{1,11} + C_{66}u_{1,22} + C_{55}u_{1,33} + 2C_{16}u_{1,12} +$$
$$C_{16}u_{2,11} + C_{26}u_{2,22} + C_{45}u_{2,33} + (C_{12} + C_{66})u_{2,12} \tag{196}$$
$$+ (C_{36} + C_{45})u_{3,23} + (C_{13} + C_{55})u_{3,31} = \rho \ddot{u}_1$$

Similarly, the $\ddot{u}_2$ and $\ddot{u}_3$ equations become

$$C_{16}u_{1,11} + C_{26}u_{1,22} + C_{45}u_{1,33} + (C_{12} + C_{66})u_{1,12} +$$
$$C_{66}u_{2,11} + C_{22}u_{2,22} + C_{44}u_{2,33} + 2C_{26}u_{2,12} \tag{197}$$
$$+ (C_{23} + C_{44})u_{3,23} + (C_{36} + C_{45})u_{3,31} = \rho \ddot{u}_2$$

$$(C_{36} + C_{45})u_{1,23} + (C_{13} + C_{55})u_{1,31}$$
$$+ (C_{23} + C_{45})u_{2,23} + (C_{36} + C_{45})u_{2,31} \tag{198}$$
$$+ C_{55}u_{3,11} + C_{44}u_{3,22} + C_{33}u_{3,33} + 2C_{45}u_{3,12} = \rho \ddot{u}_3$$

## 2.5.4 Rotated Compliance Matrix

A similar argument can be employed to find the rotated compliance matrix **S**. Recall the strain−stress matrix relation Eq. (39) and write it in local coordinates, i.e.,

$$\varepsilon' = \mathbf{S}' \, \sigma' \quad \text{(strain−stress matrix relation in local coordinates)} \tag{199}$$

Use Eqs. (180), (186) to express $\sigma'$, $\varepsilon'$ of Eq. (199) in terms of $\sigma$, $\varepsilon$ and get

$$\mathbf{T}^{-t}\varepsilon = \mathbf{S}' \, \mathbf{T}\sigma \tag{200}$$

Multiply Eq. (200) by $\mathbf{T}^t$ and obtain

$$\varepsilon = \mathbf{T}^t \mathbf{S}' \, \mathbf{T}\sigma \tag{201}$$

Use Eq. (39) to write the strain−stress relation in global coordinates, i.e.,

$$\varepsilon = \mathbf{S} \, \sigma \quad \text{(strain − stress matrix relation in global coordinates)} \tag{202}$$

Comparison of Eqs. (201), (202) yields the rotated compliance matrix **S**, i.e.,

$$\mathbf{S} = \mathbf{T}^t \mathbf{S}' \, \mathbf{T} \tag{203}$$

After performing the matrix operations, Eq. (203) yields the rotated compliance matrix **S** in the form

$$\mathbf{S} = \begin{bmatrix} S_{11} & S_{12} & S_{13} & 0 & 0 & S_{16} \\ S_{12} & S_{22} & S_{23} & 0 & 0 & S_{26} \\ S_{13} & S_{23} & S_{33} & 0 & 0 & S_{36} \\ 0 & 0 & 0 & S_{44} & S_{45} & 0 \\ 0 & 0 & 0 & S_{45} & S_{55} & 0 \\ S_{16} & S_{26} & S_{36} & 0 & 0 & S_{66} \end{bmatrix} \tag{204}$$

The strain–stress relation of Eq. (43) is written as

$$
\begin{Bmatrix} \varepsilon_{11} \\ \varepsilon_{22} \\ \varepsilon_{33} \\ 2\varepsilon_{23} \\ 2\varepsilon_{31} \\ 2\varepsilon_{12} \end{Bmatrix} =
\begin{bmatrix}
S_{11} & S_{12} & S_{13} & 0 & 0 & S_{16} \\
S_{12} & S_{22} & S_{23} & 0 & 0 & S_{26} \\
S_{13} & S_{23} & S_{33} & 0 & 0 & S_{36} \\
0 & 0 & 0 & S_{44} & S_{45} & 0 \\
0 & 0 & 0 & S_{45} & S_{55} & 0 \\
S_{16} & S_{26} & S_{36} & 0 & 0 & S_{66}
\end{bmatrix}
\begin{Bmatrix} \sigma_{11} \\ \sigma_{22} \\ \sigma_{33} \\ \sigma_{23} \\ \sigma_{31} \\ \sigma_{12} \end{Bmatrix}
\tag{205}
$$

The compliance matrix **S** of Eq. (204) also displays monoclinic symmetry.

## 2.5.5 Note on the Use of Closed-Form Expression in the C and S matrices

Note: Closed-form expressions for the elements of the rotated stiffness and compliance matrices, **C** and **S**, are possible and can be found in some textbooks. However, these closed-form expressions contain quite elaborate formulae; if manual coding of these formulae is attempted, then hard-to-trace errors may be inadvertently introduced. Hence, the use of such closed-form complicated formulae is not recommended.

## 2.5.6 Proof of $\mathbf{RTR}^{-1} = \mathbf{T}^{-t}$ in 3D

In Section 5.2, Eq. (187), we cited without proof the formula $\mathbf{RTR}^{-1} = \mathbf{T}^{-t}$ for the 3D case. Here we will perform this proof building onto the proof given in Section 4.5 for the 2D case. As in Section 4.5, we will use the convenient shorthand notations $s$ and $c$, i.e.,

$$
s = \sin\theta \quad c = \cos\theta \quad s^2 + c^2 = \sin^2\theta + \cos^2\theta = 1
\tag{206}
$$

Use Eq. (206) to write the 3D rotation matrix Eq. (179) as

$$
\mathbf{T} =
\begin{bmatrix}
c^2 & s^2 & & & & 2sc \\
s^2 & s^2 & & & & -2sc \\
& & 1 & & & \\
& & & c & -s & \\
& & & s & c & \\
-sc & sc & & & & c^2 - s^2
\end{bmatrix}
\quad \text{(3D rotation matrix)}
\tag{207}
$$

Recall the 2D rotation matrix of Eq. (156) and use the subscript 2D to differentiate it from the 3D rotation matrix **T** of Eq. (207), i.e.,

$$
\mathbf{T}_{2D} =
\begin{bmatrix}
c^2 & s^2 & 2sc \\
s^2 & c^2 & -2sc \\
-sc & sc & c^2 - s^2
\end{bmatrix}
\quad \text{(2D rotation matrix)}
\tag{208}
$$

Recall Eqs. (158), (160) which give the inverse and its transpose for the 2D rotation matrix, i.e.,

$$\mathbf{T}_{2D}^{-1} = \begin{bmatrix} c^2 & s^2 & -2sc \\ s^2 & c^2 & 2sc \\ sc & -sc & c^2-s^2 \end{bmatrix} \qquad \mathbf{T}_{2D}^{-t} = \begin{bmatrix} c^2 & s^2 & sc \\ s^2 & c^2 & -sc \\ -2sc & 2sc & c^2-s^2 \end{bmatrix} \qquad (209)$$

Recall Eq. (176) which gives the rotation relations for shear stresses $\sigma_{23}, \sigma_{31}$ and denote by $\mathbf{T}_{23-31}$ the corresponding $2 \times 2$ rotation matrix, i.e.,

$$\mathbf{T}_{23-31} = \begin{bmatrix} c & -s \\ s & c \end{bmatrix} \quad \text{(rotation matrix for } \sigma_{23}, \sigma_{31} \text{ shear stresses)} \qquad (210)$$

The inverse of Eq. (210) and its transpose are

$$\mathbf{T}_{23-31}^{-1} = \begin{bmatrix} c & s \\ -s & c \end{bmatrix} \qquad \mathbf{T}_{23-31}^{-t} = \begin{bmatrix} c & -s \\ s & c \end{bmatrix} = \mathbf{T}_{23-31} \qquad (211)$$

In view of the above, it is apparent that the 3D rotation matrix $\mathbf{T}$ of Eq. (207) is made up of three independent parts: the 2D part given by Eq. (208); the 23−31 part given by Eq. (210); and the 33 part, which is an identity. Since these three parts are decoupled and independent, their effect on the calculation of $\mathbf{RTR}^{-1}$ can be treated independently. For the 2D part, we have already proven in Section 4.5, Eq. (162), that

$$\mathbf{R}_{2D}\mathbf{T}_{2D}\mathbf{R}_{2D}^{-1} = \mathbf{T}_{2D}^{-t} \qquad (212)$$

The 33 part of $\mathbf{T}$ needs no processing since it is an identity. It remains to prove the 23−31 part of $\mathbf{T}$; recall the 23−31 part of the $\mathbf{R}$ matrix as

$$\mathbf{R}_{23-31} = \begin{bmatrix} 2 & 0 \\ 0 & 2 \end{bmatrix} \qquad \mathbf{R}_{23-31}^{-1} = \begin{bmatrix} 1/2 & 0 \\ 0 & 1/2 \end{bmatrix} \qquad (213)$$

Hence, the $\mathbf{RTR}^{-1}$ expression for the 23−31 part of $\mathbf{T}$ is calculated as

$$\mathbf{R}_{23-31}\mathbf{T}_{23-31}\mathbf{R}_{23-31}^{-1} = \begin{bmatrix} 2 & 0 \\ 0 & 2 \end{bmatrix}\begin{bmatrix} c & -s \\ s & c \end{bmatrix}\begin{bmatrix} 1/2 & 0 \\ 0 & 1/2 \end{bmatrix} = \begin{bmatrix} 2c & -2s \\ 2s & 2c \end{bmatrix}\begin{bmatrix} 1/2 & 0 \\ 0 & 1/2 \end{bmatrix}\begin{bmatrix} c & -s \\ s & c \end{bmatrix} = \mathbf{T}_{23-31}$$
$$(214)$$

Substitution of Eq. (211) into Eq. (214) yields

$$\mathbf{R}_{23-31}\mathbf{T}_{23-31}\mathbf{R}_{23-31}^{-1} = \mathbf{T}_{23-31}^{-t} \qquad (215)$$

In view of Eqs. (212), (215) and considering that the 33 part is an identity, it is apparent that the complete 3D rotation matrix **T** also satisfies the relation.

$$\mathbf{RTR}^{-1} = \mathbf{T}^{-t} \qquad \text{QED} \qquad (216)$$

## 2.6 PROBLEMS AND EXERCISES

Problems and exercises and worked out examples are given in the Instructor Manual posted on the publisher's web site.

## References

[1] Anon. (1989) "Composites II: Material Selection and Applications", Materials Engineering Institute course 31, ASM International, Materials Park, OH.

[2] Barbero, E. J. (1999) *Introduction to Composite Materials Design*, Taylor & Francis, New York, NY, 1999.

[3] Hyer, M. W. (2009) *Stress Analysis of Fiber-Reinforced Composite Materials*, 2nd Ed., DEStech Pub., Lancaster, PA, 2009.

[4] Jones, R. M. (1999) *Mechanics of Composite Materials*, 2nd Ed., Taylor & Francis.

[5] Mallick, P. K. (1993) *Fiber-Reinforced Composites*, Marcel Dekker, New York, NY.

[6] Hashin, Z.; Rosen, B. W. (1964) "The Elastic Moduli of Fiber-Reinforced Materials", *Journal of Applied Mechanics*, 223−230, Jun. 1964.

[7] Tsai, S. W.; Hahn, H. T. (1980) *Introduction to Composite Materials*, Technomic, Lancaster, PA.

[8] Herakovich, C. T. (1998) *Mechanics of Fibrous Composites*, Wiley, New York, NY, 1998.

[9] Tsai, S. W. (1987) *Composites Design*, 3rd Ed., Think Composites Pub., Dayton, OK, 1987.

[10] Tsai, S. W. (1992) *Theory of Composites Design*, Think Composites Pub., Dayton, OH, 1992.

[11] Boresi, A. P.; Schmidt, R. J.; Sidebottom, O. M. (1993) *Advanced Mechanics of Materials*, 5th Ed., Wiley, New York, NY, 1993.

[12] Reuter, R. C., Jr. (1971) "Concise Property Transformation Relations for an Anisotropic Lamina", *Journal of Composite Materials*270−272 , Apr. 1971.

[13] Tsai, S. W. (1966)  Nov. *Mechanics of Composite Materials, Part II, Theoretical Aspects*, Air Force Materials Laboratory, WPAFB, Dayton, OH.

[14] Kassapoglou, C. (2010) *Design and Analysis of Composite Structures with Applications to Aerospace Structures*, 1st Ed., Wiley, Hoboken, NJ.

# 3.1 INTRODUCTION

This chapter deals with the study of composite vibration, which may be used for structural health monitoring (SHM) applications. We will analyze the vibration of thin laminated composite plates (Figure 1). Though each lamina in the laminate may have different orientations and hence different stiffness matrices, their mass density is assumed more or less the same; hence the mid surface of the plate is considered to be placed centroidally. The top and bottom faces of the plate are free, hence the surface tractions are zero. This means that the z-direction stress and the surface shear stresses are zero at the plate surface, i.e., $\sigma_{zz}(\pm h/2) = 0$, $\sigma_{zx}(\pm h/2) = 0$, $\sigma_{yz}(\pm h/2) = 0$.

Because the plate is thin, the z-direction stress $\sigma_{zz}$ is assumed zero everywhere throughout the thickness, i.e., $\sigma_{zz} \equiv 0$. Plane-stress conditions apply in each ply of the composite layup. In our analysis which follow Refs. [2−4], we are interested in the strains $\varepsilon_{xx}$, $\varepsilon_{yy}$, $\varepsilon_{xy}$, $\varepsilon_{xz}$, $\varepsilon_{yz}$ and corresponding stresses $\sigma_{xx}$, $\sigma_{yy}$, $\sigma_{xy}$, $\sigma_{xz}$, $\sigma_{yz}$.

## 3.1.1 Displacements for Axial−Flexural Vibration of Composite Plates

Axial−flexural vibration occurs when the in-plane and out-of-plane motions occur simultaneously. Assume the $x, y, z$ displacements of the plate mid surface are $u_0, v_0, w_0$,

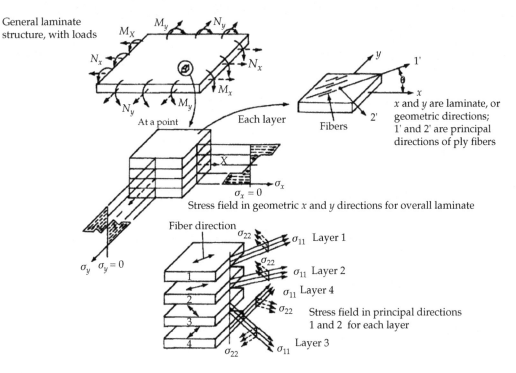

General laminate structure, with loads

x and y are laminate, or geometric directions; 1' and 2' are principal directions of ply fibers

At a point    Each layer    Fibers

Stress field in geometric x and y directions for overall laminate

Fiber direction

$\sigma_{22}$    $\sigma_{11}$  Layer 1
$\sigma_{22}$    $\sigma_{11}$  Layer 2
$\sigma_{11}$  Layer 4
$\sigma_{22}$    Stress field in principal directions 1 and 2 for each layer
Layer 3
$\sigma_{22}$    $\sigma_{11}$

$\sigma_y$  $\sigma_y = 0$

$\sigma_x = 0$    $\sigma_x$

**FIGURE 1**    Stresses and loads definitions in a laminated composite plate [1].

respectively. Under the Love–Kirchhoff plate bending theory, the displacements $u, v, w$ of any point at location $z$ are given by

$$u = u_0 - z\frac{\partial w_0}{\partial x}$$

$$v = v_0 - z\frac{\partial w_0}{\partial y} \tag{1}$$

$$w = w_0$$

## 3.1.2  Stress Resultants

Integration of stresses across the thickness gives the force stress resultants $N_x, N_y, N_{xy}, N_{xz}, N_{yz}$ (vertical forces per unit length, a.k.a. line forces) and moment stress resultants $M_x, M_y, M_{xy}$ (moments per unit length, a.k.a. line moments) as shown in Figure 2, i.e.,

$$N_x = \int_{-h/2}^{h/2} \sigma_{xx}dz \qquad N_{xy} = \int_{-h/2}^{h/2} \sigma_{xy}dz \qquad M_x = \int_{-h/2}^{h/2} \sigma_{xx}zdz$$

$$N_y = \int_{-h/2}^{h/2} \sigma_{yy}dz \qquad N_{xz} = \int_{-h/2}^{h/2} \sigma_{xz}dz \qquad M_y = \int_{-h/2}^{h/2} \sigma_{yy}zdz \tag{2}$$

$$N_{yz} = \int_{-h/2}^{h/2} \sigma_{yz}dz \qquad M_{xy} = \int_{-h/2}^{h/2} \sigma_{xy}zdz$$

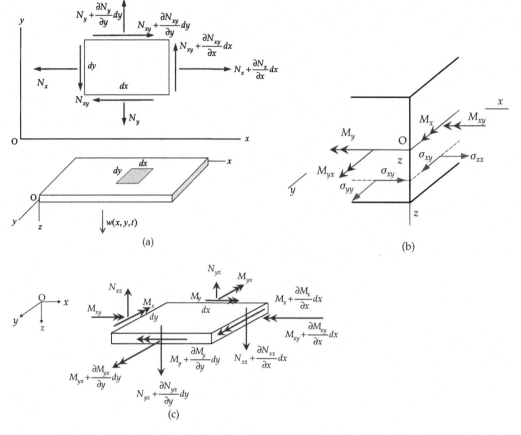

**FIGURE 2**   Infinitesimal plate element in Cartesian coordinates for the analysis of axial–flexural vibration of rectangular plates: (a) plate coordinates; (b) definition of moments and stresses; (c) stress resultants on the infinitesimal element.

## 3.2  EQUATIONS OF MOTION IN TERMS OF STRESS RESULTANTS

### 3.2.1  Derivation of Equations of Motion from Free Body Diagram

Free body analysis of the infinitesimal plate element $dx\,dy$ of Figure 2 yields

$$\text{In-plane } x\text{-direction forces:} \quad \frac{\partial N_x}{\partial x} + \frac{\partial N_{yx}}{\partial y} = m\ddot{u}_0 \tag{3}$$

$$\text{In-plane } y\text{-direction forces:} \quad \frac{\partial N_y}{\partial y} + \frac{\partial N_{xy}}{\partial x} = m\ddot{v}_0 \tag{4}$$

$$\text{Out-of-plane forces:} \quad \frac{\partial N_{xz}}{\partial x} + \frac{\partial N_{yz}}{\partial y} = m\ddot{w}_0 \tag{5}$$

$$M_x \text{ moments:} \quad \frac{\partial M_x}{\partial x} + \frac{\partial M_{yx}}{\partial y} - N_{xz} = 0 \tag{6}$$

$$M_y \text{ moments:} \quad \frac{\partial M_y}{\partial y} + \frac{\partial M_{xy}}{\partial x} - N_{yz} = 0 \tag{7}$$

where $m$ is the mass per unit area of the plate given by

$$m = \int_{-h/2}^{h/2} \rho\, dz \quad \text{(mass per unit area of the plate)} \tag{8}$$

Rotary inertia effects are ignored and hence no inertia terms appear in the moment relations Eqs. (6), (7); the $M_z$ moment equation about the out-of-plane axis $z$ is not needed. Equations (6), (7) give the shear forces $N_{xz}, N_{yz}$ in terms of the bending moments $M_x, M_{xy}, M_{yx}, M_y$, i.e.,

$$N_{xz} = \frac{\partial M_x}{\partial x} + \frac{\partial M_{yx}}{\partial y} \tag{9}$$

$$N_{yz} = \frac{\partial M_y}{\partial y} + \frac{\partial M_{xy}}{\partial x} \tag{10}$$

Upon differentiation, Eqs. (9), (10) yield

$$\frac{\partial N_{xz}}{\partial x} = \frac{\partial^2 M_x}{\partial x^2} + \frac{\partial^2 M_{yx}}{\partial x \partial y} \tag{11}$$

$$\frac{\partial N_{yz}}{\partial y} = \frac{\partial^2 M_y}{\partial y^2} + \frac{\partial^2 M_{xy}}{\partial x \partial y} \tag{12}$$

Substitution of Eqs. (11), (12) into Eq. (5) yields

$$\frac{\partial^2 M_x}{\partial x^2} + 2\frac{\partial^2 M_{xy}}{\partial x \partial y} + \frac{\partial^2 M_y}{\partial y^2} = m\ddot{w}_0 \tag{13}$$

## 3.2.2 Derivation of Axial–Flexural Equations from Stress Equations of Motion

Recall the stress equations of motion with the $\sigma_{zz} = 0$ assumption, i.e.,

$$\frac{\partial \sigma_{xx}}{\partial x} + \frac{\partial \sigma_{xy}}{\partial y} + \frac{\partial \sigma_{xz}}{\partial z} = \rho \ddot{u} \tag{14}$$

$$\frac{\partial \sigma_{xy}}{\partial x} + \frac{\partial \sigma_{yy}}{\partial y} + \frac{\partial \sigma_{yz}}{\partial z} = \rho \ddot{v} \tag{15}$$

$$\frac{\partial \sigma_{zx}}{\partial x} + \frac{\partial \sigma_{zy}}{\partial y} = \rho \ddot{w} \tag{16}$$

### 3.2.2.1 Integration of $u$-Equation of Motion

Integration of Eq. (14) with respect to $z$ gives

$$\int_{-h/2}^{h/2} \left( \frac{\partial \sigma_{xx}}{\partial x} + \frac{\partial \sigma_{xy}}{\partial y} + \frac{\partial \sigma_{xz}}{\partial z} \right) dz = \int_{-h/2}^{h/2} \rho \ddot{u} dz \tag{17}$$

Expansion of the integral in Eq. (17) gives

$$\frac{\partial}{\partial x} \int_{-h/2}^{h/2} \sigma_{xx} dz + \frac{\partial}{\partial y} \int_{-h/2}^{h/2} \sigma_{xy} dz + \int_{-h/2}^{h/2} \frac{\partial \sigma_{xz}}{\partial z} dz = \int_{-h/2}^{h/2} \rho \ddot{u} dz \tag{18}$$

The last integral in Eq. (18) is discarded because it can be evaluated exactly and its value is zero due to free surface conditions at the upper and lower faces of the plate, i.e.,

$$\int_{-h/2}^{h/2} \frac{\partial \sigma_{xz}}{\partial z} dz = \sigma_{xz} \Big|_{-h/2}^{h/2} = \sigma_{xz}(\tfrac{h}{2}) - \sigma_{xz}(-\tfrac{h}{2}) = 0 - 0 = 0 \quad \text{(free surface conditions)} \tag{19}$$

Substitution of Eq. (19) into Eq. (18) and use of Eqs. (1), (2), (8) yields

$$\frac{\partial N_x}{\partial x} + \frac{\partial N_{xy}}{\partial y} = \int_{-h/2}^{h/2} \rho \left( \ddot{u}_0 - z \frac{\partial \ddot{w}_0}{\partial x} \right) dz = \ddot{u}_0 \int_{-h/2}^{h/2} \rho dz - \frac{\partial \ddot{w}_0}{\partial x} \int_{-h/2}^{h/2} \rho z dz = m \ddot{u}_0 \tag{20}$$

The second integral in Eq. (20) vanishes because it is the first moment of inertia about a centroidal axis. Hence, Eq. (20) becomes

$$\frac{\partial N_x}{\partial x} + \frac{\partial N_{xy}}{\partial y} = \rho h \ddot{u}_0 \tag{21}$$

Note that Eq. (21) is identical with Eq. (3).

### 3.2.2.2 Integration of $v$-Equation of Motion

Integration of Eq. (15) with respect to $z$ gives

$$\int_{-h/2}^{h/2} \left( \frac{\partial \sigma_{xy}}{\partial x} + \frac{\partial \sigma_{yy}}{\partial y} + \frac{\partial \sigma_{yz}}{\partial z} \right) dz = \int_{-h/2}^{h/2} \rho \ddot{v} dz \tag{22}$$

Expansion of the integral in Eq. (22) gives

$$\frac{\partial}{\partial x} \int_{-h/2}^{h/2} \sigma_{xy} dz + \frac{\partial}{\partial y} \int_{-h/2}^{h/2} \sigma_{yy} dz + \int_{-h/2}^{h/2} \frac{\partial \sigma_{yz}}{\partial z} dz = \int_{-h/2}^{h/2} \rho \ddot{v} dz \tag{23}$$

The last integral in Eq. (23) is discarded because it can be evaluated exactly and its value is zero due to free surface conditions at the upper and lower faces of the plate, i.e.,

$$\int_{-h/2}^{h/2} \frac{\partial \sigma_{yz}}{\partial z} dz = \sigma_{yz} \Big|_{-h/2}^{h/2} = \sigma_{yz}(\tfrac{h}{2}) - \sigma_{yz}(-\tfrac{h}{2}) = 0 - 0 = 0 \quad \text{(free surface conditions)} \tag{24}$$

Substitution of Eq. (24) into Eq. (23) and use of Eqs. (1), (2), (8) yields

$$\frac{\partial N_{xy}}{\partial x} + \frac{\partial N_y}{\partial y} = \int_{-h/2}^{h/2} \rho\left(\ddot{v}_0 - z\frac{\partial \ddot{w}_0}{\partial y}\right)dz = \ddot{v}_0\int_{-h/2}^{h/2}\rho dz - \frac{\partial \ddot{w}_0}{\partial y}\int_{-h/2}^{h/2}\rho z dz = m\ddot{v}_0 \tag{25}$$

The second integral in Eq. (25) vanishes because it is the first moment of inertia about a centroidal axis. Hence, Eq. (20) becomes

$$\frac{\partial N_{xy}}{\partial x} + \frac{\partial N_y}{\partial y} = \rho h \ddot{v}_0 \tag{26}$$

Note that Eq. (26) is identical with Eq. (4).

### 3.2.2.3 Integration of $w$-Equation of Motion

Integration of Eq. (16) with respect to $z$ gives

$$\int_{-h/2}^{h/2}\left(\frac{\partial \sigma_{xz}}{\partial x} + \frac{\partial \sigma_{yz}}{\partial y}\right)dz = \int_{-h/2}^{h/2}\rho \ddot{w} dz \tag{27}$$

Expansion of the integral in Eq. (27) and use of Eqs. (2), (8) yield

$$\frac{\partial N_{xz}}{\partial x} + \frac{\partial N_{yz}}{\partial y} = m\ddot{w} \tag{28}$$

#### 3.2.2.3.1  Calculation of Out-of-Plane Shear Resultant $N_{xz}$

To get $N_{xz}$, multiply Eq. (14) by $z$ and integrate with respect to $z$ to get

$$\int_{-h/2}^{h/2}\frac{\partial \sigma_{xx}}{\partial x}z dz + \int_{-h/2}^{h/2}\frac{\partial \sigma_{xy}}{\partial y}z dz + \int_{-h/2}^{h/2}\frac{\partial \sigma_{xz}}{\partial z}z dz = \int_{-h/2}^{h/2}\rho\ddot{u}\, z dz \tag{29}$$

or

$$\frac{\partial}{\partial x}\int_{-h/2}^{h/2}\sigma_{xx}z dz + \frac{\partial}{\partial y}\int_{-h/2}^{h/2}\sigma_{xy}z dz + \int_{-h/2}^{h/2}\frac{\sigma_{xz}}{\partial z}z dz = \int_{-h/2}^{h/2}\rho\ddot{u}\, z dz \tag{30}$$

Use of Eqs. (1), (2) into Eq. (30) gives

$$\frac{\partial M_x}{\partial x} + \frac{\partial M_{xy}}{\partial y} + \int_{-h/2}^{h/2}\left(z\frac{\partial \sigma_{xz}}{\partial z}\right)dz = \ddot{u}_0\int_{-h/2}^{h/2}\rho z dz - \frac{\partial \ddot{w}_0}{\partial x}\int_{-h/2}^{h/2}\rho z^2 dz = 0 \tag{31}$$

The first integral in the right-hand side of Eq. (31) vanishes because it is the first moment of inertia about a centroidal axis. The second integral is the second moment of inertia due to $x$-motion, $I_x$, and its multiplication by $\partial \ddot{w}_0/\partial x$ represents rotary inertia effects, which are ignored under the Love–Kirchhoff plate bending theory. Also note the expression

$$\frac{\partial(z\sigma_{xz})}{\partial z} = \frac{\partial z}{\partial z}\sigma_{xz} + z\frac{\partial \sigma_{xz}}{\partial z} \tag{32}$$

Upon rearrangement, Eq. (32) yields the term under the integral sign in the left-hand side of Eq. (31), i.e.,

$$z \frac{\partial \sigma_{xz}}{\partial z} = \frac{\partial(z\sigma_{xz})}{\partial z} - \sigma_{xz} \tag{33}$$

Substitution of Eq. (33) into Eq. (31) yields

$$\frac{\partial M_x}{\partial x} + \frac{\partial M_{xy}}{\partial y} + \int_{-h/2}^{h/2} \left( \frac{\partial(z\sigma_{xz})}{\partial z} - \sigma_{xz} \right) dz = 0 \tag{34}$$

The first term of the integral is zero due to free surface conditions at the upper and lower faces of the plate, i.e.,

$$\int_{-h/2}^{h/2} \frac{\partial(z\sigma_{xz})}{\partial z} dz = z\sigma_{xz} \Big|_{-h/2}^{h/2} = 0 \quad \text{(free surface conditions)} \tag{35}$$

The second term of the integral yields $N_{xz}$ in virtue of Eq. (2), i.e.,

$$N_{xz} = \int_{-h/2}^{h/2} \sigma_{xz} dz \tag{36}$$

Substitution of Eqs. (35), (36) into Eq. (34) gives

$$\frac{\partial M_x}{\partial x} + \frac{\partial M_{xy}}{\partial y} - N_{xz} = 0 \tag{37}$$

Upon solution, Eq. (37) yields

$$N_{xz} = \frac{\partial M_x}{\partial x} + \frac{\partial M_{xy}}{\partial y} \tag{38}$$

### 3.2.2.3.2 Calculation of Out-of-Plane Shear Resultant $N_{yz}$

In a similar way, multiply Eq. (15) by z and integrate with respect to z to get

$$\int_{-h/2}^{h/2} \frac{\partial \sigma_{xy}}{\partial x} z dz + \int_{-h/2}^{h/2} \frac{\partial \sigma_{yy}}{\partial y} z dz + \int_{-h/2}^{h/2} \frac{\partial \sigma_{yz}}{\partial z} z dz = \int_{-h/2}^{h/2} \ddot{v} \rho z dz \tag{39}$$

or

$$\frac{\partial}{\partial x} \int_{-h/2}^{h/2} \sigma_{xy} z dz + \frac{\partial}{\partial y} \int_{-h/2}^{h/2} \sigma_{yy} z dz + \int_{-h/2}^{h/2} \frac{\sigma_{yz}}{\partial z} z dz = \int_{-h/2}^{h/2} \rho \ddot{v} z dz \tag{40}$$

Use Eqs. (1), (2) into Eq. (40) to get

$$\frac{\partial M_{xy}}{\partial x} + \frac{\partial M_y}{\partial y} + \int_{-h/2}^{h/2} \frac{\partial \sigma_{yz}}{\partial z} z dz = \ddot{v}_0 \int_{-h/2}^{h/2} \rho \, z dz - \frac{\partial \ddot{w}_0}{\partial y} \int_{-h/2}^{h/2} \rho z^2 dz = 0 \tag{41}$$

The first integral in the right-hand side of Eq. (41) vanishes because it is the first moment of inertia about a centroidal axis. The second integral is the second moment of inertia due to $y$-motion, $I_y$, and its multiplication by $\partial \ddot{w}_0 / \partial y$ represents rotary inertia effects, which are ignored under the Love–Kirchhoff plate bending theory. Also note the expression

$$\frac{\partial(z\sigma_{yz})}{\partial z} = \frac{\partial z}{\partial z}\sigma_{yz} + z\frac{\partial\sigma_{yz}}{\partial z} \tag{42}$$

Upon rearrangement, Eq. (42) yields the term under the integral sign in the left-hand side of Eq. (41), i.e.,

$$z\frac{\partial\sigma_{yz}}{\partial z} = \frac{\partial(z\sigma_{yz})}{\partial z} - \sigma_{yz} \tag{43}$$

Substitution of Eq. (43) into Eq. (41) yields

$$\frac{\partial M_{xy}}{\partial x} + \frac{\partial M_y}{\partial y} + \int_{-h/2}^{h/2}\left(\frac{\partial(z\sigma_{yz})}{\partial z} - \sigma_{yz}\right)dz = 0 \tag{44}$$

The first term of the integral in Eq. (44) is zero due to free surface conditions at the upper and lower faces of the plate, i.e.,

$$\int_{-h/2}^{h/2}\frac{\partial(z\sigma_{yz})}{\partial z}dz = z\sigma_{yz}\Big|_{-h/2}^{h/2} = 0 \quad \text{(free surfaces conditions)} \tag{45}$$

The second term of the integral in Eq. (44) gives $N_{yz}$ in virtue of Eq. (2), i.e.,

$$\int_{-h/2}^{h/2}\sigma_{yz}dz = N_{yz} \tag{46}$$

Substitution of Eqs. (45), (46) into Eq. (44) gives

$$\frac{\partial M_{xy}}{\partial x} + \frac{\partial M_y}{\partial y} - N_{yz} = 0 \tag{47}$$

Upon solution, Eq. (47) yields

$$N_{yz} = \frac{\partial M_{xy}}{\partial x} + \frac{\partial M_y}{\partial y} \tag{48}$$

### 3.2.2.3.3  The $w$-Equation of Motion in Terms of Moment Stress Resultants
Substitution of Eqs. (38), (48) into the left-hand side of Eq. (28) gives

$$\frac{\partial N_{xz}}{\partial x} + \frac{\partial N_{yz}}{\partial y} = \frac{\partial}{\partial x}\left(\frac{\partial M_x}{\partial x} + \frac{\partial M_{xy}}{\partial y}\right) + \frac{\partial}{\partial y}\left(\frac{\partial M_{xy}}{\partial x} + \frac{\partial M_y}{\partial y}\right) = \frac{\partial^2 M_x}{\partial x^2} + 2\frac{\partial^2 M_{xy}}{\partial x \partial y} + \frac{\partial^2 M_y}{\partial y^2} \tag{49}$$

Substitution of Eq. (49) into Eq. (28) yields the equation of motion in terms of moments, i.e.,

$$\frac{\partial^2 M_x}{\partial x^2} + 2\frac{\partial^2 M_{xy}}{\partial x \partial y} + \frac{\partial^2 M_y}{\partial y^2} = m\ddot{w} \tag{50}$$

Note that Eq. (50) is identical with Eq. (13).

### 3.2.3  Summary of Equations of Motion in Terms of Stress Resultants

Equations (3), (4), (13) are the equations of motion in terms of stress resultants, i.e.,

$$\frac{\partial N_x}{\partial x} + \frac{\partial N_{xy}}{\partial y} = m\ddot{u}_0 \quad (u\text{-equation of motion}) \tag{51}$$

$$\frac{\partial N_{xy}}{\partial x} + \frac{\partial N_y}{\partial y} = m\ddot{v}_0 \quad (v\text{-equation of motion}) \tag{52}$$

$$\frac{\partial^2 M_x}{\partial x^2} + 2\frac{\partial^2 M_{xy}}{\partial x \partial y} + \frac{\partial^2 M_y}{\partial y^2} = m\ddot{w}_0 \quad (w\text{-equation of motion}) \tag{53}$$

It is remarkable that the axial and flexural motions seem to be uncoupled since the moment stress resultants do not appear in the $u$ and $v$ equations, nor the axial force resultants appear in the $w$ equation. However, it will be shown in the subsequent derivation that coupling between in-plane and out-of-plane motions exist when the equations of motion are expressed in terms of displacements.

Equations (51)–(53) represent the equation of motion in terms of moment stress resultants. For vibration analysis, we need to obtain the equations of motion in terms of displacements. In order to obtain the equation of motion in terms of displacements, we need to express the stress resultants in terms of displacements. To do so, we will proceed in the following steps:

(a) Calculate the strains in terms of displacements
(b) Use the stress–strain relations to express the stresses in terms of displacements
(c) Substitute the stresses into the stress resultant expressions and integrate to get the stress resultants in terms of displacements.

Step (a) is a kinematic analysis and does not depend on the material properties. This step will be performed up-front before getting into the specifics of composite materials. Steps (b) and (c) are, however, dependent on the material properties. When performing these steps, we will use the stiffness matrices of each of the layers making up the composite laminate.

## 3.2.4 Strains in Terms of Displacements

The strains of interest are

$$\varepsilon_{xy} = \frac{1}{2}\left(\frac{\partial u}{\partial y} + \frac{\partial v}{\partial x}\right)$$

$$\varepsilon_{xx} = \frac{\partial u}{\partial x}$$

$$\varepsilon_{xz} = \frac{1}{2}\left(\frac{\partial u}{\partial z} + \frac{\partial w}{\partial x}\right)$$

$$\varepsilon_{yy} = \frac{\partial v}{\partial y}$$

$$\varepsilon_{yz} = \frac{1}{2}\left(\frac{\partial v}{\partial z} + \frac{\partial w}{\partial y}\right)$$

(54)

Substitution of Eq. (1) into the $\varepsilon_{xx}, \varepsilon_{yy}$ parts of Eq. (54) gives

$$\varepsilon_{xx} = \frac{\partial u}{\partial x} = \frac{\partial}{\partial x}\left(u_0 - z\frac{\partial w_0}{\partial x}\right) = \frac{\partial u_0}{\partial x} - z\frac{\partial^2 w_0}{\partial x^2}$$

(55)

$$\varepsilon_{yy} = \frac{\partial v}{\partial y} = \frac{\partial}{\partial y}\left(v_0 - z\frac{\partial w_0}{\partial y}\right) = \frac{\partial v_0}{\partial y} - z\frac{\partial^2 w_0}{\partial y^2}$$

Substitution of Eq. (1) into the $\varepsilon_{xy}, \varepsilon_{xz}, \varepsilon_{yz}$ parts of Eq. (54) gives

$$\varepsilon_{xy} = \frac{1}{2}\left(\frac{\partial u}{\partial y} + \frac{\partial v}{\partial x}\right) = \frac{1}{2}\left[\frac{\partial}{\partial y}\left(u_0 - z\frac{\partial w_0}{\partial x}\right) + \frac{\partial}{\partial x}\left(v_0 - z\frac{\partial w_0}{\partial y}\right)\right] = \frac{1}{2}\left(\frac{\partial u_0}{\partial y} + \frac{\partial v_0}{\partial x}\right) - z\frac{\partial^2 w_0}{\partial x \partial y}$$  (56)

$$\varepsilon_{xz} = \frac{1}{2}\left(\frac{\partial u}{\partial z} + \frac{\partial w}{\partial x}\right) = \frac{1}{2}\left[\frac{\partial}{\partial z}\left(u_0 - z\frac{\partial w_0}{\partial x}\right) + \frac{\partial w_0}{\partial x}\right]$$
$$= \frac{1}{2}\left[\cancel{\frac{\partial u_0}{\partial z}} - \cancel{\frac{\partial w_0}{\partial x}} - z\frac{\partial^2 w_0}{\partial x \partial z} + \cancel{\frac{\partial w_0}{\partial x}}\right] = -\frac{1}{2}z\frac{\partial^2 w_0}{\partial x \partial z}$$

(57)

$$\varepsilon_{yz} = \frac{1}{2}\left(\frac{\partial v}{\partial z} + \frac{\partial w}{\partial y}\right) = \frac{1}{2}\left[\frac{\partial}{\partial z}\left(v_0 - z\frac{\partial w_0}{\partial y}\right) + \frac{\partial w_0}{\partial y}\right]$$
$$= \frac{1}{2}\left[\cancel{\frac{\partial v_0}{\partial z}} - \cancel{\frac{\partial w_0}{\partial y}} - z\frac{\partial^2 w_0}{\partial y \partial z} + \cancel{\frac{\partial w_0}{\partial y}}\right] = -\frac{1}{2}z\frac{\partial^2 w_0}{\partial y \partial z}$$

(58)

### 3.2.5 Strains in Terms of Mid-Surface Strains and Curvatures

Equations (55), (56), (57), (58) can be written more compactly using the mid-surface strains and curvatures defined as follows:

$$\varepsilon_x^0 = \frac{\partial u_0}{\partial x}$$

$$\varepsilon_y^0 = \frac{\partial v_0}{\partial y} \qquad \text{(mid-surface strains)} \qquad (59)$$

$$\gamma_{xy}^0 = \frac{\partial u_0}{\partial y} + \frac{\partial v_0}{\partial x}$$

$$\kappa_x = -\frac{\partial^2 w_0}{\partial x^2}$$

$$\kappa_y = -\frac{\partial^2 w_0}{\partial y^2} \qquad \text{(mid-surface curvatures)} \qquad (60)$$

$$\kappa_{xy} = -2\frac{\partial^2 w_0}{\partial x \partial y}$$

Substitution of Eqs. (59), (60) into Eqs. (55), (56) (57), (58) yields

$$\varepsilon_{xx} = \varepsilon_x^0 + z\,\kappa_x$$
$$\varepsilon_{yy} = \varepsilon_y^0 + z\,\kappa_y \qquad (61)$$
$$\gamma_{xy} = 2\varepsilon_{xy} = \gamma_{xy}^0 + z\,\kappa_{xy}$$

Equation (61) can be expressed in matrix form as

$$\begin{Bmatrix} \varepsilon_{xx} \\ \varepsilon_{yy} \\ \gamma_{xy} \end{Bmatrix} = \begin{Bmatrix} \varepsilon_x^0 \\ \varepsilon_y^0 \\ \gamma_{xy}^0 \end{Bmatrix} + \begin{Bmatrix} \kappa_x \\ \kappa_y \\ \kappa_{xy} \end{Bmatrix} z \qquad (62)$$

or, in compact form,

$$\varepsilon = \varepsilon^0 + z\,\kappa \qquad (63)$$

where

$$\varepsilon = \begin{Bmatrix} \varepsilon_{xx} \\ \varepsilon_{yy} \\ \gamma_{xy} \end{Bmatrix} \quad \varepsilon^0 = \begin{Bmatrix} \varepsilon_x^0 \\ \varepsilon_y^0 \\ \gamma_{xy}^0 \end{Bmatrix} \quad \kappa = \begin{Bmatrix} \kappa_x \\ \kappa_y \\ \kappa_{xy} \end{Bmatrix} \qquad (64)$$

# 3.3 VIBRATION EQUATIONS FOR AN ANISOTROPIC LAMINATED COMPOSITE PLATE

So far, we have obtained expressions of strains in terms of displacements without paying attention to the material properties of the plate. At this stage, we need to bring in the material properties equations that relate stresses to strains. Using the material-specific stress–strain relations, we will relate stresses to displacements and then proceed to integrate the stresses across the thickness to get the stress resultants in terms of displacements. In this way, we will achieve what is commonly known as the composite lamination theory (CLT). After that, we will get the equations of motion in terms of displacements and then the vibration equations of the laminated composite plate.

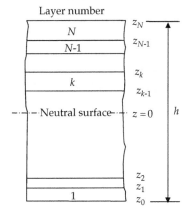

**FIGURE 3**    Coordinates definition for an $N$-ply composites laminate.

## 3.3.1 Stress–Strain Relations for an Anisotropic Laminated Composite Plate

For an anisotropic laminated composite plate with $n = 1, ..., N$ layers (Figure 3), the stress–strain relations are prescribed individually for each layer, i.e.,

$$\begin{Bmatrix} \sigma_{xx} \\ \sigma_{yy} \\ \sigma_{xy} \end{Bmatrix}_n = \begin{bmatrix} Q_{11} & Q_{12} & Q_{16} \\ Q_{12} & Q_{22} & Q_{26} \\ Q_{16} & Q_{26} & Q_{66} \end{bmatrix}_n \begin{Bmatrix} \varepsilon_{xx} \\ \varepsilon_{yy} \\ \gamma_{xy} \end{Bmatrix}_n \quad n = 1, ..., N \tag{65}$$

or, compactly,

$$\boldsymbol{\sigma}_n = \mathbf{Q}_n \boldsymbol{\varepsilon}_n \quad n = 1, ..., N \tag{66}$$

where $\mathbf{Q}_n$ is the reduced stiffness matrix of the $n$-th composite layer in global coordinates, i.e.,

$$\mathbf{Q}_n = \begin{bmatrix} Q_{11} & Q_{12} & Q_{16} \\ Q_{12} & Q_{22} & Q_{26} \\ Q_{16} & Q_{26} & Q_{66} \end{bmatrix}_n \qquad n = 1, ..., N \qquad (67)$$

## 3.3.2 Stresses in Terms of Mid-Surface Strains and Curvatures for an Anisotropic Laminated Composite Plate

For an anisotropic laminated composite plate with $n = 1, ..., N$ layers, substitution of Eq. (62) into the stress–strain relations Eq. (65) yields

$$\begin{Bmatrix} \sigma_{xx} \\ \sigma_{yy} \\ \sigma_{xy} \end{Bmatrix}_n = \begin{bmatrix} Q_{11} & Q_{12} & Q_{16} \\ Q_{12} & Q_{22} & Q_{26} \\ Q_{16} & Q_{26} & Q_{66} \end{bmatrix}_n \begin{Bmatrix} \varepsilon_x^0 \\ \varepsilon_y^0 \\ \gamma_{xy}^0 \end{Bmatrix} + z \begin{bmatrix} Q_{11} & Q_{12} & Q_{16} \\ Q_{12} & Q_{22} & Q_{26} \\ Q_{16} & Q_{26} & Q_{66} \end{bmatrix}_n \begin{Bmatrix} \kappa_x \\ \kappa_y \\ \kappa_{xy} \end{Bmatrix} \qquad n = 1, ..., N \qquad (68)$$

Alternatively, substitution of Eq. (63) into Eq. (66) gives the compact expression

$$\sigma_n = \mathbf{Q}_n \, \varepsilon^0 + z \, \mathbf{Q}_n \kappa \qquad n = 1, ..., N \qquad (69)$$

## 3.3.3 Stress Resultants in Terms of Mid-Surface Strains and Curvatures for an Anisotropic Laminated Composite Plate

### 3.3.3.1 Matrix Representation of Stresses and Stress Resultants

Recall from Eq. (2) the stress resultants $N_x, N_y, N_{xy}, M_x, M_y, M_{xy}$ and arrange them in matrix form, i.e.,

$$\mathbf{N} = \begin{Bmatrix} N_x \\ N_y \\ N_{xy} \end{Bmatrix} = \begin{Bmatrix} \int_{-h/2}^{h/2} \sigma_{xx} dz \\ \int_{-h/2}^{h/2} \sigma_{yy} dz \\ \int_{-h/2}^{h/2} \sigma_{xy} dz \end{Bmatrix} \qquad \mathbf{M} = \begin{Bmatrix} M_x \\ M_y \\ M_{xy} \end{Bmatrix} = \begin{Bmatrix} \int_{-h/2}^{h/2} \sigma_{xx} z dz \\ \int_{-h/2}^{h/2} \sigma_{yy} z dz \\ \int_{-h/2}^{h/2} \sigma_{xy} z dz \end{Bmatrix} \qquad (70)$$

Propagate the integration sign outside the matrix notation and express the force stress resultants as

$$\begin{Bmatrix} N_x \\ N_y \\ N_{xy} \end{Bmatrix} = \int_{-h/2}^{h/2} \begin{Bmatrix} \sigma_{xx} \\ \sigma_{yy} \\ \sigma_{xy} \end{Bmatrix} dz \quad \text{(force stress resultants)} \tag{71}$$

$$\begin{Bmatrix} M_x \\ M_y \\ M_{xy} \end{Bmatrix} = \int_{-h/2}^{h/2} \begin{Bmatrix} \sigma_{xx} \\ \sigma_{yy} \\ \sigma_{xy} \end{Bmatrix} z\,dz \quad \text{(moment stress resultants)} \tag{72}$$

Define the stress and strain column matrices

$$\boldsymbol{\sigma} = \begin{Bmatrix} \sigma_{xx} \\ \sigma_{yy} \\ \sigma_{xy} \end{Bmatrix} \quad \boldsymbol{\varepsilon} = \begin{Bmatrix} \varepsilon_{xx} \\ \varepsilon_{yy} \\ \gamma_{xy} \end{Bmatrix} \tag{73}$$

where $\gamma_{xy}$ is the engineering shear strain defined as

$$\gamma_{xy} = 2\varepsilon_{xy} \quad \text{(engineering shear strain)} \tag{74}$$

Using the matrix notations of Eqs. (70), (73), write Eqs. (71), (72) compactly as

$$\mathbf{N} = \int_{-h/2}^{h/2} \boldsymbol{\sigma}\,dz \quad \text{(force stress resultants)} \tag{75}$$

$$\mathbf{M} = \int_{-h/2}^{h/2} \boldsymbol{\sigma}\,z\,dz \quad \text{(moment stress resultants)} \tag{76}$$

### 3.3.3.2 Stress Resultants through Stress Integration across the Thickness of an Anisotropic Laminated Composite Plate (ABD Matrices)

For an anisotropic laminated composite plate with $n = 1, ..., N$ layers, substitution of Eq. (69) into Eqs. (75), (76) yields

$$\mathbf{N} = \sum_{n=1}^{N} \int_{z_{n-1}}^{z_n} \mathbf{Q}_n \boldsymbol{\varepsilon}^0 dz + \sum_{n=1}^{N} \int_{z_{n-1}}^{z_n} \mathbf{Q}_n \boldsymbol{\kappa} z\,dz \tag{77}$$

$$\mathbf{M} = \sum_{n=1}^{N} \int_{z_{n-1}}^{z_n} \mathbf{Q}_n \boldsymbol{\varepsilon}^0 z\,dz + \sum_{n=1}^{N} \int_{z_{n-1}}^{z_n} \mathbf{Q}_n \boldsymbol{\kappa} z^2 dz \tag{78}$$

After propagating the constant terms outside the integration sign and the summation sign, Eqs. (77), (78) become

$$\mathbf{N} = \sum_{n=1}^{N} \left( \mathbf{Q}_n \int_{z_{n-1}}^{z_n} dz \right) \boldsymbol{\varepsilon}^0 + \left( \sum_{n=1}^{N} \mathbf{Q}_n \int_{z_{n-1}}^{z_n} z\, dz \right) \boldsymbol{\kappa} \tag{79}$$

$$\mathbf{M} = \left( \sum_{n=1}^{N} \mathbf{Q}_n \int_{z_{n-1}}^{z_n} z\, dz \right) \boldsymbol{\varepsilon}^0 + \left( \sum_{n=1}^{N} \mathbf{Q}_n \int_{z_{n-1}}^{z_n} z^2\, dz \right) \boldsymbol{\kappa} \tag{80}$$

Evaluation of the integrals in Eqs. (79), (80) gives

$$\int_{z_{n-1}}^{z_n} dz = z_n - z_{n-1}$$
$$\int_{z_{n-1}}^{z_n} z\, dz = \tfrac{1}{2}\left(z_n^2 - z_{n-1}^2\right) \tag{81}$$
$$\int_{z_{n-1}}^{z_n} z^2\, dz = \tfrac{1}{3}\left(z_n^3 - z_{n-1}^3\right)$$

Substitution of Eq. (81) into Eqs. (79), (80) yields

$$\mathbf{N} = \left[ \sum_{n=1}^{N} (z_n - z_{n-1})\mathbf{Q}_n \right] \boldsymbol{\varepsilon}^0 + \left[ \sum_{n=1}^{N} \tfrac{1}{2}(z_n^2 - z_{n-1}^2)\mathbf{Q}_n \right] \boldsymbol{\kappa} \tag{82}$$

$$\mathbf{M} = \left[ \sum_{n=1}^{N} \tfrac{1}{2}(z_n^2 - z_{n-1}^2)\mathbf{Q}_n \right] \boldsymbol{\varepsilon}^0 + \left[ \sum_{n=1}^{N} \tfrac{1}{3}(z_n^3 - z_{n-1}^3)\mathbf{Q}_n \right] \boldsymbol{\kappa} \tag{83}$$

Equations (82), (83) can be written compactly in the "ABD form", i.e.,

$$\begin{Bmatrix} \mathbf{N} \\ \mathbf{M} \end{Bmatrix} = \begin{bmatrix} \mathbf{A} & \mathbf{B} \\ \mathbf{B} & \mathbf{D} \end{bmatrix} \begin{Bmatrix} \boldsymbol{\varepsilon}^0 \\ \boldsymbol{\kappa} \end{Bmatrix} \tag{84}$$

where the ABD matrices are

$$\mathbf{A} = \begin{bmatrix} A_{11} & A_{12} & A_{16} \\ A_{12} & A_{22} & A_{26} \\ A_{16} & A_{26} & A_{66} \end{bmatrix} = \left[ \sum_{n=1}^{N} (z_n - z_{n-1})\mathbf{Q}_n \right] = \sum_{n=1}^{N} (z_n - z_{n-1}) \begin{bmatrix} Q_{11} & Q_{12} & Q_{16} \\ Q_{12} & Q_{22} & Q_{26} \\ Q_{16} & Q_{26} & Q_{66} \end{bmatrix}_n \tag{85}$$

$$\mathbf{B} = \begin{bmatrix} B_{11} & B_{12} & B_{16} \\ B_{12} & B_{22} & B_{26} \\ B_{16} & B_{26} & B_{66} \end{bmatrix} = \left[ \sum_{n=1}^{N} \tfrac{1}{2}(z_n^2 - z_{n-1}^2)\mathbf{Q}_n \right] = \tfrac{1}{2}\sum_{n=1}^{N} (z_n^2 - z_{n-1}^2) \begin{bmatrix} Q_{11} & Q_{12} & Q_{16} \\ Q_{12} & Q_{22} & Q_{26} \\ Q_{16} & Q_{26} & Q_{66} \end{bmatrix}_n \tag{86}$$

$$\mathbf{D} = \begin{bmatrix} D_{11} & D_{12} & D_{16} \\ D_{12} & D_{22} & D_{26} \\ D_{16} & D_{26} & D_{66} \end{bmatrix} = \left[ \sum_{n=1}^{N} \tfrac{1}{3}(z_n^3 - z_{n-1}^3)\mathbf{Q}_n \right] = \tfrac{1}{3}\sum_{n=1}^{N} (z_n^3 - z_{n-1}^3) \begin{bmatrix} Q_{11} & Q_{12} & Q_{16} \\ Q_{12} & Q_{22} & Q_{26} \\ Q_{16} & Q_{26} & Q_{66} \end{bmatrix}_n \tag{87}$$

The longhand correspondent of Eq. (84) is

$$
\begin{Bmatrix} N_x \\ N_y \\ N_{xy} \\ M_x \\ M_y \\ M_{xy} \end{Bmatrix} = \begin{bmatrix} A_{11} & A_{12} & A_{16} & B_{11} & B_{12} & B_{16} \\ A_{12} & A_{22} & A_{26} & B_{12} & B_{22} & B_{26} \\ A_{16} & A_{26} & A_{66} & B_{16} & B_{26} & B_{66} \\ B_{11} & B_{12} & B_{16} & D_{11} & D_{12} & D_{16} \\ B_{12} & B_{22} & B_{26} & D_{12} & D_{22} & D_{26} \\ B_{16} & B_{26} & B_{66} & D_{16} & D_{26} & D_{66} \end{bmatrix} \begin{Bmatrix} \varepsilon_x^0 \\ \varepsilon_y^0 \\ \gamma_{xy}^0 \\ \kappa_x \\ \kappa_y \\ \kappa_{xy} \end{Bmatrix} \tag{88}
$$

## 3.3.4 Equations of Motion in Terms of Displacements for an Anisotropic Laminated Composite Plate

In order to derive the equations of motion in terms of displacements, recall Eqs. (59), (60) and substitute them into Eq. (84); in expanded form, one gets

$$
\begin{Bmatrix} N_x \\ N_y \\ N_{xy} \end{Bmatrix} = \begin{bmatrix} A_{11} & A_{12} & A_{16} \\ A_{12} & A_{22} & A_{26} \\ A_{16} & A_{26} & A_{66} \end{bmatrix} \begin{Bmatrix} \dfrac{\partial u_0}{\partial x} \\[2mm] \dfrac{\partial v_0}{\partial y} \\[2mm] \dfrac{\partial u_0}{\partial y} + \dfrac{\partial v_0}{\partial x} \end{Bmatrix} - \begin{bmatrix} B_{11} & B_{12} & B_{16} \\ B_{12} & B_{22} & B_{26} \\ B_{16} & B_{26} & B_{66} \end{bmatrix} \begin{Bmatrix} \dfrac{\partial^2 w_0}{\partial x^2} \\[2mm] \dfrac{\partial^2 w_0}{\partial y^2} \\[2mm] 2\dfrac{\partial^2 w_0}{\partial x \partial y} \end{Bmatrix} \tag{89}
$$

$$
\begin{Bmatrix} M_x \\ M_y \\ M_{xy} \end{Bmatrix} = \begin{bmatrix} B_{11} & B_{12} & B_{16} \\ B_{12} & B_{22} & B_{26} \\ B_{16} & B_{26} & B_{66} \end{bmatrix} \begin{Bmatrix} \dfrac{\partial u_0}{\partial x} \\[2mm] \dfrac{\partial v_0}{\partial y} \\[2mm] \dfrac{\partial u_0}{\partial y} + \dfrac{\partial v_0}{\partial x} \end{Bmatrix} - \begin{bmatrix} D_{11} & D_{12} & D_{16} \\ D_{12} & D_{22} & D_{26} \\ D_{16} & D_{26} & D_{66} \end{bmatrix} \begin{Bmatrix} \dfrac{\partial^2 w_0}{\partial x^2} \\[2mm] \dfrac{\partial^2 w_0}{\partial y^2} \\[2mm] 2\dfrac{\partial^2 w_0}{\partial x \partial y} \end{Bmatrix} \tag{90}
$$

Recall Eqs. (51), (52), (53) representing equations of motion in terms of stress resultants, i.e.,

$$
\frac{\partial N_x}{\partial x} + \frac{\partial N_{xy}}{\partial y} = m\ddot{u}_0 \qquad \text{(} u\text{-equation of motion)} \tag{91}
$$

$$
\frac{\partial N_{xy}}{\partial x} + \frac{\partial N_y}{\partial y} = m\ddot{v}_0 \qquad \text{(} v\text{-equation of motion)} \tag{92}
$$

$$
\frac{\partial^2 M_x}{\partial x^2} + 2\frac{\partial^2 M_{xy}}{\partial x \partial y} + \frac{\partial^2 M_y}{\partial y^2} = m\ddot{w}_0 \quad \text{(} w\text{-equation of motion)} \tag{93}
$$

Expand Eqs. (89), (90) and write

$$N_x = A_{11}\frac{\partial u_0}{\partial x} + A_{12}\frac{\partial v_0}{\partial y} + A_{16}\left(\frac{\partial u_0}{\partial y} + \frac{\partial v_0}{\partial x}\right) - B_{11}\frac{\partial^2 w_0}{\partial x^2} - B_{12}\frac{\partial^2 w_0}{\partial y^2} - 2B_{16}\frac{\partial^2 w_0}{\partial x \partial y} \quad (94)$$

$$N_y = A_{12}\frac{\partial u_0}{\partial x} + A_{22}\frac{\partial v_0}{\partial y} + A_{26}\left(\frac{\partial u_0}{\partial y} + \frac{\partial v_0}{\partial x}\right) - B_{12}\frac{\partial^2 w_0}{\partial x^2} - B_{22}\frac{\partial^2 w_0}{\partial y^2} - 2B_{26}\frac{\partial^2 w_0}{\partial x \partial y} \quad (95)$$

$$N_{xy} = A_{16}\frac{\partial u_0}{\partial x} + A_{26}\frac{\partial v_0}{\partial y} + A_{66}\left(\frac{\partial u_0}{\partial y} + \frac{\partial v_0}{\partial x}\right) - B_{16}\frac{\partial^2 w_0}{\partial x^2} - B_{26}\frac{\partial^2 w_0}{\partial y^2} - 2B_{66}\frac{\partial^2 w_0}{\partial x \partial y} \quad (96)$$

$$M_x = B_{11}\frac{\partial u_0}{\partial x} + B_{12}\frac{v_0}{\partial y} + B_{16}\left(\frac{\partial u_0}{\partial y} + \frac{\partial v_0}{\partial x}\right) - D_{11}\frac{\partial^2 w_0}{\partial x^2} - D_{12}\frac{\partial^2 w_0}{\partial y^2} - 2D_{16}\frac{\partial^2 w_0}{\partial x \partial y} \quad (97)$$

$$M_y = B_{12}\frac{\partial u_0}{\partial x} + B_{22}\frac{\partial v_0}{\partial y} + B_{26}\left(\frac{\partial u_0}{\partial y} + \frac{\partial v_0}{\partial x}\right) - D_{12}\frac{\partial^2 w_0}{\partial x^2} - D_{22}\frac{\partial^2 w_0}{\partial y^2} - 2D_{26}\frac{\partial^2 w_0}{\partial x \partial y} \quad (98)$$

$$M_{xy} = B_{16}\frac{\partial u_0}{\partial x} + B_{26}\frac{\partial v_0}{\partial y} + B_{66}\left(\frac{\partial u_0}{\partial y} + \frac{\partial v_0}{\partial x}\right) - D_{16}\frac{\partial^2 w_0}{\partial x^2} - D_{26}\frac{\partial^2 w_0}{\partial y^2} - 2D_{66}\frac{\partial^2 w_0}{\partial x \partial y} \quad (99)$$

Use Eqs. (94), (95), (96) to calculate the following derivatives:

$$\frac{\partial N_x}{\partial x} = A_{11}\frac{\partial^2 u_0}{\partial x^2} + A_{12}\frac{\partial^2 v_0}{\partial x \partial y} + A_{16}\left(\frac{\partial^2 u_0}{\partial x \partial y} + \frac{\partial^2 v_0}{\partial x^2}\right) - B_{11}\frac{\partial^3 w_0}{\partial x^3} - B_{12}\frac{\partial^3 w_0}{\partial x \partial y^2} - 2B_{16}\frac{\partial^3 w_0}{\partial x^2 \partial y} \quad (100)$$

$$\frac{\partial N_{xy}}{\partial y} = A_{16}\frac{\partial^2 u_0}{\partial x \partial y} + A_{26}\frac{\partial^2 v_0}{\partial y^2} + A_{66}\left(\frac{\partial^2 u_0}{\partial y^2} + \frac{\partial^2 v_0}{\partial x \partial y}\right) - B_{16}\frac{\partial^3 w_0}{\partial x^2 \partial y} - B_{26}\frac{\partial^3 w_0}{\partial y^3} - 2B_{66}\frac{\partial^3 w_0}{\partial x \partial y^2} \quad (101)$$

$$\frac{\partial N_{xy}}{\partial x} = A_{16}\frac{\partial^2 u_0}{\partial x^2} + A_{26}\frac{\partial^2 v_0}{\partial x \partial y} + A_{66}\left(\frac{\partial^2 u_0}{\partial x \partial y} + \frac{\partial^2 v_0}{\partial x^2}\right) - B_{16}\frac{\partial^3 w_0}{\partial x^3} - B_{26}\frac{\partial^3 w_0}{\partial x \partial y^2} - 2B_{66}\frac{\partial^3 w_0}{\partial x^2 \partial y} \quad (102)$$

$$\frac{\partial N_y}{\partial y} = A_{12}\frac{\partial^2 u_0}{\partial x \partial y} + A_{22}\frac{\partial^2 v_0}{\partial y^2} + A_{26}\left(\frac{\partial^2 u_0}{\partial y^2} + \frac{\partial^2 v_0}{\partial x \partial y}\right) - B_{12}\frac{\partial^3 w_0}{\partial x^2 \partial y} - B_{22}\frac{\partial^3 w_0}{\partial y^3} - 2B_{26}\frac{\partial^3 w_0}{\partial x \partial y^2} \quad (103)$$

$$\frac{\partial^2 M_x}{\partial x^2} = B_{11}\frac{\partial^3 u_0}{\partial x^3} + B_{12}\frac{\partial^3 v_0}{\partial x^2 \partial y} + B_{16}\left(\frac{\partial^3 u_0}{\partial x^2 \partial y} + \frac{\partial^3 v_0}{\partial x^3}\right) - D_{11}\frac{\partial^4 w_0}{\partial x^4} - D_{12}\frac{\partial^4 w_0}{\partial x^2 \partial y^2} - 2D_{16}\frac{\partial^4 w_0}{\partial x^3 \partial y}$$
$$(104)$$

$$\frac{\partial^2 M_{xy}}{\partial x \partial y} = B_{16}\frac{\partial^3 u_0}{\partial x^2 \partial y} + B_{26}\frac{\partial^3 v_0}{\partial x \partial y^2} + B_{66}\left(\frac{\partial^3 u_0}{\partial x \partial y^2} + \frac{\partial^3 v_0}{\partial x^2 \partial y}\right) - D_{16}\frac{\partial^4 w_0}{\partial x^3 \partial y} - D_{26}\frac{\partial^4 w_0}{\partial x \partial y^3} - 2D_{66}\frac{\partial^4 w_0}{\partial x^2 \partial y^2}$$
$$(105)$$

$$\frac{\partial^2 M_y}{\partial y^2} = B_{12}\frac{\partial^3 u_0}{\partial x \partial y^2} + B_{22}\frac{\partial^3 v_0}{\partial y^3} + B_{26}\left(\frac{\partial^3 u_0}{\partial y^3} + \frac{\partial^3 v_0}{\partial x \partial y^2}\right) - D_{12}\frac{\partial^4 w_0}{\partial x^2 \partial y^2} - D_{22}\frac{\partial^4 w_0}{\partial y^4} - 2D_{26}\frac{\partial^4 w_0}{\partial x \partial y^3}$$
$$(106)$$

Substitution of Eqs. (100), (101) into Eq. (91), of Eqs. (102), (103) into Eq. (92), and of Eqs. (104), (105), (106) into Eq. (93) yields

$$
\begin{aligned}
& A_{11}\frac{\partial^2 u_0}{\partial x^2} + 2A_{16}\frac{\partial^2 u_0}{\partial x \partial y} + A_{66}\frac{\partial^2 u_0}{\partial y^2} + A_{16}\frac{\partial^2 v_0}{\partial x^2} + (A_{12}+A_{66})\frac{\partial^2 v_0}{\partial x \partial y} + A_{26}\frac{\partial^2 v_0}{\partial y^2} \\
& \quad - B_{11}\frac{\partial^3 w_0}{\partial x^3} - 3B_{16}\frac{\partial^3 w_0}{\partial x^2 \partial y} - (B_{12}+2B_{66})\frac{\partial^3 w_0}{\partial x \partial y^2} - B_{26}\frac{\partial^3 w_0}{\partial y^3} = m\ddot{u}_0
\end{aligned}
\tag{107}
$$

$$
\begin{aligned}
& A_{16}\frac{\partial^2 u_0}{\partial x^2} + (A_{12}+A_{66})\frac{\partial^2 u_0}{\partial x \partial y} + A_{26}\frac{\partial^2 u_0}{\partial y^2} + A_{66}\frac{\partial^2 v_0}{\partial x^2} + 2A_{26}\frac{\partial^2 v_0}{\partial x \partial y} + A_{22}\frac{\partial^2 v_0}{\partial y^2} \\
& \quad - B_{16}\frac{\partial^3 w_0}{\partial x^3} - (B_{12}+2B_{66})\frac{\partial^3 w_0}{\partial x^2 \partial y} - 3B_{26}\frac{\partial^3 w_0}{\partial x \partial y^2} - B_{22}\frac{\partial^3 w_0}{\partial y^3} = m\ddot{v}_0
\end{aligned}
\tag{108}
$$

$$
\begin{aligned}
& D_{11}\frac{\partial^4 w_0}{\partial x^4} + 4D_{16}\frac{\partial^4 w_0}{\partial x^3 \partial y} + 2(D_{12}+2D_{66})\frac{\partial^4 w_0}{\partial x^2 \partial y^2} + 4D_{26}\frac{\partial^4 w_0}{\partial x \partial y^3} + D_{22}\frac{\partial^4 w_0}{\partial y^4} \\
& \quad - B_{11}\frac{\partial^3 u_0}{\partial x^3} - 3B_{16}\frac{\partial^3 u_0}{\partial x^2 \partial y} - (B_{12}+2B_{66})\frac{\partial^3 u_0}{\partial x \partial y^2} - B_{26}\frac{\partial^3 u_0}{\partial y^3} \\
& \quad - B_{16}\frac{\partial^3 v_0}{\partial x^3} - (B_{12}+2B_{66})\frac{\partial^3 v_0}{\partial x^2 \partial y} - 3B_{26}\frac{\partial^3 v_0}{\partial x \partial y^2} - B_{22}\frac{\partial^3 v_0}{\partial y^3} + m\ddot{w}_0 = 0
\end{aligned}
\tag{109}
$$

### 3.3.5 Vibration Frequencies and Modeshapes of an Anisotropic Laminated Composite Plate

Equations (107)–(109) are the equations of coupled axial–flexural vibration of a laminated composite plate. To calculate natural frequencies and modeshapes, assume time-harmonic motion of the form

$$
\begin{Bmatrix} u^0 \\ v^0 \\ w^0 \end{Bmatrix} = \begin{Bmatrix} \hat{u} \\ \hat{v} \\ \hat{w} \end{Bmatrix} e^{i\omega t}
\tag{110}
$$

where the vibration amplitudes $\hat{u}, \hat{v}, \hat{w}$ depend only on the space variables $x, y$. Substitution of Eq. (110) into Eqs. (107), (108), (109) yields

$$
\begin{aligned}
& A_{11}\frac{\partial^2 \hat{u}}{\partial x^2} + 2A_{16}\frac{\partial^2 \hat{u}}{\partial x \partial y} + A_{66}\frac{\partial^2 \hat{u}}{\partial y^2} + A_{16}\frac{\partial^2 \hat{v}}{\partial x^2} + (A_{12}+A_{66})\frac{\partial^2 \hat{v}}{\partial x \partial y} + A_{26}\frac{\partial^2 \hat{v}}{\partial y^2} \\
& \quad - B_{11}\frac{\partial^3 \hat{w}}{\partial x^3} - 3B_{16}\frac{\partial^3 \hat{w}}{\partial x^2 \partial y} - (B_{12}+2B_{66})\frac{\partial^3 \hat{w}}{\partial x \partial y^2} - B_{26}\frac{\partial^3 \hat{w}}{\partial y^3} + \omega^2 m\hat{u} = 0
\end{aligned}
\tag{111}
$$

$$A_{16}\frac{\partial^2 \hat{u}}{\partial x^2} + (A_{12} + A_{66})\frac{\partial^2 \hat{u}}{\partial x \partial y} + A_{26}\frac{\partial^2 \hat{u}}{\partial y^2} + A_{66}\frac{\partial^2 \hat{v}}{\partial x^2} + 2A_{26}\frac{\partial^2 \hat{v}}{\partial x \partial y} + A_{22}\frac{\partial^2 \hat{v}}{\partial y^2}$$

$$- B_{16}\frac{\partial^3 \hat{w}}{\partial x^3} - (B_{12} + 2B_{66})\frac{\partial^3 \hat{w}}{\partial x^2 \partial y} - 3B_{26}\frac{\partial^3 \hat{w}}{\partial x \partial y^2} - B_{22}\frac{\partial^3 \hat{w}}{\partial y^3} + \omega^2 m\hat{v} = 0 \tag{112}$$

$$D_{11}\frac{\partial^4 \hat{w}}{\partial x^4} + 4D_{16}\frac{\partial^4 \hat{w}}{\partial x^3 \partial y} + 2(D_{12} + 2D_{66})\frac{\partial^4 \hat{w}}{\partial x^2 \partial y^2} + 4D_{26}\frac{\partial^4 \hat{w}}{\partial x \partial y^3} + D_{22}\frac{\partial^4 \hat{w}}{\partial y^4}$$

$$- B_{11}\frac{\partial^3 \hat{u}}{\partial x^3} - 3B_{16}\frac{\partial^3 \hat{u}}{\partial x^2 \partial y} - (B_{12} + 2B_{66})\frac{\partial^3 \hat{u}}{\partial x \partial y^2} - B_{26}\frac{\partial^3 \hat{u}}{\partial y^3} \tag{113}$$

$$- B_{16}\frac{\partial^3 \hat{v}}{\partial x^3} - (B_{12} + 2B_{66})\frac{\partial^3 \hat{v}}{\partial x^2 \partial y} - 3B_{26}\frac{\partial^3 \hat{v}}{\partial x \partial y^2} - B_{22}\frac{\partial^3 \hat{v}}{\partial y^3} - \omega^2 m\hat{w} = 0$$

Equations (111)–(113) represent a homogeneous system of partial differential equations in the space variables $x, y$. Solution is possible only for certain frequencies $\omega_j$, which are the natural frequencies of the composite structure subject to certain boundary conditions. For each eigenvalue $\omega_j$, one finds a solution $U_j, V_j, W_j$ of Eqs. (111)–(113) representing the modeshape of the composite structure corresponding to the natural frequency $\omega_j$. The solution $U_j, V_j, W_j$ has one degree of indeterminacy, i.e., it can be scaled up and down by an arbitrary scale factor. The actual scale factor to be used in computations is determined through a modes normalization procedure.

Equations (111)–(113) do not accept closed-form solution. Solutions can be obtained numerically using various approximation methods such as Rayleigh–Ritz, Galerkin, finite element method (FEM), etc.

## 3.4 VIBRATION EQUATIONS FOR AN ISOTROPIC PLATE

In this section, we will connect the vibration analysis performed in Section 3 for an anisotropic laminated composite plate with the vibration analysis results customarily obtained for an isotropic plate. We will repeat here, for an isotropic plate, the analysis steps developed for an anisotropic laminated composite plate in Sections 3.1–3.5 and will show that we can recover the customary plate vibration equations.

Equations (51)–(53) represent the equation of motion in terms of force and moment stress resultants. For vibration analysis, we need to obtain the equations of motion in terms of displacements. In order to obtain the equation of motion in terms of displacements, we need to express the force and moment stress resultants in terms of displacements. First, we will calculate the strains in terms of displacements; next, we will use the stress–strain relations to express the stresses in terms of displacements; finally, we will substitute the stresses into the stress resultant expressions to get the stress resultants in terms of displacements.

### 3.4.1 Isotropic Stress–Strain Relations

For an isotropic plate, the strain–stress relations are

$$\varepsilon_{xx} = \frac{1}{E}\sigma_{xx} + \frac{-\nu}{E}\sigma_{yy}$$

$$\varepsilon_{yy} = \frac{-\nu}{E}\sigma_{xx} + \frac{1}{E}\sigma_{yy}$$

and

$$\varepsilon_{xy} = \frac{1}{2G}\sigma_{xy}$$

$$\varepsilon_{xz} = \frac{1}{2G}\sigma_{xz}$$

$$\varepsilon_{yz} = \frac{1}{2G}\sigma_{yz}$$

$$(114)$$

where $G = E/2(1 + \nu)$. Solution of Eq. (114) yields the stress–strain relations, i.e.,

$$\sigma_{xx} = \frac{E}{1 - \nu^2}\left(\varepsilon_{xx} + \nu\varepsilon_{yy}\right)$$

$$\sigma_{yy} = \frac{E}{1 - \nu^2}\left(\nu\varepsilon_{xx} + \varepsilon_{yy}\right)$$

$$\sigma_{xy} = 2G\varepsilon_{xy}$$
$$\sigma_{xz} = 2G\varepsilon_{xz}$$
$$\sigma_{yz} = 2G\varepsilon_{yz}$$

$$(115)$$

The in-plane components of Eq. (115) can be arranged in matrix form as

$$\begin{Bmatrix} \sigma_{xx} \\ \sigma_{yy} \\ \sigma_{xy} \end{Bmatrix} = \frac{E}{1 - \nu^2}\begin{bmatrix} 1 & \nu & \\ \nu & 1 & \\ & & G \end{bmatrix}\begin{Bmatrix} \varepsilon_{xx} \\ \varepsilon_{yy} \\ \gamma_{xy} \end{Bmatrix} \tag{116}$$

Define the isotropic reduced stiffness matrix $\mathbf{Q}^{isotropic}$, i.e.,

$$\mathbf{Q}^{isotropic} = \frac{E}{1 - \nu^2}\begin{bmatrix} 1 & \nu & \\ \nu & 1 & \\ & & G \end{bmatrix} \qquad G = \frac{E}{2(1 + \nu)} \tag{117}$$

Equations (73), (117) allow us to express Eq. (116) compactly as

$$\boldsymbol{\sigma} = \mathbf{Q}^{isotropic}\boldsymbol{\varepsilon} \tag{118}$$

### 3.4.2 Stresses in Terms of Mid-Surface Strains and Curvatures for an Isotropic Plate

For an isotropic plate, substitution of Eq. (62) into Eq. (116) yields

$$\begin{Bmatrix} \sigma_{xx} \\ \sigma_{yy} \\ \sigma_{xy} \end{Bmatrix} = \frac{E}{1 - \nu^2}\begin{bmatrix} 1 & \nu & \\ \nu & 1 & \\ & & G \end{bmatrix}\begin{Bmatrix} \varepsilon_x^0 \\ \varepsilon_y^0 \\ \gamma_{xy}^0 \end{Bmatrix} + z\frac{E}{1 - \nu^2}\begin{bmatrix} 1 & \nu & \\ \nu & 1 & \\ & & G \end{bmatrix}\begin{Bmatrix} \kappa_x \\ \kappa_y \\ \kappa_{xy} \end{Bmatrix} \tag{119}$$

Alternatively, substitution of Eq. (63) into Eq. (118) gives the compact expression

$$\sigma = \mathbf{Q}^{isotropic} \, \varepsilon^0 + z \, \mathbf{Q}^{isotropic} \kappa \qquad (120)$$

### 3.4.3  Stress Resultants for an Isotropic Plate

For an isotropic plate, substitution of Eq. (120) into Eqs. (75), (76) yields

$$\mathbf{N} = \int_{-h/2}^{h/2} \mathbf{Q}^{isotropic} \, \varepsilon^0 dz + \int_{-h/2}^{h/2} \mathbf{Q}^{isotropic} \kappa z dz \qquad (121)$$

$$\mathbf{M} = \int_{-h/2}^{h/2} \mathbf{Q}^{isotropic} \, \varepsilon^0 z dz + \int_{-h/2}^{h/2} \mathbf{Q}^{isotropic} \kappa z^2 dz \qquad (122)$$

Propagating the constant terms outside the integral sign, we write Eqs. (121), (122) as

$$\mathbf{N} = \mathbf{Q}^{isotropic} \, \varepsilon^0 \int_{-h/2}^{h/2} dz + \mathbf{Q}^{isotropic} \kappa \int_{-h/2}^{h/2} z dz \qquad (123)$$

$$\mathbf{M} = \mathbf{Q}^{isotropic} \, \varepsilon^0 \int_{-h/2}^{h/2} z dz + \mathbf{Q}^{isotropic} \kappa \int_{-h/2}^{h/2} z^2 dz \qquad (124)$$

Evaluation of the integrals in Eqs. (123), (124) gives

$$\int_{-h/2}^{h/2} dz = h$$

$$\int_{-h/2}^{h/2} z dz = \tfrac{1}{2} \left[ \left( \frac{h}{2} \right)^2 - \left( -\frac{h}{2} \right)^2 \right] = 0 \qquad (125)$$

$$\int_{-h/2}^{h/2} z^2 dz = \tfrac{1}{3} \left[ \left( \frac{h}{2} \right)^3 - \left( -\frac{h}{2} \right)^3 \right] = \tfrac{h^3}{12}$$

Substitution of Eq. (125) into Eqs. (123), (124) yields

$$\mathbf{N} = h \mathbf{Q}^{isotropic} \, \varepsilon^0 \qquad (126)$$

$$\mathbf{M} = \frac{h^3}{12} \mathbf{Q}^{isotropic} \kappa \qquad (127)$$

Equations (126), (127) indicate that, for an isotropic plate, the in-plane stress resultants contained in $\mathbf{N}$ depend only on the in-plane deformation strains contained in $\varepsilon^0$, whereas the out-of-plane stress resultants contained in $\mathbf{M}$ depend only on the curvatures contained in $\kappa$ and representing out-of-plane deformation.

### 3.4.3.1 ABD Matrices for an Isotropic Plate

For an isotropic plate, the use of Eqs. (117), (125) into Eqs. (85), (86), (87) yields

$$\mathbf{A}^{isotropic} = \frac{Eh}{1 - \nu^2} \begin{bmatrix} 1 & \nu & \\ \nu & 1 & \\ & & G \end{bmatrix} \tag{128}$$

$$\mathbf{B}^{isotropic} = \mathbf{0} \tag{129}$$

$$\mathbf{D}^{isotropic} = \frac{Eh^3}{12(1 - \nu^2)} \begin{bmatrix} 1 & \nu & \\ \nu & 1 & \\ & & G \end{bmatrix} \tag{130}$$

## 3.4.4  Equations of Motion in Terms of Displacements for an Isotropic Plate

In order to derive the equations of motion in terms of displacements, recall Eqs. (59), (60) and substitute them into the expanded form of Eqs. (126), (127) to get

$$\begin{Bmatrix} N_x \\ N_y \\ N_{xy} \end{Bmatrix} = \frac{Eh}{1 - \nu^2} \begin{bmatrix} 1 & \nu & \\ \nu & 1 & \\ & & G \end{bmatrix} \begin{Bmatrix} \dfrac{\partial u_0}{\partial x} \\[2mm] \dfrac{\partial v_0}{\partial y} \\[2mm] \dfrac{\partial u_0}{\partial y} + \dfrac{\partial v_0}{\partial x} \end{Bmatrix} \tag{131}$$

$$\begin{Bmatrix} M_x \\ M_y \\ M_{xy} \end{Bmatrix} = \frac{Eh^3}{12(1 - \nu^2)} \begin{bmatrix} 1 & \nu & \\ \nu & 1 & \\ & & G \end{bmatrix} \begin{Bmatrix} \dfrac{\partial^2 w_0}{\partial x^2} \\[2mm] \dfrac{\partial^2 w_0}{\partial y^2} \\[2mm] 2\dfrac{\partial^2 w_0}{\partial x \partial y} \end{Bmatrix} \tag{132}$$

Denote by $D$ is the *flexural stiffness* of an isotropic plate, i.e.,

$$D = \frac{Eh^3}{12(1 - \nu^2)} \quad \text{(flexural stiffness of an isotropic plate)} \tag{133}$$

Expansion of Eqs. (131), (132) gives

$$N_x = \frac{Eh}{1 - \nu^2} \left( \frac{\partial u_0}{\partial x} + \nu \frac{\partial v_0}{\partial y} \right) \tag{134}$$

$$N_y = \frac{Eh}{1 - \nu^2} \left( \nu \frac{\partial u_0}{\partial x} + \frac{\partial v_0}{\partial y} \right) \tag{135}$$

$$N_{xy} = Gh\left(\frac{\partial u_0}{\partial y} + \frac{\partial v_0}{\partial x}\right) = \frac{Eh}{2(1+\nu)}\left(\frac{\partial u_0}{\partial y} + \frac{\partial v_0}{\partial x}\right) \tag{136}$$

$$M_x = -D\left(\frac{\partial^2 w_0}{\partial x^2} + \nu\frac{\partial^2 w_0}{\partial y^2}\right) \tag{137}$$

$$M_y = -D\left(\nu\frac{\partial^2 w_0}{\partial x^2} + \frac{\partial^2 w_0}{\partial y^2}\right) \tag{138}$$

$$M_{xy} = -(1-\nu)D\frac{\partial^2 w_0}{\partial x \partial y} \tag{139}$$

It is apparent that the stress resultants related to axial and flexural motions are uncoupled, i.e., $N_x, N_y, N_{xy}$ depend only on the axial displacements $u_0, v_0$ whereas $M_x, M_y, M_{xy}$ depend only on the flexural displacement $w$. Recall Eqs. (51), (52), (53) representing equations of motion in terms of stress resultants, i.e.,

$$\frac{\partial N_x}{\partial x} + \frac{\partial N_{xy}}{\partial y} = m\ddot{u}_0 \qquad (u\text{-equation of motion}) \tag{140}$$

$$\frac{\partial N_{xy}}{\partial x} + \frac{\partial N_y}{\partial y} = m\ddot{v}_0 \qquad (v\text{-equation of motion}) \tag{141}$$

$$\frac{\partial^2 M_x}{\partial x^2} + 2\frac{\partial^2 M_{xy}}{\partial x \partial y} + \frac{\partial^2 M_y}{\partial y^2} = m\ddot{w}_0 \quad (w\text{-equation of motion}) \tag{142}$$

where $m = \rho h$. Using Eqs. (134) through (139) calculate the derivatives needed in Eqs. (140), (141), (142), i.e.,

$$\frac{\partial N_x}{\partial x} = \frac{Eh}{1-\nu^2}\left(\frac{\partial^2 u_0}{\partial x^2} + \nu\frac{\partial^2 v_0}{\partial x \partial y}\right) \tag{143}$$

$$\frac{\partial N_{xy}}{\partial y} = \frac{Eh(1-\nu)}{2(1-\nu^2)}\left(\frac{\partial^2 u_0}{\partial y^2} + \frac{\partial^2 v_0}{\partial x \partial y}\right) \tag{144}$$

$$N_{xy} = \frac{Eh}{2(1+\nu)}\left(\frac{\partial u_0}{\partial y} + \frac{\partial v_0}{\partial x}\right)$$

$$\frac{\partial N_{xy}}{\partial x} = \frac{Eh}{2(1+\nu)}\left(\frac{\partial^2 u_0}{\partial x \partial y} + \frac{\partial^2 v_0}{\partial x^2}\right) \tag{145}$$

$$\frac{\partial N_{xy}}{\partial x} = \frac{Eh(1-\nu)}{2(1-\nu^2)}\left(\frac{\partial^2 u_0}{\partial x \partial y} + \frac{\partial^2 v_0}{\partial x^2}\right)$$

$$\frac{\partial N_y}{\partial y} = \frac{Eh}{1 - \nu^2} \left( \nu \frac{\partial^2 u_0}{\partial x \partial y} + \frac{\partial^2 v_0}{\partial y^2} \right) \tag{146}$$

$$\frac{\partial^2 M_x}{\partial x^2} = -D \left( \frac{\partial^4 w_0}{\partial x^4} + \nu \frac{\partial^4 w_0}{\partial x^2 \partial y^2} \right) \tag{147}$$

$$2 \frac{\partial^2 M_{xy}}{\partial x \partial y} = -2(1 - \nu)D \frac{\partial^4 w_0}{\partial x^2 \partial y^2} \tag{148}$$

$$\frac{\partial^2 M_y}{\partial y^2} = -D \left( \nu \frac{\partial^4 w_0}{\partial x^2 \partial y^2} + \frac{\partial^4 w_0}{\partial y^4} \right) \tag{149}$$

Substitution of Eqs. (143) through (149) into Eq. (140) through (142) yields

$$\frac{Eh}{1 - \nu^2} \left( \frac{\partial^2 u_0}{\partial x^2} + \frac{(1 - \nu)}{2} \frac{\partial^2 u_0}{\partial y^2} + \frac{(1 + \nu)}{2} \frac{\partial^2 v_0}{\partial x \partial y} \right) = m \ddot{u}_0$$

$$\frac{Eh}{1 - \nu^2} \left( \frac{(1 + \nu)}{2} \frac{\partial^2 u_0}{\partial x \partial y} + \frac{(1 - \nu)}{2} \frac{\partial^2 v_0}{\partial x^2} + \frac{\partial^2 v_0}{\partial y^2} \right) = m \ddot{v}_0 \tag{150}$$

$$D \left( \frac{\partial^4 w_0}{\partial x^4} + 2 \frac{\partial^4 w_0}{\partial x^2 \partial y^2} + \frac{\partial^4 w_0}{\partial y^4} \right) + \rho h \ddot{w}_0 = 0 \tag{151}$$

It is apparent that, for isotropic materials, the axial and flexural motions are uncoupled and Eqs. (150) and (151) can be solved independently.

## 3.4.5 Vibration Frequencies and Modeshapes of an Isotropic Plate

Equations (150), (151) represent the equations for in-plane (axial) and out-of-plane (flexural) vibration of an isotropic plate. Equations (150), (151) are uncoupled and hence can be solved independently.

### 3.4.5.1 Axial Vibration of an Isotropic Plate

To calculate natural frequencies and modeshapes for axial vibration of an isotropic plate, assume time-harmonic in-plane motion of the form

$$\begin{Bmatrix} u^0 \\ v^0 \end{Bmatrix} = \begin{Bmatrix} \hat{u} \\ \hat{v} \end{Bmatrix} e^{i \omega t} \tag{152}$$

where the vibration amplitudes $\hat{u}, \hat{v}$ depend only on the space variables $x, y$. Substitution of Eq. (152) into Eq. (150) yields

$$\frac{Eh}{1-\nu^2}\left[\frac{\partial^2 \hat{u}}{\partial x^2} + \frac{(1-\nu)}{2}\frac{\partial^2 \hat{u}}{\partial y^2} + \frac{(1+\nu)}{2}\frac{\partial^2 \hat{v}}{\partial x \partial y}\right] + \omega^2 m\hat{u} = 0$$

$$\frac{Eh}{1-\nu^2}\left[\frac{(1+\nu)}{2}\frac{\partial^2 \hat{u}}{\partial x \partial y} + \frac{(1-\nu)}{2}\frac{\partial^2 \hat{v}}{\partial x^2} + \frac{\partial^2 \hat{v}}{\partial y^2}\right] + \omega^2 m\hat{v} = 0 \tag{153}$$

Equation (153) represents a coupled system of homogeneous partial differential equations in $\hat{u}, \hat{v}$. Since the system is homogeneous, solution is possible only for certain frequencies $\omega_j$, which are the in-plane (axial) natural frequencies of the isotropic plate subject to certain boundary conditions. For each eigenvalue $\omega_j$, one finds a solution $U_j, V_j$ of Eq. (153) representing the axial (in-plane) modeshape of the plate for the natural frequency $\omega_j$. The solution $U_j, V_j$ has one degree of indeterminacy, i.e., it can be scaled up and down by an arbitrary scale factor. The actual scale factor to be used in computations is determined through a modes normalization procedure.

Equation (153) does not accept closed-form solution. Solutions can be obtained numerically using various approximation methods such as Rayleigh–Ritz, Galerkin, FEM, etc.

### 3.4.5.2 Flexural Vibration of an Isotropic Plate

To calculate natural frequencies and modeshapes for flexural vibration of an isotropic plate, assume time-harmonic out-of-plane motion of the form

$$w = \hat{w}\, e^{i\omega t} \tag{154}$$

where the vibration amplitude $\hat{w}$ depends only on the space variables $x, y$. Substitution of Eq. (154) into Eq. (151) yields

$$D\left(\frac{\partial^4 \hat{w}}{\partial x^4} + 2\frac{\partial^4 \hat{w}}{\partial x^2 \partial y^2} + \frac{\partial^4 \hat{w}}{\partial y^4}\right) - \omega^2 \rho h \hat{w} = 0 \tag{155}$$

Equation (155) is a homogeneous partial differential equation for $\hat{w}$. Since the equation is homogeneous, solution is possible only for certain frequencies $\omega_j$, which are the out-of-plane (flexural) natural frequencies of the isotropic plate subject to certain boundary conditions. For each eigenvalue $\omega_j$, one finds a solution $W_j$ of Eq. (155) representing the flexural modeshape for the natural frequency $\omega_j$. The solution $W_j$ has one degree of indeterminacy, i.e., it can be scaled up and down by an arbitrary scale factor. The actual scale factor to be used in computations is determined through a modes normalization procedure.

Equation (155) does not accept closed-form solution except in some very limited cases, such as simply supported on all four edges, or simply supported on two opposite edges with any conditions on the other two edges (Young, 1950). In all other cases, solutions can be obtained numerically using various approximation methods such as Rayleigh–Ritz, Galerkin, FEM, etc.

## 3.5 SPECIAL CASES

### 3.5.1 Symmetric Laminates

Symmetric laminates are such that the layup is symmetric with respect to the neutral surface. The symmetry must be in both geometry and material properties. If the layup is symmetric, then it would consist of symmetrically-placed pairs of equal-thickness layers that satisfy the following conditions:

(i) The two layers making up a symmetric pair have the same material properties and same orientation of the material principal axes, i.e., both layers have the same global-axes stiffness matrix $\overline{\mathbf{Q}}$.

(ii) Both layers have the same thickness.

(iii) The layers that make up a symmetric pair are placed symmetrically about the neutral surface, i.e., if one layer is placed at a certain distance $z$ above the neutral surface, its pair is placed at $-z$, i.e., at the same distance below the neutral surface.

If the number of layers in a layup is an odd number, then the layer that straddles the neutral surface can be considered as a pair of half-thickness layers and that the satisfaction of the above conditions can be enforced (Figure 4).

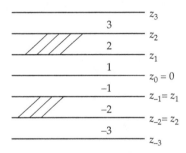

**FIGURE 4**   Numbering scheme in a symmetric layup composite.

Under these assumptions, the layup can be considered always to have an even number of layers, i.e., $N = 2N^*$ with the layer index running from $-N^*$ to $+N^*$. The $k^{*\text{th}}$ ply lies between $z_{k^*-1}$ and $z_{k^*}$, $k^* = -N^* + 1,\ \dots\ ,\ -1,\ +1,\ \dots\ ,\ +N^*$, such that the first $z$ coordinate is $z_{-N^*}$ corresponding to the bottom surface of the plate and the last is $z_{+N^*}$ corresponding to the top surface of the plate. The plate neutral surface is placed at $z_0 = 0$. Thus, condition (ii) above can be expressed as

$$\mathbf{Q}_{-k^*} = \mathbf{Q}_{+k^*}, \quad z_{-k^*} = -z_{k^*} \tag{156}$$

Equations (85), (86), (87) that define the A, B, D matrices are processed as follows. Recall Eq. (85) and change the index from $k$ to $k^*$ to get

$$\mathbf{A} = \sum_{n=1}^{N} (z_n - z_{n-1})\overline{\mathbf{Q}}_n = \sum_{n^* = -N^* + 1}^{+N^*} (z_{n^*} - z_{n^*-1})\overline{\mathbf{Q}}_{n^*} \tag{157}$$

Group the terms for the $n^*$ and $-n^*$ layers, i.e.,

$$A = \sum_{n^*=1}^{N^*} (z_{-n^*+1} - z_{-n^*})\overline{Q}_{-n^*} + (z_{n^*} - z_{n^*-1})\overline{Q}_{n^*} \tag{158}$$

Use Eq. (156) into Eq. (158) to get

$$A = \sum_{n^*=1}^{N^*} (-z_{n^*-1} + z_{n^*})\overline{Q}_{n^*} + (z_{n^*} - z_{n^*-1})\overline{Q}_{n^*}$$

$$= \sum_{n^*=1}^{N^*} (-z_{n^*-1} + z_{n^*} + z_{n^*} - z_{n^*-1})\overline{Q}_{n^*} = 2\sum_{n^*=1}^{N^*} (z_{n^*} - z_{n^*-1})\overline{Q}_{n^*} \tag{159}$$

where $N^* = N/2$. Now, recall Eq. (86) and express it as

$$B = \tfrac{1}{2}\sum_{n=1}^{N} (z_n^2 - z_{n-1}^2)\overline{Q}_k = \tfrac{1}{2}\sum_{n^*=-N^*+1}^{+N^*} (z_{n^*}^2 - z_{n^*-1}^2)\overline{Q}_{n^*} \tag{160}$$

Group the terms for the $n^*$ and $-n^*$ layers, i.e.,

$$B = \sum_{n^*=1}^{N^*} (z_{-n^*+1}^2 - z_{-n^*}^2)\overline{Q}_{-n^*} + (z_{n^*}^2 - z_{n^*-1}^2)\overline{Q}_{n^*} \tag{161}$$

Use Eq. (156) into Eq. (161) to get

$$B = \sum_{n^*=1}^{N^*} (z_{n^*-1}^2 - z_{n^*}^2)\overline{Q}_{n^*} + (z_{n^*}^2 - z_{n^*-1}^2)\overline{Q}_{n^*} = \sum_{n^*=1}^{N^*} (z_{n^*-1}^2 - z_{n^*}^2 + z_{n^*}^2 - z_{n^*-1}^2)\overline{Q}_{n^*} = 0 \tag{162}$$

where $N^* = N/2$. Finally, recall Eq. (87) and express it as

$$D = \tfrac{1}{3}\sum_{n=1}^{N} (z_n^3 - z_{n-1}^3)\overline{Q}_k = \tfrac{1}{3}\sum_{n^*=-N^*+1}^{+N^*} (z_{n^*}^3 - z_{n^*-1}^3)\overline{Q}_{n^*} \tag{163}$$

Group the terms for the $n^*$ and $-n^*$ layers, i.e.,

$$D = \tfrac{1}{3}\sum_{n^*=1}^{N^*} (z_{-n^*+1}^3 - z_{-n^*}^3)\overline{Q}_{-n^*} + (z_{n^*}^3 - z_{n^*-1}^3)\overline{Q}_{n^*} \tag{164}$$

Use Eq. (156) into Eq. (164) to get

$$D = \tfrac{1}{3}\sum_{k^*=1}^{N^*} (-z_{k^*-1}^3 + z_{k^*}^3)\overline{Q}_{k^*} + (z_{k^*}^3 - z_{k^*-1}^3)\overline{Q}_{k^*}$$

$$= \tfrac{1}{3}\sum_{k^*=1}^{N^*} (-z_{k^*-1}^3 + z_{k^*}^3 + z_{k^*}^3 - z_{k^*-1}^3)\overline{Q}_{k^*} = \tfrac{2}{3}\sum_{k^*=1}^{N^*} (z_{k^*}^3 - z_{k^*-1}^3)\overline{Q}_{k^*} \tag{165}$$

Equation (162) indicates that the axial–flexural coupling matrix **B** vanishes if the layup is symmetric. This is important in practice because the axial–flexural coupling represented by matrix **B** can lead to significant warping due to the residual stresses generated by differential thermal expansion during cure. By not having axial–flexural coupling, a symmetric laminate avoids the occurrence of such unwanted warping.

### 3.5.2 Isotropic Laminates

If the plies are made up of the same isotropic material then the ply stiffness matrix is the same for all the plies, $\overline{\mathbf{Q}}_k = \overline{\mathbf{Q}}_0$, $k = 1, ..., N$ and the **A**, **B**, **D** matrices take the form

$$\mathbf{A} = \sum_{k=1}^{N} \overline{\mathbf{Q}}_k (z_k - z_{k-1}) = \overline{\mathbf{Q}}_0 \sum_{k=1}^{N} (z_k - z_{k-1}) = \overline{\mathbf{Q}}_0 (z_N - z_1) = \overline{\mathbf{Q}}_0 \left( \frac{h}{2} - \left( -\frac{h}{2} \right) \right) = \overline{\mathbf{Q}}_0 h \quad (166)$$

$$\mathbf{B} = \frac{1}{2} \sum_{k=1}^{N} \overline{\mathbf{Q}}_k (z_k^2 - z_{k-1}^2) = \frac{1}{2} \overline{\mathbf{Q}}_0 \sum_{k=1}^{N} (z_k^2 - z_{k-1}^2) = \frac{1}{2} \overline{\mathbf{Q}}_0 \sum_{k=1}^{N} (z_N^2 - z_1^2)$$

$$= \frac{1}{2} \overline{\mathbf{Q}}_0 \sum_{k=1}^{N} \left[ \left( \frac{h}{2} \right)^2 - \left( -\frac{h}{2} \right)^2 \right] = \mathbf{0} \quad (167)$$

$$\mathbf{D} = \frac{1}{3} \sum_{k=1}^{N} \overline{\mathbf{Q}}_k (z_k^3 - z_{k-1}^3) = \frac{1}{3} \overline{\mathbf{Q}}_0 \sum_{k=1}^{N} (z_k^3 - z_{k-1}^3) = \frac{1}{3} \overline{\mathbf{Q}}_0 \sum_{k=1}^{N} (z_N^3 - z_1^3)$$

$$= \frac{1}{3} \overline{\mathbf{Q}}_0 \sum_{k=1}^{N} \left[ \left( \frac{h}{2} \right)^3 - \left( -\frac{h}{2} \right)^3 \right] = \frac{h^3}{12} \overline{\mathbf{Q}}_0 \quad (168)$$

## 3.6 PROBLEMS AND EXERCISES

Problems and exercises and worked out examples are given in the Instructor Manual posted on the publisher's website.

## References

[1] Anon. (1989) "Composites II: Material Selection and Applications", Materials Engineering Institute course 31, ASM International, Materials Park, OH.
[2] Young, D. (1950) "Vibration of Rectangular Plates by the Ritz Method", *Journal of Applied Mechanics*, Vol. 17, 448–453, 1950.
[3] Whitney, J. M. (1987) *Structural Analysis of Laminated Anisotropic Plates*, Technomic Pub. Co. Inc., Lancaster, PA, 1987.
[4] Jones, R. M. (1999) *Mechanics of Composite Materials*, 2nd Ed., Taylor & Francis, Philadelphia, PA.

# Guided Waves in Thin-Wall Composite Structures

# 4.1 INTRODUCTION

## 4.1.1 Overview

This chapter is devoted to the discussion of guided waves in composite materials as a mean for detection of internal damage. Damage detection techniques for composite structures are based on the study of complicated wave mechanisms; they rely strongly on the use of predictive modeling tools. The response method and the modal method are the two most used inspection techniques. In the response method, the reflection and transmission characteristics of the structure are examined. In the modal method, the standing-wave properties of the system are evaluated. Both methods make use of the matrix formulation, which describes elastic waves in layered media with arbitrary number of layers.

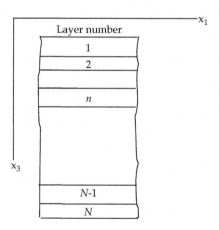

**FIGURE 1** *N*-ply laminated composite plate model for the study of ultrasonic guided waves.

Guided waves are especially important for structural health monitoring (SHM) because they can travel at large distances in structures with only little energy loss. Thus, they enable the SHM of large areas from a single location. Guided waves have the important property that they remain confined inside the walls of a thin-wall structure and hence can travel over large distances. In addition, guided waves can also travel inside curved walls.

These properties make them well suited for the ultrasonic inspection of aircraft, missiles, pressure vessels, oil tanks, pipelines, etc.

## 4.1.2 Problem Setup

Consider a composite laminate (Figure 1) made up of $N$ unidirectional composite plies of various orientations with respect to the global coordinate system $Ox_1x_2x_3$, i.e., each ply was laid up with a different rotation $\theta$ about the vertical axis $x_3$. The relations between the local stiffness matrix $\mathbf{C}'$ in material axes $Ox_1'x_2'x_3'$ and the global stiffness matrix $\mathbf{C}$ were described in Chapter 1 and will not be repeated here. For the present discussion, we will only use the global stiffness matrix $\mathbf{C}$ which is assumed to be specified for each ply. Thus, the laminate is defined by the global stiffness matrices and thicknesses of the plies, i.e.,

$$\mathbf{C}_1, ..., \mathbf{C}_n, ..., \mathbf{C}_N \quad \text{global stiffness matrices of the plies}$$
$$h_1, ..., h_n, ..., h_N \quad \text{thickness values of the plies} \qquad n = 1, ..., N \qquad (1)$$

Our objective will be to analyze the propagation of guided ultrasonic waves in the laminated composite plate depicted in Figure 1. This analysis is rather elaborate, as wave reflection and transmission at each layer interface has to be considered across the composite thickness. In the case of an isotropic plate, the dispersion curves can be derived from the solution of the Rayleigh–Lamb equation. In the case of composite materials, it is not possible to find a closed-form solution of the dispersion curves; several numerical methods have been proposed to solve the problem (GMM, TMM, SMM, etc.).

## 4.1.3 State of the Art in Modeling Guided-Wave Propagation in Laminated Composites

A review of the matrix techniques for modeling ultrasonic waves in multilayered isotropic media was given by Lowe [1]. A comprehensive presentation of the theory of guided-wave propagation in laminated composites was given by Rokhlin et al. [2].

In a bulk anisotropic material, wave propagation is governed by the Christoffel equation which yields three pairs of $\pm$ wavespeeds and thus six wave polarization directions. In a layered structure, the field equations for the displacements and stresses in a layer are expressed as the superposition of the fields of the bulk waves (a.k.a. partial waves) present within the layer. The boundary conditions at the interfaces between the layers are governed by Snell's law which requires that all the partial waves share the same $k_1 = \xi$ wavenumber component in the direction of guided-wave propagation along the laminate. Snell's law at the interface between two adjacent layers requires that all the interacting partial waves must share the same frequency and spatial properties in the $x_1$ direction. Thus, the angles of incidence, transmission, and reflection of the partial waves interacting at the interface between two adjacent layers are interrelated by Snell's law and only certain combination of wavenumbers of the partial waves may exist as given by the solution of a constrained Christoffel equation. The analysis of the layers is usually performed under $z$-invariant conditions that simplify the analysis to two dimensions although all the three displacements are present. The approach is to solve the constrained Christoffel equation to find the allowable partial waves and then express the field equations for displacements and stresses in terms of the polarization vectors of these partial waves factored by their contribution coefficients. Subsequently, the layers in the laminate are assembled by satisfying the displacement continuity and tractions balance at the interfaces between adjacent layers as well as the boundary conditions at the top and bottom faces of the laminate. Several methods of achieving this have been developed so far; the transfer matrix method (TMM), global matrix method (GMM), as well as the stiffness transfer method (STM) will be briefly discussed next.

The transfer matrix method (TMM) for wave propagation in multilayered isotropic media was initially derived by Thomson [3] and improved by Haskell [4]. In the Thomson–Haskell formulation, the displacements and the stresses at the bottom of a layer are described in terms of those at the top of the layer through a transfer matrix (TM) written in terms of the partial waves existing in the each layer. After propagating the TM process through all the layers, one obtains a compact matrix that represents the behavior of the complete system and relates the boundary conditions at the bottom of the laminate to those at the top. However, this process may show numerical instability for large layer thickness and high excitation frequencies due to the intrinsic poor numerical conditioning of the TM process due to a combination of several factors involving the presence of decaying evanescent waves at the interface.

Nayfeh [5,6] extended the Thomson–Haskell formulation to the case of anisotropic materials and composites of anisotropic layers. The constrained Christoffel equation are set up in each anisotropic composite layer and the allowable partial waves are found. The field equations for displacements and stresses are set up in terms of the polarization vectors of these partial waves factored by their contribution coefficients. Substitution of the formal solution in the wave equation and imposing the existence of a nontrivial solution lead to the formal solution for the displacements and stresses in a $6 \times 6$ matrix form as functions of partial-wave contribution factors. A transfer matrix (TM) formulation is set up to relate the displacements and tractions at the bottom surface of the layer to those at the top surface. Imposition of displacements and tractions continuity at the interfaces between two layers allows one to transfer from one layer to the next. Through a chain process, the displacements and tractions at the bottom surface of the laminate are eventually expressed in terms of those at the top surface; this is achieved in the form of an aggregate TM that reflects the sequential multiplication of the transfer matrices of each layer.

To obtain the dispersion curves, one uses the traction-free boundary conditions at the top and bottom faces of the laminate. The imposition of these traction-free boundary conditions collapses the aggregate TM formulation into a homogenous linear system that depends nonlinearly on the frequency $\omega$ and wavenumber $\xi$ of the guided wave which is propagating along the laminate. The eigenvalues of this homogeneous system are $(\xi, \omega)$ combinations that make the system matrix singular. These $(\xi, \omega)$ combinations generate the dispersion curves of the guided waves traveling in the composite laminate.

The TM method has the advantage that the resulting eigenvalue problem is at most of size $M = 6$, corresponding to the fact that three tractions have to be made zero at the top and other three at the bottom of the laminate. However, the practical application of the TM method runs into numerical difficulties at high $(\xi, \omega)$ values due to a combination of several factors involving the presence of decaying evanescent waves at the interface.

An alternative to the Thomson−Haskell formulation is the global matrix method (GMM) formulation proposed by Knopoff [7]. In the GMM formulation, all the equations of all the layers of the structure are considered. Sets of equations are set up at each interface to express displacement and tractions continuity. Boundary condition equations are also added for the top and bottom faces of the laminate. The solution is carried out on the global matrix (GM), addressing all the equations concurrently. Imposition of traction-free boundary conditions at the top and bottom faces of the laminate yield again a nonlinear eigenvalue problem in $(\xi, \omega)$, but of a much larger size than that obtained in the TM method; for $N$ layers in the laminate, the size of the GM problem is $M \times N$, where $M = 6$ corresponds to the number of unknowns in each layer. The GMM approach was used by Pavlakovic and Lowe in the commercial code DISPERSE [8].

Rokhlin and Wang [9,10], studied the numerical instability of the TM and addressed it through the layer stiffness matrix (SM). In this method, a layer SM is used to replace the layer TM. The stiffness matrix method (SMM) relates the stresses at the top and the bottom of the layer with the displacements at the top and bottom layer; the terms in the matrix have only exponentially decaying terms and the matrix had the same dimension and simplicity as the TM. The SM for a layer is calculated through a recursive algorithm that uses the SM of the previous layers. The SM method is unconditionally stable.

Rokhlin and Wang [9,10] also recommend that, for each layer, the local coordinate origin is settled at the top of the $j$-th layer for waves propagating along the $-x_3$ direction and at the bottom of the $j$-th layer for waves propagating along the $x_3$ direction. This selection of coordinate system is important for minimizing the numerical overflow of the exponential terms at large $fd$ values. In this way, the exponential terms are normalized and the nonhomogeneous exponentials are equal to one at the interface and decay toward the opposite surface of the layer.

## 4.1.4  Chapter Layout

The analysis of guided-wave propagation in a composite laminate will proceed step by step. First, the propagation of bulk waves in an infinite anisotropic medium will be studied. Next, the constrained propagation of such waves in an isolated composite ply will be analyzed. Finally, the propagation of ultrasonic waves in a stack of composite plies tightly joined together to form a composite laminate will be developed.

## 4.2  WAVE PROPAGATION IN BULK COMPOSITE MATERIAL—CHRISTOFFEL EQUATIONS

Assume plane harmonic wave of frequency $\omega$ propagating with speed $v$ in the direction $\vec{n} = n_1 \vec{e}_1 + n_2 \vec{e}_2 + n_3 \vec{e}_3$ (Figure 2). Define the wavenumber $k$ as

$$k = \omega/v \quad \text{(wavenumber)} \tag{2}$$

Also define the wave vector $\vec{k}$ as $\vec{k} = k\,\vec{n}$. Hence, write

$$
\begin{aligned}
\vec{k} &= k\,\vec{n} \\
\vec{k} &= k_1\vec{e}_1 + k_2\vec{e}_2 + k_3\vec{e}_3
\end{aligned}
\qquad
\begin{aligned}
k_1 &= kn_1 \\
k_2 &= kn_2 \\
k_3 &= kn_3
\end{aligned}
\quad\text{(wave vector)}
\tag{3}
$$

The quantities $k_1, k_2, k_3$ are the directional wavenumber and represent the projections of the wavenumber vector $\vec{k}$ in the $x_1, x_2, x_3$ directions. The particle displacement can be written in the form

$$
\vec{u}(\vec{r}, t) = \vec{\hat{u}}\, e^{i(\vec{k}\cdot\vec{r}-\omega t)} = \vec{\hat{u}}\, e^{i(k_1 x_1 + k_2 x_2 + k_3 x_3 - \omega t)}
\tag{4}
$$

Equation (4) can be written as

$$
\begin{Bmatrix} u_1 \\ u_2 \\ u_3 \end{Bmatrix} = \begin{Bmatrix} \hat{u}_1 \\ \hat{u}_2 \\ \hat{u}_3 \end{Bmatrix} e^{i(k_1 x_1 + k_2 x_2 + k_3 x_3 - \omega t)}
\tag{5}
$$

where $\hat{u}_1, \hat{u}_2, \hat{u}_3$ are complex displacement magnitudes in the $x_1, x_2, x_3$ directions. A shorthand formulation of Eq. (5) is obtained through tensor notations, i.e.,

$$
u_m = \hat{u}_m\, e^{i(k_i x_i - \omega t)} \quad m = 1, 2, 3
\tag{6}
$$

where the repeated index $m$ signifies Einstein implied summation.

## 4.2.1 Equation of Motion in Terms of Displacements

Recall in tensor notations the equation of motion in terms of displacements, i.e.,

$$
c_{ijlm} u_{m,lj} = \rho \ddot{u}_i \quad i, j = 1, 2, 3
\tag{7}
$$

where the differentiation shorthand $()_{m,l} = \partial()_m / \partial x_l$, $()_{m,lj} = \partial^2 ()_m / \partial x_l \partial x_j$ was applied.

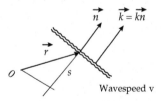

Wavespeed $v$

FIGURE 2    Plane wave propagating in an anisotropic composite material.

Differentiation of Eq. (6) yields

$$
\begin{aligned}
u_{m,l} &= ik_l\hat{u}_m\, e^{i(k_i x_i - \omega t)} = ik_l u_m \\
u_{m,lj} &= (ik_l)(ik_j)\hat{u}_m\, e^{i(k_i x_i - \omega t)} = -k_l k_j u_m \\
\ddot{u}_m &= -\omega^2 u_m
\end{aligned}
\tag{8}
$$

Substitution of Eq. (8) into Eq. (7) and cancelation of the negative sign give

$$c_{ijlm}k_lk_ju_m = \rho\omega^2 u_i \quad i = 1,2,3 \tag{9}$$

Recall Eq. (3) and write $k_l = kn_l$, $k_j = kn_j$. Use Eq. (2) to write $\omega^2 = k^2 v^2$. Upon substitution in Eq. (9) and division by $k^2$, we get

$$c_{ijlm}n_ln_ju_m = \rho v^2 u_i \quad i = 1,2,3 \tag{10}$$

Using Eq. (6) and dividing by $e^{i(k_ix_i-\omega t)}$ yields Eq. (10) only in terms of the displacement amplitudes $\hat{u}_m$, i.e.,

$$c_{ijlm}n_ln_j\hat{u}_m = \rho v^2 \hat{u}_i \quad i = 1,2,3 \tag{11}$$

## 4.2.2 Christoffel Equation for Bulk Composites

Equation (11) is a homogeneous linear algebraic equation in $\hat{u}_i$, known as *Christoffel equation*. Expansion of implied summations allows us to calculate

$$\Gamma_{im} = \sum_{j=1}^{3}\sum_{l=1}^{3} c_{ijlm}n_ln_j \quad i,m = 1,2,3 \tag{12}$$

The symmetry properties of the stiffness tensor $c_{ijlm}$ imply that the acoustic tensor is also symmetric, i.e.,

$$\Gamma_{im} = \Gamma_{mi} \quad i,m = 1,2,3 \tag{13}$$

*Proof of Eq. (13):* Recall the symmetry properties of the stiffness sensor, i.e.,

$$c_{ijlm} = c_{jilm} = c_{ijml} = c_{lmij} = c_{jiml} \tag{14}$$

Write $\Gamma_{mi} = c_{mjli}n_jn_l$ and do index permutations and interchanges using Eq. (14) to get

$$\Gamma_{mi} = c_{mjli}n_jn_l = c_{limj}n_jn_l = c_{iljm}n_jn_l \underset{\substack{j\to l \\ l\to j}}{=} c_{ijlm}n_ln_j = \Gamma_{im} \quad \text{QED} \tag{15}$$

The numbers $\Gamma_{im}$ $(i,m = 1,2,3)$ are elements of the *acoustic tensor*, i.e.,

$$\mathbf{\Gamma} = \begin{bmatrix} \Gamma_{11} & \Gamma_{12} & \Gamma_{13} \\ \Gamma_{12} & \Gamma_{22} & \Gamma_{23} \\ \Gamma_{13} & \Gamma_{23} & \Gamma_{33} \end{bmatrix} \quad \text{(acoustic tensor)} \tag{16}$$

Denote by $\hat{\mathbf{u}}$ the column vector of displacement amplitudes, i.e.,

$$\hat{\mathbf{u}} = \begin{Bmatrix} \hat{u}_1 \\ \hat{u}_2 \\ \hat{u}_3 \end{Bmatrix} \tag{17}$$

Using Eqs. (16), (17), we write Eq. (11) as an algebraic eigenvalue problem, i.e.,

$$\begin{bmatrix} \Gamma_{11} & \Gamma_{12} & \Gamma_{13} \\ \Gamma_{12} & \Gamma_{22} & \Gamma_{23} \\ \Gamma_{13} & \Gamma_{23} & \Gamma_{33} \end{bmatrix} \begin{Bmatrix} \hat{u}_1 \\ \hat{u}_2 \\ \hat{u}_3 \end{Bmatrix} = \rho v^2 \begin{Bmatrix} \hat{u}_1 \\ \hat{u}_2 \\ \hat{u}_3 \end{Bmatrix} \tag{18}$$

or, compactly,

$$\mathbf{\Gamma}\hat{\mathbf{u}} = \lambda\hat{\mathbf{u}} \quad \lambda = \rho v^2 \tag{19}$$

Equation (19) is a $3 \times 3$ algebraic eigenvalue problem. Its solution yields three eigenvalues $\lambda_I, \lambda_{II}, \lambda_{III}$ and the corresponding three mutually orthogonal eigenvectors $\hat{\mathbf{u}}_I, \hat{\mathbf{u}}_{II}, \hat{\mathbf{u}}_{III}$. For each eigenvalue, one calculates the corresponding wavespeeds as

$$^{\pm}v_I = \pm\sqrt{\lambda_I/\rho} \quad ^{\pm}v_{II} = \pm\sqrt{\lambda_{II}/\rho} \quad ^{\pm}v_{III} = \pm\sqrt{\lambda_{III}/\rho} \tag{20}$$

Note that the sign $\pm$ signifies that the wave is propagating forward and backward, respectively. Hence, three pairs of forward/backward waves exist and each pair has its own polarization $\hat{\mathbf{u}}$ which is mutually orthogonal to the polarization of the other two waves. In total, we have six independent solutions $^{-}v_I, {}^{+}v_I, {}^{-}v_{II}, {}^{+}v_{II}, {}^{-}v_{III}, {}^{+}v_{III}$.

*Pure waves*: A pure wave is obtained when the wave propagation direction is chosen such that one of the resulting polarization eigenvectors comes out to be aligned to the chosen wave propagation direction. This is an inverse problem and the outcome is only known after the calculations are performed for all possible propagation directions. In brief, pure waves may exist in composite materials but their existence is not guaranteed. This situation is in contrast to that of isotropic materials in which pure waves can always be generated, i.e., by choosing either pressure or shear polarization.

## 4.3  GUIDED WAVES IN A COMPOSITE PLY

The analysis of guided waves in a composite ply (lamina) will be studied as a superposition of partial waves that satisfy the generalized Snell's law (coherence condition). The allowable partial waves will be obtained as a solution of the constrained Christoffel equation. The top and bottom free-surface conditions will be imposed on the finite-thickness lamina and the wavespeed dispersion curves will be obtained.

### 4.3.1  Guided Wave as a Superposition of Partial Waves

Consider a composite lamina (ply) as depicted in Figure 3. Assume a guided wave of angular frequency $\omega$ is propagating along the composite lamina in the $x_1$ direction with wavenumber $\xi$ and phase velocity $v = \omega/\xi$. The guided wave that propagates in the $x_1$ direction is the result of the superposition of several partial waves that propagate in both the $x_1$ and $x_3$ directions. This wave propagation is contained in the vertical plane $x_1 O x_3$ and the problem is $x_2$ invariant.

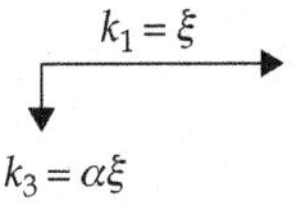

$$k_1 = \xi$$

$$k_3 = \alpha\xi$$

**FIGURE 3**   Partial wave propagating in a composite lamina.

The partial waves reflect and refract at the layer boundaries according to its boundary conditions. The result is a steady-state propagation in the $x_1$ direction of thickness-wise standing waves. Since the problem is $x_2$ invariant, the wave vector has components only in $x_1$ and $x_3$ directions and takes the form $\vec{k} = k_1 \vec{e}_1 + k_3 \vec{e}$. Hence, the particle displacement can be written as

$$\begin{Bmatrix} u_1 \\ u_2 \\ u_3 \end{Bmatrix} = \begin{Bmatrix} \hat{u}_1 \\ \hat{u}_2 \\ \hat{u}_3 \end{Bmatrix} e^{i(k_1 x_1 + k_3 x_3 - \omega t)} = \hat{\mathbf{u}}\, e^{i(k_1 x_1 + k_3 x_3 - \omega t)} \tag{21}$$

where $\hat{u}_1, \hat{u}_2, \hat{u}_3$ are complex displacement magnitudes in the $x_1, x_2, x_3$ directions and $\hat{\mathbf{u}}$ is a column vector defined as

$$\hat{\mathbf{u}} = \begin{Bmatrix} \hat{u}_1 \\ \hat{u}_2 \\ \hat{u}_3 \end{Bmatrix} \tag{22}$$

Note that the components of the particle displacement exist in all three directions $x_1, x_2, x_3$ although the wave propagation is contained in the vertical plane $x_1 O x_3$ and is invariant in the $x_2$ direction.

## 4.3.2 Coherence Condition—Generalized Snell's Law

The partial waves that make up the guided wave are solutions of the Christoffel equation (see Section 2). Our task is to find these solutions under the assumptions stated above. A common characteristic of the partial waves is that they all have the same wavenumber in the $x_1$ direction; this wavenumber is the same wavenumber $\xi$ as that of the resulting guided wave. This property results from imposing the coherence condition at the layer boundaries (a.k.a. generalized Snell's law). Hence, the directional wavenumbers of these partial waves take the form

$$\begin{aligned} k_1 &= \xi \\ k_3 &= \xi\alpha \end{aligned} \quad \text{(directional wavenumbers)} \tag{23}$$

where the parameter $\alpha$ is the ratio between the wavenumbers in the $x_3$ and $x_1$ directions, i.e.,

$$\alpha = k_3/k_1 \tag{24}$$

The problem to be solved has now become: "For a given $\omega$ and $\xi$, i.e., for a known $v$, find the value of the parameter $\alpha$ by solving the Christoffel equation". Since, as shown in Section 2, Christoffel equation has six roots (three pairs of forward/backward waves), this process will yield all the partial waves needed to build up the guided-wave solution.

### 4.3.3  Christoffel Equation for a Lamina

The Christoffel equation for a lamina is set up as follows: Substitute Eq. (23) into Eq. (21) and use $\omega = \xi v$ to write

$$\mathbf{u} = \begin{Bmatrix} u_1 \\ u_2 \\ u_3 \end{Bmatrix} = \begin{Bmatrix} \hat{u}_1 \\ \hat{u}_2 \\ \hat{u}_3 \end{Bmatrix} e^{i(k_1 x_1 + k_3 x_3 - \omega t)} = \begin{Bmatrix} \hat{u}_1 \\ \hat{u}_2 \\ \hat{u}_3 \end{Bmatrix} e^{i\xi(x_1 + \alpha x_3 - vt)} \tag{25}$$

Recall the lamina stiffness matrix given by

$$\mathbf{C} = \begin{bmatrix} C_{11} & C_{12} & C_{13} & 0 & 0 & C_{16} \\ C_{12} & C_{22} & C_{23} & 0 & 0 & C_{26} \\ C_{13} & C_{23} & C_{33} & 0 & 0 & C_{36} \\ 0 & 0 & 0 & C_{44} & C_{45} & 0 \\ 0 & 0 & 0 & C_{45} & C_{55} & 0 \\ C_{16} & C_{26} & C_{36} & 0 & 0 & C_{66} \end{bmatrix} \tag{26}$$

Recall Eq. (7), the equation of motion in terms of displacements, i.e.,

$$c_{ijlm} u_{m,lj} = \rho \ddot{u}_i \quad i,j = 1,2,3 \tag{27}$$

where $c_{ijlm}$, $i,j,l,m = 1,2,3$ is the stiffness tensor. Equation (27) can be converted from tensor notation to Voigt matrix notation and expanded to yield the equations of motion in terms of the elements $C_{pq}$, $p,q = 1,...,6$ of the stiffness matrix $\mathbf{C}$ of Eq. (26), i.e.,

$$C_{11}\frac{\partial^2 u_1}{\partial x_1^2} + C_{66}\frac{\partial^2 u_1}{\partial x_2^2} + C_{55}\frac{\partial^2 u_1}{\partial x_3^2} + 2C_{16}\frac{\partial^2 u_1}{\partial x_1 \partial x_2}$$
$$+ C_{16}\frac{\partial^2 u_2}{\partial x_1^2} + C_{26}\frac{\partial^2 u_2}{\partial x_2^2} + C_{45}\frac{\partial^2 u_2}{\partial x_3^2} + (C_{12}+C_{66})\frac{\partial^2 u_2}{\partial x_1 \partial x_2} \tag{28}$$
$$+ (C_{13}+C_{55})\frac{\partial^2 u_3}{\partial x_1 \partial x_3} + (C_{36}+C_{45})\frac{\partial^2 u_3}{\partial x_2 \partial x_3} = \rho\frac{\partial^2 u_1}{\partial t^2}$$

$$C_{16}\frac{\partial^2 u_1}{\partial x_1^2} + C_{26}\frac{\partial^2 u_1}{\partial x_2^2} + C_{45}\frac{\partial^2 u_1}{\partial x_3^2} + (C_{12}+C_{66})\frac{\partial^2 u_1}{\partial x_1 \partial x_2}$$
$$+ C_{66}\frac{\partial^2 u_2}{\partial x_1^2} + C_{22}\frac{\partial^2 u_2}{\partial x_2^2} + C_{44}\frac{\partial^2 u_2}{\partial x_3^2} + 2C_{26}\frac{\partial^2 u_2}{\partial x_1 \partial x_2} \tag{29}$$
$$+ (C_{36}+C_{45})\frac{\partial^2 u_3}{\partial x_1 \partial x_3} + (C_{23}+C_{44})\frac{\partial^2 u_3}{\partial x_2 \partial x_3} = \rho\frac{\partial^2 u_2}{\partial t^2}$$

$$(C_{13} + C_{55})\frac{\partial^2 u_1}{\partial x_1 \partial x_3} + (C_{36} + C_{45})\frac{\partial^2 u_1}{\partial x_2 \partial x_3}$$

$$+ (C_{36} + C_{45})\frac{\partial^2 u_2}{\partial x_1 \partial x_3} + (C_{23} + C_{44})\frac{\partial^2 u_2}{\partial x_2 \partial x_3} \tag{30}$$

$$+ C_{55}\frac{\partial^2 u_3}{\partial x_1^2} + C_{44}\frac{\partial^2 u_3}{\partial x_2^2} + C_{33}\frac{\partial^2 u_3}{\partial x_3^2} + 2C_{45}\frac{\partial^2 u_3}{\partial x_1 \partial x_2} = \rho\frac{\partial^2 u_3}{\partial t^2}$$

Calculate the derivatives of Eq. (25), i.e.,

$$\frac{\partial u_j}{\partial x_1} = \hat{u}_j \frac{\partial}{\partial x_1}e^{i\xi(x_1 + \alpha x_3 - vt)} = i\xi\hat{u}_j \ e^{i\xi(x_1 + \alpha x_3 - vt)} = i\xi u_j$$

$$\frac{\partial u_j}{\partial x_2} = \hat{u}_j \frac{\partial}{\partial x_2}e^{i\xi(x_1 + \alpha x_3 - vt)} = 0$$

$$\frac{\partial u_j}{\partial x_3} = \hat{u}_j \frac{\partial}{\partial x_1}e^{i\xi(x_1 + \alpha x_3 - vt)} = i\xi\alpha\hat{u}_j \ e^{i\xi(x_1 + \alpha x_3 - vt)} = i\xi\alpha u_j \qquad j = 1,2,3 \tag{31}$$

$$\frac{\partial u_j}{\partial t} = \hat{u}_j \frac{\partial}{\partial t}e^{i\xi(x_1 + \alpha x_3 - vt)} = -i\xi v\hat{u}_j \ e^{i\xi(x_1 + \alpha x_3 - vt)} = -i\xi v u_j$$

Hence,

$$\frac{\partial^2 u_j}{\partial x_1^2} = -\xi^2 u_j \qquad \frac{\partial^2 u_j}{\partial x_1 \partial x_3} = -\xi^2 \alpha u_j$$

$$\frac{\partial^2 u_j}{\partial x_2^2} = 0 \qquad \frac{\partial^2 u_j}{\partial x_1 \partial x_2} = 0 \qquad \frac{\partial^2 u_j}{\partial t^2} = -\xi^2 v^2 u_j \tag{32}$$

$$\frac{\partial^2 u_j}{\partial x_3^2} = -\xi^2 \alpha^2 u_j \qquad \frac{\partial^2 u_j}{\partial x_2 \partial x_3} = 0$$

Substitution of Eqs. (31), (32) into Eqs. (28) through (30) and division by $-\xi^2$ yields

$$\left(C_{11} + C_{55}\alpha^2\right)u_1 + \left(C_{16} + C_{45}\alpha^2\right)u_2 + (C_{13} + C_{55})\alpha u_3 = \rho v^2 u_1$$
$$\left(C_{16} + C_{45}\alpha^2\right)u_1 + \left(C_{66} + C_{44}\alpha^2 - \rho v^2\right)u_2 + (C_{36} + C_{45})\alpha u_3 = \rho v^2 u_2 \tag{33}$$
$$(C_{13} + C_{55})\alpha u_1 + (C_{36} + C_{45})\alpha u_2 + \left(C_{55} + C_{33}\alpha^2 - \rho v^2\right)u_3 = \rho v^2 u_3$$

Further simplification of Eq. (33) through division by the common factor $e^{i\xi(x_1 + \alpha x_3 - vt)}$ yields, upon rearrangement,

$$\left(C_{11} - \rho v^2 + C_{55}\alpha^2\right)\hat{u}_1 + \left(C_{16} + C_{45}\alpha^2\right)\hat{u}_2 + (C_{13} + C_{55})\alpha\hat{u}_3 = 0$$
$$\left(C_{16} + C_{45}\alpha^2\right)\hat{u}_1 + \left(C_{66} - \rho v^2 + C_{44}\alpha^2\right)\hat{u}_2 + (C_{36} + C_{45})\alpha\hat{u}_3 = 0 \tag{34}$$
$$(C_{13} + C_{55})\alpha\hat{u}_1 + (C_{36} + C_{45})\alpha\hat{u}_2 + \left(C_{55} - \rho v^2 + C_{33}\alpha^2\right)\hat{u}_3 = 0$$

Equation (34) is Christoffel equation for the situation considered here. In matrix notations, we write Eq. (34) as

$$
\begin{bmatrix}
(C_{11} - \rho v^2) + C_{55}\alpha^2 & C_{16} + C_{45}\alpha^2 & (C_{13} + C_{55})\alpha \\
C_{16} + C_{45}\alpha^2 & (C_{66} - \rho v^2) + C_{44}\alpha^2 & (C_{36} + C_{45})\alpha \\
(C_{13} + C_{55})\alpha & (C_{36} + C_{45})\alpha & (C_{55} - \rho v^2) + C_{33}\alpha^2
\end{bmatrix}
\begin{Bmatrix} \hat{u}_1 \\ \hat{u}_2 \\ \hat{u}_3 \end{Bmatrix}
= \begin{Bmatrix} 0 \\ 0 \\ 0 \end{Bmatrix}
\tag{35}
$$

where $\hat{u}$ is the column vector defined in Eq. (22). Equation (35) is a homogeneous system of linear algebraic equations with $\alpha$ as a parameter. Unlike Eq. (19), it cannot be posed as an algebraic eigenvalue problem and therefore requires a more elaborate solution route. Nontrivial solutions of Eq. (35) are obtained for the values of $\alpha$ which bring to zero the system determinant; these values are the system eigenvalues. For each eigenvalue $\alpha$, there will be a corresponding nontrivial solution $\hat{u}$ of Eq. (35) called the eigenvector. To find the eigenvalues of Eq. (35), one imposes the zero-determinant condition, i.e.,

$$
\begin{vmatrix}
(C_{11} - \rho v^2) + C_{55}\alpha^2 & C_{16} + C_{45}\alpha^2 & (C_{13} + C_{55})\alpha \\
C_{16} + C_{45}\alpha^2 & (C_{66} - \rho v^2) + C_{44}\alpha^2 & (C_{36} + C_{45})\alpha \\
(C_{13} + C_{55})\alpha & (C_{36} + C_{45})\alpha & (C_{55} - \rho v^2) + C_{33}\alpha^2
\end{vmatrix} = 0
\tag{36}
$$

Expansion of the determinant in Eq. (36) yields the bicubic equation

$$
A_6\alpha^6 + A_4\alpha^4 + A_2\alpha^2 + A_0 = 0
\tag{37}
$$

where

$$
A_6 = C_{33}C_{44}C_{55} - C_{33}C_{45}^2
\tag{38}
$$

$$
\begin{aligned}
A_4 &= (C_{55}C_{44} - C_{45}^2)(C_{55} - \rho v^2) + C_{55}C_{33}(C_{66} - \rho v^2) + C_{44}C_{33}(C_{11} - \rho v^2) \\
&\quad - 2C_{16}C_{45}C_{33} + 2(C_{45} + C_{36})(C_{13} + C_{55})C_{45} - (C_{13} + C_{55})^2 C_{44} - (C_{45} + C_{36})^2 C_{55}
\end{aligned}
\tag{39}
$$

$$
\begin{aligned}
A_2 &= C_{33}(C_{11} - \rho v^2)(C_{66} - \rho v^2) + C_{44}(C_{11} - \rho v^2)(C_{55} - \rho v^2) + C_{55}(C_{66} - \rho v^2)(C_{55} - \rho v^2) \\
&\quad - (C_{11} - \rho v^2)(C_{45} + C_{36})^2 - (C_{66} - \rho v^2)(C_{13} + C_{55})^2 - 2(C_{55} - \rho v^2)C_{16}C_{45} \\
&\quad + 2C_{16}(C_{45} + C_{36})(C_{13} + C_{55}) - C_{16}^2 C_{33}
\end{aligned}
\tag{40}
$$

$$
A_0 = \left[(C_{11} - \rho v^2)(C_{66} - \rho v^2) - C_{16}^2\right](C_{55} - \rho v^2)
\tag{41}
$$

Since Eq. (37) is bicubic, it accepts three pairs of $\pm$ solutions, $\pm\alpha_I$, $\pm\alpha_{II}$, $\pm\alpha_{III}$. The corresponding eigenvalues and eigenvectors are

$$
\boldsymbol{\alpha} = \begin{Bmatrix} \alpha^{(1)} \\ \alpha^{(2)} \\ \alpha^{(3)} \\ \alpha^{(4)} \\ \alpha^{(5)} \\ \alpha^{(6)} \end{Bmatrix}
= \begin{Bmatrix} +\alpha_I \\ -\alpha_I \\ +\alpha_{II} \\ -\alpha_{II} \\ +\alpha_{III} \\ -\alpha_{III} \end{Bmatrix}
\quad
\begin{aligned}
\alpha^{(1)} &= +\alpha_I \\
\alpha^{(2)} &= -\alpha_I \\
\alpha^{(3)} &= +\alpha_{II} \\
\alpha^{(4)} &= -\alpha_{II} \\
\alpha^{(5)} &= +\alpha_{III} \\
\alpha^{(6)} &= -\alpha_{III}
\end{aligned}
\quad \text{(eigenvalues)}
\tag{42}
$$

$$
\mathbf{U} = \begin{bmatrix} \mathbf{U}^{(1)} & \mathbf{U}^{(2)} & \mathbf{U}^{(3)} & \mathbf{U}^{(4)} & \mathbf{U}^{(5)} & \mathbf{U}^{(6)} \end{bmatrix} \quad \text{(eigenvectors)}
\tag{43}
$$

Note that $\mathbf{U}$ is a $3 \times 6$ matrix containing 3-long eigenvector columns. Substitution of Eqs. (42), (43) into Eq. (25) and factoring out $e^{i\xi(x_1 - vt)}$ yields, by superposition, the complete wave, i.e.,

$$\mathbf{u} = \hat{\mathbf{u}}(x_3)e^{i\xi(x_1 - vt)} = \left( \sum_{j=1}^{6} \eta_j \mathbf{U}^{(j)} e^{i\xi\alpha^{(j)}x_3} \right) e^{i\xi(x_1 - vt)} \tag{44}$$

where $\eta_j$, $j = 1, ..., 6$, are the partial-wave participation factors. The $x_3$-dependent part of Eq. (44) can be rearranged as

$$\hat{\mathbf{u}}(x_3) = \mathbf{B}^u(x_3)\, \boldsymbol{\eta} \tag{45}$$

where $\mathbf{B}^u(x_3)$ is the $x_3$-dependent field matrix for the displacements and $\boldsymbol{\eta}$ is a 6-long column containing the partial-wave wavenumber ratios, i.e.,

$$\boldsymbol{\eta} = \left\{ \eta_1 \quad \eta_2 \quad \eta_3 \quad \eta_4 \quad \eta_5 \quad \eta_6 \right\}^T \tag{46}$$

The $x_3$-dependent field matrix for the displacements, $\mathbf{B}^u(x_3)$, is given by

$$\mathbf{B}^u(x_3) = \left[ \mathbf{b}_u^{(1)}(x_3) \quad \mathbf{b}_u^{(2)}(x_3) \quad \mathbf{b}_u^{(3)}(x_3) \quad \mathbf{b}_u^{(4)}(x_3) \quad \mathbf{b}_u^{(5)}(x_3) \quad \mathbf{b}_u^{(6)}(x_3) \right] \tag{47}$$

where

$$\mathbf{b}_u^{(j)}(x_3) = \mathbf{U}^{(j)} e^{+i\xi\alpha^{(j)}x_3} \tag{48}$$

### 4.3.4 Stresses

#### 4.3.4.1 Stress–Displacement Relation

Recall the stress–displacement relations, i.e.,

$$\begin{Bmatrix} \sigma_{11} \\ \sigma_{22} \\ \sigma_{33} \\ \sigma_{23} \\ \sigma_{31} \\ \sigma_{12} \end{Bmatrix} = \begin{bmatrix} C_{11} & C_{12} & C_{13} & C_{14} & C_{15} & C_{16} \\ C_{12} & C_{22} & C_{23} & C_{24} & C_{25} & C_{26} \\ C_{13} & C_{23} & C_{33} & C_{34} & C_{35} & C_{36} \\ C_{14} & C_{24} & C_{34} & C_{44} & C_{45} & C_{46} \\ C_{15} & C_{25} & C_{35} & C_{45} & C_{55} & C_{56} \\ C_{16} & C_{26} & C_{36} & C_{46} & C_{56} & C_{66} \end{bmatrix} \begin{Bmatrix} u_{1,1} \\ u_{2,2} \\ u_{3,3} \\ u_{2,3} + u_{3,2} \\ u_{3,1} + u_{1,3} \\ u_{1,2} + u_{2,1} \end{Bmatrix} \tag{49}$$

where the differentiation shorthand $u_{2,3} = \partial u_2 / \partial x_3$, etc. was applied.

#### 4.3.4.2 Stress–Displacement Relations under $x_2$-Invariant Conditions

Under $x_2$-invariant conditions, we have

$$u_{i,2} = 0 \quad i = 1, 2, 3 \quad (x_2\text{-invariant conditions}) \tag{50}$$

Substitution of Eq. (50) into Eq. (49) yields the stress–displacement relations under $x_2$-invariant conditions, i.e.,

$$
\begin{Bmatrix} \sigma_{11} \\ \sigma_{22} \\ \sigma_{33} \\ \sigma_{23} \\ \sigma_{31} \\ \sigma_{12} \end{Bmatrix} = \begin{bmatrix} C_{11} & C_{12} & C_{13} & C_{14} & C_{15} & C_{16} \\ C_{12} & C_{22} & C_{23} & C_{24} & C_{25} & C_{26} \\ C_{13} & C_{23} & C_{33} & C_{34} & C_{35} & C_{36} \\ C_{14} & C_{24} & C_{34} & C_{44} & C_{45} & C_{46} \\ C_{15} & C_{25} & C_{35} & C_{45} & C_{55} & C_{56} \\ C_{16} & C_{26} & C_{36} & C_{46} & C_{56} & C_{66} \end{bmatrix} \begin{Bmatrix} u_{1,1} \\ 0 \\ u_{3,3} \\ u_{2,3} \\ u_{3,1} + u_{1,3} \\ u_{2,1} \end{Bmatrix} \quad (x_2\text{-invariant conditions}) \quad (51)
$$

### 4.3.4.3 Stresses in a Monoclinic Lamina under $x_2$-Invariant Conditions

Recall the stiffness matrix for a monoclinic lamina, i.e.,

$$
\mathbf{C} = \begin{bmatrix} C_{11} & C_{12} & C_{13} & 0 & 0 & C_{16} \\ C_{12} & C_{22} & C_{23} & 0 & 0 & C_{26} \\ C_{13} & C_{23} & C_{33} & 0 & 0 & C_{36} \\ 0 & 0 & 0 & C_{44} & C_{45} & 0 \\ 0 & 0 & 0 & C_{45} & C_{55} & 0 \\ C_{16} & C_{26} & C_{36} & 0 & 0 & C_{66} \end{bmatrix} \quad (52)
$$

Substitution of Eq. (52) into Eq. (51) yields

$$
\begin{Bmatrix} \sigma_{11} \\ \sigma_{22} \\ \sigma_{33} \\ \sigma_{23} \\ \sigma_{31} \\ \sigma_{12} \end{Bmatrix} = \begin{bmatrix} C_{11} & C_{12} & C_{13} & 0 & 0 & C_{16} \\ C_{12} & C_{22} & C_{23} & 0 & 0 & C_{26} \\ C_{13} & C_{23} & C_{33} & 0 & 0 & C_{36} \\ 0 & 0 & 0 & C_{44} & C_{45} & 0 \\ 0 & 0 & 0 & C_{45} & C_{55} & 0 \\ C_{16} & C_{26} & C_{36} & 0 & 0 & C_{66} \end{bmatrix} \begin{Bmatrix} u_{1,1} \\ 0 \\ u_{3,3} \\ u_{2,3} \\ u_{3,1} + u_{1,3} \\ u_{2,1} \end{Bmatrix} \quad (x_2\text{-invariant conditions}) \quad (53)
$$

### 4.3.4.4 Boundary Tractions for a Monoclinic Lamina

In the study of guided-wave propagation in a lamina, we need to calculate the tractions on the upper and lower faces of lamina. Since the lamina faces have the normal vector $\vec{e}_3$, the tractions are actually the normal stress $\sigma_{33}$ and the two shear stresses $\sigma_{23}$, $\sigma_{31}$. Hence, to calculate the tractions, we are going to find the expressions of the stresses $\sigma_{33}$, $\sigma_{23}$, $\sigma_{31}$ at a generic location inside the lamina and then evaluate them at the upper and lower faces of the lamina. Equation (53) gives

$$
\begin{aligned}
\sigma_{33} &= C_{13}u_{1,1} + C_{33}u_{3,3} + C_{36}u_{2,1} \\
\sigma_{23} &= C_{44}u_{2,3} + C_{45}(u_{3,1} + u_{1,3}) \quad (x_2\text{-invariant conditions}) \\
\sigma_{31} &= C_{45}u_{2,3} + C_{55}(u_{3,1} + u_{1,3})
\end{aligned} \quad (54)
$$

Now, we have to express Eq. (54) in terms of the wave propagation expression Eq. (44).

### 4.3.4.5 Boundary Tractions in Terms of Wave Propagation

Substitution of the wave solution Eq. (44) into the boundary tractions Eq. (54). First, calculate the required derivatives, i.e.,

$$u_{i,1} = i\xi \hat{u}_i(x_3)\, e^{i\xi(x_1-vt)} = i\xi u_{i,1} \quad i = 1,2,3 \tag{55}$$

$$u_{i,3} = i\xi \hat{u}_{i,3}(x_3)\, e^{i\xi(x_1-vt)} = \left( \sum_{j=1}^{6} i\xi \alpha^{(j)} \eta_j U_i^{(j)} e^{i\xi\alpha^{(j)}x_3} \right) e^{i\xi(x_1-vt)} \quad i = 1,2,3 \tag{56}$$

Elimination of the common factor $e^{i\xi(x_1-vt)}$ and rearrangement allows us to write Eqs. (55), (56) as

$$\hat{u}_{i,1} = i\xi \sum_{j=1}^{6} \eta_j U_i^{(j)} e^{i\xi\alpha^{(j)}x_3} \quad i = 1,2,3 \tag{57}$$

$$\hat{u}_{i,3} = i\xi \sum_{j=1}^{6} \alpha^{(j)} \eta_j U_i^{(j)} e^{i\xi\alpha^{(j)}x_3} \quad i = 1,2,3 \tag{58}$$

Substitution of Eqs. (57), (58) into Eq. (54) yields

$$\hat{\sigma}_{33} = C_{13}\hat{u}_{1,1} + C_{33}\hat{u}_{3,3} + C_{36}\hat{u}_{2,1}$$

$$= C_{13}i\xi \sum_{j=1}^{6} \eta_j U_1^{(j)} e^{i\xi\alpha^{(j)}x_3} + C_{33}i\xi \sum_{j=1}^{6} \alpha^{(j)} \eta_j U_3^{(j)} e^{i\xi\alpha^{(j)}x_3} + C_{36}i\xi \sum_{j=1}^{6} \eta_j U_2^{(j)} e^{i\xi\alpha^{(j)}x_3}$$

$$= i\xi \sum_{j=1}^{6} (C_{13}U_1^{(j)} + C_{33}\alpha^{(j)}U_3^{(j)} + C_{36}U_2^{(j)})\eta_j e^{i\xi\alpha^{(j)}x_3} \tag{59}$$

$$\hat{\sigma}_{23} = C_{44}\hat{u}_{2,3} + C_{45}(\hat{u}_{3,1} + \hat{u}_{1,3})$$

$$= C_{44}i\xi \sum_{j=1}^{6} \alpha^{(j)} \eta_j U_2^{(j)} e^{i\xi\alpha^{(j)}x_3} + C_{45}i\xi \sum_{j=1}^{6} \eta_j U_3^{(j)} e^{i\xi\alpha^{(j)}x_3} + C_{45}i\xi \sum_{j=1}^{6} \alpha^{(j)} \eta_j U_1^{(j)} e^{i\xi\alpha^{(j)}x_3}$$

$$= i\xi \sum_{j=1}^{6} \left( C_{44}\alpha^{(j)}U_2^{(j)} + C_{45}U_3^{(j)} + C_{45}\alpha^{(j)}U_1^{(j)} \right)\eta_j e^{i\xi\alpha^{(j)}x_3} \tag{60}$$

$$\hat{\sigma}_{31} = C_{45}\hat{u}_{2,3} + C_{55}(\hat{u}_{3,1} + \hat{u}_{1,3})$$

$$= C_{45}i\xi \sum_{j=1}^{6} \alpha^{(j)} \eta_j U_2^{(j)} e^{i\xi\alpha^{(j)}x_3} + C_{55}i\xi \sum_{j=1}^{6} \eta_j U_3^{(j)} e^{i\xi\alpha^{(j)}x_3} + C_{55}i\xi \sum_{j=1}^{6} \alpha^{(j)} \eta_j U_1^{(j)} e^{i\xi\alpha^{(j)}x_3}$$

$$= i\xi \sum_{j=1}^{6} \left( C_{45}\alpha^{(j)}U_2^{(j)} + C_{55}U_3^{(j)} + C_{55}\alpha^{(j)}U_1^{(j)} \right)\eta_j e^{i\xi\alpha^{(j)}x_3} \tag{61}$$

Using Eqs. (59)–(61) we write the stress column vector $\hat{\sigma}(x_3)$ as

$$\hat{\sigma}(x_3) = \begin{Bmatrix} \hat{\sigma}_{33} \\ \hat{\sigma}_{23} \\ \hat{\sigma}_{31} \end{Bmatrix} = \mathbf{B}^{\sigma}(x_3)\,\eta \tag{62}$$

where $\eta$ is given by Eq. (46) and

$$\mathbf{B}^{\sigma}(x_3) = \begin{bmatrix} \mathbf{b}_{\sigma}^{(1)}(x_3) & \mathbf{b}_{\sigma}^{(2)}(x_3) & \mathbf{b}_{\sigma}^{(3)}(x_3) & \mathbf{b}_{\sigma}^{(4)}(x_3) & \mathbf{b}_{\sigma}^{(5)}(x_3) & \mathbf{b}_{\sigma}^{(6)}(x_3) \end{bmatrix} \tag{63}$$

and

$$\mathbf{b}_{\sigma}^{(j)}(x_3) = i\xi \begin{Bmatrix} (C_{13}U_1^{(j)} + C_{33}\alpha^{(j)}U_3^{(j)} + C_{36}U_2^{(j)}) \\ \left(C_{44}\alpha^{(j)}U_2^{(j)} + C_{45}U_3^{(j)} + C_{45}\alpha^{(j)}U_1^{(j)}\right) \\ \left(C_{45}\alpha^{(j)}U_2^{(j)} + C_{55}U_3^{(j)} + C_{55}\alpha^{(j)}U_1^{(j)}\right) \end{Bmatrix} e^{i\xi\alpha^{(j)}x_3} \tag{64}$$

### 4.3.5  State Vector and Field Matrix

Define the state vector $\mathbf{z}$ as the vertical concatenation of displacement and stress vectors.

$$\mathbf{z}(x_3) = \begin{Bmatrix} \hat{\mathbf{u}}(x_3) \\ \hat{\sigma}(x_3) \end{Bmatrix} \tag{65}$$

Combining Eqs. (45), (62), we write the state vector $\mathbf{z}$ as

$$\mathbf{z}(x_3) = \begin{Bmatrix} \hat{\mathbf{u}}(x_3) \\ \hat{\sigma}(x_3) \end{Bmatrix} = \begin{Bmatrix} \mathbf{B}^{u}(x_3)\,\eta \\ \mathbf{B}^{\sigma}(x_3)\eta \end{Bmatrix} = \begin{bmatrix} \mathbf{B}^{u}(x_3) \\ \mathbf{B}^{\sigma}(x_3) \end{bmatrix} \eta = \mathbf{B}(x_3)\eta \tag{66}$$

where the field matrix $\mathbf{B}(x_3)$ is given by

$$\mathbf{B}(x_3) = \begin{bmatrix} \mathbf{B}^{u}(x_3) \\ \mathbf{B}^{\sigma}(x_3) \end{bmatrix} \tag{67}$$

It is apparent that the only unknowns in Eq. (66) are the partial-wave participation factors contained in $\eta$. These six unknowns are going to be obtained through the imposition of the boundary conditions at the upper and lower faces of the lamina.

### 4.3.6  Dispersion Curves

#### 4.3.6.1  Boundary Conditions at Upper and Lower Faces of the Lamina

Assume tractions-free boundary conditions at the upper and lower faces of the laminate, i.e.,

$$\hat{\sigma}(0) = \mathbf{0} \quad \text{(upper free surface of the composite laminate)} \tag{68}$$

$$\hat{\sigma}(h) = \mathbf{0} \quad \text{(lower free surface of the composite laminate)} \tag{69}$$

Substitution of Eq. (62) into Eqs. (68), (69) yields

$$\begin{bmatrix} \mathbf{B}^\sigma(0) \\ \mathbf{B}^\sigma(h) \end{bmatrix} \boldsymbol{\eta} = \mathbf{0} \tag{70}$$

Equation (70) can be written as a system of linear algebraic equation, i.e.,

$$\mathbf{D}\boldsymbol{\eta} = \mathbf{0} \tag{71}$$

where

$$\mathbf{D} = \begin{bmatrix} \mathbf{B}^\sigma(0) \\ \mathbf{B}^\sigma(h) \end{bmatrix} \tag{72}$$

### 4.3.6.2 Search for the Solution

Equation (71) is a homogenous equation that depends on the $x_1$-wave number $\xi$, the $x_1$-wavespeed $v$, the partial-wave eigenvalues $\alpha^{(j)}$, $j = 1, ..., 6$ and eigenvectors (partial-wave polarization vectors) $U^{(j)}$, $j = 1, ..., 6$. In their turn, the partial-wave eigenvalues $\alpha^{(j)}$ and eigenvectors $U^{(j)}$ depend on the $x_1$-wavespeed $v$. Hence, Eq. (71) depends only on the $x_1$-wave number $\xi$, the $x_1$-wavespeed $v$. Since $\xi = \omega/v$, one can say that Eq. (71) depends on the $x_1$-wave number $\xi$ and the excitation frequency $\omega$. In fact any two of the $\xi, \omega, v$ triad can be chosen as independent variables, i.e., either $(\xi, \omega)$, or $(\omega, v)$, or even $(\xi, v)$.

The search for the solution consists in finding the values of the two independent variables, say $(\xi, v)$, which make the equation system Eq. (71) singular. For these values, a solution $\boldsymbol{\eta}$ exists, admittedly with one degree of indeterminacy. The condition for the equation system Eq. (71) to be singular is that its determinant be zero, i.e.,

$$f(\xi, \omega) = \det(\mathbf{D}) \tag{73}$$

Once a solution $(\xi, v)$ is found, the corresponding frequency $\omega$ can be found from $\omega = \xi v$. By assembling all the possible solutions, one may plot the wavespeed dispersion curves $(\xi, v)$ or $(\omega, v)$ as desired.

### 4.3.6.3 Modeshapes

For every solution of Eq. (73) $(\xi, v)_k$, $k = 1, 2, ...,$ one calculates the modeshape as follows: use Eq. (71) which is singular for the found value of $(\xi, v)_k$ to get the vector $\boldsymbol{\eta}_k$ that satisfies the condition

$$\mathbf{D}(\xi_k, v_k)\boldsymbol{\eta}_k = \mathbf{0} \tag{74}$$

Since $\boldsymbol{\eta}_k$ has one degree of indeterminacy, an appropriate normalization procedure should be applied. The extraction of $\boldsymbol{\eta}_k$ could be achieved by evaluating $\mathbf{D}(\xi_k, v_k)$ and the using an eigenvalue algorithm to find the vector $\boldsymbol{\eta}_k$ corresponding to a zero-valued eigenvalue of $\mathbf{D}(\xi_k, v_k)$. Since $\boldsymbol{\eta}_k$ has one degree of indeterminacy, an appropriate normalization procedure should be applied. The extraction of $\boldsymbol{\eta}_k$ could be achieved by evaluating $\mathbf{D}(\xi_k, v_k)$ and the using an eigenvalue algorithm to find the vector $\boldsymbol{\eta}_k$ corresponding to a zero-valued eigenvalue of $\mathbf{D}(\xi_k, v_k)$. Then, evaluate the field matrix $\mathbf{B}(x_3)$ of Eq. (67) for the particular pair $(\xi, v)_k$ to get

$$\mathbf{B}_k(x_3) = \mathbf{B}(x_3;\ \xi_k, v_k) \tag{75}$$

Next, use the vector $\boldsymbol{\eta}_k$ into Eq. (66) to calculate the state vector $\mathbf{z}_k$ at various desired locations $x_3$ across the thickness in order to calculate and plot the displacement and stress modeshape, i.e.,

$$\mathbf{z}_k(x_3) = \mathbf{B}_k(x_3)\boldsymbol{\eta}_k \tag{76}$$

or

$$\begin{Bmatrix} \hat{\mathbf{u}}_k(x_3) \\ \hat{\boldsymbol{\sigma}}_k(x_3) \end{Bmatrix} = \begin{Bmatrix} \mathbf{B}_k^u(x_3)\,\boldsymbol{\eta}_k \\ \mathbf{B}_k^\sigma(x_3)\boldsymbol{\eta}_k \end{Bmatrix} \quad \rightarrow \quad \begin{aligned} \hat{\mathbf{u}}_k(x_3) &= \mathbf{B}_k^u(x_3)\,\boldsymbol{\eta}_k \\ \hat{\boldsymbol{\sigma}}_k(x_3) &= \mathbf{B}_k^\sigma(x_3)\boldsymbol{\eta}_k \end{aligned} \tag{77}$$

## 4.4 GUIDED-WAVE PROPAGATION IN A LAMINATED COMPOSITE

Assume a guided wave of frequency $\omega$, wavespeed $v$, wavenumber $\xi = \omega/v$ propagating along the $x_1$ direction in an $N$-layer laminated composite. Each layer of the laminate has an arbitrary orientation, hence its own stiffness matrix $\mathbf{C}_n$; the layer thickness is $h_n$ where $n = 1, ..., N$.

Assume tractions-free boundary conditions at the upper and lower faces of the laminate, i.e.,

$$\hat{\boldsymbol{\sigma}}_1(0) = \mathbf{0} \quad \text{(upper free surface of the composite laminate)} \tag{78}$$

$$\hat{\boldsymbol{\sigma}}_N(h_N) = \mathbf{0} \quad \text{(lower free surface of the composite laminate)} \tag{79}$$

Strong connection is assumed between the layers of laminate such that displacement continuity and stress balance conditions apply, i.e.,

$$\begin{aligned} \hat{\mathbf{u}}_n(0) &= \hat{\mathbf{u}}_{n-1}(h_{n-1}) \\ \hat{\boldsymbol{\sigma}}_n(0) &= \hat{\boldsymbol{\sigma}}_{n-1}(h_{n-1}) \end{aligned} \quad n = 2, ..., N \tag{80}$$

We note that imposition of Eqs. (78)–(80) generates a system of $6N$ linear algebraic equations that need to be solved for the $6N$ unknown partial-wave participation factors contained in the column vectors $\boldsymbol{\eta}_1, ..., \boldsymbol{\eta}_N$ corresponding to the $1, ..., N$ layers. Once the partial-wave participation factors are determined for each layer, the wave displacements and stresses can be determined at any location in the layer using Eq. (66).

It is apparent that the resulting system of $6N$ linear algebraic equations in the unknowns $\boldsymbol{\eta}_1, ..., \boldsymbol{\eta}_N$ is a homogeneous system which admits solutions only for specific values of the parameters $\omega$ and $v$, which are common for all the layers.

To solve for the unknown partial-wave participation factors $\boldsymbol{\eta}_1, ..., \boldsymbol{\eta}_N$, several methods have been proposed:

- Global matrix method (GMM)
- Transfer matrix method (TMM)
- Stiffness matrix method (SMM)

These methods are discussed next.

## 4.4.1 Global Matrix Method (GMM)

Use Eqs. (66), (67) to write Eqs. (78), (79), (80) as

$$\mathbf{B}_1^\sigma(0)\ \boldsymbol{\eta}_1 = \mathbf{0} \quad \text{(upper free surface)} \tag{81}$$

$$\mathbf{B}_n(h_n)\ \boldsymbol{\eta}_n - \mathbf{B}_{n+1}(0)\ \boldsymbol{\eta}_{n+1} = \mathbf{0} \quad n = 1, ..., N-1 \tag{82}$$

$$\mathbf{B}_N^\sigma(h_N)\ \boldsymbol{\eta}_N = \mathbf{0} \quad \text{(lower free surface)} \tag{83}$$

Equations (81), (82), (83) can be combined and written in matrix form as

$$
\begin{bmatrix}
\mathbf{B}_1^\sigma(0) & & & & & & \\
& \ddots & \ddots & & & & \\
& & \mathbf{B}_{n-1}(h_{n-1}) & -\mathbf{B}_n(0) & & & \\
& & & \mathbf{B}_n(h_n) & -\mathbf{B}_{n+1}(0) & & \\
& & & & \mathbf{B}_{n+1}(h_{n+1}) & -\mathbf{B}_{n+2}(0) & \\
& & & & & \ddots & \ddots \\
& & & & & & \mathbf{B}_N^\sigma(h_N)
\end{bmatrix}
\begin{Bmatrix}
\boldsymbol{\eta}_1 \\
\cdots \\
\boldsymbol{\eta}_{n-1} \\
\boldsymbol{\eta}_n \\
\boldsymbol{\eta}_{n+1} \\
\boldsymbol{\eta}_{n+2} \\
\cdots \\
\boldsymbol{\eta}_N
\end{Bmatrix}
=
\begin{Bmatrix}
0 \\
\cdots \\
0 \\
0 \\
0 \\
0 \\
\cdots \\
0
\end{Bmatrix}
\tag{84}
$$

Note that the top and bottom matrix elements in Eq. (84), $\mathbf{B}_1^\sigma(0)$ and $\mathbf{B}_N^\sigma(h_N)$, respectively, are $3 \times 6$ matrices, whereas the rest of the matrices are $6 \times 6$ matrices. Hence, the complete matrix is of size $6N \times 6N$.

Equation (84) is a $6N \times 6N$ homogenous algebraic system with $\xi$ and $v$ (or $\omega$, since $\omega = v\xi$) as parameters. Nontrivial solutions of a homogeneous algebraic system are possible if the system determinant is zero. Hence, one has to search for particular combinations of $\xi$ and $v$ (or $v$ and $\omega$; or $\xi$ and $\omega$) which annuls the system determinant, i.e., solve the equation

$$D(\xi, v) = 0 \tag{85}$$

where

$$
D(\xi, v) =
\begin{vmatrix}
\mathbf{B}_1^\sigma(0) & & & & & \\
& \ddots & \ddots & & & \\
& & \mathbf{B}_{n-1}(h_{n-1}) & -\mathbf{B}_n(0) & & \\
& & & \mathbf{B}_n(h_n) & -\mathbf{B}_{n+1}(0) & \\
& & & & \mathbf{B}_{n+1}(h_{n+1}) & -\mathbf{B}_{n+2}(0) \\
& & & & \ddots & \ddots \\
& & & & & \mathbf{B}_N^\sigma(h_N)
\end{vmatrix}
\tag{86}
$$

Once an eigenvalue pair $(\xi, v)$ has been found, the unknown partial-wave participation factors $\boldsymbol{\eta}_1, ..., \boldsymbol{\eta}_N$ can be obtained by solution of Eq. (84) which would now have a rank deficient matrix and hence can be solved with one degree on indeterminacy in the solution. This means that the solution will contain an arbitrary scale factor that should be determined through an appropriate normalization procedure. The partial-wave participation factors in each layer are then used in Eq. (45) to calculate the mode shapes as described in Section 3.6.3.

## 4.4.2 Transfer Matrix Method (TMM)

The transfer matrix method (TMM) makes use of the fact that the global matrix in Eq. (84) is banded, i.e., one layer only depends on its immediately adjacent neighbors, the layer above and the layer below it. This property can be used to reduce the size of the problem from $6N \times 6N$ to only $6 \times 6$. The TMM approach proceeds as follows.

Write the state vectors at the beginning and the end of a layer by recalling Eq. (66) and evaluating it at $x_3 = 0$ and at $x_3 = h$, i.e.,

$$\mathbf{z}(0) = \mathbf{B}(0)\,\boldsymbol{\eta} \tag{87}$$

$$\mathbf{z}(h) = \mathbf{B}(h)\,\boldsymbol{\eta} \tag{88}$$

Eliminate $\boldsymbol{\eta}$ between Eqs. (87), (88) by writing

$$\boldsymbol{\eta} = \mathbf{B}^{-1}(0)\,\mathbf{z}(0) \tag{89}$$

$$\mathbf{z}(h) = \mathbf{B}(h)\,\mathbf{B}^{-1}(0)\,\mathbf{z}(0) \tag{90}$$

Define the following notations with $n = 1, ..., N$ being the layer number:

$$\mathbf{z}_n^{\mathrm{T}} = \mathbf{z}_n(0) \quad \text{(state vector at the top of the layer } n\text{)} \tag{91}$$

$$\mathbf{z}_n^{\mathrm{B}} = \mathbf{z}_n(h_n) \quad \text{(state vector at the bottom of the layer } n\text{)} \tag{92}$$

$$\mathbf{A}_n = \mathbf{B}_n(h_n)\,\mathbf{B}_n^{-1}(0) \quad \text{(layer transfer matrix)} \tag{93}$$

Substitute Eqs. (91)–(93) into Eq. (90) to write the layer transfer matrix relation for the layer number $n-1$ as

$$\mathbf{z}_{n-1}^{\mathrm{B}} = \mathbf{A}_{n-1}\,\mathbf{z}_{n-1}^{\mathrm{T}} \quad n = 2, ..., N \tag{94}$$

Apply displacement continuity and stress balance conditions Eq. (80) between layers $n-1$ and $n$, i.e.,

$$\begin{aligned}\hat{\mathbf{u}}_n(0) &= \hat{\mathbf{u}}_{n-1}(h_{n-1}) \\ \hat{\boldsymbol{\sigma}}_n(0) &= \hat{\boldsymbol{\sigma}}_{n-1}(h_{n-1})\end{aligned} \quad n = 2, ..., N \tag{95}$$

or, in state vector form,

$$\mathbf{z}_n^{\mathrm{T}} = \mathbf{z}_{n-1}^{\mathrm{B}} \quad n = 2, ..., N \tag{96}$$

Substitute Eq. (94) into Eq. (96) to write the transfer from the top of layer $n-1$ to the top of layer, $n$, i.e.,

$$\mathbf{z}_n^{\mathrm{T}} = \mathbf{A}_{n-1}\,\mathbf{z}_{n-1}^{\mathrm{T}} \quad n = 2, ..., N \tag{97}$$

Repeated application of Eq. (97) for $n = 2, ..., N$ and use of Eq. (94) with $n = N$ yield a relation between the state vector $\mathbf{z}_1^T$ at top of first layer and the state vector $\mathbf{z}_n^B$ at the bottom of the last layer, i.e.,

$$\mathbf{z}_N^B = \mathbf{A}_N...\mathbf{A}_{n+1}\mathbf{A}_n...\mathbf{A}_1 \, \mathbf{z}_1^T \tag{98}$$

Define the overall transfer matrix $\mathbf{A}$ as

$$\mathbf{A} = \mathbf{A}_N...\mathbf{A}_{n+1}\mathbf{A}_n...\mathbf{A}_1 = \prod_{n=1}^{N} \mathbf{A}_n \tag{99}$$

Use Eq. (99) to write Eq. (98) as

$$\mathbf{z}_N^B = \mathbf{A} \, \mathbf{z}_1^T \tag{100}$$

Equation (100) expresses the state vector $\mathbf{z}_N^B$ at the bottom face of the composite laminate in terms of the state vector $\mathbf{z}_1^T$ at the top face of the laminate. All that remains to be done is to impose the boundary conditions Eqs. (78), (79) at the top and bottom faces of the laminate. In order to do so, rewrite Eq. (100) such as to show explicitly the displacement and stress parts, i.e.,

$$\begin{Bmatrix} \hat{\mathbf{u}}_N^B \\ \hat{\boldsymbol{\sigma}}_N^B \end{Bmatrix} = \begin{bmatrix} \mathbf{A}_{uu} & \mathbf{A}_{u\sigma} \\ \mathbf{A}_{\sigma u} & \mathbf{A}_{\sigma\sigma} \end{bmatrix} \begin{Bmatrix} \hat{\mathbf{u}}_1^T \\ \hat{\boldsymbol{\sigma}}_1^T \end{Bmatrix} \tag{101}$$

where $\mathbf{A}_{uu}, \mathbf{A}_{u\sigma}, \mathbf{A}_{\sigma u}, \mathbf{A}_{\sigma\sigma}$ are submatrices obtained through the partition of $\mathbf{A}$. The boundary conditions at the top and bottom of the laminate Eqs. (78), (79) imply

$$\hat{\boldsymbol{\sigma}}_1^T = \mathbf{0} \tag{102}$$

$$\hat{\boldsymbol{\sigma}}_N^B = \mathbf{0} \tag{103}$$

Substitution of Eqs. (102), (103) into Eq. (101) yields

$$\begin{Bmatrix} \hat{\mathbf{u}}_N^B \\ \mathbf{0} \end{Bmatrix} = \begin{bmatrix} \mathbf{A}_{uu} & \mathbf{A}_{u\sigma} \\ \mathbf{A}_{\sigma u} & \mathbf{A}_{\sigma\sigma} \end{bmatrix} \begin{Bmatrix} \hat{\mathbf{u}}_1^T \\ \mathbf{0} \end{Bmatrix} \tag{104}$$

Expansion of Eq. (104) gives

$$\hat{\mathbf{u}}_N^B = \mathbf{A}_{uu}\hat{\mathbf{u}}_1^T \tag{105}$$

$$\mathbf{0} = \mathbf{A}_{\sigma u}\hat{\mathbf{u}}_1^T \tag{106}$$

Equation (106) can be solved for $\hat{\mathbf{u}}_1^T$. Note that Eq. (106) is a $3 \times 3$ homogenous algebraic system with $\xi$ and $v$ (or $\omega$, since $\omega = v\xi$) as parameters. Nontrivial solutions of a homogeneous algebraic system are possible if the system determinant is zero. Hence, one has to search for particular combinations of $\xi$ and $v$ (or $v$ and $\omega$; or $\xi$ and $\omega$) which annul the system determinant, i.e., solve the equation

$$D(\xi, v) = 0 \tag{107}$$

where

$$D(\xi, v) = |\mathbf{A}_{\sigma u}| \tag{108}$$

Once an eigenvalue pair $(\xi, v)$ has been found, the unknown $\hat{\mathbf{u}}_1^T$ can be obtained by solution of Eq. (106) which would now have a rank deficient matrix and hence can be solved with one degree on indeterminacy in the solution. This means that the solution will contain an arbitrary scale factor that should be determined through an appropriate normalization procedure.

Knowledge of $\hat{\mathbf{u}}_1^T$ and use of Eq. (102) allows us to write $\mathbf{z}_1(0)$, i.e.,

$$\mathbf{z}_1(0) = \begin{Bmatrix} \hat{\mathbf{u}}_1(0) \\ \hat{\sigma}_1(0) \end{Bmatrix} = \begin{Bmatrix} \hat{\mathbf{u}}_1^T \\ \mathbf{0} \end{Bmatrix} \tag{109}$$

Hence we use Eq. (89) to find the partial-wave participation factors in the first layer, i.e.,

$$\boldsymbol{\eta}_1 = \mathbf{B}_1^{-1}(0)\ \mathbf{z}_1(0) \tag{110}$$

Substitution of $\boldsymbol{\eta}_1$ into Eq. (66) allows us to calculate the state vector $\mathbf{z}_1(x_3)$ at any location within the first layer, i.e.,

$$\mathbf{z}_1(x_3) = \mathbf{B}_1(x_3)\boldsymbol{\eta}_1 \tag{111}$$

where $x_3$ is a local coordinate inside the layer. Furthermore, we can calculate the state vector at the bottom of the first layer, i.e., $\mathbf{z}_1^B = \mathbf{z}_1(h_1)$. Now, we recall Eq. (96) and transfer to the second layer, i.e.,

$$\mathbf{z}_2^T = \mathbf{z}_1^B \quad n = 2, ..., N \tag{112}$$

Since $\mathbf{z}_2(0) = \mathbf{z}_2^T$, we can now calculate the partial-wave participation factors $\boldsymbol{\eta}_2$ in the second layer, i.e.,

$$\boldsymbol{\eta}_2 = \mathbf{B}_2^{-1}(0)\ \mathbf{z}_2(0) \tag{113}$$

The process described by Eqs. (111)–(113) is repeated until we get to the last layer at the bottom of the laminate and calculate the state vector $\mathbf{z}_N^B$ at the bottom face of the laminate as

$$\mathbf{z}_N^B = \mathbf{z}_N(h_N) = \mathbf{B}_N(h_3)\boldsymbol{\eta}_N \tag{114}$$

As a verification of calculation accuracy, the state vector $\mathbf{z}_N^B$ and the state vector $\mathbf{z}_1^T$ at the top face of the laminate should satisfy Eq. (100).

Through the process described above by Eqs. (109)–(114), we have been able to construct the complete modeshape inside the laminate.

### 4.4.3 Stiffness Matrix Method (SMM)

The TMM approach described in Section 4.2 is fast and efficient because it deals with very small matrices. However, this process becomes numerically unstable at high frequencies (large $\xi h$ values). This numerical instability is due to the numerical phenomenon known as "small differences of large numbers" which poses a problem when performing finite-mantissa computations. One way to prevent such high-frequency numerical instability is to use the SMM approach, as described next.

Recall Eq. (97) and evaluate it at top and bottom of the lamina, i.e., at $x_3 = 0$ and $x_3 = h$, respectively, such that it explicitly shows the displacement and stress contributions, i.e.,

$$\begin{Bmatrix} \hat{\mathbf{u}}^{\mathrm{T}} \\ \hat{\boldsymbol{\sigma}}^{\mathrm{T}} \end{Bmatrix} = \begin{bmatrix} \mathbf{B}^u(0) \\ \mathbf{B}^\sigma(0) \end{bmatrix} \boldsymbol{\eta} \tag{115}$$

$$\begin{Bmatrix} \hat{\mathbf{u}}^{\mathrm{B}} \\ \hat{\boldsymbol{\sigma}}^{\mathrm{B}} \end{Bmatrix} = \begin{bmatrix} \mathbf{B}^u(h) \\ \mathbf{B}^\sigma(h) \end{bmatrix} \boldsymbol{\eta} \tag{116}$$

Expand Eqs. (115), (116) to get

$$\hat{\mathbf{u}}^{\mathrm{T}} = \mathbf{B}^u(0)\boldsymbol{\eta} \tag{117}$$

$$\hat{\boldsymbol{\sigma}}^{\mathrm{T}} = \mathbf{B}^\sigma(0)\boldsymbol{\eta} \tag{118}$$

$$\hat{\mathbf{u}}^{\mathrm{B}} = \mathbf{B}^u(h)\boldsymbol{\eta} \tag{119}$$

$$\hat{\boldsymbol{\sigma}}^{\mathrm{B}} = \mathbf{B}^\sigma(h)\boldsymbol{\eta} \tag{120}$$

Rearrange Eqs. (117) through (120) to express stresses in terms of displacements, i.e.,

$$\begin{Bmatrix} \hat{\boldsymbol{\sigma}}^{\mathrm{T}} \\ \hat{\boldsymbol{\sigma}}^{\mathrm{B}} \end{Bmatrix} = \boldsymbol{\kappa} \begin{Bmatrix} \hat{\mathbf{u}}^{\mathrm{T}} \\ \hat{\mathbf{u}}^{\mathrm{B}} \end{Bmatrix} \tag{121}$$

where

$$\boldsymbol{\kappa} = \begin{bmatrix} \mathbf{B}^\sigma(0) \\ \mathbf{B}^\sigma(h) \end{bmatrix} \begin{bmatrix} (\mathbf{B}^u(0)) \\ (\mathbf{B}^u(h)) \end{bmatrix}^{-1} \quad \text{(lamina stiffness)} \tag{122}$$

*Proof of Eqs. (121), (122):* Eliminate $\boldsymbol{\eta}$ between Eqs. (117), (118) to write

$$\hat{\boldsymbol{\sigma}}^{\mathrm{T}} = \mathbf{B}^\sigma(0)(\mathbf{B}^u(0))^{-1}\hat{\mathbf{u}}^{\mathrm{T}} \tag{123}$$

Similarly, eliminate $\boldsymbol{\eta}$ between Eqs. (119), (120) to write

$$\hat{\boldsymbol{\sigma}}^{\mathrm{B}} = \mathbf{B}^\sigma(h)(\mathbf{B}^u(h))^{-1}\hat{\mathbf{u}}^{\mathrm{B}} \tag{124}$$

Combine Eqs. (123), (124) in matrix form and get

$$\begin{Bmatrix} \hat{\boldsymbol{\sigma}}^{\mathrm{T}} \\ \hat{\boldsymbol{\sigma}}^{\mathrm{B}} \end{Bmatrix} = \begin{bmatrix} \mathbf{B}^\sigma(0) \\ \mathbf{B}^\sigma(h) \end{bmatrix} \begin{bmatrix} (\mathbf{B}^u(0)) \\ (\mathbf{B}^u(h)) \end{bmatrix}^{-1} \begin{Bmatrix} \hat{\mathbf{u}}^{\mathrm{T}} \\ \hat{\mathbf{u}}^{\mathrm{B}} \end{Bmatrix} \quad \text{QED} \tag{125}$$

Next step in the SMM approach is to build up the stiffness matrix of the laminate from the stiffness matrices of the individual laminae through a recursive process, where the stiffness $\mathbf{K}_n$ of the laminate from layer 1 through layer $n$ is obtained by combining the stiffness $\mathbf{K}_{n-1}$ of the laminate from layer 1 through layer $n-1$ with the stiffness $\boldsymbol{\kappa}_n$ of the $n$-th layer, i.e.,

$$\mathbf{K}_n = \begin{bmatrix} \mathbf{K}_n^{\mathrm{TT}} & \mathbf{K}_n^{\mathrm{TB}} \\ \mathbf{K}_n^{\mathrm{BT}} & \mathbf{K}_n^{\mathrm{BB}} \end{bmatrix} = \begin{bmatrix} \mathbf{K}_{n-1}^{\mathrm{TT}} + \mathbf{K}_{n-1}^{\mathrm{TB}}(\boldsymbol{\kappa}_n^{\mathrm{TT}} - \mathbf{K}_{n-1}^{\mathrm{BB}})^{-1}\mathbf{K}_{n-1}^{\mathrm{BT}} & -\mathbf{K}_{n-1}^{\mathrm{TB}}(\boldsymbol{\kappa}_n^{\mathrm{TT}} - \mathbf{K}_{n-1}^{\mathrm{BB}})^{-1}\boldsymbol{\kappa}_n^{\mathrm{TB}} \\ \boldsymbol{\kappa}_n^{\mathrm{BT}}(\boldsymbol{\kappa}_n^{\mathrm{TT}} - \mathbf{K}_{n-1}^{\mathrm{BB}})^{-1}\mathbf{K}_{n-1}^{\mathrm{BT}} & \boldsymbol{\kappa}_n^{\mathrm{BB}} - \boldsymbol{\kappa}_n^{\mathrm{BT}}(\boldsymbol{\kappa}_n^{\mathrm{TT}} - \mathbf{K}_{n-1}^{\mathrm{BB}})^{-1}\boldsymbol{\kappa}_n^{\mathrm{TB}} \end{bmatrix} \quad (126)$$

where

$$\begin{Bmatrix} \hat{\boldsymbol{\sigma}}_1^{\mathrm{T}} \\ \hat{\boldsymbol{\sigma}}_{n-1}^{\mathrm{B}} \end{Bmatrix} = \begin{bmatrix} \mathbf{K}_{n-1}^{\mathrm{TT}} & \mathbf{K}_{n-1}^{\mathrm{TB}} \\ \mathbf{K}_{n-1}^{\mathrm{BT}} & \mathbf{K}_{n-1}^{\mathrm{BB}} \end{bmatrix} \begin{Bmatrix} \hat{\mathbf{u}}_1^{\mathrm{T}} \\ \hat{\mathbf{u}}_{n-1}^{\mathrm{B}} \end{Bmatrix} \qquad \mathbf{K}_{n-1} = \begin{bmatrix} \mathbf{K}_{n-1}^{\mathrm{TT}} & \mathbf{K}_{n-1}^{\mathrm{TB}} \\ \mathbf{K}_{n-1}^{\mathrm{BT}} & \mathbf{K}_{n-1}^{\mathrm{BB}} \end{bmatrix} \qquad (127)$$

$$\begin{Bmatrix} \hat{\boldsymbol{\sigma}}_1^{\mathrm{T}} \\ \hat{\boldsymbol{\sigma}}_n^{\mathrm{B}} \end{Bmatrix} = \begin{bmatrix} \mathbf{K}_n^{\mathrm{TT}} & \mathbf{K}_n^{\mathrm{TB}} \\ \mathbf{K}_n^{\mathrm{BT}} & \mathbf{K}_n^{\mathrm{BB}} \end{bmatrix} \begin{Bmatrix} \hat{\mathbf{u}}_1^{\mathrm{T}} \\ \hat{\mathbf{u}}_{n-1}^{\mathrm{B}} \end{Bmatrix} \qquad \mathbf{K}_n = \begin{bmatrix} \mathbf{K}_n^{\mathrm{TT}} & \mathbf{K}_n^{\mathrm{TB}} \\ \mathbf{K}_n^{\mathrm{BT}} & \mathbf{K}_n^{\mathrm{BB}} \end{bmatrix} \qquad (128)$$

$$\begin{Bmatrix} \hat{\boldsymbol{\sigma}}_n^{\mathrm{T}} \\ \hat{\boldsymbol{\sigma}}_n^{\mathrm{B}} \end{Bmatrix} = \begin{bmatrix} \boldsymbol{\kappa}_n^{\mathrm{TT}} & \boldsymbol{\kappa}_n^{\mathrm{TB}} \\ \boldsymbol{\kappa}_n^{\mathrm{BT}} & \boldsymbol{\kappa}_n^{\mathrm{BB}} \end{bmatrix} \begin{Bmatrix} \hat{\mathbf{u}}_n^{\mathrm{T}} \\ \hat{\mathbf{u}}_n^{\mathrm{B}} \end{Bmatrix} \qquad \boldsymbol{\kappa}_n = \begin{bmatrix} \boldsymbol{\kappa}_n^{\mathrm{TT}} & \boldsymbol{\kappa}_n^{\mathrm{TB}} \\ \boldsymbol{\kappa}_n^{\mathrm{BT}} & \boldsymbol{\kappa}_n^{\mathrm{BB}} \end{bmatrix} \qquad (129)$$

*Proof of Eq.* (126): Expand Eqs.(127), (129) to get

$$\hat{\boldsymbol{\sigma}}_1^{\mathrm{T}} = \mathbf{K}_{n-1}^{\mathrm{TT}}\hat{\mathbf{u}}_1^{\mathrm{T}} + \mathbf{K}_{n-1}^{\mathrm{TB}}\hat{\mathbf{u}}_{n-1}^{\mathrm{B}} \qquad (130)$$

$$\hat{\boldsymbol{\sigma}}_{n-1}^{\mathrm{B}} = \mathbf{K}_{n-1}^{\mathrm{BT}}\hat{\mathbf{u}}_1^{\mathrm{T}} + \mathbf{K}_{n-1}^{\mathrm{BB}}\hat{\mathbf{u}}_{n-1}^{\mathrm{B}} \qquad (131)$$

$$\hat{\boldsymbol{\sigma}}_n^{\mathrm{T}} = \boldsymbol{\kappa}_n^{\mathrm{TT}}\hat{\mathbf{u}}_n^{\mathrm{T}} + \boldsymbol{\kappa}_n^{\mathrm{TB}}\hat{\mathbf{u}}_n^{\mathrm{B}} \qquad (132)$$

$$\hat{\boldsymbol{\sigma}}_n^{\mathrm{B}} = \boldsymbol{\kappa}_n^{\mathrm{BT}}\hat{\mathbf{u}}_n^{\mathrm{T}} + \boldsymbol{\kappa}_n^{\mathrm{BB}}\hat{\mathbf{u}}_n^{\mathrm{B}} \qquad (133)$$

Recall the interface conditions between layers $n-1$ and $n$, i.e.,

$$\hat{\mathbf{u}}_n^{\mathrm{T}} = \hat{\mathbf{u}}_{n-1}^{\mathrm{B}} \qquad (134)$$

$$\hat{\boldsymbol{\sigma}}_n^{\mathrm{T}} = \hat{\boldsymbol{\sigma}}_{n-1}^{\mathrm{B}} \qquad (135)$$

Use Eqs. (134), (135) into Eqs. (130), (131) to get

$$\hat{\boldsymbol{\sigma}}_1^{\mathrm{T}} = \mathbf{K}_{n-1}^{\mathrm{TT}}\hat{\mathbf{u}}_1^{\mathrm{T}} + \mathbf{K}_{n-1}^{\mathrm{TB}}\hat{\mathbf{u}}_n^{\mathrm{T}} \qquad (136)$$

$$\hat{\boldsymbol{\sigma}}_n^{\mathrm{T}} = \mathbf{K}_{n-1}^{\mathrm{BT}}\hat{\mathbf{u}}_1^{\mathrm{T}} + \mathbf{K}_{n-1}^{\mathrm{BB}}\hat{\mathbf{u}}_n^{\mathrm{T}} \qquad (137)$$

Eliminate $\hat{\boldsymbol{\sigma}}_n^{\mathrm{T}}$ between Eqs. (132), (137) and write

$$\boldsymbol{\kappa}_n^{\mathrm{TT}}\hat{\mathbf{u}}_n^{\mathrm{T}} + \boldsymbol{\kappa}_n^{\mathrm{TB}}\hat{\mathbf{u}}_n^{\mathrm{B}} = \mathbf{K}_{n-1}^{\mathrm{BT}}\hat{\mathbf{u}}_1^{\mathrm{T}} + \mathbf{K}_{n-1}^{\mathrm{BB}}\hat{\mathbf{u}}_n^{\mathrm{T}} \qquad (138)$$

Solve Eq. (138) for $\hat{\mathbf{u}}_n^{\mathrm{T}}$, i.e.,

$$\hat{\mathbf{u}}_n^{\mathrm{T}} = (\boldsymbol{\kappa}_n^{\mathrm{TT}} - \mathbf{K}_{n-1}^{\mathrm{BB}})^{-1}\mathbf{K}_{n-1}^{\mathrm{BT}}\hat{\mathbf{u}}_1^{\mathrm{T}} - (\boldsymbol{\kappa}_n^{\mathrm{TT}} - \mathbf{K}_{n-1}^{\mathrm{BB}})^{-1}\boldsymbol{\kappa}_n^{\mathrm{TB}}\hat{\mathbf{u}}_n^{\mathrm{B}} \qquad (139)$$

Substitute Eq. (139) into Eqs. (133), (136) and write

$$\hat{\sigma}_1^T = \left[ \mathbf{K}_{n-1}^{TT} + \mathbf{K}_{n-1}^{TB}(\kappa_n^{TT} - \mathbf{K}_{n-1}^{BB})^{-1}\mathbf{K}_{n-1}^{BT} \right]\hat{\mathbf{u}}_1^T - \mathbf{K}_{n-1}^{TB}(\kappa_n^{TT} - \mathbf{K}_{n-1}^{BB})^{-1}\kappa_n^{TB}\hat{\mathbf{u}}_n^B \tag{140}$$

$$\hat{\sigma}_n^B = \kappa_n^{BT}(\kappa_n^{TT} - \mathbf{K}_{n-1}^{BB})^{-1}\mathbf{K}_{n-1}^{BT}\hat{\mathbf{u}}_1^T + \left[ \kappa_n^{BB} - \kappa_n^{BT}(\kappa_n^{TT} - \mathbf{K}_{n-1}^{BB})^{-1}\kappa_n^{TB} \right]\hat{\mathbf{u}}_n^B \tag{141}$$

Rearrange Eqs. (140), (141) in matrix form to get

$$\begin{Bmatrix} \hat{\sigma}_1^T \\ \hat{\sigma}_n^B \end{Bmatrix} = \begin{bmatrix} \mathbf{K}_{n-1}^{TT} + \mathbf{K}_{n-1}^{TB}(\kappa_n^{TT} - \mathbf{K}_{n-1}^{BB})^{-1}\mathbf{K}_{n-1}^{BT} & -\mathbf{K}_{n-1}^{TB}(\kappa_n^{TT} - \mathbf{K}_{n-1}^{BB})^{-1}\kappa_n^{TB} \\ \kappa_n^{BT}(\kappa_n^{TT} - \mathbf{K}_{n-1}^{BB})^{-1}\mathbf{K}_{n-1}^{BT} & \kappa_n^{BB} - \kappa_n^{BT}(\kappa_n^{TT} - \mathbf{K}_{n-1}^{BB})^{-1}\kappa_n^{TB} \end{bmatrix} \begin{Bmatrix} \hat{\mathbf{u}}_1^T \\ \hat{\mathbf{u}}_n^B \end{Bmatrix} \quad \text{QED} \tag{142}$$

The recursive application of Eq. (126) allows as to calculate the overall stiffness matrix $\mathbf{K}_N$ of the composite laminate, i.e.,

$$\mathbf{K}_N = \begin{bmatrix} \mathbf{K}_N^{TT} & \mathbf{K}_N^{TB} \\ \mathbf{K}_N^{BT} & \mathbf{K}_N^{BB} \end{bmatrix} \tag{143}$$

Using Eq. (143), we write

$$\begin{Bmatrix} \hat{\sigma}_1^T \\ \hat{\sigma}_N^B \end{Bmatrix} = \begin{bmatrix} \mathbf{K}_N^{TT} & \mathbf{K}_N^{TB} \\ \mathbf{K}_N^{BT} & \mathbf{K}_N^{BB} \end{bmatrix} \begin{Bmatrix} \hat{\mathbf{u}}_1^T \\ \hat{\mathbf{u}}_N^B \end{Bmatrix} \tag{144}$$

Recall the boundary conditions at the top and bottom of the laminate Eqs. (78), (79), i.e.,

$$\hat{\sigma}_1^T = 0 \tag{145}$$

$$\hat{\sigma}_N^B = 0 \tag{146}$$

Substitution of Eqs. (145), (146) into Eq. (144) yields

$$\begin{bmatrix} \mathbf{K}_N^{TT} & \mathbf{K}_N^{TB} \\ \mathbf{K}_N^{BT} & \mathbf{K}_N^{BB} \end{bmatrix} \begin{Bmatrix} \hat{\mathbf{u}}_1^T \\ \hat{\mathbf{u}}_N^B \end{Bmatrix} = \begin{Bmatrix} 0 \\ 0 \end{Bmatrix} \tag{147}$$

Equation (147) is a $6 \times 6$ homogenous algebraic system with $\xi$ and $v$ (or $\omega$, since $\omega = v\xi$) as parameters. Nontrivial solutions of a homogeneous algebraic system are possible if the system determinant is zero. Hence, one has to search for particular combinations of $\xi$ and $v$ (or $v$ and $\omega$; or $\xi$ and $\omega$) which annul the system determinant, i.e., solve the equation

$$D(\xi, v) = 0 \tag{148}$$

where

$$D(\xi, v) = \begin{vmatrix} \mathbf{K}_N^{TT} & \mathbf{K}_N^{TB} \\ \mathbf{K}_N^{BT} & \mathbf{K}_N^{BB} \end{vmatrix} \tag{149}$$

For each eigenvalue pair $(\xi, v)$, the system of Eq. (147) has a rank deficient matrix and hence can be solved with one degree on indeterminacy in the solution. This means that the solution will contain an arbitrary scale factor that should be determined through an appropriate normalization procedure. This means that we now know (to one degree of indeterminacy) the values

$$\hat{\mathbf{u}}_1^T = \mathbf{0} \tag{150}$$

$$\hat{\mathbf{u}}_N^B = \mathbf{0} \tag{151}$$

Knowledge of $\hat{\mathbf{u}}_1^T$ and use of Eq. (145) allows us to write $\mathbf{z}_1(0)$, i.e.,

$$\mathbf{z}_1(0) = \begin{Bmatrix} \hat{\mathbf{u}}_1(0) \\ \hat{\boldsymbol{\sigma}}_1(0) \end{Bmatrix} = \begin{Bmatrix} \hat{\mathbf{u}}_1^T \\ \mathbf{0} \end{Bmatrix} \tag{152}$$

From here, the calculation of the modeshape proceeds in the same way as for the TMM approach. Using the process described by Eqs. (109)–(114), we can construct the complete modeshape inside the laminate.

## 4.5 NUMERICAL COMPUTATION

The numerical implementation of the computational procedure outlined in this chapter may not be as straightforward as it seems. The numerical process may show numerical instability for large layer thickness and high excitation frequencies due to the intrinsic poor numerical conditioning that may be due to a combination of several factors involving the presence of decaying evanescent waves at the interfaces. An extensive treatment of these issues accompanied by step-by-step examples is given in the Instructor Manual posted on the publisher's website. The reader is encouraged to study these examples and, if questions arise, contact the author directly at victorg@sc.edu.

## 4.6 PROBLEMS AND EXERCISES

Problems and exercises and worked out examples are given in the Instructor Manual posted on the publisher's website.

## References

[1] Lowe, M. J. S. (1995) "Matrix Technique for Modeling Ultrasonic Waves in Multilayered Media", *IEEE*, July, 1995.
[2] Rokhlin, S. I.; Chimenti, D. E.; Nagy, P. B. (2011) *Physical Ultrasonics of Composites*, Oxford University Press, Oxford, UK.
[3] Thomson, W. T. (1950) "Transmission of Elastic Waves through a Stratified Solid Medium", *Journal of Applied Physics*, **Vol. 21**, 89–93.
[4] Haskell, N. A. (1953) "Dispersion of Surface Waves on Multilayer Media", *Bulletin of the Seismological Society of America*, **Vol. 43**, 17–34.
[5] Nayfeh, A. H. (1991) "The general problem of elastic wave propagation in multilayered anisotropic media", *Journal of Acoustical Society of America*, **89**(4), Pt. 1.

[6] Nayfeh, A. H. (1995) Wave propagation in layered anisotropic media with application to composites, Elsevier, Amsterdam, The Netherlands.

[7] Knopoff, L. (1964) "A Matrix Method for Elastic Waves Problems", *Bulletin of the Seismological Society of America*, **Vol. 43**, 431–438.

[8] Pavlakovic, B.; Lowe, M. (2003) *DISPERSE—A System for Generating Dispersion Curves*, version 2.0.16B, Imperial College, London, UK, July 2003.

[9] Wang, L.; Rokhlin, S. I. (2001) "Stable Reformulation of Transfer Matrix Method for Wave Propagation in Layered Anisotropic Media", *Ultrasonics*, **39**(2001), 413–424.

[10] Rokhlin, S. I.; Wang, L. (2002) "Stable Recursive Algorithm for Elastic Wave Propagation in Layered Anisotropic Media: Stiffness Matrix Method", *Journal of Acoustic Society of America*, 2002, **112**(3), 822–834.

# Damage and Failure of Aerospace Composites

# 5.1 INTRODUCTION

The damage and failure of metallic structures is relatively well understood; their in-service damage and failure occurs mostly due to fatigue cracks that propagate under cyclic loading in metallic materials. In contrast, damage in composite materials occurs in many more ways than in metals [1]. Composites fail differently under tension than they fail in compression, and the effect of fastener holes is much more complicated than in metals. In addition, the composites are prone to hidden damage from low-velocity impact (e.g., the drop of a hand tool on a wing, or large hail impact on a radome); such damage can be barely visible and may go undetected, but its effect on the degradation of the composite structure strength can be dramatic. In order to satisfy the aircraft damage tolerance requirements [2,3], one has to demonstrate that a composite aircraft structure possesses adequate residual strength at the end of service life in the presence of an assumed worst-case damage, as for example that caused by a low-velocity impact on a composite structure.

The damage in composite materials, its basic mechanisms, accumulation, and characterization, as well as the concept of damage tolerance have been studied for over three decades [4], but many of the initial questions still stand unanswered, especially in connection with composites fatigue [5]. Tension, compression, and shear are the three fundamental modes in which a composite may fail. As the composite is made up of layers of various orientations, the projection of the global stresses onto the lamina principal directions varies from lamina to lamina. As the load is increased, so do the various stresses in the laminae, and failure values may be attained in a certain lamina in a certain principal direction without the overall laminate to experience actual failure; in other words, the failure of the composite laminate is a progressive phenomenon. This progressive damage evolution is subcritical for a while, but eventually leads to ultimate failure of the composite laminate. The monitoring of subcritical damage occurrence and evolution in a composite material is one of the main objectives of composite structural health monitoring (SHM) endeavor (Table 1).

TABLE 1    Typical strength values for CFRP and GFRP composites [8]

| Property | CFRP (Mpa) | GFRP (Mpa) |
| --- | --- | --- |
| Longitudinal tension strength $X_t$ | 1500 | 1000 |
| Transverse tension strength $Y_t$ | 50 | 30 |
| Longitudinal compression strength $X_c$ | −1250 | −600 |
| Transverse compression strength $Y_c$ | −200 | −120 |
| In-plane shear strength $S_{12}$ | 100 | 70 |

## 5.2 COMPOSITES DAMAGE AND FAILURE MECHANISMS

Aerospace composites are made up of two high-performance basic constituents: fibers and matrix. A variety of high-performance fibers have been developed for aerospace applications; of these, carbon fibers have gained wide acceptance in the manufacturing of primary aerospace structures. A variety of polymeric materials have been considered for matrix usage; of these, the toughened epoxy resins have gained large acceptance in primary aerospace structures. The damage and failure of the laminated composite depends on the damage and failure of its constituent fibers and matrix as well as on the damage and failure of the various interfaces between these constituents and between the layers of the composite laminate [6].

### 5.2.1 Fiber and Matrix Stress–Strain Curves

To understand the damage and failure of a composite material, we will start with a understanding of the damage and failure behavior of its constituents, i.e., (a) the fiber and (b) the matrix. Figure 1a shows stress–strain curves for typical high-performance fibers used in aerospace composites, whereas Figure 1b shows the stress–strain curves for the high-performance polymeric matrix materials [7]. A common fiber–matrix combination that has gained acceptance in aerospace composite manufacturing is that between a high-strength carbon fiber (e.g., IM-7 in Figure 1a) and a toughened epoxy (Figure 1b).

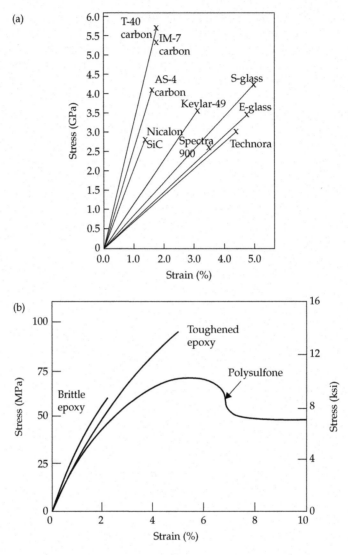

**FIGURE 1**   Stress–strain curves for composite constituents: (a) high-performance fibers; (b) high-performance polymeric matrix materials [7].

It is immediately apparent from Figure 1 that the behavior of the fiber and matrix are substantially different. The high-performance fibers have a linear stress–strain curve with an abrupt brittle ending and a relatively small strain to failure ($\varepsilon_f^{failure} \simeq 1.8\%$ for IM-7). The toughened epoxy matrix has a nonlinear stress–strain curve with progressive diminishing slope and a significantly larger strain to failure ($\varepsilon_m^{failure} \simeq 5\%$). Also apparent from Figure 1 is the extremely different tensile modulus values for the fibers ($E_f \simeq 300\text{GPa}$) and for the matrix ($E_m \simeq 2.5\text{GPa}$); hence, it is apparent that, for a given strain, the fibers carry most of the load.

## 5.2.2 Failure Modes in Unidirectional Fiber-Reinforced Composites

To understand the failure of a composite laminate, we shall first focus our attention on the failure of the unidirectional composite lamina (a.k.a. ply). Six failure modes and associated strengths can be distinguished in a unidirectional composite ply, i.e.,

- longitudinal failure in tension, strength $X_t$
- transverse failure in tension, strength $Y_t$
- longitudinal failure in compression, strength $X_c$
- transverse failure in compression, strength $Y_c$
- failure through in-plane shear, strength $S_{12}$
- failure through transverse shear, strength $S_{23}$.

Recall that the unidirectional composite lamina is transversely isotropic; this implies that the transverse and shear strengths associated with the z-direction are identical with those associated with the y-direction, i.e., $Z_t = Y_t$, $Z_c = Y_c$ and that $S_{13} = S_{12}$.

Figure 2 depicts conceptually the two tension failure modes associated with $X_t$ and $Y_t$. Longitudinal tension failure is controlled by the fibers; hence, the composite ply will fail in longitudinal tension when $X_t$ is reached because the fibers fracture. Transverse tension failure is controlled by the matrix strength and/or the strength of the fiber−matrix interface. (Transverse failure of the fibers is also possible, though it is less common). Failure of the lamina occurs in transverse tension when $Y_t$ is reached.

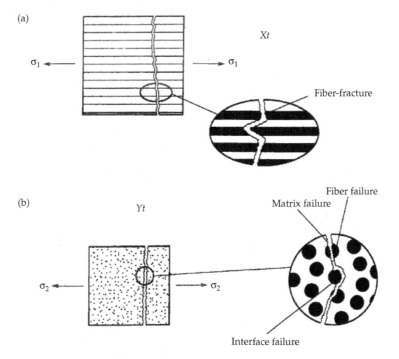

**FIGURE 2**    Tension failure modes in a unidirectional composite ply: (a) longitudinal tension failure (strength $X_t$) through fiber fracture; (b) transverse tension failure (strength $Y_t$) through matrix fracture and/or failure of the fiber−matrix interface (transverse failure of the fibers is also depicted, though this is not as common) [8].

Figure 3 depicts conceptually the two compression failure modes associated with $X_c$ and $Y_c$. Longitudinal compression strength $X_c$ in carbon-fiber-reinforced polymer (CFRP) is controlled by fiber microbuckling (or kinking) as depicted in Figure 3a. Fiber microbuckling failure is especially found when the fibers have a very small diameter and hence a small bending stiffness, such as in the case of carbon fibers. Fiber microbuckling generates kink bands with the width of 10−15 fiber diameters. The fibers in the kink band are fractured at both ends of the kink. Fiber microbuckling is promoted by initial misalignment in the fibers layout. In high-quality composites, such misalignment is kept within 1−4°.

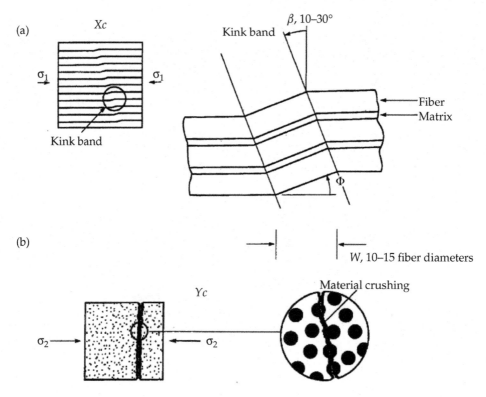

**FIGURE 3**   Compression failure modes in a unidirectional composite ply: (a) longitudinal compression failure (strength $X_c$) through fiber microbuckling leading to kink bands; (b) transverse compression failure (strength $Y_c$) through material crushing [8].

If the fibers are thicker, e.g., boron fibers, then their bending stiffness is greatly increased and fiber microbuckling may not occur. Hence, in boron-fiber-reinforced polymer (BFRP) composites, longitudinal compression failure is dominated by longitudinal crushing of the fibers.

Transverse compression failure (Figure 3b) occurs through material crushing, either matrix crushing or fiber crushing, or both. The transverse compression strength $Y_c$ is usually significantly higher than the transverse tension strength $Y_t$.

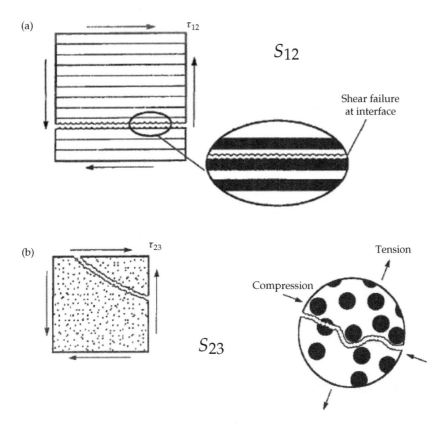

**FIGURE 4**    Shear failure modes in a unidirectional composite ply: (a) in-plane shear failure (strength $S_{12}$) dominated by the shear failure of the fiber–matrix interface; (b) transverse shear failure (strength $S_{23}$) dominated by the shear of the matrix or, sometimes, the fibers [8].

Shear failure modes of the unidirectional composite ply are depicted in Figure 4. In-plane shear failure (strength $S_{12}$) is controlled by the shear strength of the matrix, the longitudinal shear strength of the fiber, and the shear strength of the fiber–matrix interface, which, in many cases, is the weakest link (Figure 4a). The transverse shear failure (strength $S_{23}$) is dominated by the shear of the matrix or, sometimes, the transverse shear of the fibers.

## 5.3 TENSION DAMAGE AND FAILURE OF A UNIDIRECTIONAL COMPOSITE PLY

### 5.3.1 Strain-Controlled Tension Failure due to Fracture of the Fibers

When the fiber and matrix are combined to create a unidirectional composite, the failure process under tension loading will be dominated by the fiber fracture. Figure 5 shows diagrammatically that the composite failure strain is dictated by the fiber failure strain $\varepsilon_f$ which, as already discussed, is much smaller than the matrix failure strain.

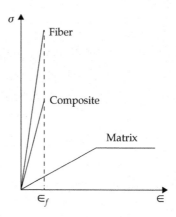

**FIGURE 5**    Failure of a unidirectional composite is controlled by the fiber fracture [7].

### 5.3.2 Statistical Effects on Unidirectional Composite Strength and Failure

The high tensile strength of high-performance fibers used in aerospace composites is generally attributed to their filamentary form in which there are statistically fewer flaws than in the bulk material. However, as in other brittle materials, the measured tensile strength of high-performance fibers shows a significant statistical spread. This means that, although the tensile modulus may be the same, the tensile ultimate strain of the fiber would vary statistically. Such variations in strength occur not just between one fiber and another, but also between different locations on the same fiber indicating that weak spots may exist along a given fiber and that a continuous fiber of considerable length has a greater chance of breaking than a shorter fiber (Figure 6).

Due to this statistical variation in strength from fiber to fiber, it is conceivable that, as a unidirectional composite is loaded into tension, the various fibers within the composite would fail progressively according to their place in the statistical distribution, first the weaker ones, then the stronger ones. As some of the fibers fail, the effective tensile modulus of the composite diminishes and the actual stress—strain curve measured during the static test of a unidirectional composite may show a diminishing slope as the strain increases.

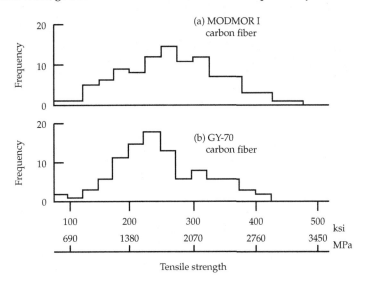

**FIGURE 6**   Example of statistical spread of carbon fiber tensile strength [7].

## 5.3.3 Shear-Lag Load Sharing between Broken Fibers

When a fiber breaks, the load is redistributed to the other fibers. In a bunch of dry fibers, such a load redistribution would be uniform to all the fibers because all the fibers in the bunch are assumed connected to a common loading point. In a composite, however, the redistribution is done through the matrix and it only affects the fibers in the immediate vicinity of the broken fiber (Figure 7). The transfer of load from the broken fiber to the adjacent fibers is by a shear-lag process in the matrix [9] and in the fiber–matrix interface/interphase [10]. As illustrated in Figure 7, this shear-lag process proceeds as follows: the stress in a broken fiber builds back up to the undisturbed level by shear transfer from the surrounding matrix, composite, and interphase region following a shear-lag expression (Ref. [9] cited in Ref. [10]), i.e.,

$$\sigma_f(x) = \frac{\overline{\sigma} E_f}{E_c}\left(1 - e^{-\eta x}\right) \quad \text{(exponential decay)} \tag{1}$$

where $x$ is measured from the fiber tip and

$$\eta = \frac{1}{r_f}\sqrt{\frac{2G_m \sqrt{v_f}}{E_f(1 - v_f)}} \quad \text{(shear-lag parameter)} \tag{2}$$

In Eqs. (1), (2), the subscripts $f, m, c$ refer to fiber, matrix, and composite, respectively, whereas $r_f$ is the radius of the fiber and $\overline{\sigma}$ is the average stress in the composite away from the broken fiber. Equation (1) shows that the stress in the broken fiber experiences an exponential buildup: as $x$ increases away from the tip of the broken fiber, the stress in the

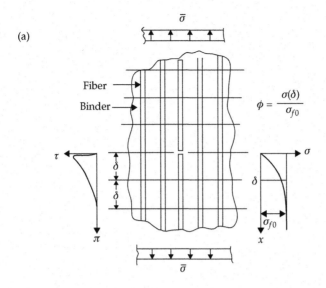

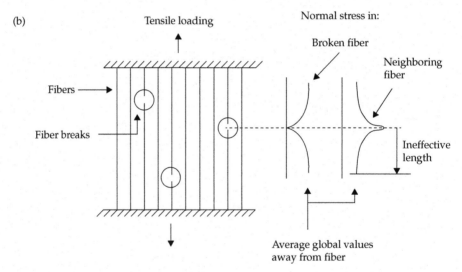

**FIGURE 7** Tensile failure model for fiber-reinforced composite: (a) schematic of the Rosen model [9]; (b) improved representation showing several failed fibers [10].

fiber is restored to the original value $\sigma_{f0} = \lim_{x \to \infty} \sigma_f$, and a distance $\Delta$ can be defined over which this restoration has reached to a value of, say, $\phi = 99\%$, i.e.,

$$\phi = \frac{\sigma_f(\Delta)}{\sigma_{f0}} = 1 - e^{-\eta\Delta} \quad \text{(stress restoration efficiency)} \tag{3}$$

Hence, an "ineffective length" $\delta = 2\Delta$ can be defined as the length of the composite zone around the two tips of a broken fiber in which the stress that used to be carried by the broken fiber had to be redistributed to the adjacent fibers. Outside the ineffective length, the broken fiber continues to perform its duties and carry its fair share of the load. In this way, a unidirectional composite will continue to carry load even after some of its fibers have broken (these early-breaking fibers are those weak spots placed at the lower end of the statistical spread). This process of load sharing will continue until the stress concentration in the adjacent fibers exceeds their load-carrying capacity.

It is apparent that the stress concentration in the adjacent fibers is directly proportional to the value of the ineffective length $\delta$ [10]: if the stress buildup occurs over a short distance, the stress concentration in the neighboring fibers is great and they tend to break quickly, causing very brittle composite behavior. If the buildup occurs over a large distance (e.g., a very compliant matrix or a very low fiber–matrix interface), the strength of each fiber is almost completely lost when a local break occurs and the load sharing process is ineffectual. Hence, an optimum combination of strength values for the fiber, matrix, and interface/interphase properties should exist that provides an intermediate condition to those two extremes that best satisfies the composite tensile strength requirements.

### 5.3.4 Fiber Pullout

Resilient fiber-reinforced polymer composites do not accommodate a clean fracture across their cross-section but rather present multiple failure sites distributed in their mass. These types of fiber–matrix material systems allow for crack deflection at the fiber–matrix interface and present the phenomenon of fiber bridging that prevents cracks from leading to catastrophic failure. When finally failing, these material systems exhibit extensive fiber sliding and pullout taking place at multiple sites. The length of the part of the fiber that is being pulled out and the intrinsic strength of the fiber are controlling the amount of the energy being dissipated during the fiber pullout process. Figure 8 presents the mechanism of fiber pullout. The pullout load, $P_0$, is transferred to the matrix through the shear stresses on the fiber–matrix interface. In a simplified analysis, one may assume that the fiber surface is subjected to constant shear $\tau_0$, which is the shear strength of the fiber–matrix interface.

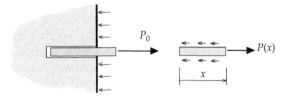

**FIGURE 8** Schematic representation of a single fiber pullout mechanism under constant interfacial shear strength $\tau_0$.

As the broken fiber is being pulled out, its axial force builds up gradually along its length through the accumulation of the shear stress $\tau_0$ applied to its surface. At a distance $x$ along the fiber, the axial force in the fiber has built up to the value $P(x)$ given by

$$P(x) = 2\pi\, r_f\, x\, \tau_0 \tag{4}$$

However, the axial stress in the fiber cannot exceed the fiber strength, i.e., $P(x)/A_f \leq X_f$, where $A_f = \pi r_f^2$ is the fiber cross-sectional area. Hence, the pullout length, $\delta$, is the value of $x$ at which the fiber strength has been reached, i.e.,

$$P(\delta) = A_f X_f = \pi r_f^2 X_f \tag{5}$$

Combining Eqs. (4), (5) gives

$$P(\delta) = \pi r_f^2 X_f = 2\pi\, r_f\, \delta\, \tau_0 \tag{6}$$

Solving Eq. (6) for $\delta$ yields the pullout length as

$$\delta = \frac{r_f X_f}{2\,\tau_0} \quad \text{(pullout length)} \tag{7}$$

It is apparent that the pullout length $\delta$ increases with fiber strength $X_f$ and decreases with interfacial strength $\tau_0$.

## 5.4 TENSION DAMAGE AND FAILURE IN A CROSS-PLY COMPOSITE LAMINATE

When subjected to axial tension, a $[0,90]_S$ cross-ply composite displays progressive failure because several damage mechanisms take place sequentially. Part of these damage mechanisms can be accounted for through the ply discount method.

### 5.4.1 Ply Discount Method

When calculating the quasi-static strength of a laminate with plies of various orientations, the common scheme [11] is to calculate the ply stresses using composite lamination theory (CLT), invoke some failure criterion to predict first ply failure (usually matrix cracking), reduce the moduli in the failed ply (usually the $E_2$ modulus perpendicular to the fibers and the in-plane shear modulus $G_{12}$), recalculate ply stresses, increase the load, test for second ply failure, and so on until "last ply failure" is predicted in the $0°$ plies. Thus, the quasi-static strength of the laminate is controlled (at least to an engineering approximation level) by the net section strength of the $0°$ plies. This scheme, commonly referred to as the ply discount method is known to provide good engineering estimates of laminate strength when edge effects do not dominate the failure process [11].

### 5.4.2 Progressive Failure of a Cross-Ply Laminate

Consider for example the $[0,90]_S$ cross-ply laminate of Figure 9 consisting of a CFRP laminate with properties given in Table 2. Figure 9a presents a visual description of the salient features, whereas Figure 9b presents a load–displacement diagram calculated with the ply discount model.

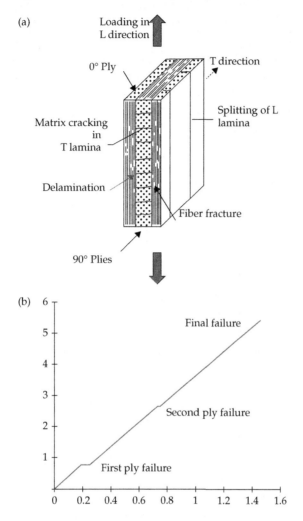

**FIGURE 9** Longitudinal tension of a 0/90 composite laminate: (a) highlight of several damage modes: matrix cracking in transverse (T) lamina; splitting of longitudinal (L) laminae; delamination between T and L laminae (after Ref. [12]); (b) load/displacement curve showing kinks and slope changes after each partial failure.

When an axial load is applied in the longitudinal (L) direction, the 0° ply is loaded along its reinforcing fibers, whereas the 90° ply is loaded across the fibers. Because the strength of the polymeric matrix is much less than that of the fibers, the across-the-fiber transverse strength of the lamina is much lower than the along-the-fiber longitudinal strength, $Y_t \ll X_t$. Hence, matrix cracking of the 90° ply occurs at an early stage in the loading cycle; this corresponds to "first ply failure" event marked on Figure 9b. Thus, the transverse

**TABLE 2**   Typical properties of a [0,90]$_S$ CFRP laminate

| *ELASTIC PROPERTIES* | | | |
|---|---|---|---|
| Longitudinal modulus | $E_L = 155$GPa | Longitudinal strength | $X_t = 1447$MPa |
| Transverse modulus | $E_T = 12.1$GPa | Transverse strength | $Y_t = 62$MPa |
| Shear modulus | $G_{LT} = 4.4$GPa | Transverse strength | $Y_t = 62$MPa |
| Poisson ratio | $\nu_{LT} = 0.248$ | Shear strength | $S = 62$MPa |
| *THERMAL EXPANSION PROPERTIES* | | | |
| Longitudinal CTE | $\alpha_L = -0.1 \times 10^{-6}$ | Transverse CTE | $\alpha_T = 24.3 \times 10^{-6}$ |
| *GEOMETRIC PROPERTIES* | | | |
| Stacking sequence | | [0, 90, 90, 0] | |
| Ply thickness | | $h = 75\,\mu m$ | |
| Specimen width | | $b = 25$mm | |
| Specimen length | | $l = 150$mm | |
| *TEMPERATURES* | | | |
| Curing temperature | | $T_1 = 200°C$ | |
| Testing temperature | | $T_1 = 20°C$ | |

plies have lost the ability to resist stress across their fibers and the corresponding transverse modulus vanishes in these plies. As a result, we see a sudden extension of the specimen, i.e., a manifestation of stiffness loss. As the tension load increases, the outer $0°$ layers crack laterally (i.e., split) due to the transverse fibers in the $90°$ layers resisting the Poisson effect. This corresponds to the "second ply failure" event marked on Figure 9b. At this stage, the $0°$ plies loose their transverse modulus as well and the load−displacement curves display another sudden displacement increase associated with the additional loss in the overall stiffness of the specimen. If further loading of the specimen is attempted, the $0°$ plies will eventually fail due to fiber fracture, at which point no load can be any longer supported. Thus, the final tension failure of a cross-ply composite under tension is controlled by the tension strength $X_t$ in the $0°$ plies. This observation also applies to other composite layups that contain plies of other fiber orientations in addition to the $0°$ plies.

One important aspect of this simple [0/90]$_S$ example is that internal damage in a composite laminate can happen at relatively low stress levels in the form of matrix cracking to be followed, at intermediate levels, by inter-ply delamination and lamina splitting. Such internal damage is not catastrophic (i.e., it does not lead to the failure of the composite) but diminishes the composite stiffness and allows for environmental effects to penetrate into to the composite material.

### 5.4.3 Interfacial Stresses at Laminate Edges and Cracks

Further damage may also occur in composite laminates in the form of delaminations between plies of dissimilar properties. Such delaminations are promoted by the interfacial shear and normal stresses that appear in the "boundary layer" at the laminate edges and near internal cracks. These interfacial stresses that appear near free surfaces are 3D effects that are not predicted by CLT, because CLT does not include plate edge effects. The CLT approach relies on the plane-stress assumption. The in-plane stresses components predicted by CLT vary from layer to layer and across the thickness of each layer, but are constant along the layers, which is correct for an infinite plate. However, if the composite plate has finite dimensions, the CLT assumptions break down near the free edges since the in-plane stresses are required to vanish along the free edges; in this case, a full 3D analysis is required.

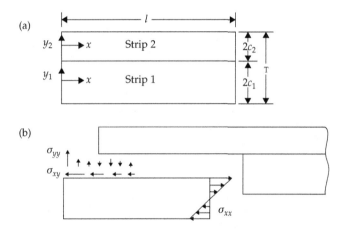

**FIGURE 10**   Elementary explanation of interlaminar stresses: (a) two-layer laminated strip made up of two dissimilar materials under differential expansion; (b) schematic of the stresses taking place near the laminate free end [13].

An elementary explanation of free-edge effects was given by Hess [13] through the analysis of a two-layer laminated strip made up of two dissimilar materials undergoing differential expansion (Figure 10a). Mechanics of materials analysis of the differential expansion predicts that each strip is under a state of combined axial–flexural loading. This axial–flexural state of loading results in a $\sigma_{xx}$ stress that varies linearly across the strip thickness as depicted at the right-hand side of Figure 10b. However, the free-surface boundary conditions at the strip ends are such that $\sigma_{xx}(y) \equiv 0$ at these locations. Hence, a small element cut out near the end of one of the strips would have a linearly varying axial–flexural $\sigma_{xx}$ stress at one end (right-hand side of the element shown in Figure 10b) but no such stress at the other end which is a free end. To keep the small element in equilibrium, other stresses are needed, specifically the normal stress $\sigma_{yy}$ and the shear stress $\sigma_{xy}$, which are confined to the vicinity of the free end. The normal stress $\sigma_{yy}$ tends to open the bond between the two strips whereas the shear stress $\sigma_{xy}$ tends to shear this bond, both promoting the formation of a crack in the bond between the two strips at this free end. Hess [13] calculated these local $\sigma_{yy}$ and $\sigma_{xy}$ stresses using a series expansion solution in terms of the eigenvalues of the Airy stress function and found that, in certain cases, the $\sigma_{yy}$ may take very large values at the free end hinting to the existence of a singularity as indicated in a related problem solved by Bogy [14].

The free-edge problem for composite laminates was solved by Pipes and Pagano [15] using a finite-differences scheme applied to the complete 3D equations of the laminate. It became apparent from the analysis of the numerical results that the "region of disturbance" in which interlaminar stresses are significant is restricted to a small region near the edges that extends over a length that is approximately equal to the laminate thickness. Pipes and Pagano [15] called this region "boundary layer" or "edge effect zone." The numerical results also showed a tendency for very large and rapid increase of the interlaminar stresses near the free edge which seemed to confirm the singularity suggestion of Hess [13].

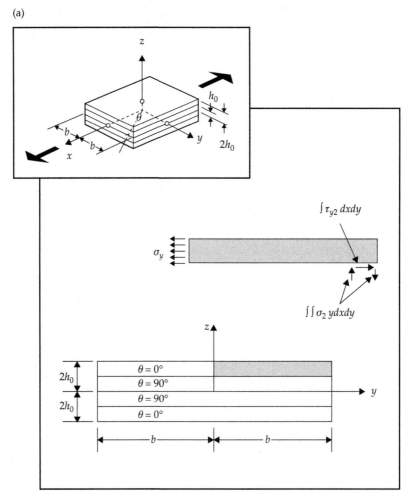

(a)

**FIGURE 11** Interlaminar stresses in a CFRP cross-ply laminate: (a) schematic diagram; (b) distribution of interlaminar shear stress; (c) distribution of the interlaminar normal stress [16].

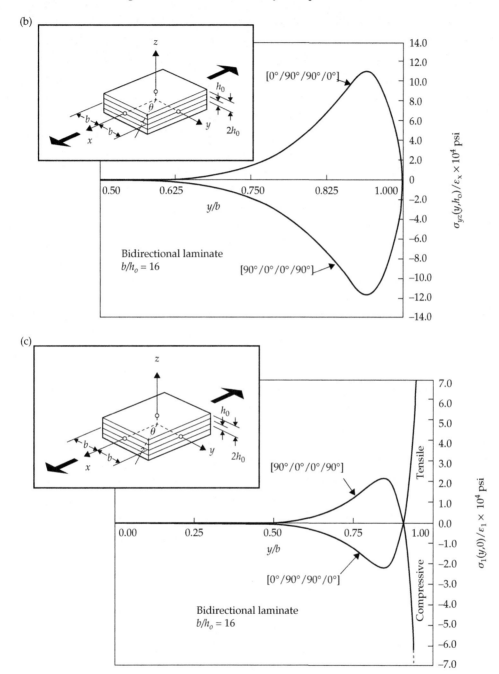

**FIGURE 11**    (Continued)

Figure 11 presents the interlaminar stresses that take place in a cross-ply laminate [16]. Figure 11a depicts a schematic representation indicating that the interlaminar shear stresses $\tau_{yz}$ and $\sigma_z$ are only present near the free edge in order to balance the effect of the in-plane stress $\sigma_y$. Figure 11b shows the distribution of the interlaminar shear stress at the interface between the 0° and 90° plies, whereas Figure 11c shows the distribution of interlaminar normal stress. The fact that the interlaminar normal stress is almost singular at the free edge is apparent. It is also remarkable to notice that the [0, 90, 90, 0] laminate stacking sequence produces tensile normal stresses that would tend to break the bond between the plies and separate them, whereas the [90, 0, 0, 90] stacking sequence would produce compressive normal stress that may have only a limited effect on the interlaminar bond.

### 5.4.4 Effect of Matrix Cracking on Interlaminar Stresses

Interlaminar stress effects similar to those appearing at the laminate free edges also appear due to matrix cracking in the transverse or off-axis layers of a laminate because these matrix cracks act as discontinuities generating local 3D disbonding stresses that promote delaminations.

If the applied stress is cyclic, as in fatigue loading, then these low-level damage states can increase and propagate further and further into the composite with each load cycle. The reinforcing fibers have high strength and good load-carrying properties, but the matrix cracking, delamination, and lamina splitting mechanisms usually lead to in-service composite structures becoming operationally unfit and requiring replacement.

## 5.5 CHARACTERISTIC DAMAGE STATE (CDS)

Reifsnider and colleagues [10,11,17,18] discovered that early damage accumulation in off-diagonal plies present an asymptotic behavior, i.e., a limit of loading is reached beyond which no more damage of this type can occur. They named this situation characteristic damage states (CDS) and postulated that CDS is a laminate property, i.e., it is completely defined by the properties of the individual plies, their thickness, and the stacking sequence of the various-orientation plies. The CDS is independent of extensive variables such as load history and environment (except as the ply properties are altered) and internal effects such as residual stresses or moisture-related stresses.

### 5.5.1 Definition of the Characteristic Damage State

The explanation of this CDS phenomenon is as follows [11]: in the transverse ply of a cross-ply laminate, cracks develop quite early in loading history and quickly stabilize to a very nearly constant pattern with a fixed spacing. The same behavior occurs for quasi-static loading and for cyclic loading (Figure 12). Under quasi-static loading, crack development occurs over a small range of load at the beginning of the loading curve and quickly stabilizes into a fixed pattern. Under cyclic loading, cracks develop during the early cycles and stabilize in a pattern with the same spacing as for the quasi-static loading. In fact, the two patterns of regular crack arrays are essentially identical regardless of load history.

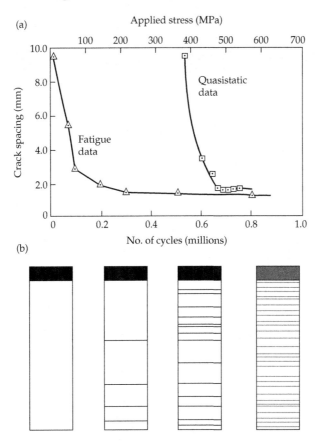

**FIGURE 12**  Characteristic damage state (CDS) concept: asymptotic behavior of crack spacing in 45° plies of a [0, 90, ±45]$_S$ CFRP laminate as a function of increasing quasi-static load or fatigue cycles (adapted after Ref. [11]); (b) experimental X-ray evidence of CDS formation (cited in Ref. [19] after Ref. [20]).

Formation of the cracks can be reasonably well anticipated by laminate analysis coupled with a common "failure theory" such as the maximum strain, Tsai-Wu, or Tsai-Hill concepts. It is possible to anticipate the number and arrangement of such cracks, information which can be used for subsequent analysis of behavior.

Similar behavior is observed for other off-axis plies. Figure 13 shows a schematic view of several typical crack patterns as seen from the edge of laminates loaded with tension–tension fatigue histories [11]. This type of transverse crack formation has received a great deal of attention and is, in comparison with other microevents, relatively well described and understood. The CDS patterns can be predicted by several methods, all of which are based in some way on local stress analysis. On the basis of several dozen successful predictions, Reifsnider et al. [11] state that the CDS can be anticipated with at least "engineering accuracy". Figure 12 shows the spacing between cracks in a −45° ply of a [0, 90, ±45]$_S$ AS3501-5 CFRP laminate as a function of quasi-static load level and cycles of loading at about two-thirds of the ultimate strength $(R = 0.1)$. The correspondence between quasi-static and fatigue effects in determining the CDS crack density is quite apparent.

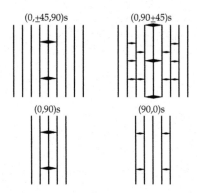

**FIGURE 13**   Schematic drawing of several typical transverse crack patterns [11].

The CDS concept is important in the context of composite strength analysis. Since it is a well-defined laminate property, it is possible to properly set boundary value problems based on the pattern geometry and expect the results to be useful for the prediction of subsequent laminate behavior.

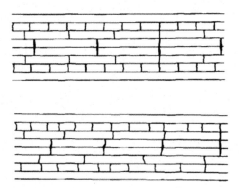

**FIGURE 14**   Schematic representation of predicted (top) and observed (bottom) characteristic damage states for a $(0, 90 \pm 45)_s$ CFRP laminate [11].

An example of a predicted and observed pattern is shown in Figure 14. The effect of stacking sequence, ply properties, and ply thickness on that pattern and others was faithfully reproduced by the analytical models in Refs. [17,18].

Table 3, reproduced from Ref. [11], provides the stresses in the individual plies of a sample laminate before and after matrix cracks form in the 90° and ±45° plies (for which $E_2$ and $G$ are then set equal to zero). The stress in the fiber direction of the 0° plies (which controls final fracture) is increased from 2631 to 2993 MPa, a jump of 14%, which is then used in a failure analysis to predict the "correct" strength (if both off-axis plies fail before laminate failure).

**TABLE 3**    Stresses in individual plies of a T300-5208 laminate composite [0, 90, ±45]$_S$ before and after matrix cracks form in the 90° and ±45° plies [11]

| | $\sigma_x$ (MPa)[a] | | $\sigma_y$ (MPa) | | $\sigma_{xy}$ (MPa) | |
|---|---|---|---|---|---|---|
| Ply | Before | After | Before | After | Before | After |
| 0 | 2631 | 2993 | −2.3 | −4.7 | 0 | 0 |
| 90 | 167 | 0 | −796 | −1000 | 0 | 0 |
| +45 | 600 | 503 | 400 | 503 | 417 | 503 |
| −45 | 600 | 503 | 400 | 503 | −417 | −503 |

[a]*Applied stress σ = 1000 MPa.*

## 5.5.2 Damage Modes That Modify Local Stress Distribution

Reifsnider and Case [10] indicate that there are a number of composite damage modes that modify the local stress distribution and hence, when they appear, strongly affect the evolution of damage (Figure 15). Transverse matrix cracking (A) appears at an early stage due to the low transverse strength of a unidirectional composite plies oriented transverse to the loading direction. Another matrix dominated failure is the longitudinal matrix cracking (C) which appears at higher loading and is due to the opposition expressed by the fibers in the transverse ply to the Poisson shrinkage during longitudinal tension loading. 3D effects near the crack sites create interlaminar debonding stresses that produce delaminations (B) between the composite plies as well as debonding around the end of a fractured fiber (E).

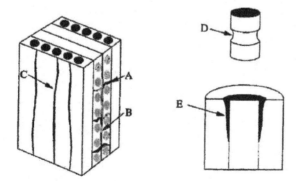

**FIGURE 15**    Damage modes modifying the local stress distribution in fiber-reinforced composites: transverse matrix cracking (A); delamination (B), longitudinal matrix cracking (C), fiber degradation (D), fiber debonding (E) [10].

### 5.5.3 Stiffness Evolution with Damage Accumulation

Stiffness is a well-defined engineering property, routinely measured, clearly interpreted, and directly involved in mechanics (stress analysis) calculations through the constitutive relations. Stiffness changes are directly related to internal stress redistributions. The idea of using stiffness change as a quantitative indicator of structural damage has received considerable attention in the literature [21,22].

In conventional built-up aerospace structures, e.g., metallic airframes with their multitude of rivets and other fasteners, it is commonly accepted that an overall "loosening" of the structure takes place during operational service. The structure starts its operational life as "tight and crisp" and progressively becomes more compliant as minute play develops in the joints especially due to cyclic loading and extreme load excursions. Such loosening of the structure due to service loads can be quantified by measuring its stiffness, i.e., the ratio between the applied load and the resulting deflection in a simple static test (e.g., loading the wing tip with a known number of standardized sandbags and measuring its resulting sag). Thus, the structural stiffness is notice to decrease with service life; however, this effect is not (up to a certain point) safety-critical since the structural strength of a built-up metallic structures is correlated with failure stress $\sigma_f$ rather than stiffness. (The loss of stiffness may reduce the natural vibration frequencies and induce unwanted in-flight aeroelastic phenomena that might eventually affect flight safety, but this would be an indirect effect.)

Loss of stiffness also occurs due to crack propagation in metallic structures and, in this case, it may be safety-critical. Consider, for example, a fracture-mechanics test coupon in which a crack starting from an initial notch is propagated through cyclic fatigue loading. As the crack grows through the accumulation of fatigue cycles, the remaining load-bearing section of the specimen diminishes resulting in higher stress and hence higher local elongation under the same load. The higher elongation translates into lower stiffness, which, in this case, relates directly to safety-critical damage since increasing crack length increases the stress intensity factor and brings the specimen closer to failure.

In high-strength fiber composite structures, the effect of stiffness loss is more dramatic than the loosening of built-up metallic structures and resembles more the case of stiffness loss due to fracture-mechanics damage. One important aspect is that, in high-strength fiber composites, the fracture-mechanics damage is diffused/distributed (e.g., a multitude of fine cracks in the matrix and in the fiber—matrix and inter-ply interfaces), whereas in metallic structures it tends to be rather concentrated at "hot-spot" sites. Another important aspect to consider is that failure of a high-performance fiber composite occurs ultimately through fiber breakage with is controlled, more or less, by the fiber failure strain $\varepsilon_f$ and the corresponding stress, i.e., $\sigma_f = E\varepsilon_f$. Hence, as shown diagrammatically in Figure 16a, a loss of composite stiffness by, say, 30%, would result in a corresponding reduction in the failure stress. If the 30% is the result of internal damage accumulated through fatigue cycling, than fatigue failure would occur at stress levels 30% less than the quasi-static failure stress$\sigma_f$, i.e. $\sigma_{fatigue} = 0.7\sigma_f$. This means that, if the specimen is loaded cyclically with an amplitude $\sigma_{fatigue} = 0.7\sigma_f$, and if the specimen fails at that amplitude, then the final stiffness $E'$ should be $0.7E$ if the strain to failure is the same $\varepsilon_f$. For long-term fatigue behavior

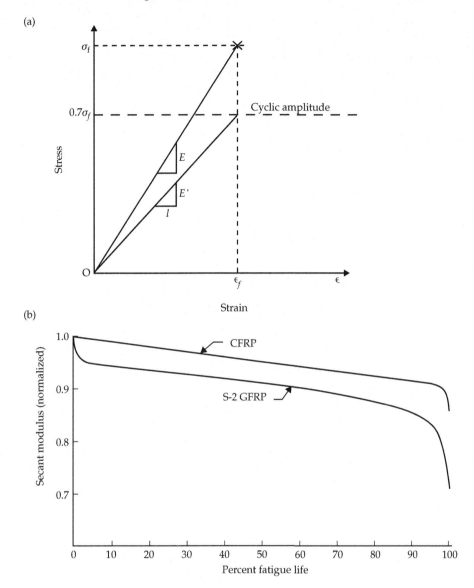

**FIGURE 16**  Stiffness evolution with damage accumulation: (a) reduction of elastic modulus due to damage accumulation results in reduction of strength for the same failure strain; (b) normalized secant modulus data recording during cyclic loading of $[0, \pm 45, 0]_S$ CFRP and $[0, 90_2]_S$ GFRP specimens [11].

where strength reductions are large, the attendant stiffness changes are also large [11], as illustrated in Figure 16b. It is thus apparent that a measurement of composite stiffness could serve as a quantitative indicator for monitoring the fatigue damage accumulation inside a high-performance composite material.

## 5.6 FATIGUE DAMAGE IN AEROSPACE COMPOSITES

Composites fatigue is substantially more complicated than that of metals [11]. One major difference is that composite materials have a fibrous nature and the fatigue mechanism is strongly influenced by fiber orientation and by the stacking sequence in the composite laminate. Another major difference is that fatigue damage may appear in various places in the laminated fiber—matrix material systems: in the matrix, in the matrix—fiber interface, in the interface between composite layers with various orientations, and, when final failure eventually occurs, in the fibers themselves.

### 5.6.1 Fatigue of Unidirectional Composites

While studying the fatigue and failure of unidirectional composites, Talreja and colleagues [19,23] indicated that damage in unidirectional composites is controlled by four typical mechanisms (Figure 17):

(a) Fiber break causing interfacial debonding and matrix crack
(b) Coalescence of matrix crack through a fiber break
(c) Fiber-bridged matrix cracking
(d) A combination of (a), (b), and (c)

When subjected to cyclic fatigue loading, unidirectional composites may behave differently depending on the load level as defined by the cyclic strain amplitude $\varepsilon_{max}$.

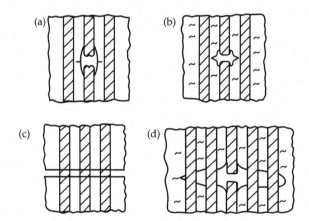

**FIGURE 17**   Typical damage mechanisms in unidirectional composites subjected to tension fatigue [23].

The $\varepsilon_{max} - \log N$ fatigue diagram of unidirectional composites under cyclic tension loading displays three characteristic regions (Figure 18):

• Region I is the horizontally extending scatter band (e.g., between 5% and 95% probability of failure) of the composite static failure strain. This region represents a lack of degradation in strength (failure strain), i.e., the underlying predominant mechanism of fiber failure is non-progressive.

- Region II is the fatigue-life scatter band, which deviates from Region I at a certain number of cycles and extends down to the fatigue limit. This region is governed by the progressive mechanism of fiber-bridged matrix cracking.
- Region III is the region of no fatigue failure (in a selected large number of cycles, say 106) lying below the fatigue limit.

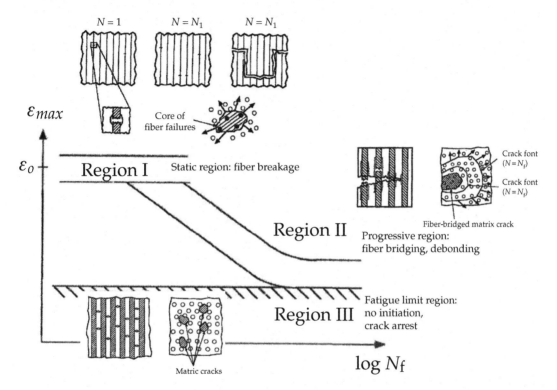

**FIGURE 18**    Fatigue diagram ($\varepsilon_{max} - \log N$) of unidirectional composites under cyclic tension loading *(collage of various diagrams in Ref. [19])*.

Region I may accommodate three damage scenarios as depicted in the upper left corner of Figure 18. The **first damage scenario in Region I** consists of fiber breaks resulting from the right-away application of a high load that produces a maximum strain lying within the scatter band of the composite static failure strain. These fiber breaks are caused by the local fiber stress (or strain) exceeding the fiber strength at those (statistically weaker) points during the first application of load. Unloading and reapplying the load would change the local fiber stresses only if an irreversible (inelastic) deformation occurs. Assume now that the matrix surrounding a fiber break undergoes only small inelastic deformation due to the constraint of the stiff fibers. This is likely if the matrix is relatively brittle, such as an epoxy. For small inelastic deformation in the matrix, the repeated application of load would cause little change in the stresses on fibers surrounding a broken fiber site.

**The second damage scenario in Region I** deals with the fibers that are likely to break in a subsequent load cycle; such fiber breaks could appear near any of the previously broken fibers—but not necessarily near all previously broken fibers—when the local stresses exceed their local strengths (see the $N = N_1$ schematic in the upper left corner of Figure 18). The important thing to note is that, due to the small cycle-to-cycle changes in the local fiber stresses and randomness of the fiber strength, it is unlikely that the fiber breakage will be a progressive mechanism in the sense that the number of fiber failures in a given location increases monotonically.

**The third damage scenario in Region I** deals with the final failure, which may result from a core of (a few) fiber failures growing unstably (see the case $N = N_f$ schematic in the upper left corner of Figure 18); this could occur in any of the potential failure sites, without preference.

The consequence is that, in Region I, the composite failure under the specified loading condition can occur at any value of the number of load cycles. Thus, Region I can be described as dominated by a non-progressive failure mechanism, with no associated strength degradation. For this reason, the scatter band in Region I is horizontal [19].

**Region II** displays a mechanism of fatigue failure which is substantially different from that of Region I, as illustrated in the right-hand side of Figure 18. When the applied fatigue load levels are below the lower bound of Region I scatter band, the composite failure is unlikely to result from a cluster of neighboring fiber breaks. Instead, a mechanism of progressive degradation will take place. As the applied load cycles increase, the matrix will undergo fatigue cracking. The matrix cracks will progress (i) by producing failure in the adjoining fibers; or (ii) by debonding the fiber—matrix interface and progressing along the fibers; or (iii) by cracking the material around the fibers. A typical fiber-bridged crack would then appear as illustrated in the right-hand side of Figure 18. Among several such cracks, the one to cause failure earliest would be the one that will be the first to undergo unstable growth [19].

**Region III** of the fatigue-life diagram can be viewed as the domain of no fatigue failure in the desired range (say $>10^6$) of load cycles. One possible fatigue failure scenario in this region is illustrated in the lower left side of Figure 18: fatigue cracks in the matrix are likely to develop, but they remain confined to parts of the cross-section which are between fibers. The "driving force" for the cracks is insufficient to advance them by fracturing and/or debonding fibers. Another likely scenario in this region is fiber-bridged matrix cracking progressing at rates too low to cause failure in the preselected large number of cycles. A "true" fatigue will exist only if a mechanism of effectively arresting matrix crack growth is available [19].

## 5.6.2 Fatigue of Cross-Ply Composite Laminate

Damage progression in a cross-ply laminated composite under fatigue loading was studied in Ref. [23]. The following stages were identified in a cross-ply composite laminate under cyclic tension (Figure 19):

1. Matrix cracking
2. Crack coupling and interfacial debonding
3. Delamination
4. Fiber breakage
5. Fracture failure of the overall composite laminate.

The evolution of damage through these phases depends on the cyclic fatigue loading level; three damage levels were identified [23]:

*Level* 1: Matrix cracks begin to develop while the fibers remain intact. Under cyclic loading, these cracks propagate until they reach a fiber/matrix interface or some other inhomogeneity. If the cyclic load is sufficiently low, these cracks will not break through the interphase and propagate any further; no more fracture surfaces will be created until the fatigue limit of the embedded fibers is reached and overall composite failure occurs.

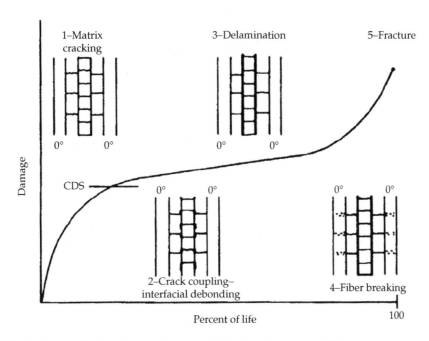

**FIGURE 19**    Evolution of fatigue damage in a cross-ply laminated composite [24].

*Level* 2: If the cyclic load is at an intermediate level, then Level 2 of damage evolution occurs. Because the load is higher, the matrix cracks may not stop at the fiber/matrix interface as they did in Phase 1. There are several things that could happen at this stage: (a) when the matrix crack reaches a fiber, the stress concentration may cause the fiber to break, forming a larger crack. (b) If the stress concentration is not sufficiently high for the fiber to break, a separation of the fiber and matrix could occur and the crack propagates along the fiber/matrix interface. As damage propagates through (a) and (b) mechanisms, it reaches new fiber/matrix interfaces and may cause further fiber fractures or fiber/matrix separation. Eventually, the degradation of the composite will be great enough to cause fracture.

*Level* 3: If the cyclic fatigue load level is near the ultimate strength, some fibers will fail immediately due to the statistical spread of their individual strengths. As the cyclic loading continues, more fibers will break throughout the composite; when enough fibers have failed in a particular area, a crack will form leading to a rapid fracture of the composite. At this level of loading, the final fracture will occur quickly and the damage progression trend, as shown in Figure 1, will not occur.

The orientation of the laminae in the composite laminate may modify the damage evolution. In a generic situation, the individual lamina is subjected to off-axis loading which is projected onto the lamina principal axes and may cause the matrix crack to form parallel to the fibers. These cracks can propagate freely down the length of the fiber either in the matrix or in the fiber/matrix interface. Debonds between the fibers and matrix eventually will occur.

The damage occurring in a composite laminate is a complex phenomenon that is governed by the processes taking in each individual lamina. A generic scenario may look as follows:

(a) Small microcracks develop in the matrix mainly between fibers that are not parallel to the loading direction.
(b) As the cyclic loading continues, the microcracks grow and become macroscopic cracks. At this point, the material has reached the CDS as marked in Figure 19.
(c) The cracks in the matrix will spread through the ply in which they began.
(d) Stress concentrations cause microcracks to develop in the plies on either side of the initial ply.
(e) Stress concentrations between plies cause local delaminations.
(f) Once delaminations have formed, damage increases rapidly up to complete failure.

Because of this progressive fatigue damage behavior, a composite structure should be periodically inspected to monitor the damage progression just as a metallic structure is currently monitored. Reference [19] states that a similar damage progression happens under quasi-static loading.

## 5.7 LONG-TERM FATIGUE BEHAVIOR OF AEROSPACE COMPOSITES

Reifsnider et al. [11] identified three distinct regions of damage development in the long-term fatigue behavior of composites. These regions are depicted as Region I, Region II, and Region III in Figure 20.

### 5.7.1 Damage Region I—Progression toward Widespread CDS

Damage Region I covers a zone in which the composite exposed to cyclic loading undergoes a "period of adjustment". Damage accumulation is fast at first and then decreases rapidly as it reaches a state of saturation and settles down into a steady-state situation that will be representative for most of the upcoming Region II.

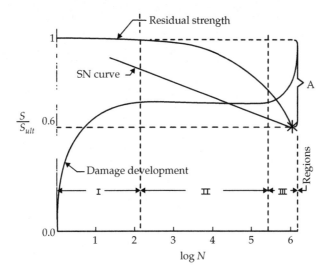

**FIGURE 20**    Long-term composite behavior diagram showing three distinct regions of development [11].

For laminates that have off-axis plies, such as the common $[0, 90, \pm 45]_S$ quasi-isotropic stacking sequence, Region I usually involves matrix cracking through the thickness of the off-axis ($90°$ or $45°$) plies parallel to the fibers and perpendicular (at least in transverse projection) to the dominant load axis (the $0°$ direction). These cracks evolve until the CDS described in Section 5 is reached. The effect of matrix cracking and CSD formation is twofold.

On the one hand, matrix cracks introduce a *reduction in the stiffness of the laminate* since the cracked plies carry less load than they did prior to cracking as predicted by the ply discount method discussed in Section 4.1. For CFRP quasi-isotropic laminates, that laminate stiffness change is on the order of 10% or somewhat less, while a comparable GFRP laminate stiffness would change by somewhat greater amounts. These stiffness changes are generally not of great engineering consequence except as they may be used in the SHM process because they affect vibration frequencies.

On the other hand, matrix cracks produce *stress redistributions*. Using the ply discount method discussed in Section 4.1, one can predict the stress redistributions associated with matrix cracks in the off-axis plies. However, these stress redistributions (and the stiffness reductions that caused them) are not, in reality, uniform. They truly exist only near the matrix cracks in the off-axis plies, whereas decreasing gradually between the cracks according to a shear-lag law. As shown in Figure 21a, a region of stress redistribution exists around the tip of cracks in the off-axis plies.

Nonetheless, no reduction in the overall residual strength of the laminate is observed since, as predicted by the ply discount method discussed in Section 4.1, the quasi-static strength of the laminate is controlled (at least to an engineering approximation level) by the net section strength of the $0°$ plies. And the net section strength of the $0°$ plies is not affected by matrix cracks in the adjacent off-axis plies (assuming that the local stress concentration is not high enough to produce a local fracture in the 0° ply). This explains why the residual strength in Region I of Figure 20 does not show a decrease in spite of CDS damage propagating throughout the off-axis plies of the composite laminate.

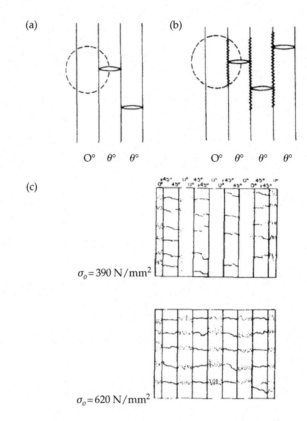

**FIGURE 21** Schematic diagrams of crack damage development: (a) in damage Region I, local zone of influence in the longitudinal ply near a matrix crack in an off-axis ply characterized by stress concentration; (b) in damage Region II, the local zone of influence near matrix cracking in an off-axis ply produces crack coupling and local delamination in the interface between the plies; (c) in damage Region III, the formation of fracture in the longitudinal 0° fibers happens with increasing load in correlation with the location of matrix cracks in the off-axis plies *(adapted from Ref. [11]).*

## 5.7.2 Damage Region II—Crack Coupling and Delamination

The stability of the off-axis crack CDS pattern is the reason for the marked decrease in damage rate between Regions I and II of Figure 20. It also accounts for the relatively flat nature of the damage growth curve in Region II. Generally speaking, the behavior in Region II is dominated by coupling and growth: matrix cracks may couple together and grow, especially along interfaces; delaminations, if present, may also grow.

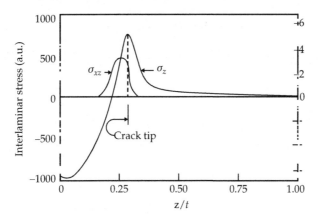

**FIGURE 22**   Distribution of interlaminar stresses near a transverse crack in the 90° ply (the interlaminar stresses act in the interface between the 0° and 90° plies) *(adapted from Ref. [11])*.

The mechanics of this region are typified by the following scenario. Figure 22 shows schematically the interlaminar stress distributions near a transverse crack in the 90° ply; these distributions are depicted as functions of the distance halfway through the thickness (the interlaminar stresses act in the interface between the 0° and 90° plies). For an uncracked material, these interlaminar stresses would not exist; but when a transverse crack forms, it creates an interlaminar tensile normal stress which tends to separate the plies, and an interlaminar shear stress, which tends to shear the plies along their interface in the neighborhood of the crack. These interlaminar tensile and shear stresses provide the basis for a mechanism of crack coupling and delamination as shown in Figure 23. The nature of the interfacial growth is influenced by locality in the specimen and by the nature of the matrix cracks which precede it.

### 5.7.2.1 Edge Cracks versus Internal Cracks

The pattern of damage shown in Figure 23 is typical of crack patterns observed at the edges of composite laminate specimens. The same interlaminar stresses also operate in the interior of laminates, but the results are frequently somewhat different from those shown in Figure 23.

Figure 24 is a schematic diagram of the type of damage commonly found in the interior of a laminate (a cross-ply laminate for this example). The details are generalizations of results obtained by X-ray radiography [11]. These results indicate that interfacial growth at a transverse matrix crack in the interior of a laminate is frequently constrained to a small region, at least in Region II, and may also involve growth of the crack along 0° fiber directions into the 0° ply thickness as well as along the ply interface plane.

### 5.7.2.2 Comparison between Region I and Region II

#### 5.7.2.2.1 Phenomenological Comparison

In Region I of Figure 20, matrix cracking of the off-axis plies was identified as the primary damage mechanism, and the internal stress redistribution associated with this damage

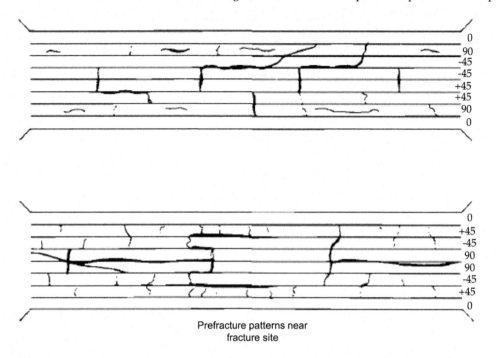

Prefracture patterns near
fracture site

**FIGURE 23**  Prefracture crack patterns showing a breakdown of CDS through coupling of transverse cracks and appearance of longitudinal cracks; schematic drawings after replicas of the specimen edges [11].

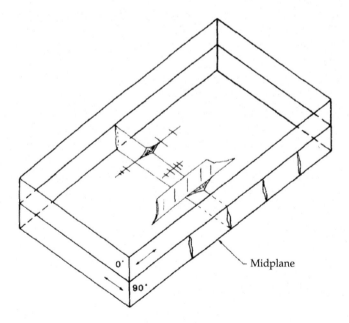

**FIGURE 24**  Schematic representation of internal interfacial damage in a cross-ply laminate [11].

mode was discussed as a means of explaining the resulting residual strength of the laminate. This was presented schematically in Figure 21a which shows a small zone of influence characterized by stress concentration in the longitudinal ply around the tip of matrix crack in the off-axis ply.

In Region II, mechanisms of crack coupling and interface separation (delamination and debonding) seem to dominate the damage development. The internal stress redistribution associated with this type of damage is characterized, in part, by the diagram in Figure 21b. This figure is a Region II version of Figure 21a and continues the local stress argument associated with matrix cracking.

### 5.7.2.2.2 Stress−Distribution Comparison

When the cracks in the off-axis plies grow along the ply interfaces to join other cracks, how is the local stress altered? Reference [11] provides a numerical example which we will discuss only qualitatively here:

- In Region I, when the ply discount analysis of progressive failure was applied and the local $E_2$ and $G$ moduli of the cracked off-axis plies were reduced to zero, it was found that the axial stress in the longitudinal $0°$ plies increased by approximately 14%. At that point, the plies were still bonded together, and the cracked off-axis plies still contributed some load sharing to the stress state in the $0°$ plies by supporting load in their off-axis fiber direction.
- In Region II, when local ply delamination/separation occurs in certain zones, the load sharing in Region I is relaxed and the longitudinal $0°$ plies are under completely uniaxial stress in that local delamination zone. The delaminated off-axis plies carry no part of the $0°$ ply load and do not constrain the transverse expansion or contraction of the $0°$ plies. In this situation, the longitudinal stress in the $0°$ plies becomes the only load-bearing stress of the laminate.

The numerical calculations presented in Ref. [11] show that the $\sigma_x$ stress in the $0°$ plies could grow, in the local delamination zone, by approximately 34%, which means an approximately 34% reduction of the laminate strength. This strength reduction is consistent with experimental results that indicate the long-term fatigue strength of composite laminates to be 30−40% of the static strength. So it would appear that crack coupling and (local) delamination or debonding (even in the interior of a specimen) near matrix cracks are viable strength reduction mechanisms for the long-term fatigue behavior of composite laminates.

## 5.7.3 Damage Region III—Damage Acceleration and Final Failure

Figure 20 indicates that the rate of damage development increases sharply near the end of life. Before concentrating on the micromechanical behavior is this Region III, one should also recall the strong correlation between damage and stiffness loss, as already discussed in Section 5.3.

### 5.7.3.1 Stiffness-Damage Correlation in Region III toward Composite End of Life

One important point should be made: stiffness decreases sharply toward the end of life in a composite under long-term fatigue loading. Evidence of this is found in the damage observations and is demonstrated in Figure 16 which shows the normalized axial stiffness as a function of the number of cycles of tensile fatigue loading ($R = 0.1$) for two laminates, one CFRP, the other GFRP. Both curves show the same generic shape suggested by the "damage development" curve in Figure 20. In the $[0, 90_2]_S$ GFRP laminate, the initial drop in modulus corresponds to 90° matrix crack (CDS) formation which, for this material, changes the laminate modulus considerably. Whereas for the $[0, \pm45, 0]_S$ CFRP specimen, the change in laminate modulus attributable to matrix crack formation is very small.

Reference [11] indicates that the idea of using stiffness change as a quantitative indicator of fatigue damage has received considerable attention in the literature. Stiffness is a well-defined engineering property, routinely measured, clearly interpreted, and directly involved in mechanics (stress analysis) calculations through the constitutive relations. Stiffness changes are directly related to internal stress redistributions, as discussed earlier and illustrated in Figure 16b. For long-term fatigue behavior where strength reductions are large, the attendant stiffness changes are also large, as demonstrated by Figure 16b. Figure 16a illustrates the fact that, for a laminate which has a quasi-static failure strength of $\sigma_f$ and which is fatigue-loaded with an amplitude of $0.7\sigma_f$ for a long period of time, if the laminate fails at that amplitude, then the final stiffness, $E'$, must be $0.7E$ or less in magnitude if the strain to failure is constant or increases. Whether the laminate is fiber dominated or not makes no difference in this argument, a fact also demonstrated by the CFRP laminate behavior in Figure 16b. But it does make a difference how carefully the modulus is measured, since the large change in Region III happens very quickly.

### 5.7.3.2 Role of Fiber Fracture in Final Failure

Reference [11] discusses the role of fiber fracture in connection with Figure 16b, especially the $[0, \pm45, 0]_S$ CFRP data. Final failure and, it appears, those events immediately preceding failure are dominated by fiber failure. In Figure 16b, the sharp drop in stiffness quite near the end of the test involves fiber fracture events, but fiber fracture occurs throughout Region III. Two major modes of fiber failure have been observed:

*Fiber failure Mode* 1 is associated with specimen edges. Near the specimen edges, there is a distinct fiber failure sequence which begins in the early stages of life. Debonds frequently serve to connect fiber fractures in a stair-step fashion. Fiber fractures develop as a function of increasing load or cycles in regions frequently associated with matrix cracks in the off-axis plies and preferentially near the specimen surfaces. Delaminations between plies are also associated with this fiber failure mode.

*Fiber failure Mode* 2 is associated with fiber fracture in the interior of laminates. When matrix cracks intersect an adjacent ply during fatigue loading, a series of local ply delaminations develop and cracks (or debonds) develop along fibers in the adjacent plies. Once the CDS cracks have been established, tiny "dendrites" of interface separation appear at right angles to the matrix crack directions, as illustrated in Figure 24. Evidence exists of fiber fractures associated with this damage mode [11]. Debonding along alternate sides of a fiber near a matrix crack in a 90° layer separated from an adjacent 0° ply has also been observed and evidence associating debonding (or longitudinal "splitting") with fiber fracture in the interior of laminates has been recorded [11].

### 5.7.4 Summary of Long-Term Fatigue Behavior of Composites

Reference [11] summarizes the long-term fatigue behavior of laminated composites by indicating that, although not all aspects of long-term fatigue-related fracture are clear, it is however clear that near the end of a long-term fatigue exposure, at stress levels on the order of 60−70% of the ultimate strength of the laminate, more damage of all types has occurred than under any type of load history. Matrix cracks have formed and generated interlaminar stresses, which initiate debonding of fibers (or splitting along fibers) and delamination between plies. And fiber failures in isolated random locations as well as in clusters around matrix cracks have occurred in the 0° laminae (and the off-axis plies). The laminate has developed preferential failure sites which are under redistributed stress states that cause elevated (and probably predominantly unidirectional) load-direction stresses in the 0° plies.

Attempts are still being made [11] to determine the nature of the final failure event itself, but the damage state described previously is certainly sufficient to cause fracture. The development of the damage state differs for different materials and stacking sequences and laminae orientations, but the previously described development is generic in the sense that it is based on observations of events in CFRP and GFRP laminates having a wide range of ply orientations and stacking sequences. Of course, the role of matrix cracking in a $[0, \pm 45, 0]_S$ CFRP specimen is not as great as it is for a $[0, 90_3]_S$ GFRP or other laminate, but these variations are a matter of degree.

## 5.8 COMPRESSION FATIGUE DAMAGE AND FAILURE IN AEROSPACE COMPOSITES

Reference [11] indicates that compressive fatigue loading produces a somewhat different scenario from tensile fatigue loading. When subjected to axial compression, the composite fails through loss of elastic stability (buckling). This buckling may take place at various scales: (a) structural; (b) composite layup; (c) composite material itself.

At global structural scale, buckling can be avoided by relative sizing the length and bending stiffness of the component such that loss of elastic stability does not occur for the given boundary conditions and operational load levels. Hence, this type of buckling will not be pursued further in the present discussion.

### 5.8.1 Compression Fatigue Delamination Damage

At the composite layup scale, the major damage mode is delamination, as shown in Figure 25. This figure shows a schematic view of the edge of a laminate that delaminates under compressive loading. The pattern of development depicted is typical of the patterns observed in cyclically-loaded $[0, \pm 45, 90_2 \pm 45, 0]_{3S}$ CFRP laminates. Several general features of the damage development are observed: the delamination generally develops preferentially from the exterior (planar) surfaces of the laminates and subsequently initiates at positions of high interlaminar stress in the interior of the thickness. This surface-to-interior initiation sequence appears to be generic for long-term damage development in compressive loading. As the cyclic compressive loading continues, the delaminations grow from the edges across the width of the coupon specimens as well as along the length of the specimens. Growth along the length is highly dependent upon the specimen geometry and test arrangement. However, growth across the width appears to be generic and can be described and predicted using strain energy release concepts [11].

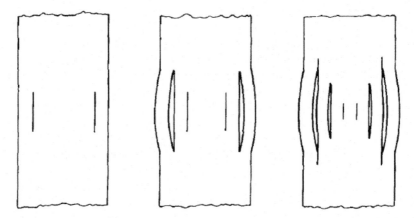

**FIGURE 25**   Schematic diagram of the development of delamination during cyclic compressive fatigue loading (edge view of laminate) [11].

## 5.8.2  Compression Fatigue Local Microbuckling Damage

At the local scale, the composite material itself may fail under compression through the microbuckling mechanism (Figure 26). The high-strength fibers encased in the polymeric matrix can be viewed as beams on elastic foundation, where the elastic support is provided by the matrix stiffness [25]. Under axial compression, such a beam on elastic foundation would eventually buckle and take an undulatory shape (Figure 26a). The compressive stress values at which this buckling occurs are dictated by the fiber bending stiffness and matrix compression stiffness. As the compressive load is further increased, the microbuckling is further exacerbated until local failure occurs in the form of a *kink band* (Figure 26b).

For a given composite material system with a certain fiber/matrix combination, the microbuckling compressive strength is fixed and cannot be altered through structural design. In order to modify the microbuckling compressive strength, one has to address the very constituents of the composite material system. For example, the thicker boron fibers have a higher compression buckling strength than the thinner carbon fibers; for this reason, boron fiber composites may be preferred in places where material compression strength is the critical factor.

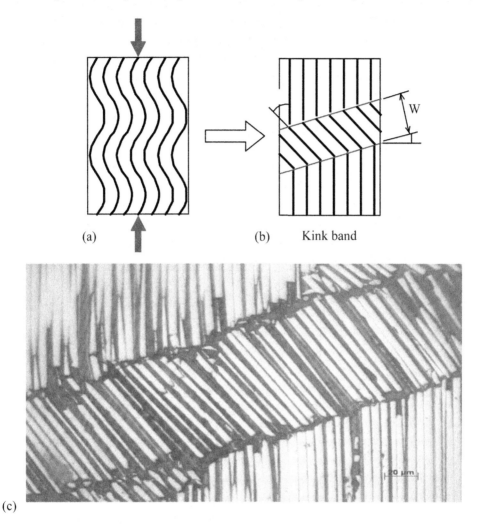

**FIGURE 26**  Compression damage of fiber composites through microbuckling: (a) undulations of buckled fibers; (b) kink band local failure schematic [25]; (c) micrograph of kink band formation in T800/924C carbon fiber composite [26].

The growth of fiber microbuckling damage zone in composites subjected to cyclic compressive loading has a behavior similar to the growth of a tension crack growth in sheet metal, although the underlying mechanisms are radically different. Reference [27] performed an extensive study of compression damage growth in composites through fiber microbuckling in the stress concentration zones of hole and found that fracture-mechanics concepts that were initially developed for sheet metal structures in tension could be extended to the study of composites in compression using the cohesive

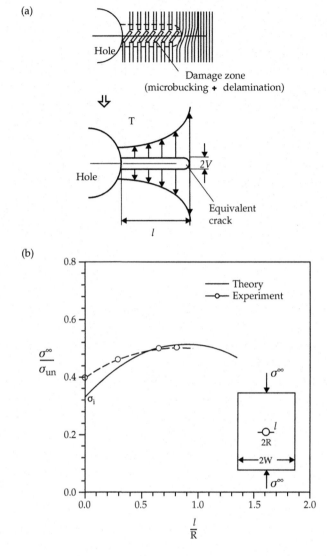

**FIGURE 27** Growth of compression damage zone through fiber microbuckling and delamination near a composite hole: (a) schematic of the mechanism taking place in the stress concentration zones near the hole; (b) stable growth curve of the compression damage zone [27].

zone model (Figure 27a). The gradual lateral extension of the microbuckling plus delamination damage zones from a composite hole under compression loading resembles closely the lateral extension of butterfly cracks from a metallic hole under tension (Figure 27b).

## 5.8.3 Compression Fatigue Damage under Combined Tension-Compression Loading

Another aspect that needs to be considered is the combined loading in tension and in compression during fatigue cycling. Under combined tension-compression cyclic loading, the tension damage modes discussed in Section 7 will interact synergistically with the damage modes discussed here to produce a "worst-case" situation that is distinct in its severity.

## 5.9  OTHER COMPOSITE DAMAGE TYPES

### 5.9.1 Fastener Hole Damage in Composites

Mechanical fasteners used in riveted and bolted joints are prevalent in metallic aircraft structures, where they offer a rapid and convenient method of assembling large structures from smaller components. The load-bearing mechanisms of metallic joints are well understood and easily predicted. The use of mechanical fasters in composite structures is also allowed, but this comes with significant strength and fatigue penalties. Nonetheless, mechanical fasteners are still widely used in the construction of composite and/or hybrid structures, especially when load transfer has to be achieved between composite and metallic components.

A typical example is illustrated in Figure 28a, where the load from a composite wing skin is transferred into an aluminum metallic bracket through a 9-bolt junction. When in service, each hole in the composite skin would be subjected to tension and/or compression loading that may, under certain circumstances, promote damage initiation and damage progression.

**Under tension** (Figure 28b), the composite joint may fail in three major modes: (i) tear failure; (ii) bearing failure; and (iii) shear-out failure. Of these, the tear failure is unlikely to happen because the fiber reinforcement is strongest in tension. The shear-out failure would happen if the fibers are predominant in the tension direction; shear-out failure can be counteracted through design by the addition of 45° reinforcement. The bearing failure is more difficult to prevent because it is a compression-type loading that has to be taken up by the polymeric matrix and by the fibers under compression. Bearing failure may occur through matrix crushing, or fiber microbuckling, or both.

**Under compression** (Figure 28c), the composite joint may fail in three major modes: (i) overall buckling of the component; (ii) local buckling of the region weakened by the hole; (iii) fiber microbuckling at the areas of highest compression strength. The overall compression buckling can be prevented by proper component design. Local buckling and fiber microbuckling may also be prevented by design, but damage accumulation during cyclic loading would eventually weaken it.

The use of mechanical fasteners in fiber composites is somehow counterintuitive, but expedient. The very premise of fiber composites is to have the load carried through the high-strength fibers embedded in a relatively weak polymeric matrix. This type of load-carrying capability benefits from a smooth and continuous load "flow" and is adverse to sudden changes in material properties and geometries such as those imposed by a fastener

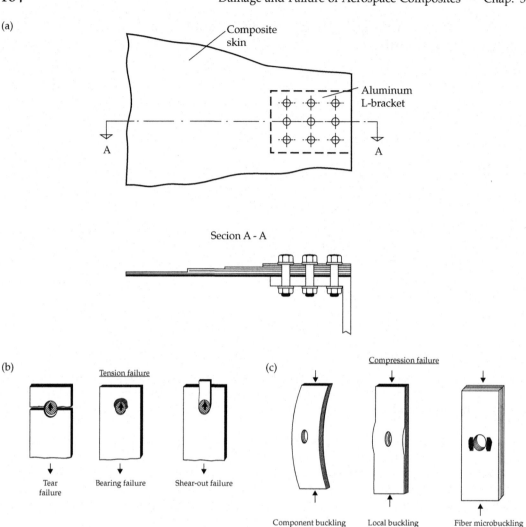

**FIGURE 28** Composite damage due to fastener holes: (a) typical bolted junction between a composite wing skin and metallic connection; (b) tension failure modes; compression failure modes.

hole. Ideally, composite joints should be done through adhesive bonding with gradual transition from one component into the next. However, mechanical fasteners are often used for a variety of reasons and one has to assess the consequences of such a design decision: fasteners holes drilled in the composite structures produce sudden discontinuities, interrupt the fiber flow, and act as stress concentrators. They also act as crack and delamination initiators due to microdamage introduced during the hole drilling process. Special attention can be given to creating stress-free holes by designing local reinforcement; damage-free holes can also be manufactured with special tooling. However, these aspects come with added cost and may not be always implementable in practice.

## 5.9.2 Impact Damage in Composites

Composite aerospace structures are prone to a particular type of damage that is not critical in metallic aerospace structures, i.e., **low-velocity impact damage**. Such damage may occur during manufacturing or in service due to, say, a hand tool being dropped onto a thin-wall composite part. When such an impact happens on a conventional metallic structural part, either the part is not damaged at all, or, if it is damaged, then it shows clearly as an indent or scratch. In composite structures, a similar impact may damage the structure without leaving any visible marks on the surface (the so-called **barely visible damage, BVD**). In this case, the impact result takes the form of **delaminations** in the composite layup. (A more drastic impact may also show spalling on the back side, while having no visible marks on the front side).

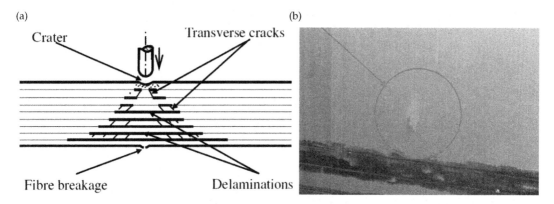

**FIGURE 29**    Impact damage effects on laminated composites: (a) schematics of various damage mechanisms [28]; (b) magnified photo of barely visible impact damaged on a composite.

Delamination due to barely visible impact damage may not have a large effect on the tension strength of the composite, but it can significantly **diminish the composite compression strength** (delaminated plies have a much weaker buckling resistance than the same plies solidly bonded together). The component buckling strength and the local buckling strength may be both affected; when a fastening hole is present, as depicted in Figure 28c, this effect may be even worse. For this reason, manufacturing companies place a strong emphasis on testing the **open-hole compression strength after impact** of their composite structures.

Worst-case impact damage is defined as the damage caused by an impact event (e.g., a 1-in hemispherical impactor) at the lesser of the following two energy levels: (a) 100 ft-lb, or (b) energy to cause a visible dent (0.1-in deep).

A schematic of the various damage mechanics that takes place in a laminated composite under low-velocity impact is shown in Figure 29a; they can be summarized as [28]: (a) front face damage (crater); (b) transverse cracks; (c) delaminations; (d) back face fiber

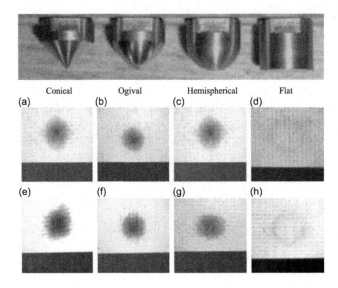

**FIGURE 30**   Effect of impactor size and type on the resulting damage in a laminated composite. *(adapted after Ref.* [29]*).*

breakage. In practical instances, the front face damage may be barely visible. The impactor shape is also important and it may have a significant effect on damage amplitude (Figure 30).

### 5.9.3 Composite Sandwich Damage

The composite sandwich construction consists of rigid thin high-strength composite faces (CFRP or GFRP) that are adhesively bonded on the top and bottom of a relatively thick low-density core (honeycomb or rigid foam) as illustrated in Figure 31a. The CFRP sandwich panels are much more susceptible to impact damage than GFRP sandwich panels, and the predominant type of damage is different: fiber breaking for CFRP sandwiches and core crushing for GFRP sandwich. Reference [30] describes the impact damage types in GFRP composite sandwich structures (e.g., an aircraft radome). Shearography is used to detect barely visible impact damage in a GFRP/foam and GFRP/honeycomb sandwich specimens impacted with 10 J using a drop weight resulting in a barely visible localized surface indentation.

Figure 31b shows the various types of damage that can happen in a composite sandwich under impact, i.e.,

- Front skin damage: (a) skin indentation; (b) skin cracking; (c) resin crushing; (d) fiber breaking; (e) delamination
- Foam core damage: (a) core crushing; (b) core cracking
- Interface damage: debonding at the skin/core interface.

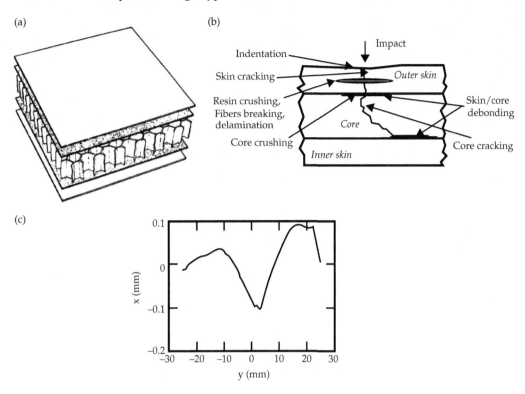

**FIGURE 31** Impact damage on composite/foam sandwich: (a) typical composite sandwich construction; (b) various types of possible sandwich damage; (c) indentation profile after a 10-J impact on a GFRP/foam core sandwich [30].

Figure 31c shows the mapped indentation profile on a GFRP/foam core sandwich after a 10-J impact. The indentation depth and the crater ridge swelling can be as high as 0.1 mm.

### 5.9.3.1 Skin Damage

On the front face (impact side) of the impacted sandwich panel, an indentation occurs. This type of damage, invisible to the naked eye for very low impact energy, becomes barely visible at an energy threshold which depends on the materials used, the structure arrangement, and the attachment of the coupon to the sample holder of the impact machine. For larger energy, the indentation is easily visible and its detection does not need the use of sophisticated nondestructive evaluation (NDE) techniques. It is crucial to be able to detect the BVD phenomenon, because it can be accompanied by a severe internal damage reducing significantly the local mechanical strength. The ability of existing NDE techniques and future SHM methods to detect and characterize such type of damage is a criterion for selecting the more efficient ones. In the present case, the indentation obtained is barely visible, even for a 10-J impact, as shown in Figure 31b. The maximum of the depression of the surface is less than 0.1 mm over an extent of 40 mm. In fact, in radome sandwich structures, due to the GFRP nature of the skin, visual inspection can sometimes detect the damage by a slight change in the color of the surface. The surface alterations, which, by themselves, have no influence on the mechanical strength of the structure, can be accompanied by more critical damage such as matrix cracking, matrix crushing, fiber breaking, and delamination. The occurrence of such types of damage strongly depends on the nature of the composite skin.

### 5.9.3.2 Interface Damage

In parallel with skin damage, the skins/core interfaces can also be damaged; this occurs by debonding. For CFRP skins and foam core, debonds do not seem to appear in spite of the presence of delaminations and fiber breaking of the front face skin. As shown in Figure 31a, debonding of the core/rear-face skin interface may also occur; this would happen at higher level of impact energy.

### 5.9.3.3 Core Damage

In the case of lightweight foam, two types of damage take place: core cracking, in relation with skin cracking, and core crushing located underneath debonds.

## 5.9.4 Damage in Adhesive Composite Joints

Adhesive composite joints offer unquestionable manufacturing and cost advantages over mechanical fasteners. In addition, adhesive joining of composite parts seems more appropriate than drilling holes, as discussed in Section 9. However, the reliability of adhesive bonding cannot always be guaranteed due to a variety of reasons. Hence, delamination damage in composite joints is probably the most frequently occurring damage and the consequences may be quite severe. If the joint is part of a fail-safe construction, then the occurrence of delamination reduces the stiffness of the structure and hence the load-carrying capability of the structure. However, if the joint is not in a fail-safe construction, then the consequences of the joint failure are dramatic. SHM of adhesive composite joints has been considered by various authors; for example, Ref. [31] describes the monitoring of a delamination in a composite T-joint (Figure 32).

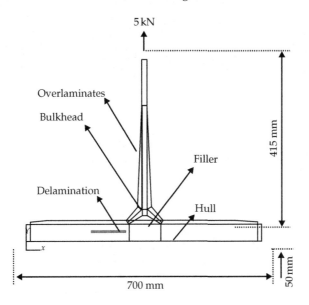

**FIGURE 32**    Schematic of a composite T-joint susceptible to delamination failure [31].

## 5.10 FABRICATION DEFECTS VERSUS IN-SERVICE DAMAGE

Distinction should be made between the defects that may occur during composites manufacturing and the in-service damage that could occur during the composites usage. The inspection and/or monitoring method of choice at the time of manufacture may be different from the inspection and/or monitoring method of choice for in-service assessments.

### 5.10.1 Fabrication Defects

The two major categories of polymer composite used in aerospace are laminates and sandwich structures. Common fabrication defects found in aerospace composites are improper curing, voids and porosity, inclusions, delaminations, unbond areas, disbonds, cracks, fiber-to-resin ratio deviations, and wrinkles (wavy or out-of-plane ply). Sandwich composites may also display core crushing and fluid ingress. *Delaminations* are areas of separation between the layers of a laminate composite or between the faces and the core of a composite sandwich. *Disbonds* are areas in which two adherends have separated at the bondline. In contrast, unbond areas are areas in which the two adherends or prepreg layers failed to ever bond. *Voids and porosity* are produced by entrapment of air bubbles or by gases produced during the chemical reaction of setting the polymeric matrix. *Cracks* are fractures in a composite lamina that typically extend throughout its thickness. In this respect, cracks in a composite lamina are similar to cracks in a metallic plate. In contrast, delaminations are cracks that extend parallel to the lamina, typically between two laminae. *Foreign* inclusions may happen inadvertently during manufacturing, such as forgetting to remove a peel ply during the layup process. *Core crush* is a defect that may happen during the manufacturing of a sandwich composite when the honeycomb core may get crushed due to excessive pressure during cure.

The purpose of inspection and quality assurance (QA) procedures is to detect such manufacturing defects and to take appropriate repair or replace measures. Common nondestructive inspection (NDI) and nondestructive testing (NDT) methods used in aerospace composite manufacturing include [32] visual inspection, tap testing, through-transmission ultrasonics, pulse-echo ultrasonics, bond testers, radiography, thermography, shearography, and electromagnetic inspection. In many cases, the form of the structure to be examined determines the types of features of concern and the NDI techniques that may be applied to detect and measure them.

## 5.10.2 In-service Damage

In-service damage experienced by aerospace composites can be categorized in (a) normal "wear and tear", a.k.a. fatigue and degradation, and (b) accidental damage. In the first category, the most common damage is matrix microcracking (a.k.a. crazing), disbonding, progressive delamination, stiffness loss, and eventual fracture and failure. In the second category, one finds impact damage, impact delaminations, lightning burns, and core degradation of composite sandwich. Note that we also view as damage the progressive in-service degradation of the composite strength due to normal long-term fatigue under operational loads and environmental exposure although this degradation process is more gradual than the sudden occurrence of accidental damage.

*Impact damage* is an internal damage of the composite that may be caused by bird collision in flight, by hale, by collision during docking or other ground maneuvers, or by dropping of a hand tool. Impact damage is typically marked by delaminations, fiber breaking, and matrix cracking. The *bare visible impact damage* (BVID) that may be caused by a tool drop during ground maintenance may lead to serious in-service consequences though it is very hard to detect. *In-service delaminations* inside the composite may occur due to unexpected out-of-plane stresses that were not considered during design. *Disbonding* in composite/composite or composite/metal adhesive joints may occur due to unexpected out-of-plane stresses or to environmental degradation, e.g., in a composite/metal joint. *Lightning burns* appear in an area of the composite that has been subjected to the high temperatures experienced during a lightning discharge resulting in degradation and decomposition of the polymeric matrix. *Core degradation* appears in sandwich composites, e.g., the disbond and/or corrosion of the honeycomb core due to water ingress.

In-service damage may be detected during the routine inspection and maintenance using NDE methods such as visual inspection, tap testing, through-transmission ultrasonics, pulse-echo ultrasonics, bond testers, radiography, thermography, shearography, and electromagnetic testing. However, the application of NDE methods requires taking the aircraft out of service and hence cannot be done often without major disruption of the aircraft normal usage. The possibility exists that composite damage may occur and progress undetected between scheduled NDE inspections; this may have grave safety-critical consequences. Hence, the on-demand or even continuous detection and monitoring of in-service damage without taking the aircraft out of service constitutes one of the major objectives of SHM. However, once the in-service damage has been found with SHM method, the NDE techniques may be called upon to obtain a more precise characterization of its intensity and severity.

## 5.11  WHAT COULD SHM SYSTEMS AIM TO DETECT?

The previous sections have shown that there is a multitude of damage types that could appear in aerospace composites. Then, it seems natural to ask the following question:

- Which are the most important (most critical) damage types that should be on the top of the list when developing new SHM methods/systems?

This is hard question to answer; we could try to develop a tentative answer by exclusion. For example, an SHM system could not substitute the work of production NDI, which should ensure that the aircraft that leaves the factory door is free of any detectable defects. Hence, we will assume that the composite structure starts in pristine condition and builds up internal damage during its in-service operation. Two situations are possible:

(a) Normal operational life under in-service loads without damaging events
(b) Shortened operational life due to BVD events superposed on normal operational loads.

In the former case, the composite will undergo a gracious diminishing of its remaining strength due to progressive damage accumulation produced by the normal effect of in-service operational loads. In this case (Figure 33a), the end of operational life occurs when the remaining strength has diminished to the level of the applied loading.

In the latter case (Figure 33b), the operational life is shortened because BVD events occurred and produced stepwise decreases in the remaining strength which add to the gradual degradation of remaining strength induced by progressive damage accumulation during normal operational loads. If the damage occurs during service, the current strength would be reduced in a step, as shown in Figure 33b; in that case, it is like dropping down to a remaining strength curve for a higher load [33]. Alternatively, if the initial strength is reduced, then the remaining strength curve is moved down initially, driving the intersection point with the life curve to the left (Figure 33c).

Hence, it is apparent that an SHM system/methodology capable of detecting BVD events would be of great assistance and would significantly increase the safety of an aerospace composite structure. A cursory examination of the specialized literature indicates that, indeed, the development of SHM methods and systems that would detect BVD events has captured the attention of a large proportion of the SHM investigators. Besides the detection of BVD events, the detection of the induced BVD size and location as well as subsequent monitoring of its growth has also received considerable attention.

The detection and monitoring of the appearance and growth of the progressive damage due to normal operational loads has also received attention, though to a lesser degree than the detection of BVD events.

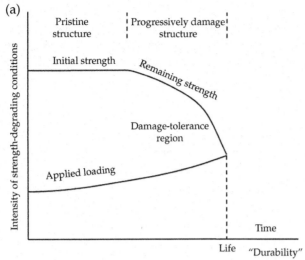

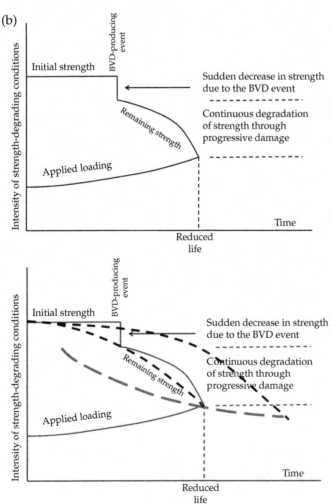

**FIGURE 33** In-service life of an aerospace composite structure: (a) normal operational life with progressive damage growth under in-service loads [10]; (b) shortened operational life due to the superposed effect of a one-time event that only leaves BVD; (c) overlap of (a) and (b) showing the life-shortening effect [33].

Overall, SHM should contribute to lower the maintenance costs of composite structures for which deterministic damage events, types, and limit sizes are difficult to predict. SHM would also facilitate the introduction of new composite materials and development of new composite structures by reducing the uncertainty component of the aircraft design cycle.

## 5.12  SUMMARY AND CONCLUSIONS

This chapter has addressed the topic of damage in composite materials. This has proven to be a much more difficult topic than that of damage in metallic structures, which, by comparison, is relatively well understood. In contrast, damage in composite materials occurs in many more ways than in metals. Composites fail differently under tension than they fail in compression, and the effect of fastener holes is much more complicated than in metals. In addition, the composites are prone to hidden damage from low-velocity impact (e.g., the drop of a hand tool on a wing, or large hail impact on a radome); such damage can be barely visible and may go undetected, but its effect on the degradation of the composite structure strength can be dramatic.

The chapter has started with a review of the composites damage and failure basic mechanisms from stress–strain behavior of the constituents through the failure modes of the unidirectional lamina. Focus was then placed on tension damage: first, the damage and failure of unidirectional composites was analyzed including fracture of fiber, statistical effects and spread, shear-lag load redistribution, and fiber pullout mechanisms. Next, the tension damage and failure of cross-ply composite laminates were considered. The ply discount method that partly explains the progressive failure of a cross-ply laminate was briefly discussed and illustrated. Other aspects that play an important role in the tension failure of cross-ply laminates were discussed next. Interfacial stresses at laminate edges and at cracks and the effect of matrix cracking on the interlaminar stresses were reviewed. The CDS concept was presented and discussed in coordination with the discussion of stiffness evolution with damage accumulation.

The next couple of sections covered fatigue damage in aerospace composites starting with the fatigue of unidirectional plies and then discussing the fatigue of cross-ply laminates. The long-term behavior of composites was described in terms of three major regions, with an initial progression toward widespread CDS taking place in Region I, followed by a period of crack coupling and delamination (Region II) with damage acceleration toward end-of-life final failure taking place in Region III.

Compression fatigue damage in composite, which is fundamentally different from tension fatigue damage, was covered next. Compression fatigue is dominated by fiber microbuckling and local delamination. It was found that the growth of fiber microbuckling damage zone in composites subjected to cyclic compressive loading has a behavior similar to the growth of a tension crack growth in sheet metal, although the underlying mechanisms are radically different. The gradual lateral extension of the microbuckling plus delamination damage zones from a composite hole under compression loading resembles closely the lateral extension of butterfly cracks from a metallic hole under tension.

Other composite damage types were also discussed: fastener hole damage, impact damage of composites, damage of composite sandwich structures, damage in adhesive composite joints, etc.

A discussion of the difference between fabrication defects and in-service damage was performed in order to distinguish between the relative role and scope of NDI and SHM, inasmuch the former is focused more on fabrication defects whereas the latter would deal more with in-service damage that may appear between scheduled NDE inspections. However, once the in-service damage has been found with SHM method, the NDE techniques may be called upon to obtain a more precise characterization of its intensity and severity. Hence, this chapter ends with a discussion of what type of damage could SHM systems aim to detect.

# References

[1] Soutis, C.; Beaumont, P. W. R. (2005) *Multi-scale Modelling of Composite Material Systems: The Art of Predictive Damage Modelling*, Elsevier Woodhead Publishing Ltd, New York, NY.

[2] MIL-STD-1530B, "Standard Practice: Aircraft Structural Integrity Program (ASIP)", United State Air Force Aeronautical Systems Division, WPAFB, Dayton, OH, USA MIL STD 1530B, 2004.

[3] Baker, A.; Dutton, S.; Kelly, D. (2004) *Composite Materials for Aircraft Structures*, AIAA Education Series, AIAA, Reston, VA.

[4] Reifsnider, K. L. (1982) *Damage in Composite Materials: Basic Mechanisms, Accumulation, Tolerance, and Characterization*, ASTM SP 775, ASTM International, Philadelphia, PA.

[5] Talreja, R. (1981) "Fatigue of Composite Materials: Damage Mechanisms and Fatigue-life Diagrams", *Proc. Royal Soc. London* Series A, Vol. 378, No. 1775, 1981 pp. 461-475.

[6] Gdoutos, E. E.; Pilakoutas, K.; Rodopoulos, C. A. (2000) *Failure Analysis of Industrial Composite Materials*, McGraw-Hill, New York, NY, 2000.

[7] Mallick, P. K. (1993) *Fiber-Reinforced Composites*, Marcel Dekker, New York, NY.

[8] Hyer, M. W. (2009) *Stress Analysis of Fiber-Reinforced Composite Materials*, DEStech Pub. Inc., Lancaster, PA.

[9] Rosen, W. B. (1964) "Tensile Failure of Fibrous Composites', *AIAA Journal*, **2**(11), 1985–1991, Nov. 1964.

[10] Reifsnider, K. L.; Case, S. W. (2002) *Damage Tolerance and Durability of Material Systems*, Wiley Interscience, New York, NY.

[11] Reifsnider, K. L.; Schulte, K.; Duke, J. C. (1983) "Long-Term Fatigue Behavior of Composite Materials". in *Long Term Behavior of Composites*, T. K. O'Brien (Ed.), ASTM STP 813, ASTM International, Philadelphia, PA, 1983.

[12] Kashtalyan, M.; Soutis, C. (2005) "Analysis of Composite Laminates with Intra- and Interlaminar Damage", *Progress in Aerospace Sciences*, **41**(2), 152–173, 2005.

[13] Hess, M. S. (1969) "The End Problem for a Laminated Elastic Strip—II. Differential Expansion Stresses", *Journal of Composite Materials*, **3**, 630–641, Oct. 1969.

[14] Bogy, D. B. (1968) "Edge-bonded Dissimilar Orthogonal Elastic Wedges under Normal and Shear Loadind", *Journal of Applied Mechanics*, **35**, 460–466, 1968.

[15] Pipes, R. B.; Pagano, N. J. (1970) "Interlaminar Stresses in Composite Laminates Under Uniform Axial Extension", *Journal of Composite Materials*, **4**, 538–548, Oct. 1970.

[16] Whitney, J. M. (1990) "Lamination Theory and Free Edge Stress Analysis". in *Chapter 5.4 in* Delaware Composites Design Encyclopedia—Vol. 5/Design Studies, K. T. Kedward, J. M. Whitney (Eds.), Technomic Pub. Co., Lancaster, PA, 1990.

[17] Reifsnider, K. L.; Talug, A. (1980) "Analysis of Fatigue Damage in Composite Laminates", *International Journal of Fatigue*3–11, Jan. 1980.

[18] Reifsnider, K. L.; Highsmith, A. L. (1981) "Characteristic Damage States: A New Approach to Representing Fatigue Damage in Composite Laminates,". in *Materials: Experimentation and Design in Fatigue*, F. Sherratt, J. B. Sturgeon (Eds.), Westbury House, Guilford, England, pp. 246–260, 1981.

[19] Talreja, R.; Singh, C. V. (2012) *Damage and Failure of Composite Materials*, Cambridge Univ. Press, New York, NY.

[20] Wang, A. S. D. (1984) "Fracture Mechanics of Sublaminate Cracks in Composite Materials", *Journal of Composites, Technology and Research*, **6**(2), July 1984, http://dx.doi.org/10.1520/CTR10817J.

[21] O'Brien, T. K.; Reifsnider, K. L. (1977) "Fatigue Damage: Stiffness/Strength Comparisons for Composite Materials,", *Journal of Testing and Evaluation*, **5**(5), .384–393, 1977.

[22] Reifsnider, K.L.; Highsmith, A.L. (1981) "The Relationship of Stiffness Changes in Composite Laminates to Fracture-Related Damage Mechanisms", *Proceedings, Second USA-USSR Symposium on Fracture of Composite Materials*, Lehigh University, Bethleham, PA, 9–12 March 1981.

[23] Talreja, R. (1990) "Internal Variable Damage Mechanics of Composite Materials". in *Yielding, Damage, and Failure of Anisotropic Solids*, J. P. Boehler, A. Sawczuk (Eds.), Wiley, London, pp. 509–533.

[24] Reifsnider, K. L.; Henneke, E. G.; Stinchcomb, W. W.; Duke, J. C. (1982) "Damage mechanics and NDE of Composite Laminates". in *Mechanics of Composite Materials; Recent Advances—Proceedings of the 1982 IUTAM Symposium on Mechanics of Composite Materials*, Zvi Hashin, Carl Herakovich (Eds.), Pergamon Press, pp. 399–420, 1983.

[25] Berbinau, P.; Soutis, C.; Guz, I. A. (1999) "Compressive Failure of 0 Unidirectional CFRP Laminates by Fibre Microbuckling", *Composites Science and Technology*, **59**(9), 1451–1455, 1999.

[26] Soutis, C.; Lee, J. (2008) "Scaling Effects in Notched Carbon Fibre/Epoxy Composites Loaded in Compression", *J Mater Sci*, **43**(20), 6593–6598, 2008.

[27] Soutis, C.; Curtis, P. T. (2000) "A Method for Predicting the Fracture Toughness of CFRP Laminates Failing by Fibre Microbuckling", *Composites: Part A*, **31**, 733–740, 2000.

[28] Azouaoui, K.; Azari, Z. (2010) "Evaluation of Impact Fatigue Damage in Glass/Epoxy Composite Laminate", *International Journal of FatigueU*, **32**(2), 443–452, 2010.

[29] Mitrevski, T.; Marshall, I. H.; Thomson, R. S.; Jones, R. (2006) "Low-Velocity Impacts on Preloaded GFRP Specimens with Various Impactor Shapes", *Composite Structures*, **76**, 209–217, 2006.

[30] Balageas, D.; Bourasseau, S.; Dupont, M.; Bocherens, E.; Dewynter-Marty, V.; Ferdinand, P. (2000) "Comparison Between Non-Destructive Evaluation Techniques and Integrated Fiber Optic Health Monitoring Systems for Composite Sandwich Structures", *Journal of Intelligent Material Systems and Structures*, **11**(6), 426–437, 2000, Available from: http://dx.doi.org/10.1106/mfm1-c5ft-6bm4-afud.

[31] Kesavan, A.; John, S.; Herszberg, I. (2008) "Strain-Based Structural Health Monitoring of Complex Composite Structures", *Structural Health Monitoring—An International Journal*, **7**(3), 203–213, Sep. 2008, Available from: http://dx.doi.org/10.1177/1475921708090559.

[32] Bossi, R. H.; Giurgiutiu, V. (2014) "Nondestructive Testing of Damage in Aerospace Composites'. in *Ch. 15 in Polymer Composites in Aerospace Industry*, P. Irving, C. Soutis (Eds.), Elsevier Woodhead Pub., 2014.

[33] Reifsnider, K.L. (2014) Personal Communication, Oct. 2014.

# Piezoelectric Wafer Active Sensors

*Structural Health Monitoring of Aerospace Composites*
DOI: http://dx.doi.org/10.1016/B978-0-12-409605-9.00006-4
     **177**     

## 6.1 INTRODUCTION

Piezoelectric wafer active sensors (PWAS) are inexpensive and easy to use low-profile transducers that have been used extensively in guided-waves structural health monitoring (SHM) of aerospace composites. PWAS transducers are made of thin piezoceramic wafers electrically poled in the thickness direction. Typical PWAS are 7-mm squares or disks with a 0.2-mm thickness (Figure 5). PWAS can be bonded to the structure with strain-gage installation methodology. They have also been experimentally inserted between the layers of a composite layup, but this option has raised some structural integrity issues that are still being examined. When electrically excited with an ultrasonic voltage, the PWAS expand and contract inducing in-plane strain which produces guided waves into a thin-wall structure. Guided waves can travel large distances thus facilitating damage detection over a large area of the structure. Surface-mounted PWAS transducers undergo in-plane extension and expansion. They induce into the structure both symmetric (quasi-axial) and antisymmetric (quasi-flexural) guided waves because their surface placement creates an off-axis effect. The PWAS transducers are called differently by various authors: "piezos", "piezo wafers", "PZTs", etc. In this book, we will consistently call them "PWAS" (same in singular and in plural) or "PWAS transducers".

### 6.1.1 SMART Layer™ and SMART Suitcase™

Fu-Kuo Chang and coworkers were among the first to identify the opportunity for impact detection with a piezo wafers network, as well as using the piezo wafers as both transmitters and receivers of guided Lamb waves for composite damage detection. Reference [1] describes a proposed built-in damage diagnostics system for composite structures aimed at detecting damaging events and monitoring the in-service structural integrity of the composites (Figure 1). The proposed system consisted of two major diagnostic processes: passive sensing diagnosis (PSD) and active sensing diagnosis (ASD). The PSD process utilizes the measurements done by the sensors to identify damage-inducing event. The ASD process uses diagnostic signals sent by the actuators and received by the sensors to diagnose the change in structural integrity. The PSD and ASD concepts discussed in the previous section were captured in the US Patent 6,370,964 [2] which was licensed to Acellent Technologies, Inc. The acronym "SMART", which stands for "Stanford Multi-Actuator Receiver Transduction", was introduced. The appropriate hardware and software was developed and commercialized under the trademarks SMART Layer™ and SMART Suitcase™ [3–8] (Figure 2). The SMART Layer™ is a thin dielectric film with a network of piezoelectric wafer transducers intended for embedding inside composite structures or mounting on the surface of existing structures. The SMART Layer™ fabrication process is based on the flexible printed circuit technique used in the electronics industry, with modifications to accommodate the composite manufacturing process. The major processing steps involve printing and etching a conductor pattern onto a dielectric substrate, laminating a dielectric cover layer for electrical insulation, and mounting arrays of piezoceramic disks on the circuit [5]. The commercially available SMART Layer™ technology [4,5] has facilitated the wide use of this type of transducers in both academic and industrial SHM research.

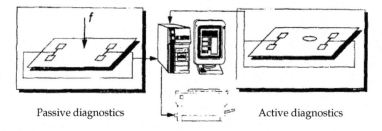

Passive diagnostics                                     Active diagnostics

**FIGURE 1** Damage diagnostics system for composite materials using a network of piezoelectric wafer sensor–actuator transducers [1].

For composite structures, one option is to embed the SMART Layer™ into the structure itself during the manufacturing stage. In this case, the SMART Layer™ is treated as an extra ply that can be placed between composite plies during composite layup process. After co-curing in an autoclave, the resulting composite structure would have a highly integrated network of piezoceramics that can be used to send and receive diagnostic signals for monitoring the structure. Since the SMART Layer™ has temperature tolerance exceeding 400°F, it can be co-cured with a wide range of composite materials. Preliminary test showed that embedding a SMART Layer™ inside composite materials does not significantly alter the composite manufacturing process [5]. A 36-in × 30-in composite panel with an embedded SMART Layer™ is shown in Figure 3.

The SMART Layer™ can also be mounted on the structural surface including both metallic and composite structures. In this case, the SMART Layer™ is bonded onto the structural surfaces. The SMART Layer™ manufactured by Acellent Technologies, Inc. (www.acellent. com) comes with an epoxy film adhesive added onto one side of the layer for bonding to metals. Users simply peel off the backing film and attach the SMART Layer™ onto the structural structure. The epoxy film adhesive can be selected to cure either at elevated temperature or at room temperature depending on the type of application [5].

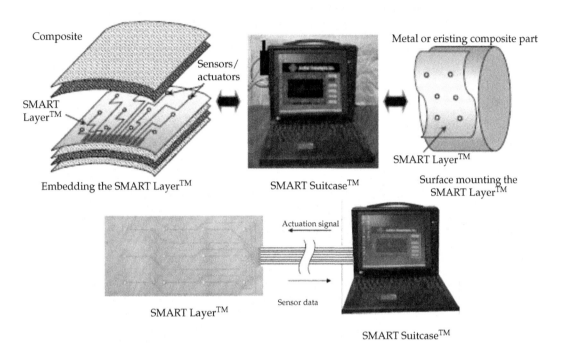

**FIGURE 2**   SMART Layer™ and SMART Suitcase™ products from Acellent Technologies, Inc. [5].

The SMART Suitcase™ is a portable diagnostic instrument that has multiple sensor/actuator (I/O) channels to interface specifically with the SMART Layer™ as illustrated in Figure 2. It has the built-in capability to drive the piezo actuators embedded on the SMART Layer™ and record measurements from adjoining piezo sensors. It can store the sensory data and perform real-time data analysis. The current SMART Suitcase™ model has the capability to interface with up to 30 piezo actuators/sensors. It drives the piezo actuators with a specific preprogrammed diagnostic waveform, while recording the output from neighboring piezo sensors. The SMART Suitcase™ is designed as a PC-based portable instrument that has the built-in capability to generate a specific waveform for structural diagnostics and to collect sensor data with high sampling rate and resolution. It also has multichannel capability to accommodate a network of piezos. A schematic diagram of the SMART Suitcase™ system is shown in Figure 4. They include a diagnostic waveform generator, an actuator power amplifier, a multichannel switching matrix, a sensor signal filter and amplifier board, a sensor data acquisition board, and devices for data storage and processing.

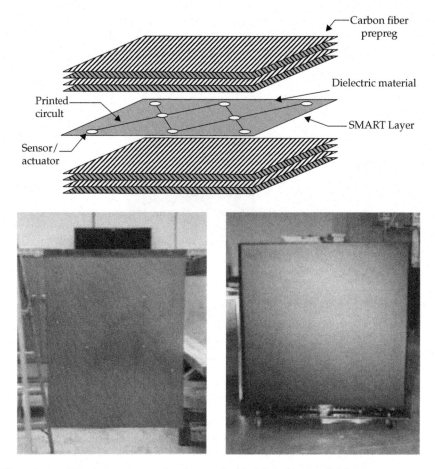

**FIGURE 3**    A 36-in × 30-in composite panel with an embedded SMART Layer™ [3].

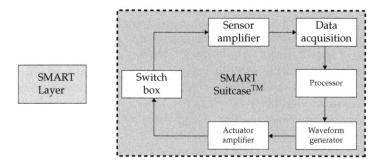

**FIGURE 4**    Schematic of the SMART Suitcase™ hardware system [5].

### 6.1.2  Advantages of PWAS Transducers

PWAS have been extensively used for SHM demonstrations because they convert directly electric energy into elastic energy and vice versa and thus require very simple instrumentation: effective measurements of composite impact waves and guided-waves transmission/reception have been achieved with experimental setups consisting of no more than a signal generator, a digitizing oscilloscope, and a PC [9] (Figure 5).

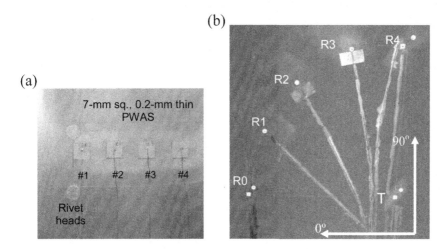

**FIGURE 5**    PWAS installation: (a) 7-mm square PWAS on a metallic structure; (b) 7-mm round and square PWAS on a composite structure.

An extensive description of PWAS transducers principles, methodology, and use in SHM applications is given in Ref. [9]. In this chapter, we will attempt to cover the main highlights; the reader wishing further details is kindly referred to this comprehensive reference [9].

## 6.2  PWAS CONSTRUCTION AND OPERATIONAL PRINCIPLES

PWAS construction and operational principles is covered extensively in chapter 7 of Ref. [9]. PWAS transducers are small, lightweight, and relatively low-cost devices based on the piezoelectric principle that couples the electrical and mechanical variables in the material (mechanical strain, $S_{ij}$, mechanical stress, $T_{kl}$, electrical field, $E_k$, and electrical displacement $D_j$) in the form:

$$S_{ij} = s^E_{ijkl} T_{kl} + d_{kij} E_k$$
$$D_j = d_{jkl} T_{kl} + \varepsilon^T_{jk} E_k$$

(1)

**TABLE 1**  Piezoelectric wafer properties (APC-850)

| Property | Symbol | Value |
|---|---|---|
| Compliance, in-plane | $s_{11}^E$ | $15.30 \cdot 10^{-12}\,\text{Pa}^{-1}$ |
| Compliance, thickness wise | $s_{33}^E$ | $17.30 \cdot 10^{-12}\,\text{Pa}^{-1}$ |
| Dielectric constant | $\varepsilon_{33}^T$ | $\varepsilon_{33}^T = 1750\varepsilon_0$ |
| Thickness-wise induced-strain coefficient | $d_{33}$ | $400 \cdot 10^{-12}\,\text{m/V}$ |
| In-plane induced-strain coefficient | $d_{31}$ | $-175 \cdot 10^{-12}\,\text{m/V}$ |
| Coupling factor, parallel to electric field | $k_{33}$ | 0.72 |
| Coupling factor, transverse to electric field | $k_{31}$ | 0.36 |
| Poisson ratio | $\nu$ | 0.35 |
| Density | $\rho$ | $7700\,\text{kg/m}^3$ |
| Sound speed | $c$ | $2900\,\text{m/s}$ |

Note: $\varepsilon_0 = 8.85 \times 10^{-12}\,\text{F/m}$.

where $s_{ijkl}^E$ is the mechanical compliance of the material measured at zero electric field ($E = 0$), $\varepsilon_{jk}^T$ is the dielectric permittivity measured at zero mechanical stress ($T = 0$), and $d_{kij}$ represents the piezoelectric coupling effect. Typical values of these constants are given in Table 1. The direct piezoelectric effect converts the stress applied to the sensor into electric charge. Similarly, the converse piezoelectric effect produces strain when a voltage is applied to the sensor.

The electrical field $E_3$ is applied across the PWAS thickness in the $x_3$ direction. The prevalent piezoelectric coupling is between in-plane strain in the $x_1$ and $x_2$ directions and out-of-plane electrical field in the $x_3$ direction. At ultrasonic frequencies, PWAS can sense and excite guided Lamb waves traveling long distances along the thin-wall shell structures of aircraft and space vehicles (Figure 6).

When permanently attached into the structure, PWAS provide bidirectional energy transduction from the electronics into the structure and from the structure back into the electronics. As shown in Figure 7, PWAS transducers can serve several purposes [9]: (a) high-bandwidth strain sensors; (b) high-bandwidth wave exciters and receivers; (c) resonators; (d) embedded modal sensors with the electromechanical (E/M) impedance method. By application types, PWAS transducers can be used for (i) **active sensing of far-field damage** using pulse-echo, pitch-catch, and phased-array methods, (ii) **active sensing of near-field damage** using high-frequency E/M impedance method and thickness-gage mode, and (iii) **passive sensing of damage-generating events** through detection of low-velocity impacts and acoustic emission (AE) at the tip of advancing cracks. By using Lamb

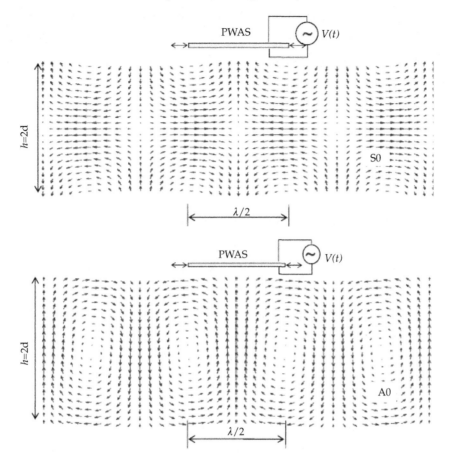

**FIGURE 6** Typical structure of S0 and A0 Lamb waves in a thin-wall structure and their coupling with PWAS transducers: (a) symmetric S0 mode; (b) antisymmetric A0 mode.

waves in a thin-wall structure, one can detect structural anomaly, i.e., cracks, corrosions, delaminations, and other damage. Because of the physical, mechanical, and piezoelectric properties of PWAS transducers, they act as both transmitters and receivers of Lamb waves traveling through the structure. Upon excitation with an electric signal, the PWAS generate Lamb waves in a thin-wall structure. The generated Lamb waves travel through the structure and are reflected or diffracted by the structural boundaries, discontinuities, and damage. The reflected or diffracted waves arrive at the PWAS where they are transformed into electric signals.

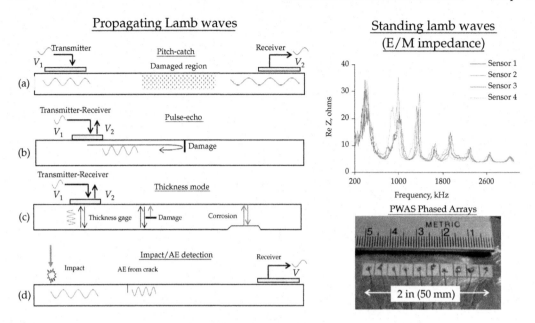

**FIGURE 7**   Use of PWAS transducers for damage detection with propagating and standing guided waves in thin-wall structures: (a) pitch-catch; (b) pulse-echo; (c) thickness mode; (d) impact and acoustic emission (AE) detection; (e) electromechanical (E/M) impedance; (f) PWAS phased array.

## 6.3  COUPLING BETWEEN THE PWAS TRANSDUCER AND THE MONITORED STRUCTURE

PWAS transducers can be used as embedded ultrasonic transducers that act as both exciters and detectors of elastic waves. PWAS couple their in-plane extension and contraction with the in-plane elastic strain of the elastic waves on the structural surface. The in-plane PWAS motion is excited by the applied oscillatory voltage through the $d_{31}$ piezoelectric effect. The PWAS action as ultrasonic transducers is fundamentally different from that of conventional ultrasonic transducers. Conventional ultrasonic transducers activate the structure through surface tapping and apply vibrational pressure to the structural surface. PWAS act through surface pinching and are strain-coupled with the structural surface. Because the PWAS are bonded to the structural surface, the strain transmission between the PWAS and the structure is affected only by the shear-lag effect in the adhesive bonding layer. This type of surface coupling imparts to PWAS a better efficiency and predictive consistency in transmitting and receiving ultrasonic Lamb and Rayleigh waves than conventional ultrasonic transducers that act through gel coupling. Rectangular-shaped PWAS with high length-to-width ratio can generate unidirectional waves. Circular PWAS excite omnidirectional waves that propagate in circular wave fronts. Unidirectional and omnidirectional wave propagations are both illustrated in Figure 8. Omnidirectional waves can also be generated by square PWAS, although their pattern is somehow irregular in the PWAS proximity. At far enough distance $(r \gg a)$, the wave front generated by square PWAS is practically identical with that generated by circular PWAS.

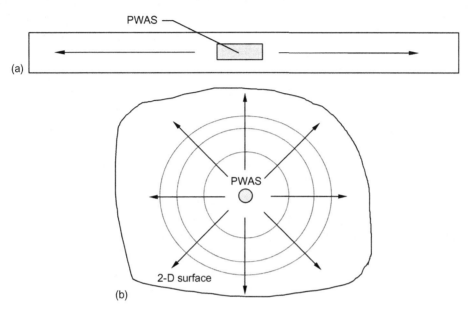

**FIGURE 8**   Elastic waves generated by a PWAS in a structure: (a) unidirectional Lamb waves generated by a PWAS in a 1D structure; (b) circular-crested Lamb waves generated by a PWAS in a 2D structure.

This section will review how surface-mounted PWAS transducers couple with the structure through the bonding adhesive. An extensive treatment of this topic is given in chapter 8 of Ref. [9] to which the reader is referred to for a detailed analysis; here, we will just recall the main highlights. The shear-lag solutions that apply to 1D and circular PWAS will be recalled from chapter 8 of Ref. [9]. It will be shown that these shear-lag solutions can be simplified using the effective pin-force and effective line-force concepts which yield an effective PWAS size that it is usually smaller than the physical PWAS size, thus taking into account the transmission losses that take place in the shear layer. In the extreme when the bonding layer is either very thin or very stiff (or both), the effective pin-force and line-force models will yield, through asymptotic convergence, the "ideal-bonding conditions" in which all the load transfers take place at the PWAS boundary.

## 6.3.1 1D Analysis of PWAS Coupling

Figure 9 shows a thin-wall structure of thickness $t$, unit width $(b = 1m)$ and elastic modulus $E$, having a PWAS of thickness $t_a$ and elastic modulus $E_a$ attached to its upper surface through a bonding layer of thickness $t_b$ and shear modulus $G_b$. The PWAS length is $l_a$ whereas its half-length is $a = l_a/2$. The structural half-thickness is $d = t/2$. Note that the analysis is done per unit width with the motion constraint in the $Oxy$ plane. This assumption may correspond to a strip PWAS of length $l_a$ by infinite width attached to an infinite structure. Plane strain effects are ignored but they can be introduced through the $1 - \nu^2$ correction. It should also be noted that our analysis does not address stress concentration and stress singularity effects that appear at the PWAS ends because we are interested in the overall effects of the coupling between the PWAS and the structure.

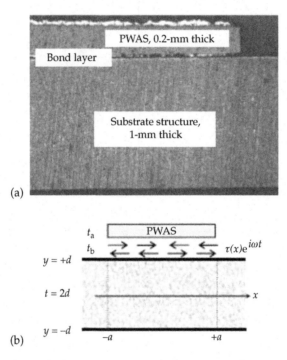

**FIGURE 9** Interaction between the PWAS and the structure: (a) micrograph picture of an actual PWAS installed on a 1-mm plate shown a very thin bond layer; (b) schematic model showing the bonding-layer interfacial shear stress, $\tau(x)$.

Upon application of an electric voltage $V$, the PWAS experiences an induced strain $\varepsilon_{ISA}$ given by $\varepsilon_{ISA} = d_{31}V/t_a$. The induced strain is transmitted to the structure through the bonding-layer interfacial shear stress $\tau$. For harmonic varying excitation, the shear stress has the expression $\tau(x, t) = \tau(x)e^{i\omega t}$.

## 6.3.1.1 Shear-Lag Solution for 1D Coupling

According to chapter 8 of Ref. [9], the 1D shear-lag solution is

$$\varepsilon_a(x) = \frac{\alpha}{\alpha + \psi}\varepsilon_{ISA}\left(1 + \frac{\psi}{\alpha}\frac{\cosh\Gamma x}{\cosh\Gamma a}\right) \qquad \text{(PWAS actuation strain)} \qquad (2)$$

$$\sigma_a(x) = -\frac{\psi}{\alpha + \psi}E_a\varepsilon_{ISA}\left(1 - \frac{\cosh\Gamma x}{\cosh\Gamma a}\right) \qquad \text{(PWAS stress)} \qquad (3)$$

$$u_a(x) = \frac{\alpha}{\alpha + \psi}\varepsilon_{ISA}a\left(\frac{x}{a} + \frac{\psi}{\alpha}\frac{\sinh\Gamma x}{(\Gamma a)\cosh\Gamma a}\right) \qquad \text{(PWAS displacement)} \qquad (4)$$

$$\tau(x) = \frac{t_a}{a}\frac{\psi}{\alpha + \psi}E_a\varepsilon_{ISA}\left(\Gamma a\frac{\sinh\Gamma x}{\cosh\Gamma a}\right) \qquad \text{(interfacial shear stress in bonding layer)} \qquad (5)$$

$$\varepsilon(x) = \frac{\alpha}{\alpha + \psi}\varepsilon_{ISA}\left(1 - \frac{\cosh\Gamma x}{\cosh\Gamma a}\right) \qquad \text{(structure strain at the surface)} \qquad (6)$$

$$\sigma(x) = \frac{\alpha}{\alpha + \psi}E\varepsilon_{ISA}\left(1 - \frac{\cosh\Gamma x}{\cosh\Gamma a}\right) \qquad \text{(structure stress)} \qquad (7)$$

$$u(x) = \frac{\alpha}{\alpha + \psi}\varepsilon_{ISA}a\left(\frac{x}{a} - \frac{\sinh\Gamma x}{(\Gamma a)\cosh\Gamma a}\right) \qquad \text{(structure displacement at the surface)} \qquad (8)$$

where

$$\psi = \frac{Et}{E_a t_a} \qquad \text{(relative stiffness coefficient)} \qquad (9)$$

$$\Gamma^2 = \frac{G_b}{E_a}\frac{1}{t_a t_b}\frac{\alpha + \psi}{\psi} \qquad \text{(shear-lag parameter)} \qquad (10)$$

These equations apply for $x \in [-a, +a]$. Outside the $[-a, +a]$ interval, the strain and stress variables are zero, whereas the displacements maintain a constant value. Note that Eqs. (2), (3) indicate that $\sigma_a \neq E_a\varepsilon_a$. The effect of the PWAS is transmitted to the structure through the interfacial shear stress of the bonding layer. A small shear stress value in the bonding layer produces a gradual transfer of strain from the PWAS to the structure, whereas a large shear stress produces a rapid transfer. Because the PWAS ends are stress free, the buildup of strain takes place at the ends, and it is more rapid when the shear stress is more intense. For large values of $\Gamma a$, the shear transfer process becomes concentrated toward the PWAS ends.

The shear-lag parameter, $\Gamma$, plays a very important role in determining the distribution of $\varepsilon_a$, $\varepsilon$, $\tau$ along the span of the PWAS, i.e., over the range $x \in [-a, +a]$. Figure 10a presents the strain distribution in the structure and PWAS, whereas Figure 10b presents the shear stress distribution for a bond thickness range $t_b = 1$, 10, 100 $\mu m$ (further details of this numerical example are given in section 8.2.2 of Ref. [9]).

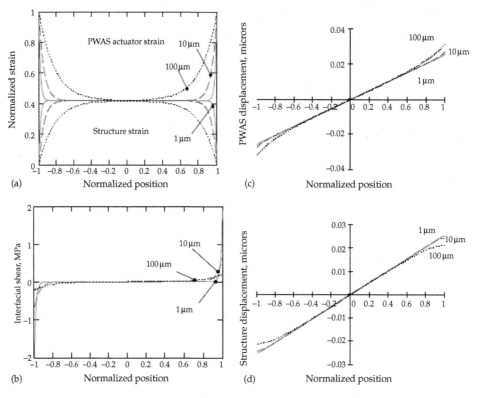

**FIGURE 10**  Variation of shear-lag transfer mechanism with bond thickness: (a) strain distribution in the PWAS and in the structure; (b) interfacial shear stress distribution; (c) displacement distribution in the PWAS; (d) displacement distribution in the structure (bond thickness $t_b$ = 1, 10, 100 $\mu m$).

### 6.3.1.2 Ideal-Bonding Solution: Pin-Force Model

It is apparent from the previous section that a relatively thick bonding layer produces a slow transfer over the entire span of the PWAS (the "100 $\mu m$" curves in Figure 10), whereas a thin bonding layer produces a very rapid transfer (the "1 $\mu m$" curves in Figure 10). The shear-lag analysis indicated that, as the bond thickness decreases, $\Gamma a$ increases. The shear stress transfer becomes concentrated over some infinitesimal distances at the ends of the PWAS actuator. In the limit, as $\Gamma a \to \infty$, all the load transfer can be assumed to take place at the PWAS actuator ends. This leads to the concept of **ideal bonding** (also known as the **pin-force model**), in which all the load transfer is assumed to take place over an infinitesimal region at the PWAS ends, and the induced-strain action is assumed to consist of a pair of concentrated forces applied at the ends (Figure 11). It should also be noted that our analysis does not address stress concentration and stress singularity effects that appear at the PWAS ends because we are interested in the overall effects of the coupling between the PWAS and the structure. This situation, as depicted in

Figure 11, can be described mathematically with the help of the Dirac delta function $\delta(x)$ which satisfies the condition

$$\int_{-\infty}^{+\infty} \delta(x)\,dx = 1 \qquad \text{(Dirac delta function)} \qquad (11)$$

The Dirac function has the localization property

$$\int_{-\infty}^{+\infty} f(x)\delta(x - x_0)dx = f(x_0) \qquad \text{(Dirac function localization property)} \qquad (12)$$

Under these assumptions, Eqs. (2)–(8) take the simple forms

$$\varepsilon_a(x) = \frac{\alpha}{\alpha + \psi}\varepsilon_{ISA}[H(x + a) - H(x - a)] \qquad \text{(PWAS actuation strain)} \qquad (13)$$

$$\sigma_a(x) = -\frac{\psi}{\alpha + \psi}\varepsilon_{ISA}[H(x + a) - H(x - a)] \qquad \text{(stress in the PWAS)} \qquad (14)$$

$$u_a(x) = \frac{\alpha}{\alpha + \psi}\varepsilon_{ISA}x(H(x + a) - H(x - a)) \qquad \text{(displacement in the PWAS)} \qquad (15)$$

$$\tau(x) = \frac{\psi}{\alpha + \psi}t_a E_a \varepsilon_{ISA}(-\delta(x + a) + \delta(x - a)) \qquad \text{(interfacial shear stress in bonding layer)} \qquad (16)$$

$$F(x) = \frac{\psi}{\alpha + \psi}t_a E_a \varepsilon_{ISA}(-H(x + a) + H(x - a)) \qquad \text{(interfacial shear force in bonding layer)} \qquad (17)$$

$$\varepsilon(x) = \frac{\alpha}{\alpha + \psi}\varepsilon_{ISA}[H(x + a) - H(x - a)] \qquad \text{(structure strain at the surface)} \qquad (18)$$

$$\sigma(x) = \frac{\alpha}{\alpha + \psi}E\varepsilon_{ISA}[H(x + a) - H(x - a)] \qquad \text{(structure stress at the surface)} \qquad (19)$$

$$u(x) = \frac{\alpha}{\alpha + \psi}\varepsilon_{ISA}x(H(x + a) - H(x - a)) \qquad \text{(structure displacement at the surface)} \qquad (20)$$

where $H(x)$ and $\delta(x)$ are the Heaviside step function and the Dirac delta function, respectively. Note that, under the ideal-bonding assumption, the strains in the PWAS and at the surface of the structure are equal. However, the stresses are still different. Equations (16) and (17) can be written in the compact form

$$\tau(x) = a\tau_a[-\delta(x + a) + \delta(x - a)] \qquad \text{(shear stress distribution in ideal-bonding)} \qquad (21)$$

$$F(x) = F_a[H(x + a) - H(x - a)] \qquad \text{(force distribution in the pin-force model)} \qquad (22)$$

where

$$\tau_a = \frac{\psi}{\alpha + \psi}\frac{t_a}{a}E_a\varepsilon_{ISA} \qquad \text{(shear stress for ideal-bonding condition)} \qquad (23)$$

$$F_a = a\tau_a \qquad \text{(pin-end forces per unit width)} \qquad (24)$$

Equation (24) represents the pin forces, $F_a = a\tau_a$, applied by the PWAS to the structure. These forces are localized at the PWAS ends (Figure 11a). The pin-force model is convenient for obtaining simple solutions that represent a first-order approximation to the PWAS–structure interaction. Note that this extreme situation implies that the shear stress reaches very large values over diminishing areas at the PWAS ends.

The axial force and bending moment associated with the ideal-bonding assumption (pin-force model) are

$$N_a = F_a \qquad \text{(axial force per unit width)} \tag{25}$$

$$M_a = F_a d = F_a \frac{t}{2} \qquad \text{(bending moment per unit width)} \tag{26}$$

The axial force and bending moment described by Eqs. (25) and (26) represent the excitation induced by the PWAS into the plate under the ideal-bonding hypothesis.

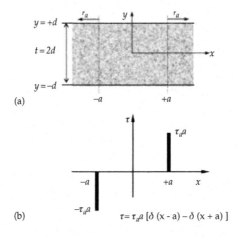

**FIGURE 11**  Pin-force model: (a) surface shear distribution; (b) direct strain induced in the structure at the upper structural surface.

### 6.3.1.3 Effective Pin-Force Model for Nonideal Bonding

The pin-force equations developed under the ideal-bonding conditions can also be used for nonideal bonding conditions through the effective shear stress concept (Figure 12). Consider an effective pin force per unit width $F_e = a\tau_e$ which, by acting at position $a_e$ through a Dirac delta function $\delta(x - a_e)$, would have the same effect as the distributed shear $\tau(x)$, i.e.,

$$F_e \, \delta(x - a_e) \quad \Leftrightarrow \quad \tau(x), \qquad F_e = a\tau_e \quad \text{(effective line force)} \tag{27}$$

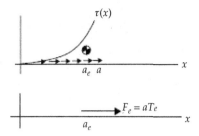

**FIGURE 12**   Effective shear stress concept.

By "same effect" we mean that the integrals of $\tau(x)$ and $x\,\tau(x)$ for the distributed model and the pin-force model should be equal. Thus we can determine the magnitude $F_e = a\tau_e$ and location $a_e$ of the effective shear force. According to section 8.2.4 of Ref. [9], one gets

$$\tau_e = \frac{1}{a}\int_0^\infty \tau(x)dx = C\left(1 - \frac{1}{\cosh\Gamma a}\right), \quad C = \frac{\psi}{\alpha + \psi}\frac{t_a}{a}E_a\varepsilon_{ISA} \quad \text{(effective shear stress)} \quad (28)$$

$$a_e = \frac{\displaystyle\int_0^\infty \tau(x)x\,dx}{\displaystyle\int_0^\infty \tau(x)dx} = a\left(\frac{1 - \dfrac{\sinh\Gamma a}{\Gamma a\cosh\Gamma a}}{1 - \dfrac{1}{\cosh\Gamma a}}\right) \quad \text{(location of the effective shear stress)} \quad (29)$$

*Asymptotic behavior*: As the bonding layer becomes very thin ($t_b \to 0$), or very stiff ($G_b \to \infty$), or both, the shear-lag constant becomes very large ($\Gamma \to \infty$) and the effective values $\tau_e$, $a_e$ tend toward the ideal-bonding values $\tau_a$, $a$. Recall the shear stress for ideal-bonding condition as given by Eq. (23), i.e.,

$$\tau_a = \frac{\psi}{\alpha + \psi}\frac{t_a}{a}E_a\varepsilon_{ISA} \quad \text{(shear stress for ideal-bonding condition)} \quad (30)$$

Applying the limit $\Gamma \to \infty$ to Eqs. (28), (29) yields

$$\lim_{\Gamma\to\infty}\tau_e = C\lim_{\Gamma\to\infty}\left(1 - \frac{1}{\cosh\Gamma a}\right) = C\left(1 - \lim_{\Gamma\to\infty}\frac{1}{\cosh\Gamma a}\right) = C = \frac{\psi}{\alpha + \psi}\frac{t_a}{a}E_a\varepsilon_{ISA} = \tau_a \quad (31)$$

$$\lim_{\Gamma\to\infty}a_e = a\lim_{\Gamma\to\infty}\left(\frac{1 - \dfrac{\sinh\Gamma a}{\Gamma a\cosh\Gamma a}}{1 - \dfrac{1}{\cosh\Gamma a}}\right) = a\left(\frac{1 - \lim_{\Gamma\to\infty}\dfrac{\sinh\Gamma a}{\Gamma a\cosh\Gamma a}}{1 - \lim_{\Gamma\to\infty}\dfrac{1}{\cosh\Gamma a}}\right) = a \quad (32)$$

## 6.3.2 Shear-Layer Analysis for a Circular PWAS

Consider a circular PWAS (Figure 13) of radius $a$ and thickness $t$ undergoing piezoelectric excitation with induced strain $\varepsilon_{ISA} = d_{31}V/t_a$ in all directions. The problem is axisymmetric, i.e., $\partial(\cdot)/\partial\theta = 0$, with null circumferential displacement, i.e., $u_\theta = 0$. Since the radial displacement $u_r$ is the only displacement, we can simplify notations, drop the subscript $r$, and simply write $u$ instead of $u_r$. The strain−displacement relations in polar coordinates reduce to

$$\varepsilon_r = u', \qquad \varepsilon_\theta = u/r \tag{33}$$

Note that although the circumferential displacement $u_\theta$ is zero, i.e., $u_\theta = 0$, the circumferential strain $\varepsilon_\theta$ is nonzero, which implies that the circumferential stress $\sigma_\theta$ is also nonzero.

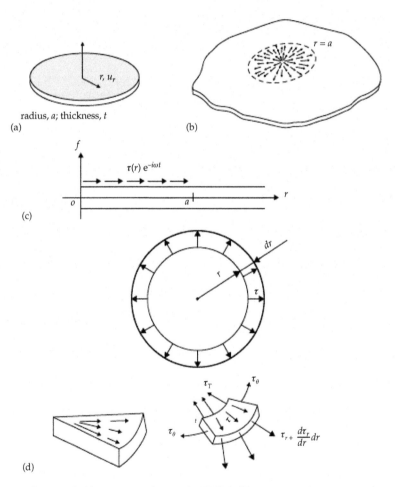

**FIGURE 13**   Circular PWAS: (a) geometry of a circular PWAS; (b) axisymmetric excitation shear stress applied to the plate top surface; (c) radial section showing the surface shear stress; (d) surface shear stress interaction.

### 6.3.2.1 Shear-Lag Solution for Circular PWAS

The shear-lag solution is developed in terms of modified Bessel functions $I_0(x)$, $I_1(x)$, as documented fully in section 8.4.1 of Ref. [9]. The resulting expression of the shear-lag stress in the bonding layer between a circular PWAS and the structural substrate is given as

$$\tau(r) = C\, I_1(\Gamma r), \qquad r \le a \tag{34}$$

where

$$C = \left[ \Gamma I_0(\Gamma a) \left( \frac{1 - \nu_a}{E_a t_a} + \alpha \frac{1 - \nu}{E t} \right) - \left( \frac{(1 - \nu_a)^2}{E_a t_a} + \alpha \frac{(1 - \nu)^2}{E t} \right) \frac{I_1(\Gamma a)}{a} \right]^{-1} \Gamma^2 \varepsilon_{ISA} \tag{35}$$

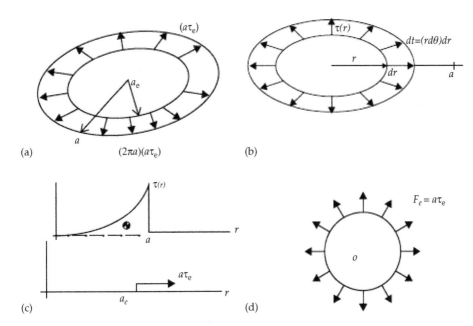

FIGURE 14    Effective shear excitation from a circular PWAS: (a) problem setup; (b) integration setup; (c) effective shear stress concept; (d) ideal-bonding approximation.

### 6.3.2.2 Effective Line-Force Model for a Circular PWAS

The pin-force model used in the 1D shear-lag analysis of Sections 3.1.2, 3.1.3 can be extended to the circular PWAS analysis as a circular line-force model (Figure 14). The effect of the circular line force of intensity $F_e = a\tau_e$ acting over the circumference $2\pi a$ (Figure 14a) should be the same as the integrated effect of the shear stress $\tau(r)$ acting over the whole area of PWAS (Figure 14b). We represent the effective shear stress with the help of the delta function $\delta(r)$ in the form

$$F_e \frac{a}{r} \delta(r - a_e), \quad F_e = a\tau_e \qquad \text{(effective line force)} \tag{36}$$

According to section 8.4.2 of Ref. [9], the effective shear stress $\tau_e$ and it location $a_e$ are given by

$$\tau_e = \frac{C}{a^2}\int_0^a I_1(\Gamma r)r\,dr, \quad a_e = \frac{\displaystyle\int_0^a I_1(\Gamma r)r^2\,dr}{\displaystyle\int_0^a I_1(\Gamma r)r\,dr} \tag{37}$$

*Note*: As indicated in section 8.4.2 of Ref. [9], closed-form solution exists for the integral $\int_0^a I_1(\Gamma r)r^2\,dr$ since $\int_0^a I_1(z)z^2\,dz = z^2 I_0(z) - 2z I_1(z)$. No closed-form solution could yet be found for the integral.

### 6.3.2.3 Ideal-Bonding Solution for a Circular PWAS

As the bonding layer becomes very thin ($t_b \to 0$), or very stiff ($G_b \to \infty$), or both, the shear-lag constant $\Gamma$ becomes very large ($\Gamma \to \infty$) and the effective values $\tau_e$ and $a_e$ tend toward the ideal-bonding values $\tau_a$ and $a$, i.e.,

$$\begin{cases} \tau_e & \xrightarrow[\Gamma \to \infty]{} \quad \tau_a = \dfrac{1}{a^2}\lim_{\Gamma \to \infty} C\displaystyle\int_0^a I_1(\Gamma r)r\,dr \\[2mm] a_e & \xrightarrow[\Gamma \to \infty]{} \quad a \\[2mm] F_e = a\,\tau_e & \xrightarrow[\Gamma \to \infty]{} F_a = a\,\tau_a \\[2mm] \tau(r) & \xrightarrow[\Gamma \to \infty]{} \quad F_a\dfrac{a}{r}\delta(r-a) = a^2\tau_a\dfrac{\delta(r-a)}{r} \end{cases} \quad \text{(ideal-bonding condition)} \tag{38}$$

In virtue of Eq. (38), the ideal-bonding shear stress distribution for a circular PWAS can be expressed as

$$\tau(r) = a^2\tau_a\frac{\delta(r-a)}{r} \quad \text{(ideal-bonding shear stress distribution for a circular PWAS)} \tag{39}$$

It is apparent from Eq. (39) that, under ideal-bonding condition, the load transfer between the circular PWAS and the structural substrate takes through a line force acting on the PWAS rim.

## 6.4 TUNING BETWEEN PWAS TRANSDUCERS AND STRUCTURAL GUIDED WAVES

A remarkable property of PWAS transducers is their ability to tune into the various guided Lamb-wave modes traveling into the structure on which they are attached. This tuning depends on PWAS size and on the excitation frequency. Under electric excitation, the transmitter PWAS transducers undergo oscillatory contractions and expansions, which are transferred to the structure through the bonding layer in order to excite Lamb waves into the structure. In the same time, the receiver PWAS transducers are able to convert acoustic energy of the ultrasonic waves back into an electric signal. Several factors influence the interaction between the PWAS transducers and the guided waves in the structure: PWAS length, excitation frequency, wavelength of the guided wave, thickness of the bonding layer, etc. By varying these parameters in a judicious way, one can enhance the excitation of certain Lamb-wave modes while suppressing other Lamb-wave modes, thus enhancing the SHM process. An extensive treatment of this topic is given in chapter 11 of Ref. [9] to which the reader is referred to for a detailed analysis; here, we will just recall the main highlights.

## 6.4.1 Lamb-Wave Tuning with Linear PWAS Transducers

In the first level of analysis, we consider straight-crested Lamb waves interacting with infinitely wide PWAS of length $l_a = 2a$. The PWAS width is parallel to the wave front whereas its length is parallel to the wave propagation direction. Under these assumptions, the problem is $z$ invariant and the wave propagation is one-dimensional (1D analysis). Recall Figure 6 showing the coupling between PWAS and two Lamb modes, S0 and A0 in a plate of thickness $h = 2d$. It is apparent from Figure 6 that maximum coupling between the PWAS and the Lamb wave would occur when the PWAS length is an odd multiple of the half-wavelength. Since different Lamb modes have different wavelengths, which vary with frequency, the opportunity arises for selectively exciting various Lamb modes at various frequencies, i.e., making the PWAS **tune** into one or another Lamb mode.

### 6.4.1.1 Solution of Lamb-Wave PWAS Tuning with Shear Lag in the Bonding Layer

The analysis presented in section 11.4 of Ref. [9] considers as input the surface shear distribution generated by the PWAS transducer and determines the Lamb-wave response in the structure. In such analysis, the choice of shear distribution is very important. As indicated in Section 3, this shear stress distribution is nonuniform, with high values at the PWAS ends. Hence, we will have to assume a generic shear stress distribution model that may even be frequency dependence, i.e.,

$$\tau(x,t) = \tau(x)e^{-i\omega t} \tag{40}$$

Section 11.4 of Ref. [9] performs the analysis of this problem using the space-domain Fourier transform, followed by solution in the Fourier domain and return to the physical domain through inverse space-domain Fourier transform assisted by the residue theorem. This process yields the strain wave evaluated at the top surface of the plate, i.e.,

$$\varepsilon_x(x,t) = \frac{1}{2\mu}\sum_{m=0}^{M_S} \frac{\tilde{\tau}(\xi_m^S)N_S(\xi_m^S)}{D_S'(\xi_m^S)} e^{i(\xi_j^S x - \omega t)} + \frac{1}{2\mu}\sum_{m=0}^{M_A} \frac{\tilde{\tau}(\xi_m^A)N_A(\xi_m^A)}{D_A'(\xi_m^A)} e^{i(\xi_m^A x - \omega t)} \tag{41}$$

$$u_x(x,t) = \frac{1}{2\mu}\sum_{m=0}^{M_S} \frac{1}{i\xi_m^S} \frac{\tilde{\tau}(\xi_m^S)N_S(\xi_m^S)}{D_S'(\xi_m^S)} e^{i(\xi_m^S x - \omega t)} + \frac{1}{2\mu}\sum_{m=0}^{M_A} \frac{1}{i\xi_m^A} \frac{\tilde{\tau}(\xi_m^A)N_A(\xi_m^A)}{D_A'(\xi_m^A)} e^{i(\xi_m^A x - \omega t)} \tag{42}$$

where $\tilde{\tau}(\xi)$ is the space-domain Fourier transform of $\tau(x)$ and

$$D_S = \left(\xi^2 - \eta_S^2\right)^2 \cos\eta_P d \sin\eta_S d + 4\xi^2 \eta_P \eta_S \sin\eta_P d \cos\eta_S d \tag{43}$$

$$D_A = \left(\xi^2 - \eta_S^2\right)^2 \sin\eta_P d \cos\eta_S d + 4\xi^2 \eta_P \eta_S \cos\eta_P d \sin\eta_S d \tag{44}$$

$$\begin{aligned} N_S &= \xi\eta_S\left(\xi^2 + \eta_S^2\right)\cos\eta_P d \cos\eta_S d \\ N_A &= -\xi\eta_S\left(\xi^2 + \eta_S^2\right)\sin\eta_P d \sin\eta_S d \end{aligned} \tag{45}$$

$$\eta_P^2 = \frac{\omega^2}{c_P^2} - \xi^2, \quad \eta_S^2 = \frac{\omega^2}{c_S^2} - \xi^2 \tag{46}$$

In Eq. (46), $\xi$ is the wavenumber in the $x$-direction, $c_P^2 = (\lambda + 2\mu)/\rho$ and $c_S^2 = \mu/\rho$ are the longitudinal (pressure) and transverse (shear) wavespeeds, $\lambda$ and $\mu$ are Lamé constants, and $\rho$ is the mass density of the plate. Superscripts S and A signify the symmetric and antisymmetric Lamb-wave modes. At a given value of $\omega$ in a given plate, there are $m = 0, 1, ..., M_S$ symmetric Lamb-wave modes and $m = 0, 1, ..., M_A$ antisymmetric Lamb-wave modes. The expressions $D'_S$ and $D'_A$ are derivatives of $D_S$ and $D_A$ with respect to $\xi$ evaluated at the corresponding $\xi_m$ poles. Note that $D_S$ and $D_A$ are the left-hand sides of the Rayleigh–Lamb equation for symmetric and antisymmetric Lamb-wave modes, respectively, as given by Eqs. (6.101), (6.115) of Ref. [9].

### 6.4.1.2 Lamb-Wave PWAS Tuning at Low Frequencies

At low frequencies ($fd \to 0$), only two propagating Lamb-wave modes exist, S0 and A0 (one evanescent mode, A1, also exists). In this case, the general solution of Eq. (41) has only two propagating terms, i.e.,

$$\begin{aligned} \varepsilon_x(x, t) &= \frac{1}{2\mu} \frac{\tilde{\tau}(\xi_0^S) N_S(\xi_0^S)}{D'_S(\xi_0^S)} e^{i(\xi_0^S x - \omega t)} \\[2mm] &\quad + \frac{1}{2\mu} \frac{\tilde{\tau}(\xi_0^A) N_A(\xi_0^A)}{D'_A(\xi_0^A)} e^{i(\xi_0^A x - \omega t)} \end{aligned} \qquad \text{(low frequency)} \tag{47}$$

$$\begin{aligned} u_x(x, t) &= -\frac{a^2 \tau_a}{\mu} \frac{\sin\xi_0^S a}{\xi_0^S a} \frac{N_S(\xi_0^S)}{D'_S(\xi_0^S)} e^{i(\xi_0^S x - \omega t)} \\[2mm] &\quad - \frac{a^2 \tau_a}{\mu} \frac{\sin\xi_0^A a}{\xi_0^A a} \frac{N_A(\xi_0^A)}{D'_A(\xi_0^A)} e^{i(\xi_0^A x - \omega t)} \end{aligned} \qquad \text{(low frequency)} \tag{48}$$

As the frequency-thickness product $fd$ approaches zero ($fd \to 0$), the behavior of the S0 and A0 modes approaches the behavior of the axial and flexural waves and we find that

$$\frac{N_A(\xi_0^A)}{D'_A(\xi_0^A)} \bigg/ \frac{N_S(\xi_0^S)}{D'_S(\xi_0^S)} \xrightarrow[fd \to 0]{} \frac{3}{2} \tag{49}$$

Indeed, when we plot the expressions contained in Eq. (49), we find (Figure 15) that the ratio depicted in Eq. (49) approaches indeed the limit $3/2$ as $fd \to 0$ since it takes the values 1.491 at $fd = 1.0$ kHz-mm and 1.5001 at $fd = 0.5$ kHz-mm. This indicates that, at low frequencies when the Lamb waves can be approximated by axial and flexural waves, the flexural solution is stronger by a factor of $3/2$ than the axial solution.

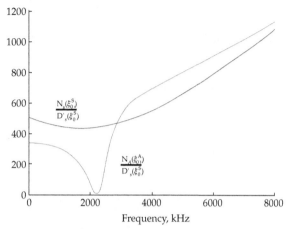

**FIGURE 15**    Behavior of the functions $\dfrac{N_S(\xi_0^S)}{D_S'(\xi_0^S)}$ and $\dfrac{N_A(\xi_0^A)}{D_A'(\xi_0^A)}$ in a 1-mm ($d = 0.5$ mm) 2024 aluminum plate.

### 6.4.1.3 Ideal-Bonding Solution for Lamb-Wave PWAS Tuning

In the case of ideal bonding, the shear stress in the bonding layer is concentrated at the ends and one can use the pin-force model of Eq. (21)

$$\tau(x) = a\tau_a[\delta(x - a) - \delta(x + a)] \tag{50}$$

The Fourier transform of Eq. (50) is

$$\tilde{\tau} = a\tau_a[-2i\sin\xi a] \tag{51}$$

Substitution of Eq. (51) into Eqs. (41), (42) yields

$$\varepsilon_x(x,t) = -i\frac{a\tau_a}{\mu}\sum_{m=0}^{M_S}\left(\sin\xi_m^S a\right)\frac{N_S(\xi_m^S)}{D_S'(\xi_m^S)}e^{i(\xi_m^S x - \omega t)} - i\frac{a\tau_a}{\mu}\sum_{m=0}^{M_A}\left(\sin\xi_m^A a\right)\frac{N_A(\xi_m^A)}{D_A'(\xi_m^A)}e^{i(\xi_m^A x - \omega t)} \tag{52}$$

$$u_x(x,t) = -\frac{a^2\tau_a}{\mu}\sum_{m=0}^{M_S}\frac{\sin\xi_m^S a}{\xi_m^S a}\frac{N_S(\xi_m^S)}{D_S'(\xi_m^S)}e^{i(\xi_m^S x - \omega t)} - \frac{a^2\tau_a}{\mu}\sum_{m=0}^{M_A}\frac{\sin\xi_m^A a}{\xi_m^A a}\frac{N_A(\xi_m^A)}{D_A'(\xi_m^A)}e^{i(\xi_m^A x - \omega t)} \tag{53}$$

Equation (52) can also be used for less-than-ideal bonding conditions in which the shear stress $\tau$ varies with $x$ as $\tau(x)$ by replacing $\tau_a$ and $a$ with their effective values $\tau_e$ and $a_e$.

At low frequencies, only two propagating Lamb-wave modes exist, S0 and A0, and the general solution Eqs. (52), (53) has only two terms, i.e.,

$$\varepsilon_x(x,t) = -i\frac{a\tau_a}{\mu}\left(\sin\xi_0^S a\right)\frac{N_S(\xi_0^S)}{D'_S(\xi_0^S)}e^{i(\xi_0^S x-\omega t)}$$

$$\qquad\qquad\qquad\qquad\qquad\qquad\qquad\qquad\text{(low frequency)}\qquad\qquad(54)$$

$$-i\frac{a\tau_a}{\mu}\left(\sin\xi_0^A a\right)\frac{N_A(\xi_0^A)}{D'_A(\xi_0^A)}e^{i(\xi_0^A x-\omega t)}$$

$$u_x(x,t) = -\frac{a^2\tau_a}{\mu}\frac{\sin\xi_0^S a}{\xi_0^S a}\frac{N_S(\xi_0^S)}{D'_S(\xi_0^S)}e^{i(\xi_0^S x-\omega t)}$$

$$\qquad\qquad\qquad\qquad\qquad\qquad\qquad\qquad\text{(low frequency)}\qquad\qquad(55)$$

$$-\frac{a^2\tau_a}{\mu}\frac{\sin\xi_0^A a}{\xi_0^A a}\frac{N_A(\xi_0^A)}{D'_A(\xi_0^A)}e^{i(\xi_0^A x-\omega t)}$$

Equations (52)−(55) contain the tuning function

$$F_\varepsilon(\xi a) = \sin\xi a \qquad \text{(tuning function for strain)}\qquad\qquad(56)$$

$$F_u(\xi a) = \frac{\sin\xi a}{\xi a} \qquad \text{(tuning function for displacement)}\qquad\qquad(57)$$

The tuning functions depend on frequency $f$ since $\xi(\omega) = \omega/c(\omega)$ and $\omega = 2\pi f$. The behavior of the tuning function of Eq. (56) is such that it displays maxima when the PWAS length $l_a = 2a$ equals an odd multiple of the half-wavelength $\lambda/2 = \pi/\xi$ and minima when it equals an even multiple of the half-wavelength.

A complex pattern of such maxima and minima may evolve when several Lamb modes coexist, each with its own different wavelength. However, frequencies can be found when the response is dominated by certain modes that can be preferentially excited through **mode tuning**. An additional factor must be considered besides wavelength tuning, i.e., the relative mode amplitude at the top plate surface. This factor is contained in the values taken by the functions $N_S/D'_S$, $N_A/D'_A$. Hence, it is conceivable that some modes may have little surface amplitude, while others may have larger surface amplitudes at a given frequency. For illustration, Figure 16 shows a plot of the individual tuning behavior of S0 and A0 modes in the 0−1000kHz range.

### 6.4.1.4 Experimental Verification of the Lamb-Wave PWAS-Tuning Phenomenon

The Lamb-wave PWAS-tuning phenomenon has been extensively verified through experimental testing as reported in Ref. [9] and elsewhere. One of the earlier experiments identified a "sweet spot" at which the dispersive A0 mode could be rejected and the much less dispersive S0 mode would remain the only Lamb-wave mode present in the structure (Figure 17a). Such a situation is very advantageous in damage detection with pulse-echo method because the wave packet of the S0 mode is much more compact than the dispersed package of the A0 mode and hence easier to interpret, as illustrated in Section 5 dealing with wave propagation SHM and Section 6 dealing with PWAS phased array.

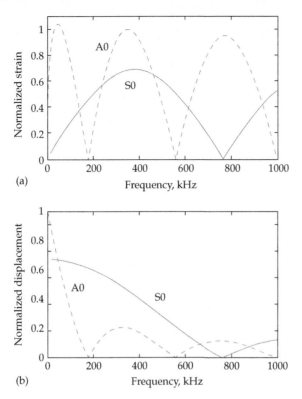

**FIGURE 16**    Predicted Lamb-wave response on the top surface of a 1-mm aluminum plate under a 7-mm PWAS excitation.

The tuning principles have been found experimentally to also apply to composite structures (Figure 17b) although the expressions in Eq. (52) should be extended to also include quasi-SH modes. Figure 17b shows tuning between a PWAS and the guided waves present in a CFRP composite: quasi-A0, quasi-S0, quasi-SH0 waves. It is remarkable that the quasi-A0 is very strong at low frequencies below 100 kHz, but disappears almost entirely at frequencies beyond 200 kHz, where only the quasi-S0 and quasi-SH0 seem to be present. The quasi-SH0 guided wave seems to have a tuning maximum around 350 kHz, whereas the quasi-S0 guided wave has a persistently strong response between 250 and 500 kHz.

## 6.4.2  Lamb-Wave Tuning with Circular PWAS

The Lamb-wave PWAS-tuning principles also apply to circular PWAS, only that the analysis is more elaborate and involves Bessel functions, Hankel functions, and the Hankel transform.

(a)

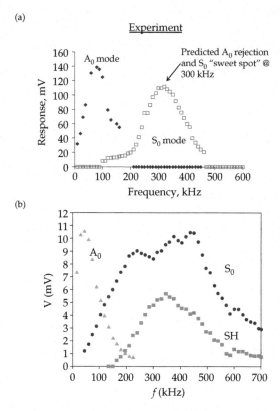

(b)

FIGURE 17   Experimental frequency-tuning results: (a) maximum S0 response at around 300 kHz due to A0 rejection in a 1-mm aluminum plate [9]; (b) similar rejection of A0 mode in a CFRP composite plate [10].

### 6.4.2.1  General Solution for Circular Lamb-Wave Tuning

An extensive treatment of this topic is given in section 11.5 of Ref. [9] to which the reader is referred to for a detailed analysis. The result of this analysis is that the Lamb-wave response to a circular PWAS excitation takes the form

$$u_r(r)\big|_{z=d} = -\frac{\pi i}{2\mu}\sum_{m=0}^{M_S}\frac{\tilde{\tau}(\xi_m^S)N_S(\xi_j^S)}{D_S'(\xi_m^S)}H_1^{(1)}(\xi_m^S r)e^{-i\omega t} - \frac{\pi i}{2\mu}\sum_{m=0}^{M_A}\frac{\tilde{\tau}(\xi_m^A)N_A(\xi_m^A)}{D_A'(\xi_m^A)}H_1^{(1)}(\xi_m^A r)e^{-i\omega t} \qquad (58)$$

where $H_1^{(1)}$ is the Hankel function of the first kind and order 1. The corresponding radial strain on the top surface of the plate can be derived through the differentiation of Eq. (58), i.e.,

$$\varepsilon_r(r,t)\big|_{z=d} = \frac{\partial u_r(r,t)\big|_{z=d}}{\partial r} \qquad (59)$$

Recall the Hankel functions differentiation formula

$$\frac{d}{dr}H_1^{(1)}(\xi r) = \xi H_0^{(1)}(\xi r) - \frac{H_1^{(1)}(\xi r)}{r} \qquad (60)$$

In view of Eq. (60), it is apparent that the radial strain of Eq. (59) will have components of both $H_0^{(1)}$ and $H_1^{(1)}$, i.e.,

$$
\begin{aligned}
\varepsilon_r(r)\big|_{z=d} = & -\frac{\pi i}{2\mu} \sum_{m=0}^{M_S} \frac{\tilde{\tau}(\xi_m^S) N_S(\xi_m^S)}{D_S'(\xi_m^S)} \left( \xi_m^S H_0^{(1)}(\xi_m^S r) - \frac{H_1^{(1)}(\xi_m^S r)}{r} \right) e^{-i\omega t} \\
& -\frac{\pi i}{2\mu} \sum_{j=0}^{M_A} \frac{\tilde{\tau}(\xi_m^A) N_A(\xi_m^A)}{D_A'(\xi_m^A)} \left( \xi_m^A H_0^{(1)}(\xi_m^A r) - \frac{H_1^{(1)}(\xi_m^A r)}{r} \right) e^{-i\omega t}
\end{aligned}
\tag{61}
$$

### 6.4.2.2 Ideal-Bonding Solution for Circular Lamb-Waves Tuning

In the case of ideal bonding of a circular PWAS that expands and contracts radially, the shear transfer in the bonding layer is concentrated on the PWAS outer contour in the form of a radially acting horizontal line force as discussed in Section 3.2.3. The radial shear stress applied to the plate top surface is given by Eq. (39), i.e.,

$$
\tau(r) = a^2 \tau_a \frac{\delta(r-a)}{r}
\tag{62}
$$

The $J_1$ Hankel transform of Eq. (62) is

$$
\tilde{\tau}(\xi)_{J_1} = \int_0^\infty r\left[ \tau_a a^2 \frac{1}{r} \delta(r-a) \right] J_1(\xi r)dr = \tau_a a^2 \int_0^\infty \delta(r-a) J_1(\xi r)dr = \tau_a a^2 J_1(\xi a)
\tag{63}
$$

Substitution of Eq. (63) into Eq. (58) gives the radial displacement on the top surface of the plate in the case of an ideally bonded circular PWAS, i.e.,

$$
\begin{aligned}
u_r(r)\big|_{z=d} = & -\pi i \frac{a^2 \tau_a}{2\mu} \sum_{m=0}^{M_S} \frac{J_1(\xi_m^S a) N_S(\xi_m^S)}{D_S'(\xi_m^S)} H_1^{(1)}(\xi_m^S r) e^{-i\omega t} \\
& -\pi i \frac{a^2 \tau_a}{2\mu} \sum_{m=0}^{M_A} \frac{J_1(\xi_m^A a) N_A(\xi_m^A)}{D_A'(\xi_m^A)} H_1^{(1)}(\xi_m^A r) e^{-i\omega t}
\end{aligned}
\tag{64}
$$

where $J_1$ is the Bessel function of the first kind and order 1, and $H_1^{(1)}$ is the Hankel function of the first kind and order 1. The summation is taken over all the symmetric, $\xi^S$, and anti-symmetric, $\xi^A$, Lamb-wave modes that exist at a given frequency $\omega$ in a given plate. The expressions $D_S'$ and $D_A'$ are derivatives of $D_S$ and $D_A$ with respect to $\xi$ evaluated at the corresponding $\xi^S$ and $\xi^A$ poles.

The corresponding radial strain on the top surface of the plate can be derived from the differentiation of Eq. (64) using Eq. (60) or by substitution of Eq. (63) into Eq. (61); by either route, one gets

$$
\begin{aligned}
\varepsilon_r(r)\big|_{z=d} = & -\pi i \frac{a^2 \tau_a}{2\mu} e^{-i\omega t} \sum_{m=0}^{M_S} \frac{J_1(\xi_m^S a) N_S(\xi_m^S)}{D_S'(\xi_m^S)} \left( \xi_m^S H_0^{(1)}(\xi_m^S r) - \frac{H_1^{(1)}(\xi_m^S r)}{r} \right) \\
& -\pi i \frac{a^2 \tau_a}{2\mu} e^{-i\omega t} \sum_{m=0}^{M_A} \frac{J_1(\xi_m^A a) N_A(\xi_m^A)}{D_A'(\xi_m^A)} \left( \xi_m^A H_0^{(1)}(\xi_m^A r) - \frac{H_1^{(1)}(\xi_m^A r)}{r} \right)
\end{aligned}
\tag{65}
$$

The expressions in Eqs. (64), (65) can also be used for less-than-ideal bonding conditions in which the shear stress $\tau$ varies with $r$ as $\tau(r)$ by replacing $\tau_a$ and $a$ with their effective values $\tau_e$ and $a_e$.

## 6.5 WAVE PROPAGATION SHM WITH PWAS TRANSDUCERS

Figure 7a illustrates the **pitch-catch method**. An electric signal applied at the transmitter PWAS generates, through piezoelectric transduction, elastic waves that travel into the structure and are captured at the receiver PWAS. As long as the structural region between the transmitter and receiver is in pristine condition, the received signal will be consistently the same; if the structure becomes damaged, then the received signal will be modified. Comparison between the historically stored signals and the currently read signal will indicate when changes (e.g., damage) take place in the structure. The pitch-catch method may be applied to situations in which the damage is diffuse and/or distributed such as corrosion in metals or degradation in composites. By extension of the pitch-catch method to several pitch-catch pairs in a network of PWAS ("sparse array") placed around a structural region of interest, one achieves **ultrasonic tomography** through a round-robin process. The processing of all the collected data during the round-robin process yields an image of the monitored region indicating the damage area.

Figure 7b illustrates the **pulse-echo** method. In this case, the same PWAS transducer acts as both transmitter and receiver. A tone-burst signal applied to the PWAS generates an elastic wave packet that travels through the structure and reflects at structural boundaries and at cracks and abrupt discontinuities. In a pristine structure, only boundary reflections are present whereas in a damaged structure, reflections from cracks also appear. By comparing historical signals, one can identify when new reflections appear due to the presence of damage. This comparison may be facilitated by the differential signal method.

Figure 7c illustrates the use of PWAS transducers in **thickness mode**. The thickness mode is usually excited at much higher frequencies than the guided-wave modes discussed in the previous two paragraphs. For example, the thickness mode for a 0.2-mm PWAS is excited at around 12 MHz, whereas the guided-wave modes are excited at tens and hundreds of kHz. When operating in thickness mode, the PWAS transducer can act as a thickness gage. In metallic structures, thickness mode measurements allow the detection of damage that affects the structural thickness, e.g., corrosion, which can be detected from that side of the structure, which is outside from the corrosive environment. In composite structures, thickness mode measurements may detect cracks which are parallel to the surface, such as delaminations. A limitation of the thickness mode approach is that detection can only be made directly under the PWAS location, or in its proximity. In this respect, this method is rather localized, which may be alright for monitoring well-defined critical areas, but insufficient for large area monitoring.

Figure 7d illustrates the detection of **impacts and AE events**. In this case, the PWAS transducer is operating as a passive receiver of the elastic waves generated by impacts or by AE events. By placing several PWAS transducers in a network configuration around a given structural area, one can set up a "listening" system that would monitor if impact damage or AE events take place. Because the PWAS is self-energized through piezoelectric transduction, the listening system can stay in a low-energy dormant mode until a triggering by the PWAS wakes it up. The signals recorded by the PWAS network can be processed to yield the location and amplitude of the impact and/or AE event.

An extensive treatment of this topic is given in chapter 13 of Ref. [9] to which the reader is referred to for a detailed analysis; here, we will just recall the main highlights.

## 6.5.1 Pitch-Catch Guided-Wave Propagation SHM

Cracks in metallic structures typically run perpendicular to wall surface. A fully developed crack will cover the whole thickness (through-thickness crack) and will produce a tear of the metallic material. In conventional nondestructive evaluation (NDE), metallic structure cracks are detected with ultrasonic or eddy current probes that have pointwise capabilities. Intensive manual scanning is required for crack detection. The aim of embedded pitch-catch NDE is to detect cracks in metallic structures using guided waves transmitted from one location and received at a different location. The analysis of the change in guided-wave shape, phase, and amplitude should yield indications about the crack presence and extension.

Reference [11] presents an embedded pitch-catch method for the detection of cracks in metallic structures using an array of 12-mm diameter piezoceramic disks (Figure 18). A 5-count 300-kHz smoothed tone-burst applied to the transmitter (T) produced an omnidirectional Lamb wave into the plate. The receiver (R) detects a wave that is modified by the crack presence. The crack was grown through cyclic loading; measurements were taken at various crack length. The readings at zero crack length were taken as the baseline. The scatter wave was defined as the difference between the current received wave and the baseline wave. It was shown that the scatter amplitude increases linearly with the crack length.

The pitch-catch guided-wave propagation method for damage detection is also known as acousto-ultrasonics because, as indicated in Ref. [12], the method combines an excitation similar to that used in ultrasonics with an analysis of the detected signal similar to that used in acoustic emission monitoring.

## 6.5.2 Pulse-Echo Guided-Wave Propagation SHM

Pulse-echo wave propagation SHM can be illustrated by considering a 914-mm × 14-mm × 1.6-mm narrow-strip specimen with an 8-mm transverse crack. Note that the crack covers only approximately 57% of the specimen width. The wave propagation in the specimen was modeled with the finite element method (FEM). Two forms of the elastic wave propagation were studied: flexural waves and axial waves. To attain wave excitation, we applied prescribed harmonic displacements to the nodes delimiting the contour of the PWAS. Consistent with the physical phenomenon, the displacement applied to nodes representing opposite ends of the PWAS had to be in opposite direction. This ensures that the net effect on the structure is self-equilibrating. To generate

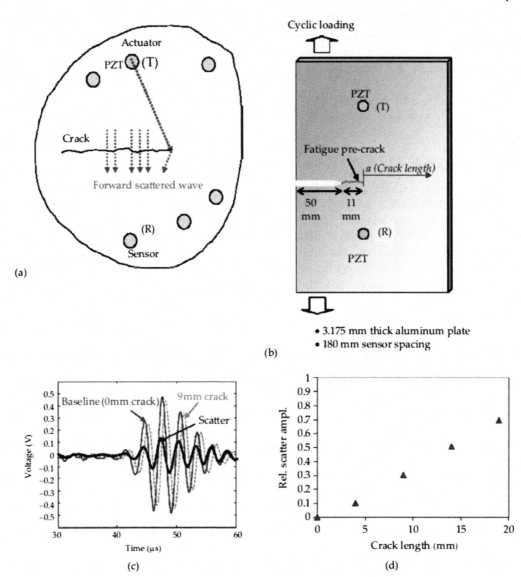

**FIGURE 18**   Crack detection in metallic plate with the embedded pitch-catch method: (a) conceptual configuration; (b) experimental setup; (c) received waves and scatter wave; (d) linear variation of scatter amplitude with crack length [11].

axial waves, we applied nodal translations. To generate flexural waves, we applied nodal rotations. The detection of the elastic waves followed the same general principle as that applied to wave generation. The variables of interest were the differences between the displacements at the opposing ends of the PWAS, i.e., the $\Delta u$ for axial waves and $\Delta w'$ for flexural waves.

The applied loads were varied in the time domain according to a Hanning-windowed tone 5-count burst. Time marching FEM solution produced the time response of the strip specimen, allowing us to follow the wave propagation patterns.

### 6.5.2.1 Simulation of Axial Waves

Figure 19 shows FEM simulation of axial waves in the strip specimen excited with 100-kHz 5-count Hanning-windowed axial tone-burst applied at the left-hand side end. The patterns of dilatation and contraction (in-plane motion) are shown. The wave was captured after traveling for 50 μs. Figure 19a gives an overall view, whereas Figure 19b gives a magnified detail of the first quarter of the strip specimen. One notes that the number of peaks in the wave is greater than the burst count of 5 because both the incident wave and the wave reflected from the left-hand side end of the strip specimen are superposed in this wave front.

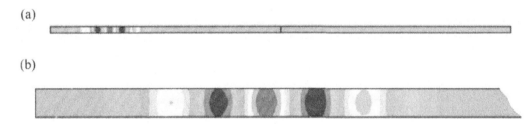

(a)

(b)

**FIGURE 19**    FEM simulation of axial waves in the 914-mm × 14-mm × 1.6-mm aluminum alloy strip specimen excited with 100-kHz 5-count Hanning-windowed axial burst at left-hand side end. The wave is captured after traveling for 50 μs: (a) overall view; (b) details of first quarter of the strip specimen. The number of peaks in the wave is greater than five because both the incident wave and the wave reflected from the left-hand side end of the strip specimen are superposed in this wave front.

Figure 20 shows FEM simulation of the pulse-echo method for damage detection. The process described above was applied to (a) a pristine specimen and (b) to a specimen with an 8-mm long through-thickness transverse crack is simulated in the center of the strip specimen. The PWAS placed at the left-hand side end was used to send a 100-kHz 5-count Hanning-windowed axial burst and to receive the elastic wave responses from the strip specimen. In a strip specimen without crack, only the initial excitation signal and the reflection from the right-hand side end of the specimen appear (Figure 20a). In a specimen with the 8-mm long through-thickness transverse crack at mid-length, reflection (echo) from this crack also appears (Figure 20b). As the crack length increases, the amplitude of the reflection increases (Figure 20c).

### 6.5.2.2 Simulation of Flexural Waves

Figure 21 shows FEM simulation of flexural waves in the strip specimen excited with a 100-kHz 5-count Hanning-windowed flexural tone-burst at the left-hand side end. The peaks and valleys of flexural wave propagation are apparent. The wave was captured after traveling for 99.3 μs. Figure 21a gives an overall view, whereas Figure 21b gives a

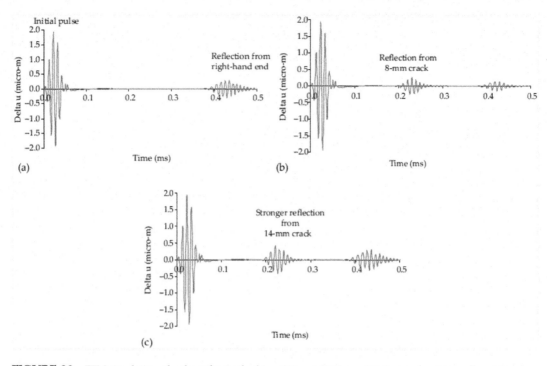

**FIGURE 20**   FEM simulation of pulse-echo method in a 914-mm × 14-mm × 1.6-mm aluminum alloy strip specimen using axial waves. A 100-kHz 5-count Hanning-windowed axial burst was applied at the left-hand side end: (a) strip specimen without crack shows only the reflection from the right-hand side end; (b) strip specimen with 8-mm long through-thickness transverse crack shows, in addition, the reflection from crack; (c) a longer crack (14-mm) gives a stronger reflection.

magnified detail of the first quarter of the strip specimen. One notes that the number of peaks in the wave is greater than the burst count of 5 because of dispersion effects and of the fact that both the incident wave and the wave reflected from the left-hand side end of the strip specimen are superposed in this wave front.

Figure 22 shows FEM simulation of the pulse-echo method used for damage detection. The PWAS placed at the left-hand side end was used to send a 100-kHz 5-count Hanning-windowed flexural burst and to receive the elastic wave responses. In a strip specimen without crack (Figure 22a), the initial signal and the reflection from the right-hand side end appear. If an 8-mm long through-thickness transverse crack is placed mid-span in the strip specimen, the reflection from this crack also appears (Figure 22b). As the crack length increases to 14 mm, the amplitude of the reflection increases (Figure 22c).

### 6.5.2.3  Comparison between Axial and Flexural Wave Simulation Results

The main differences between using the pulse-echo method with axial waves versus flexural waves are revealed by the comparison of Figure 20 and Figure 22. Figure 20a and Figure 22a show that, in this thin-strip specimen, the flexural wavespeed at 100 kHz is roughly half of the axial wavespeed, because the same distance of 914 mm is traveled in roughly twice the time. In addition, the flexural wave echo of Figure 22a shows a

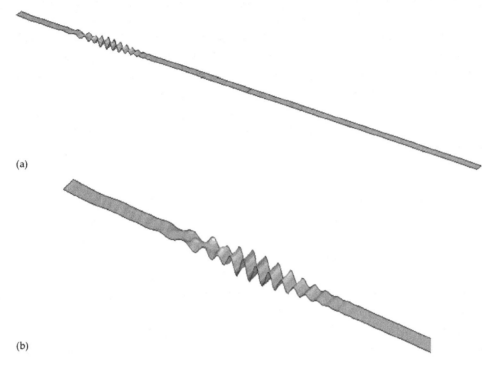

(a)

(b)

**FIGURE 21**   FEM simulation of flexure waves in the 914-mm × 14-mm × 1.6-mm aluminum alloy strip specimen excited with a 100-kHz 5-count Hanning-windowed burst at left-hand side end. The wave is captured after traveling for 99.3 µs: (a) overall view; (b) details of first quarter of the strip specimen.

dispersion pattern, whereas the axial wave echo of Figure 20a shows a much more compact pattern. This is in agreement with the dispersion curves of A0 and S0 Lamb-wave modes. At this value of the frequency-thickness product, the quasi-flexural wave mode A0 is much more dispersive than the quasi-axial wave mode S0. Figure 20b and Figure 22b show that, *for partial cracks of the same crack size* (e.g., 8 mm), *the quasi-axial wave echo is much stronger than the quasi-flexural wave echo*. Figure 20c and Figure 22c, *on the other hand, show that for a large crack* (e.g., 14 mm), *the quasi-flexural wave echo is much larger than the quasi-axial wave echo*. These observations indicate that both the quasi-axial and the quasi-flexural wave modes can offer advantages, under appropriate circumstances, for damage detection, and that both should be retained for further studies. For crack detection with the pulse-echo method, an appropriate Lamb-wave mode must be selected.

### 6.5.2.4 The Importance of High-Frequency Excitation

Although the results shown above refer mainly to one frequency (100 kHz), our wave propagation simulation efforts were performed for a variety of frequencies in the range 10–100 kHz. It was found that the lower-frequency limit were easier to simulate because at low frequency the wavelength is longer and spans several finite elements, and hence the distortion of each element is less severe. Consequently, at low frequency, we could use larger elements, i.e., a coarser mesh, and less computation time. However, low-frequency waves

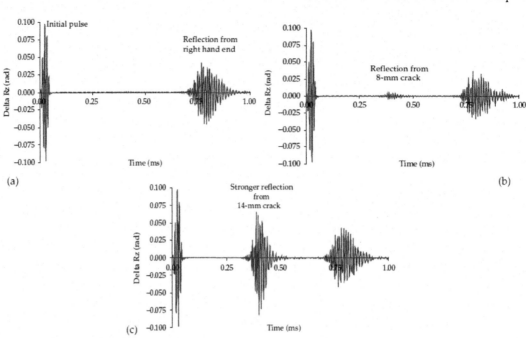

**FIGURE 22**    FEM of pulse-echo method in a 914-mm × 14-mm × 1.6-mm aluminum alloy strip specimen using flexural waves. A 100-kHz 5-count Hanning-windowed flexural burst was applied at the left-hand side end: (a) strip specimen without crack shows only the reflection from the right-hand side end; (b) strip specimen with 8-mm long through-thickness transverse crack shows, in addition, the reflection from crack; (c) a longer crack (14-mm) gives a stronger reflection.

are inappropriate for ultrasonic SHM applications. To achieve damage detection with the pulse-echo method, the timewise length of the wave packet must be much less than the time taken for the echo to return. We observed that, at low frequencies (e.g., 10 kHz), the echo signal starts to appear before the incident signal has finished developing, thus markedly impeding the damage detection process. Hence, high excitation frequencies are important. As the frequency increases, the wavelength decreases, and hence a finer finite element mesh is needed to capture the wave propagation process, which increases considerably the computational effort. After several trials, the compromised frequency of 100 kHz was selected. With this frequency, we were able to simulate successfully both axial and flexural wave patterns, and to identify defect-generated echoes as close as 100 mm from the source. For the detection of defects closer than 100 mm, higher frequencies are required.

## 6.5.3 Impact and AE Wave Propagation SHM

The pitch-catch and pulse-echo methods discussed in the previous sections are active-SHM method because they interrogate the structure in order to find its state of "health". However, the same PWAS installation can also be used for passive-SHM purposes, in which case the PWAS transducers will only "listen" to the waves generated by various events, e.g., (a) impacts that can produce damage in an aerospace composite structure and (b) AE signals that are produced when a crack propagates through a structures.

Reference [9] discusses two examples of signals obtained during an experiment in which low energy impacts (0.16-g small steel ball dropped from 50-mm height) and simulated AE events (pencil lead break) were applied to a plate instrumented with a network of 11 PWAS placed in a grid pattern. Figure 23a shows a schematic of these impact detection and simulated AE experiments indicating the location of impact and AE event relative to the PWAS transducers. Figure 23b shows the impact signals received at various PWAS transducers. Figure 23c shows the AE signals received at various PWAS transducers.

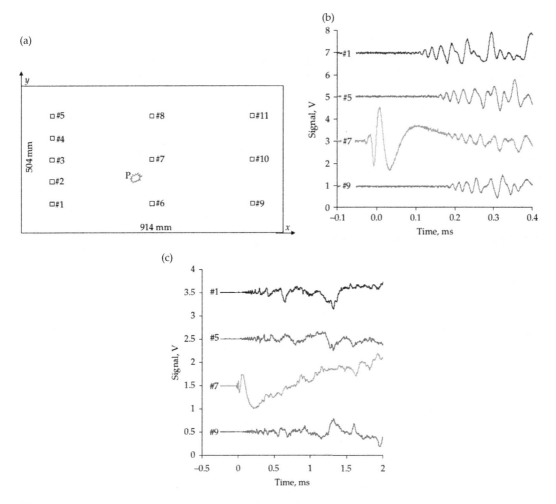

**FIGURE 23**  Impact detection and simulated AE experiments: (a) location of impact and AE event relative to the PWAS transducers; (b) impact signals received at various PWAS transducers; (c) AE signals received at various PWAS transducers.

## 6.6 PWAS PHASED ARRAYS AND THE EMBEDDED ULTRASONICS STRUCTURAL RADAR

A natural extension of the PWAS pulse-echo method is the development of a PWAS phased array (Figure 7f), which is able to scan a large area from a single location. Phased arrays were first used in radar applications because they allowed the replacement of the rotating radar dish with a fixed panel equipped with an array of transmitter—receivers which were energized with prearranged phase delays. When simultaneous signals are emitted from an array of transmitters, the constructive interference of the propagating waves creates a beam positioned broadside to the array. If prearranged phase delays are introduced in the firing of the signals of individual array elements, then the constructive interference beam can be steered to difference angular positions. Thus, an azimuth and elevation sweep can be achieved without mechanical rotation of the radar platform. The phased-array principle has gained wide utilization in ultrasonics, both for medical applications and for NDE, because ultrasonic phased arrays permit the sweeping of large volumes from a single location.

The PWAS phased arrays utilize the phase array principles to create an interrogating beam of guided waves that travel in a thin-wall structure and can sweep a large area from a single location. An extensive treatment of this topic is given in chapter 13 of Ref. [9] to which the reader is referred to for a detailed analysis; here, we will just recall the main highlights.

### 6.6.1 Phased-Array Processing Concepts

The array processing is based on previously discussed PWAS Lamb-wave propagation and tuning properties. The following assumptions are made:

1. The PWAS transducers are omnidirectional, i.e., having equal transmission and reception sensitivity in all directions. The radiation pattern of these PWAS sources is described by circular wave fronts. Far away from the source, the wave front approaches the plane-wave condition as the wave front curvature decreases. Thus, plane-wave models can be used in the far field, where the parallel-ray approximation may apply. However, the parallel-ray approximation does not hold in the near field.
2. The wave propagation direction is equivalently represented by the wavenumber vector, $\vec{k}$, and the slowness vector, $\vec{\alpha}$, where $\vec{\alpha} = \vec{k}/\omega$.
3. Propagating waves are single-mode tone-burst signals that can be described by a simple function, $f(t - \vec{\alpha} \cdot \vec{x})$, in which the space—time relationship $t - \vec{\alpha} \cdot \vec{x}$ applies, with $\vec{\alpha} = \vec{k}/\omega$. When the tone-burst is not sufficiently narrowband, dispersion may occur, and group velocity measurements will be used. If the dispersion is too large, special signal-processing methods (as described in a later chapter) can be used.
4. The superposition principle applies. This allows several propagating waves to occur simultaneously. The constructive or destructive interference of the separate wave patterns generated by each of the array elements represents the essence of the phased-array principle.
5. Common assumptions of a homogeneous, linear, lossless elastic medium are used in derivation of the phased-array equations. However, the phased-array principle can also be used in media that do not satisfy all these assumptions (e.g., composite materials) by making the parameters direction dependent.

When signals are superposed in an array, signal enhancement results. The signal enhancement within an array can be explained easily. In the simplest assumption, the signal $y_m(t)$ produced at the $m$-th sensor consists of a signal, $s(t)$, which is assumed to be more-or-less identical for all sensors, and random noise $N_m(t)$, which varies from sensor to sensor, i.e.,

$$y_m(t) = s(t) + N_m(t) \tag{66}$$

By summing the signals received by all the $M$ sensors, one obtains

$$z(t) = \sum_{m=0}^{M-1} y_m(t) = M\,s(t) + \sum_{m=0}^{M-1} N_m(t) \tag{67}$$

Equation (67) shows that while the signal, $s(t)$, is amplified $M$ times, the random noise is also summed up. The summing of several random noise signals usually results in noise reduction through mutual cancelation. Thus, a significant increase in the signal-to-noise ratio has been achieved. In phased arrays, the processing will be more complicated, because delays are used to create a beamforming effect. However, the basic noise reduction effect described by Eq. (67) will still apply.

### 6.6.1.1 Generic Delay-and-Sum Beamforming

Consider the generic geometric arrangement of Figure 24. The outputs of the array elements may be modified by the weighting factors $w_m$ such as to enhance the beam shape and reduce the side lobes. The origin of coordinate system is defined in the weighted array center, i.e., the origin is chosen such as

$$\sum w_m \vec{s}_m = 0 \tag{68}$$

Assume, as shown in Figure 24, that the target P is located at position vector $\vec{r}$. The $m$-th element of the phased array is placed at $\vec{s}_m$. The vector between the target P and the $m$-th element is $\vec{r}_m$. We define $\vec{\xi}$ as the unit vector for the direction $\vec{r}$ and $\vec{\xi}_m$ for the direction $\vec{r}_m$.

Assume that tuning between the PWAS and the Lamb waves in the structure has been achieved such that a low dispersion Lamb wave of wavespeed $c$ is dominant (details of such a tuning process were given in Chapter 10). Assume a generic pulse $f(t)$ is applied to a PWAS source located at the origin O. The wave front propagates radially from the source with wavespeed $c$. The wave front at a point P $(\vec{r})$ can be expressed as

$$f(\vec{r}, t) = \frac{1}{\sqrt{r}} f\!\left(t - \frac{r}{c}\right) \tag{69}$$

where $r = |\vec{r}|$. The wave transmitted by the $m$-th PWAS toward the target P $(\vec{r})$ is

$$y_m(t) = f(\vec{r}_m, t) = \frac{1}{\sqrt{r_m}} f\!\left(t - \frac{r_m}{c}\right) \tag{70}$$

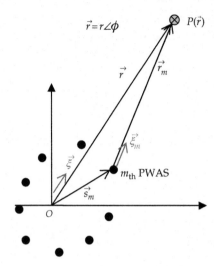

**FIGURE 24**　Geometric schematics of the $m$-th PWAS at $\vec{s}_m$ and the reflector at $P(\vec{r})$.

Equation (70) can be rearranged as

$$y_m(t) = f(\vec{r}_m, t) = \frac{1}{\sqrt{r}\sqrt{r_m/r}} f\left(t - \frac{r - r + r_m}{c}\right) = \frac{1}{\sqrt{r}} \frac{1}{\sqrt{r_m/r}} f\left(t - \frac{r}{c} + \frac{r - r_m}{c}\right) \qquad (71)$$

The delay-and-sum beamforming consists of two steps:

1. Apply a delay $\Delta_m$ and an optional weighting factor $w_m$ to the output of the $m$-th PWAS
2. Sum up the output signals of all the $M$ PWAS

This processing can be expressed as

$$z(t) = \sum_{m=0}^{M-1} w_m y_m(t - \Delta_m) \qquad (72)$$

Substitution of Eq. (71) into Eq. (72) yields

$$z(t; \vec{r}) = \sum_{m=0}^{M-1} w_m y_m(t - \Delta_m) = \sum_{m=0}^{M-1} \frac{1}{\sqrt{r}} \frac{w_m}{\sqrt{r_m/r}} f\left(t - \frac{r}{c} + \frac{r - r_m}{c} - \Delta_m\right) \qquad (73)$$

The delays $\Delta_m$ can be chosen such as to strengthen and focus the array's output beam on a particular point in space, $P(\vec{r}_P)$. For example, if the delays are chosen as to cancel the last two terms in Eq. (73), then the waves from all the array elements will be in phase when they arrive at $P(\vec{r}_P)$ and thus the signal will be strengthen almost $M$ times, i.e., if

$$\Delta_m = \frac{r_P - r_m}{c} \qquad (74)$$

then

$$z(t; \vec{r}_P) = \sum_{m=0}^{M-1} \frac{1}{\sqrt{r_P}} \frac{w_m}{\sqrt{r_m/r_P}} f\left(t - \frac{r_P}{c} + \frac{r_P - r_m}{c} - \Delta_m\right) = \sum_{m=0}^{M-1} \frac{1}{\sqrt{r_P}} \frac{w_m}{\sqrt{r_m/r_P}} f\left(t - \frac{r_P}{c}\right)$$

$$= \frac{1}{\sqrt{r_P}} f\left(t - \frac{r_P}{c}\right) \sum_{m=0}^{M-1} \frac{w_m}{\sqrt{r_m/r_P}}$$

(75)

Furthermore, the weights $w_m$ can be adjusted such as to compensate for the difference between $r_m$ and $r_P$, i.e., $w_m = \sqrt{r_m/r_P}$. In this case, one gets an exact $M$ times reinforcement of the signal, i.e.,

$$\sum_{m=0}^{M-1} \frac{w_m}{\sqrt{r_m/r_P}} = \sum_{m=0}^{M-1} 1 = M \quad \text{for} \quad w_m = \sqrt{r_m/r_P}$$

(76)

Substitution of Eq. (76) into Eq. (75) yields the fully focused signal

$$z(t)\Big|_{\text{focused on } P(\vec{r}_P)} = \frac{M}{\sqrt{r_0}} f\left(t - \frac{r_P}{c}\right) \quad \text{for} \quad w_m = \sqrt{r_m/r_P}, \quad \Delta_m = \frac{r_P - r_m}{c}$$

(77)

However, the weights $w_m$ may also be chosen in such a way as to enhance the beam shape and reduce the side lobes. Thus, we have seen how the delay-and-sum formulation of Eq. (72) allows us to calculate how an array of $M$ emitters located at positions $\vec{s}_m$, $m = 0, ..., M-1$ and fired with weights $w_m$ can focus the wave field generated toward a certain location $P(\vec{r}_P)$.

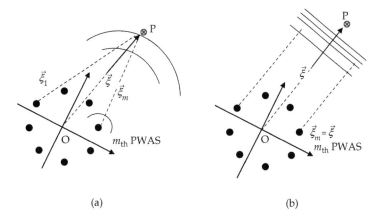

(a)                              (b)

**FIGURE 25**    Beamforming in the array's near and fields: (a) near field; (b) far field.

### 6.6.2 Beamforming Formulae for 2D PWAS Phased Arrays

We will develop harmonic wave beamforming formulae for an arbitrary PWAS phased array using the near-field full wave traveling paths (Figure 25a) and then simplify this formulation to the far-field case using the parallel-ray approximation(Figure 25b). Recall the PWAS phased array of Figure 24 containing $M$ elements defined by the position vector $\vec{s}_m$, $m = 0, 1, ..., M - 1$. The axes origin is chosen in the array centroid, i.e., Eq. (68) holds. Assume an unit amplitude harmonic wave of frequency $\omega$ emanates from a generic PWAS source located at the origin O. The wave front propagates radially with wavespeed $c$. The wave front at a point P $(\vec{r})$ can be expressed as

$$f(\vec{r}, t) = \frac{1}{\sqrt{r}} e^{i\left(\omega t - \vec{k} \cdot \vec{r}\right)} \tag{78}$$

where $\vec{k} = \vec{\xi} \, \omega/c$ is the wavenumber and $r = |\vec{r}|$. For each array element $m$, one has a vector $\vec{r}_m$ to the target P $(\vec{r})$ and a corresponding unit vector $\vec{\xi}_m$ that defines the angular direction from the $m$-th element to the target (Figure 25a). The following notations apply

$$\vec{\xi} = \frac{\vec{r}}{r}, \quad r = |\vec{r}|, \quad \vec{r} = r \angle \phi \tag{79}$$

$$\vec{r}_m = \vec{r} - \vec{s}_m, \quad r_m = |\vec{r}_m|, \quad \vec{\xi}_m = \frac{\vec{r}_m}{|\vec{r}_m|}, \quad m = 0, 1, \ldots, M - 1 \tag{80}$$

Assume all the PWAS elements in the array are fired simultaneously. The wave front coming from the $m$-th PWAS element toward the target P $(\vec{r})$, $\vec{r} = r \angle \phi$ can be written as

$$f(\vec{r}_m, t) = \frac{1}{\sqrt{r_m}} e^{j\left(\omega t - \vec{k}_m \cdot \vec{r}_m\right)} \tag{81}$$

The total wave front arriving at target P $(\vec{r})$ from all the array elements is obtained by superposition. If each source is fired with a different weight, $w_m$, then the superposition gives

$$z(\vec{r}, t) = \sum_{m=0}^{M-1} w_m f(\vec{r}_m, t) = \sum_{m=0}^{M-1} w_m \frac{1}{\sqrt{r_m}} e^{j\left(\omega t - \vec{k}_m \cdot \vec{r}_m\right)} \tag{82}$$

We will call the signal in Eq. (82) the *synthetic wave front* of the phased array. The generic Eq. (82) will be now used to study the near-field and far-field conditions. For near-field conditions, we will examine the exact traveling path, a.k.a. the *triangular algorithm*. For far-field conditions, we will study the *parallel-ray approximation*. Both conditions will still maintain a generic arbitrary phased-array arrangement. In the end, we will show that the parallel-ray approximation reduces to the 1D linear array algorithm when the generic array is reduced to an equally spaced linear array.

### 6.6.2.1 Near Field: Exact Traveling Path Analysis (Triangular Algorithm)

For the generic situation, exact traveling wave paths are used in the beamforming formulation. Equation (81) can be rewritten as

$$f(\vec{r}_m, t) = \frac{1}{\sqrt{r}\sqrt{r_m/r}} e^{j(\omega t - \vec{k}\cdot\vec{r} + \vec{k}\cdot\vec{r} - \vec{k}_m\cdot\vec{r}_m)} = \frac{1}{\sqrt{r}} e^{j(\omega t - \vec{k}\cdot\vec{r})} \frac{1}{\sqrt{r_m/r}} e^{j(\vec{k}\cdot\vec{r} - \vec{k}_m\cdot\vec{r}_m)}$$

$$= f(\vec{r}, t) \frac{1}{\sqrt{r_m/r}} e^{j(\vec{k}\cdot\vec{r} - \vec{k}_m\cdot\vec{r}_m)} \tag{83}$$

According to Eqs. (79), (80), we have

$$\vec{k}\cdot\vec{r} = \vec{k}\cdot r\vec{\xi} = r\frac{\omega}{c}\vec{\xi}\cdot\vec{\xi} = r\frac{\omega}{c}, \qquad \vec{k}_m\cdot\vec{r}_m = \vec{k}_m\cdot r_m\vec{\xi} = r_m\frac{\omega}{c}\vec{\xi}_m\cdot\vec{\xi}_m = r_m\frac{\omega}{c} \tag{84}$$

Substitution of Eq. (84) into Eq. (78) yields

$$f(\vec{r}, t) = \frac{1}{\sqrt{r}} e^{i\left(\omega t - \vec{k}\cdot\vec{r}\right)} = \frac{1}{\sqrt{r}} e^{i\left(\omega t - \omega\frac{r}{c}\right)} = \frac{1}{\sqrt{r}} e^{i\omega\left(t - \frac{r}{c}\right)} = f\left(t - \frac{r}{c}\right) \tag{85}$$

Substitution of Eqs. (84), (85) into Eq. (83) gives

$$f(\vec{r}_m, t) = f(\vec{r}, t) \frac{1}{\sqrt{r_m/r}} e^{j(\vec{k}\cdot\vec{r} - \vec{k}_m\cdot\vec{r}_m)} = f\left(t - \frac{r}{c}\right) \frac{1}{\sqrt{r_m/r}} e^{j\omega\frac{r - r_m}{c}} \tag{86}$$

Substitution of Eq. (86) into Eq. (82) yields the synthetic signal $z(\vec{r}, t)$ as

$$z(\vec{r}, t) = f\left(t - \frac{r}{c}\right) \sum_{m=0}^{M-1} w_m \frac{1}{\sqrt{r_m/r}} e^{j\omega\frac{r - r_m}{c}} \tag{87}$$

Equation (87) is made up from the multiplication of two terms. The first term, $f(t - r/c)$, is a function that does not depend on the locations or the weights of the phased-array elements or on the location of target. This term represents the individual wave signal that would be produced by a single PWAS element placed at the origin. We will leave this term alone. The second term, however, depends on the locations and weights of the phased-array elements and on the location of target. Hence, this term will change if we change the array configuration. It will also change if the target changes. We will call this second term the *beamforming function* (denoted as *BF*). In accordance with Eq. (87), the beamforming function is given by

$$BF(\mathbf{w}, \mathbf{r}, \vec{r}) = \sum_{m=0}^{M-1} w_m \frac{1}{\sqrt{r_m/r}} e^{j\omega\frac{r - r_m}{c}} \qquad \text{(beamforming function)} \tag{88}$$

where $\mathbf{w} = \{w_0, w_1, ..., w_{M-1}\}$, $\mathbf{r} = \{r_0, r_1, ..., r_{M-1}\}$. To strengthen and focus the array's output beam on a particular point in space, $P(\vec{r}_P)$, $\vec{r}_P = r_P \angle \phi_0$, we introduce the delays $\Delta_m$ and write the beamforming function of Eq. (88) as

$$BF(\mathbf{w}, \mathbf{r}, \vec{r}_P) = \sum_{m=0}^{M-1} w_m \frac{1}{\sqrt{r_m/r_P}} e^{j\omega\left(\frac{r_P - r_m}{c} - \Delta_m\right)} \tag{89}$$

Equation (89) shows that a maximum of the beamforming function $BF(\mathbf{w}, \bar{\mathbf{r}})$ can be achieved if one can make all the exponentials equal to one, which happens when the exponents are zero, i.e.,

$$e^{j\omega\left(\frac{r_P - r_m}{c} - \Delta_m\right)} = 1 \quad \text{for} \quad \frac{r_P - r_m}{c} - \Delta_m = 0 \tag{90}$$

To achieve Eq. (90), one has to apply to each element of the PWAS phased array the delay

$$\Delta_m = \frac{r - r_m}{c}, \qquad m = 0, 1, \ldots, M - 1 \tag{91}$$

When the delays of Eq. (91) are used, the beamforming function corresponding to the particular point in space, $P(\vec{r}_P)$, reaches a maximum and takes the value

$$BF(\mathbf{w}, \bar{\mathbf{r}}, \vec{r}_P) = \sum_{m=0}^{M-1} w_m \frac{1}{\sqrt{\tilde{r}_m / r_P}} \tag{92}$$

Equation (92) shows that further manipulation of the value of the beamforming function to focus on the desired point in space, $P(\vec{r}_P)$, can be achieved by adjusting the weighting factors $w_m$. One way of using this effect is to try to compensate the effect caused by the PWAS elements being placed at different locations, i.e., by taking

$$w_m = \sqrt{r_m / r} \tag{93}$$

If Eq. (93) is used, then the value of the beamforming function on the desired point in space, $P(\vec{r}_P)$, reaches the value $M$, i.e.,

$$BF(\mathbf{w}, \bar{\mathbf{r}}, \vec{r}_P) = \sum_{m=0}^{M-1} 1 = M \tag{94}$$

Substituting Eq. (94) into Eq. (87) yields the synthetic wave of the $M$-PWAS array at $P$ is under optimum conditions as

$$z(\vec{r}_P, t) = M f\left(t - \frac{r}{c}\right) \tag{95}$$

We see that the synthetic wave $z(\vec{r}, t)$ has become reinforced $M$ times relative to the individual wave signal that would be produced by a single PWAS element placed at the origin.

The process described above indicates that, with proper delays and weights, the phased array can be made to focus on a desired point in space, $P(\vec{r}_P)$, in a given direction, $\phi_0$, and at a certain distance, $r_P$. In contrast with the simplified parallel-ray algorithm, the exact algorithm presented here is able to focus both azimuthally through the angle, $\phi_0$, and radially through the presumed target location, $r_P$. For this reason, it does not depend on the commonly used parallel-ray approximation and hence can be used in the near field where the parallel algorithm fails. However, the implementation of this exact algorithm is more elaborate and requires more computational time.

### 6.6.2.2 Far Field: Parallel-Ray Approximation (Parallel Algorithm)

In the previous section, we deduced a generic beamforming formula for PWAS phased arrays that is exact but requires intensive computation. However, if the target is in the far field and the parallel-ray assumption applies, the algorithm can be simplified and the computational time can be reduced. If the target is located far away, we can assume that the rays emanating from the array elements toward the target are approximately parallel. Hence the $\vec{\xi}_m$ unit vectors become approximately equal, i.e.,

$$\vec{\xi}_m \approx \vec{\xi}, \qquad m = 0, 1, \dots, M-1 \tag{96}$$

Under these conditions, Eqs. (79) and (80) become

$$\vec{k}_m \approx \vec{\xi}\frac{\omega}{c} = \vec{k}, \quad \sqrt{r_m} \approx \sqrt{r}, \quad m = 0, 1, \dots, M-1 \tag{97}$$

Furthermore, the term $r_m/r$ in Eq. (86) becomes $r_m/r \approx 1$. The term $r - r_m$ becomes

$$r - r_m \approx \vec{\xi} \cdot \vec{s}_m \tag{98}$$

since $\vec{r} - \vec{r}_m = \vec{s}_m$, $\vec{\xi} \cdot (\vec{r} - \vec{r}_m) \approx \vec{\xi} \cdot \vec{s}_m$, and $\vec{\xi} \cdot (\vec{r} - \vec{r}_m) = \vec{\xi} \cdot \vec{r} - \vec{\xi} \cdot \vec{r}_m \approx \vec{\xi} \cdot r\vec{\xi} - \vec{\xi} \cdot r_m\vec{\xi} = r - r_m$. Substitution of Eq. (98) into Eq. (86) yields

$$f(\vec{r}_m, t) \approx f\left(t - \frac{r}{c}\right) \; e^{j\frac{\omega}{c}(\vec{\xi} \cdot \vec{s}_m)} \tag{99}$$

To generate beamforming in a direction $\phi_0$ defined by the unit vector $\vec{\xi}_0 = \vec{e}_x\cos\phi_0 + \vec{e}_y\sin\phi_0$, we apply delays $\Delta_m(\phi_0)$ and weights $w_m$. Thus, the beamforming function becomes

$$BF(\mathbf{w}, \vec{s}, \phi_0) = \sum_{m=0}^{M-1} w_m e^{j\frac{\omega}{c}\left(\vec{\xi} \cdot \vec{s}_m - \Delta_m(\phi_0)\right)} \tag{100}$$

where $\vec{s} = \{\vec{s}_0, \vec{s}_1, \dots, \vec{s}_{M-1}\}$. To achieve beamforming in the direction $\phi_0$ defined by the unit vector $\vec{\xi}_0$, one chooses the delays such as to make all the exponentials equal to one for that particular direction $\vec{\xi}_0$, i.e.,

$$\vec{\xi}_0 \cdot \vec{s}_m - \Delta_m(\phi_0) = 0 \quad \text{for} \quad \Delta_m(\phi_0) = \frac{\vec{\xi}_0 \cdot \vec{s}_m}{c} \quad \rightarrow \quad e^{j\frac{\omega}{c}\left(\vec{\xi}_0 \cdot \vec{s}_m - \Delta_m(\phi_0)\right)} = 1 \tag{101}$$

If we also choose the weights as unity, i.e., $w_m = 1$, then the beamforming function will take the value $M$, and the synthetic signal $z(\vec{r}, t)$ will become $M$ times reinforced with respect to the individual reference signal $f(\vec{r}, t)$. The weights $w_m$ may also be taken with different values such as to optimize the beamforming shape at various angles. However, as already noted, the formulation of Eq. (100) holds only when the far-field assumption is valid. However, the generic equations derived in Section 6.2.1 do not have this limitation and can be used for any situation, regardless of the target position.

## 6.6.3 Linear PWAS Phased Arrays

The generic 2D PWAS phased-array formulation presented in Section 6.2 simplifies considerably when applied to a 1D linear PWAS phased array. Linear arrays are made up of a number of elements, usually identical in size and arranged along a line, at uniform pitch (Figure 26).

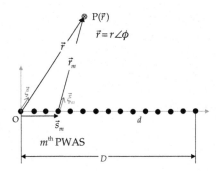

**FIGURE 26**   Schematic of an $M$-PWAS linear phased array.

Consider the linear PWAS array depicted in Figure 26. The array consists of $M$ elements, $m = 0, 1, ..., M - 1$, spaced at equal distance $d$. The $m$-th PWAS element in the array is located at position $\vec{s}_m$ given by

$$\vec{s}_m = md\,\vec{e}_x \tag{102}$$

where $\vec{e}_x$ is the unit vector in the $x$ direction, i.e., along the array. Assume a target located at point P of position vector $\vec{r} = r\vec{\xi}$. The target position vector is at an azimuth angle $\phi$ as measured from the $x$-axis. Each PWAS in the array is an omnidirectional wave source that transmits a pulse $s_T(t)$ as illustrated in Figure 26. The pulse travels with wavespeed $c$ toward the target P. The vector between the $m$-th PWAS and the target P is $\vec{r}_m = r_m \vec{\xi}_m$. The wave received at the target P from the $m$-th PWAS is

$$y_m(t) = \frac{1}{\sqrt{r}} s_T\left(t - \frac{r_m}{c}\right), \quad m = 0, 1, ..., M - 1 \tag{103}$$

where $r_m = |\vec{r}_m|$ is the distance between the $m$-th PWAS and the target P. The factor $1/\sqrt{r}$ represents the decrease in the wave amplitude due to of the omnidirectional 2D radiation. The $1/\sqrt{r}$ factor is based on the wave front energy conservation assumption. The signal received at the target P from the superposition of the effect of all the PWAS sources is

$$s_P(t) = \sum_{m=0}^{M-1} y_m(t) = \frac{1}{\sqrt{r}} \sum_{m=0}^{M-1} s_T\left(t - \frac{r_m}{c}\right) \tag{104}$$

### 6.6.3.1 Far-Field Parallel-Ray Approximation

If the target P is far away from the PWAS array, then the far-field parallel-ray approximation can be applied (Figure 27). The basic assumption of this approximation is that the source is sufficiently far away such that the position vectors $\vec{r}_m$ drawn from the array elements to the target are all parallel to the position vector $\vec{r}$ drawn from the origin to target P, i.e.,

$$\vec{r}_m \parallel \vec{r}, \qquad m = 0, 1, ..., M - 1 \tag{105}$$

Equation (105) implies that $\vec{\xi}_m = \vec{\xi}$; recall

$$\vec{\xi} = \vec{e}_x \cos\phi + \vec{e}_y \sin\phi, \qquad \vec{r} = r\vec{\xi} = \vec{e}_x r \cos\phi + \vec{e}_y r \sin\phi \tag{106}$$

Hence, $\vec{\xi}_m = \vec{\xi}$ implies

$$\vec{\xi}_m = \vec{\xi} = \vec{e}_x \cos\phi + \vec{e}_y \sin\phi, \qquad \vec{r}_m = r_m \vec{\xi} = \vec{e}_x r_m \cos\phi + \vec{e}_y r_m \sin\phi \tag{107}$$

Under these assumptions, the distance $r_m$ in Eq. (104) can be evaluated through vector operations. We start with the vector sum between the position vector $\vec{r}_m$ and the PWAS element location vector $\vec{s}_m$; according to Figure 26, we have

$$\vec{r} = \vec{r}_m + \vec{s}_m \tag{108}$$

The projection of Eq. (108) on the target direction $\vec{\xi}$ yields, upon rearrangement,

$$\vec{r}_m \cdot \vec{\xi} = \vec{r} \cdot \vec{\xi} - \vec{s}_m \cdot \vec{\xi} \tag{109}$$

Substitution of Eqs. (102), (106), (107) into Eq. (109) gives

$$r_m = r - md \; \vec{e}_x \cdot \vec{\xi} = r - md \cos\phi \tag{110}$$

Equation (110) indicates that, for the $m$-th PWAS, the distance to the target P is shorter by the value $m(d\cos\phi)$. Substitution of Eq. (110) into Eq. (104) yields

$$s_P(t) = \frac{1}{\sqrt{r}} \sum_{m=0}^{M-1} s_T\left(t - \frac{r_m}{c}\right) = \frac{1}{\sqrt{r}} \sum_{m=0}^{M-1} s_T\left(t - \frac{r - md \cos\phi}{c}\right) = \frac{1}{\sqrt{r}} \sum_{m=0}^{M-1} s_T\left(t - \frac{r}{c} + m\frac{d \cos\phi}{c}\right) \tag{111}$$

where $r/c$ is the delay due to the travel distance between the reference PWAS ($m = 0$) and the target P. Equation (111) indicates that, if all the PWAS were fired simultaneously, the signal from the $m$-th PWAS will arrive at the target P quicker by the amount

$$\delta_m(\phi) = m\frac{d\cos\phi}{c} \tag{112}$$

Substitution of Eq. (112) into Eq. (111) yields

$$s_P(t) = \frac{1}{\sqrt{r}} \sum_{m=0}^{M-1} s_T\left(t - \frac{r}{c} + \delta_m(\phi)\right) \tag{113}$$

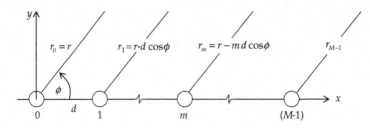

**FIGURE 27**   Far-field approximation of an $M$-element PWAS linear array with pitch $d$.

### 6.6.3.2 Firing with Time Delays

Assume that the PWAS are not fired simultaneously, but with some individual delays, $\Delta_m, m = 0, 1, ..., M - 1$. Equation (111) can be used to achieve array focusing and beamforming through judicious choice of the time delays applied to each individual array element. The wave arriving at the target P from the $m$-th PWAS is

$$y_m(t) = \frac{1}{\sqrt{r}} s_T\left(t - \frac{r_m}{c} - \Delta_m\right), \quad m = 0, 1, ..., M - 1 \tag{114}$$

where $\Delta_m$ is the delay of the $m$-th element. Correspondingly, Eq. (113) becomes

$$s_P(t) = \frac{1}{\sqrt{r}} \sum_{m=0}^{M-1} s_T\left(t - \frac{r}{c} + \delta_m(\phi) - \Delta_m\right) \tag{115}$$

### 6.6.3.3 Transmitter Beamforming

Beamforming is based on constructive interference of the waves emanating from the omnidirectional array elements. We can achieve constructive interference by making the waves from all the array elements arrive at the target P at the same time. According to Eq. (115), this desiderate can be achieved by firing the array elements with delays $\Delta_m$ that cancel the differences $\delta_m(\phi)$ in arrival time, i.e., by taking

$$\Delta_m = \delta_m(\phi) \tag{116}$$

Substitution of Eq. (116) into Eq. (115) yields the simple form

$$s_P(t) = \frac{1}{\sqrt{r}} \sum_{m=0}^{M-1} s_T\left(t - \frac{r}{c}\right) = \frac{1}{\sqrt{r}} s_T\left(t - \frac{r}{c}\right) \sum_{m=0}^{M-1} 1 = M \frac{1}{\sqrt{r}} s_T\left(t - \frac{r}{c}\right) \tag{117}$$

where the factor $M$ is due to the $M$-times summation of unity. Equation (117) shows an $M$ times increase in the signal strength received at a target P placed in the direction $\phi$ in comparison to the signal that would be received from a single transmitter PWAS. Thus, we have achieved *beamforming* in the direction $\phi$.

If the target direction $\phi$ is not known, then one has to consider sweeping the whole space by varying the beamforming angle $\phi$ through changes in the time delays. Substitution of Eq. (112) into Eq. (116) gives the time delay expression that has to be used for each given $\phi$, i.e.,

$$\Delta_m = m\frac{d}{c}\cos\phi, \quad m = 0, 1, ..., M - 1 \tag{118}$$

Since the function $\cos\phi$ is symmetric about the horizontal axis, the application of the delays $\Delta_m$ given by Eq. (118) will produce beamforming simultaneously at two angles, $\phi$ and $-\phi$. The $\pm\phi$ uncertainty did not present a problem in conventional radar because that device swept the sky at different elevation angles $\phi$ that would only vary in the range $0 - 180°$. But SHM applications of linear PWAS phased arrays often involve the placement of the array at the center of a large area. In this situation, the angle $\phi$ would be an azimuth angle that varies from $0°$ to $360°$. The $\pm\phi$ uncertainty of Eq. (118) should limit the practical range of such an application to $0 - 180°$. For example, the beamforming pattern for $\phi = 53°$ shown in Figure 28 shows a symmetric pattern with two beamforming lobes, one at $\phi = 53°$, the other at $\phi = -53°$.

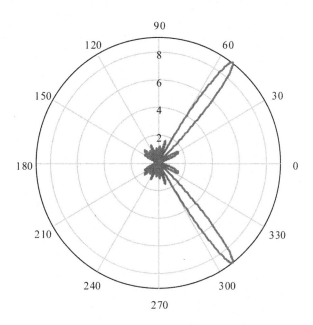

**FIGURE 28**   Calculated beamforming pattern for a 9-PWAS phased array at $\phi = 53°$.

### 6.6.3.4 Receiver Beamforming

The receiver beamforming principle is the reciprocal of the transmitter beamforming principle. If the point P is an omnidirectional source placed at azimuth $\phi$, then the signals received at the $m$-th sensor will arrive quicker by $m\,(d\cos\phi)/c$. Hence, one can synchronize the signals received at all the sensors by delaying them individually by $\Delta_m(\phi) = m(d\cos\phi)/c$.

### 6.6.3.5 Phased-Array Pulse-Echo

Assume that a target exists at azimuth $\phi$ and distance $R$. The transmitter beamformer is sweeping the azimuth in increasing angles $\phi$ by varying the delays $\Delta_m$ according to Eq. (118). An echo is received when $\phi = \phi_0$. The process through which this echo is formed is as follows. According to Eq. (117), the signal $s_P$ received at the target P from the transmitter beamformer is an $M$ times boost of the individual signals $s_T$, i.e.,

$$s_P(t) = \frac{M}{\sqrt{R}} s_T\left(t - \frac{R}{c}\right) \tag{119}$$

At the target P, the signal is backscattered with a backscatter coefficient, $A$. The radiated backscatter is further reduced by the factor $1/\sqrt{r}$ due to omnidirectional radiation and is delayed by $\delta_m(\phi_0)$. Hence, the signal received back at the $m$-th sensor in the array will be

$$y_m^R = \frac{1}{\sqrt{R}} A \frac{M}{\sqrt{R}} s_T\left(t - \frac{2R}{c} + \delta_m(\phi_0)\right), \quad m = 0, 1, ..., M - 1 \tag{120}$$

where $\delta_m(\phi_0) = m(d\cos\phi_0)/c$ The receiver beamformer applies the delays $\Delta_m(\phi_0) = m(d\cos\phi_0)/c$ and assembles the signals from all the sensors to create the receiver signal $s_R$, i.e.,

$$s_R(t) = \frac{A\,M}{R} \sum_{m=0}^{M-1} s_T\left(t - \frac{2R}{c} + \delta_m(\phi_0) - \Delta_m(\phi_0)\right) \tag{121}$$

Constructive interference between the received signals is achieved because $\Delta_m(\phi_0) = \delta_m(\phi_0) = m(d\cos\phi_0)/c$ and Eq. (121) produces an $M$-times boosted receiver signal, i.e.,

$$s_R(t) = \frac{A\,M^2}{R} s_T\left(t - \frac{2R}{c}\right) \tag{122}$$

The time delay of the received signal, $s_R(t)$, with respect to the transmit signal, $s_T(t)$, is

$$\tau = \frac{2R}{c} \tag{123}$$

Measurement of the time delay $\tau$ observed in $s_R(t)$ allows one to calculate the target range, $R = c\tau/2$.

### 6.6.3.6 Damage Detection with Tuned PWAS Phased Arrays

Once the beam steering and focusing concepts of the PWAS phased-array have been established, the detection of internal flaws and damage can be done with the pulse-echo method (Figure 29). A pulse, consisting of a smooth-windowed tone-burst of duration $t_p$, is transmitted toward the target. The target reflects the signal and creates an echo, which is detected by the PWAS phased array. By analyzing the phased-array signal in the interval $(t_p, t_p + t_0)$, one identifies the delay, $\tau$, representing the time-of-flight (TOF) taken by the wave to travel to the target and back. Knowing TOF and wavespeed allows one to precisely determine the target position.

The single most important phenomenon that enables the use of PWAS phased arrays in conjunction with multi-modal guided waves in thin-wall structures is the *PWAS Lamb-wave tuning principle*, which was described in Section 4. The PWAS Lamb-wave-tuning principle allows one to find convenient combinations of PWAS dimensions and excitation frequency that permit the preferential excitation of just one Lamb-wave mode, preferably one of minimal dispersion. In the following developments, we will assume that such tuning is possible and that a minimally dispersive Lamb wave can be tuned into. In this way, the situation depicted in Figure 29 can be achieved in spite of the generally multi-modal dispersive character of the Lamb waves.

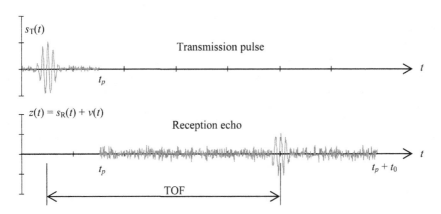

**FIGURE 29**    The pulse-echo method: (a) the transmitted pulse, $s_T(t)$; (b) the received echo, $z(t)$, consisting of the backscattered signal, $s_R(t)$, and the noise, $v(t)$. The difference between the pulse transmission and echo reception is the TOF.

## 6.6.4 Embedded Ultrasonics Structural Radar

The embedded ultrasonics structural radar (EUSR) is a concept that utilizes PWAS phased-array radar principles and ultrasonic guided waves (Lamb waves) to scan large surface areas of thin-wall structures and detect cracks and corrosion [9].

### 6.6.4.1 The EUSR Concept

In the EUSR concept, the guided Lamb waves are generated with surface-mounted PWAS that couple their in-plane motion with the in-plane particle motion of the Lamb wave as perceived at the structural surface. The guided Lamb waves stay confined inside the walls of the thin-wall structure and travel large distances with little attenuation. Thus, the target location relative to the phased-array origin is described by the radial position, $R$, and the azimuth angle, $\phi$. The EUSR algorithm works as follows. Consider a PWAS array as presented in Figure 28. Each element in the PWAS array plays the role of both transmitter and receiver. A methodology is designed to change the role of each PWAS in a round-robin fashion. The responses of the structure to all the excitation signals are collected. By applying the EUSR algorithm, an appropriate delay is applied to each signal in the data set to make them all focus on a direction denoted by angle $\phi$. When this angle $\phi$ is changed from 0° to 180°, a virtual scanning beam is formed and a large area of the structure can be interrogated.

As shown in chapter 6 of Ref. [9], guided Lamb waves can exist in a number of dispersive modes. However, through the PWAS-tuning approach discussed in Section 4, it is possible to confine the excitation to a particular Lamb-wave mode of wavelength $\lambda = c/F_c$, where $F_c$ is the carrier frequency and $c$ is the wavespeed of the tuned Lamb-wave mode. The tuning is usually done such that a very low dispersion Lamb-wave mode is selected, thus allowing us to assume quasi-constant wavespeed $c$. Hence, the smoothed tone-burst signal generated by one PWAS is assumed of the form

$$s_T(t) = s_0(t)\cos2\pi F_c t, \quad 0 < t < t_p \tag{124}$$

where $s_0(t)$ is a short-duration smoothing window that is applied to the carrier signal of frequency $F_c$ between 0 and $t_p$ (Figure 29). As in conventional phased-array radar, we assume a uniform linear array of $M$ PWAS, with each PWAS acting as a pointwise omnidirectional transmitter and receiver. The PWAS in the array are spaced at the distance $d$, which is assumed to be much smaller than the distance $r$ to a generic, far-distance point, P. Since $d < < r$, the rays joining the sensors with the point P can be assimilated with a parallel fascicle, of azimuth $\phi$ (Figure 27).

### 6.6.4.2 Practical Implementation of the EUSR Algorithm

The practical implementation of the EUSR signal generation and collection algorithms is described next. In a round-robin fashion, one active sensor at a time is activated as the transmitter. The reflected signals are received at all the sensors. The activated sensor acts in pulse-echo mode, i.e., as both transmitter and receiver; the other sensors act as passive sensors. Thus, an $M \times M$ matrix of elemental signals is generated. The elemental signals are assembled into synthetic beamforming responses using the synthetic beamformer algorithm as given by Eqs. (112), (115), (116). The delays, $\Delta_j$, are selected in such a way as to steer the interrogation beam at a certain angle, $\phi_0$. The synthetic-beam sensor responses, $w_i(t)$, synthesized for a transmitter beam with angle $\phi_0$, are assembled by the receiver beamformer into the total received signal, $s_R(t)$, using the same delays $\Delta_j$ as for the transmitter beamformer. However, to apply this method directly, one needs to know the target angle $\phi_0$. Since, in general applications, the target angle is not known, we need to use an inverse approach to determine it. Hence, we write the received signal as a function of the parameter $\phi_0$, using the array unit delay for the direction $\phi_0$ as $\delta_0(\phi_0) = (d\cos\phi_0)/c$. (To accurately implement the time shifts when the time values fall in between the fixed values of the sampled time, we have used a spline interpolation algorithm.)

A coarse estimate of the target direction is obtained by using an azimuth sweep technique, in which the beam angle, $\phi_0$, is modified until the maximum received energy is attained, i.e.,

$$\max E_R(\phi_0), \qquad E_R(\phi_0) = \int_{t_p}^{t_p+t_0} \left| s_R(t, \phi_0) \right|^2 dt \qquad (125)$$

After a coarse estimate of the target direction is found $\phi_0$, the actual round-trip TOF, $\tau_{TOF}$, is calculated using an optimal estimator, e.g., the cross-correlation between the receiver and the transmitter signals

$$y(\tau) = \int_{t_p}^{t_p+t_0} s_R(t) s_T(t - \tau) dt \qquad (126)$$

Then, the estimated $\tau_{TOF} = 2R/c$ is obtained as the value of $\tau$ where $y(\tau)$ is maximum. Hence, the estimated target distance is

$$R_{exp} = c \frac{\tau_{TOF}}{2} \qquad (127)$$

This algorithm works best for targets in the far field, for which the "parallel-rays" assumption holds. For targets in the near and intermediate fields, a more sophisticated self-focusing algorithm, that uses triangulation principles, is used. This algorithm is an outgrowth of the passive sensors target-localization methodologies. The self-focusing algorithm modifies the delay times used in each synthetic-beam response, $w_i(t)$. The total response is maximized by finding the focal point of individual responses, i.e., the common location of the defect that generated the echoes recorded at each sensor.

## 6.6.5 EUSR System Design and Experimental Validation

The EUSR system consists of three major modules: (a) the PWAS array; (b) the data acquisition (DAQ) module; and (c) the signal-processing module. A proof-of-concept EUSR system was built in the Laboratory for Active Materials and Smart Structures (LAMSS) at the University of South Carolina to evaluate the feasibility and capability of the EUSR system.

### 6.6.5.1 Experimental Setup

Three specimens were used in the experiments. These specimens were 1220-mm square panels of 1-mm thick 2024-T3 Al-clad aircraft grade sheet metal stock. One of the specimens (specimen #0) was pristine and was used to obtain baseline data. The other two specimens were manufactured with simulated cracks. The cracks were placed on a line midway between the center of the plate and its upper edge (Figure 30). The cracks were 19 mm long, 0.127 mm wide. On specimen #1, the crack was placed broadside with respect to the phase array, at coordinates $(0, 0.305 \text{ m})$, i.e., at $R = 305$ mm, $\phi_0 = 90°$. On the specimen 2, the crack was placed offside with respect to the phased array, at coordinates

$(-0.305 \text{ m}, 0.305 \text{ m})$, which corresponds to $R = 409$ mm, $\phi_0 = 136.3°$ with respect to the reference point of the PWAS array. The PWAS array was constructed from nine 7-mm square, 0.2-mm thick piezoelectric wafers (American Piezo Ceramic Inc., APC-850) placed on a straight line in the center of the plate. The sensors were spaced at pitch $d = \lambda/2$, where $\lambda = c/f$ is the wavelength of the guided wave propagating in the thin-wall structure. Since the first optimum excitation frequency for $S_0$ mode was 300kHz, and the corresponding wavespeed was $c = 5.440$ mm/$\mu s$, the wavelength was $\lambda = 18$ mm. Hence, the spacing in the PWAS array was selected as $d = 9$ mm.

The DAQ module consisted of an HP33120A arbitrary signal generator, a Tektronix TDS210 digital oscilloscope, and a portable PC with DAQ and GPIB interfaces. The HP33120A arbitrary signal generator was used to generate a 300-kHz Hanning-windowed tone-burst excitation with a 10-Hz repetition rate. Under the Hanning-windowed tone-burst excitation, one element in the PWAS array generated a Lamb-wave packet that spread out into the entire plate in an omnidirectional pattern (circular wave front). The Tektronix TDS210 digital oscilloscope, synchronized with the signal generator, collected the response signals from the PWAS array. One of the oscilloscope channels was connected to the transmitter PWAS, whereas the other was switched among the remaining elements in the PWAS array by using a digitally controlled switching unit. A LabVIEW computer program was developed to digitally control the signal switching, to record the data from the digital oscilloscope, and to generate the group of raw data files. Photographs of the experimental setup are presented in Figure 30.

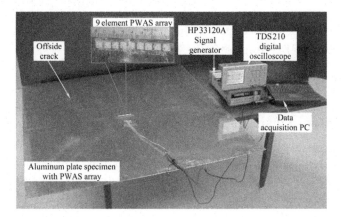

**FIGURE 30**   Experimental setup for EUSR experiment showing the plate, active sensors, and instrumentation.

### 6.6.5.2 Implementation of the EUSR Data-Processing Algorithm

The signal-processing module reads the raw data files and processes them using the EUSR algorithm. Although the EUSR algorithm is not computationally intensive, the large amount of data points in each signal made this step time consuming. Hence, we elected to save the resulting EUSR data on the PC for later retrieval and post-processing. This approach also enables other programs to access the EUSR data. Based on the EUSR algorithm, the resulting data file is a collection of signals that represent the structure response at different angles, defined by the parameter $\phi$. In other words, they represent the response when the EUSR scanning beam turned at incremental angles $\phi$.

After being processed, the data was transformed from the time domain to the 2D physical domain. Knowing the Lamb-wavespeed $c$, and using $r = ct$, the EUSR signal was transformed from voltage $V$ versus time $t$ to voltage $V$ versus distance $r$. The signal detected at angle $\phi$ was plotted on a 2D plane at angle $\phi$. Since angle $\phi$ was stepped from $0°$ to $180°$, at constant increments, the plots covered a half-space. These plots generate a 3D surface, which is a direct mapping of the structure being interrogated, with the $z$ value of the 3D surface representing the detected signal at that $(x, y)$ location (Figure 31). If we present the $z$ value on a color scale, then the 3D surface is projected onto the 2D plane, and the color of each point on the plane represents the intensity of the reflections.

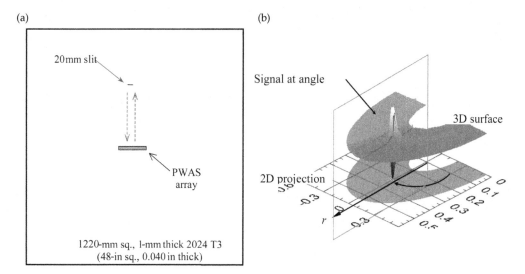

**FIGURE 31**    EUSR signal reconstruction examples: (a) schematic of the broadside crack; (b) 3D visualization of EUSR signal reconstruction of the broadside crack.

The implementation of these concepts in a graphical user interface (GUI) is presented in Figure 32. The angle sweep is performed automatically to produce the structure/defect imaging picture on the right. Manual sweep of the beam angle can be also performed with the turn knob; the signal reconstructed at the particular beam angle (here, $\phi_0 = 136°$) is shown in the lower picture. In NDE terminology, the 2D image corresponds to a C-scan, whereas the reconstructed signal would be an A-scan.

The EUSR methodology uses the full matrix of captured signals collected by the PWAS phased array to recreate a virtual sweeping of the monitored structural area. The associated image represents the reconstruction of the complete area as if the interrogating beam

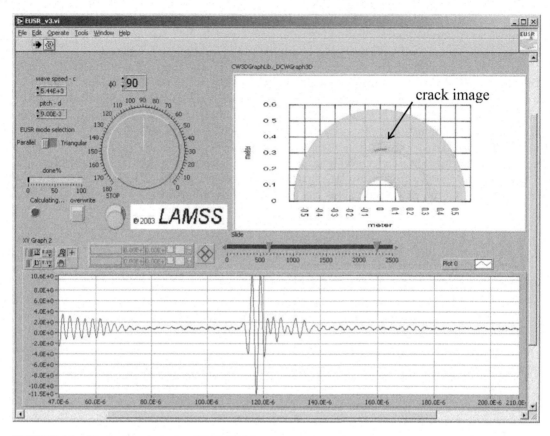

**FIGURE 32**  Graphical user interface (EUSR-GUI) front panel. The angle sweep is performed automatically to produce the structure/defect imaging picture on the right. Manual sweep of the beam angle can also be performed with the turn knob; the signal reconstructed at the particular beam angle (here, $\phi_0 = 136°$) is shown in the lower picture.

was actually sweeping it. When no damage is present, the only echoes are those arriving from the natural boundaries of the interrogated area; if damage is present, its echo reflection is imaged on the EUSR screen indicating its location in $(R, \theta)$ or $(x, y)$ coordinates. PWAS phased arrays have been used to monitor crack growth during fatigue testing [9].

## 6.7  PWAS RESONATORS

A PWAS transducer excited by an ultrasonic alternating voltage acts as a resonator, i.e., it undergoes large-amplitude vibration at certain frequencies which are its natural frequencies of vibration. An extensive treatment of this topic is given in chapter 9 of Ref. [9] to which the reader is referred to for a detailed analysis; here, we will just recall the main highlights.

## 6.7.1 Linear PWAS Resonators

Consider a slender piezoelectric wafer of length $l_a$, width $b_a$, and thickness $t_a$, with $l_a \gg b_a \gg t_a$. The wafer undergoes piezoelectric expansion induced by the thickness polarization electric field, $E_3$ (Figure 33). The electric field is produced by the application of a harmonic voltage $V(t) = \hat{V} e^{i\omega t}$ between the top and bottom surface electrodes. The electric field is assumed uniform over the piezoelectric wafer.

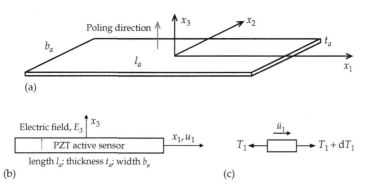

(a)

(b)

(c)

**FIGURE 33**   Schematic of a piezoelectric active sensor and infinitesimal axial element.

The analysis performed in section 9.2 of Ref. [9] shows that the admittance and impedance of a free linear PWAS resonator are given by

$$Y = i\omega C\left[1 - k_{31}^2\left(1 - \frac{1}{\phi\cot\phi}\right)\right] \tag{128}$$

$$Z = \frac{1}{i\omega C}\left[1 - k_{31}^2\left(1 - \frac{1}{\phi\cot\phi}\right)\right]^{-1} \tag{129}$$

### 6.7.1.1 Resonances and Anti-Resonances of Linear PWAS Resonators

The following conditions are considered:

- **Resonance**, when $Y \to \infty$, i.e., $Z = 0$
- **Anti-resonance**, when $Y = 0$, i.e., $Z \to \infty$

   **Electrical resonance** is associated with the situation in which a device is drawing very large currents when excited harmonically with a constant voltage at a given frequency. At resonance, the admittance becomes very large, whereas the impedance goes to zero. As the admittance becomes very large, the current drawn under constant-voltage excitation also becomes very large as $I = Y \cdot V$. In piezoelectric devices, the mechanical response at electrical resonance also becomes very large. This happens because the electromechanical coupling of the piezoelectric material transfers energy from the electrical input into the mechanical response. For these reasons, the resonance of an electrically driven piezoelectric device must be seen as an **electromechanical resonance.** A piezoelectric wafer driven at electrical resonance may undergo mechanical deterioration and even break up.

**Electrical anti-resonance** is associated with the situation in which a device under constant-voltage excitation draws almost no current. At anti-resonance, the admittance goes to zero, whereas the impedance becomes very large. Under constant-voltage excitation, this condition results in very small current being drawn from the source. In a piezoelectric device, the mechanical response at electrical anti-resonance is also very small. A piezoelectric wafer driven at the electrical anti-resonance hardly moves at all.

The **condition for electromechanical resonance** is obtained by studying the poles of $Y$, i.e., the values of $\phi$ that make $Y \to \infty$. Equation (128) reveals that $Y \to \infty$ as $\cot\phi \to 0$. This happens when $\cos\phi \to 0$, i.e., when $\phi$ is an odd multiple of $\pi/2$ as indicated below:

$$\cos\phi = 0 \quad \to \quad \phi = (2n-1)\frac{\pi}{2} \tag{130}$$

Hence, for electromechanical resonance, the angle $\phi$ can take the following values:

$$\phi^{EM} = \frac{\pi}{2}, \frac{3\pi}{2}, \frac{5\pi}{2}, \cdots \tag{131}$$

These values are the **electromechanical eigenvalues.** They are marked by superscript EM. Because $\phi = \frac{1}{2}\gamma l_a$, Eq. (131) implies that

$$\gamma l_a = \pi, 3\pi, 5\pi \tag{132}$$

Recall the definitions $\gamma = \omega/c$ and $\omega = 2\pi f$. Hence, the **electromechanical resonance frequencies** are given by the formula

$$f_n^{EM} = (2n-1)\frac{c}{2l} \tag{133}$$

It is remarkable that the electromechanical resonance frequencies do not depend on any of the electric or piezoelectric properties. They depend entirely on the speed of sound in the material and on the geometric dimensions.

### 6.7.1.2 Admittance and Impedance Formulae with Damping

The formulae of Eqs. (128), (129) predict that the admittance and the impedance become infinitely large at resonance and anti-resonance, respectively. In practice this does not happen due to internal damping and electrical loss. Hence, Eqs. (128), (129) were modified to include internal damping and electrical loss effects, i.e.,

$$\overline{Y} = i\omega\overline{C}\left[1 - \overline{k}_{31}^2\left(1 - \frac{1}{\overline{\phi}\cot\overline{\phi}}\right)\right] \quad \text{(admittance of linear PWAS resonator)} \tag{134}$$

$$\overline{Z} = \frac{1}{i\omega\overline{C}}\left[1 - \overline{k}_{31}^2\left(1 - \frac{1}{\overline{\phi}\cot\overline{\phi}}\right)\right]^{-1} \quad \text{(impedance of a linear PWAS resonator)} \tag{135}$$

where a bar over a variable signifies a complex quantity and

$$\bar{k}_{13}^2 = \frac{d_{31}^2}{\bar{s}_{11}^E \bar{\varepsilon}_{33}^T}, \qquad \overline{C} = \bar{\varepsilon}_{33}^T \frac{b_a l_a}{t_a}, \qquad \bar{c} = \sqrt{\frac{1}{\rho \bar{s}_{11}^E}}, \qquad \bar{\phi} = \frac{1}{2} \omega l_a \bar{c} \qquad (136)$$

$$\bar{s}_{11}^E = s_{11}^E(1 - i\eta), \quad \bar{\varepsilon}_{33}^T = \varepsilon_{33}^T(1 - i\delta) \qquad (137)$$

The coefficients $\eta$ and $\delta$ represent the mechanical and electrical loss factor; their values vary with the piezoceramic formulation but are usually small ($\eta, \delta < 5\%$). The use of Eq. (137) permits us to develop complex-number expressions for admittance and impedance that can be used when comparing predicted results with actual experimental data.

As the PWAS is excited at various frequencies, its admittance and impedance behave as shown in Figure 34. Outside resonances and anti-resonances, the admittance and impedance essentially behave like $Y = i\omega C$ and $Z = 1/i\omega C$ (Figure 34a). For example, the imaginary part of the admittance follows a straight line pattern outside resonances, whereas the real part is practically zero. At resonances and anti-resonances, these basic patterns are modified by the addition of a pattern of behavior specific to resonance and anti-resonance. These patterns include zigzags of the imaginary part and a sharp peak of the real part. The admittance shows zigzags of the imaginary part and peaks of the real part around the resonance frequencies (Figure 34a, left). The impedance shows the same behavior around the anti-resonance frequencies (Figure 34a, right).

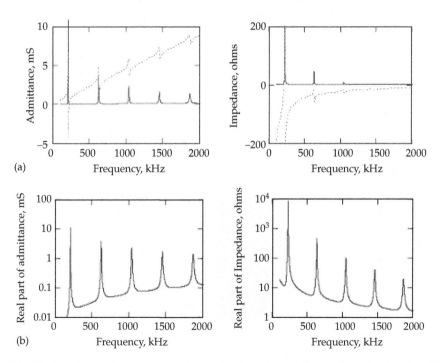

**FIGURE 34** Simulated frequency response of admittance and impedance of a piezoelectric active sensor ($l_a = 7$mm, $b_a = 1.68$mm, $t_a = 0.2$mm, APC-850 piezoceramic, $\eta = \delta = 1\%$): (a) complete plots showing both real (full line) and imaginary (dashed line) parts; (b) plots of real part only, log scale.

Figure 34b shows log scale plots of the real parts of admittance and impedance. The log-scale plots are better for graphically identifying the resonance and anti-resonance frequencies. The peaks of the admittance and impedance real part spectra can be used to measure the resonance and anti-resonance frequencies. The same information could also be extracted from the plots of the imaginary part, but this approach would be less practical. In the imaginary part plots, the resonance and anti-resonance patterns are masked by the intrinsic $Y = i\omega C$ and $Z = 1/i\omega C$ behaviors. In addition, the imaginary parts may undergo sign changes at resonances and anti-resonances; this would not allow log-scale plots to be applied, thus making the readings less precise.

One shortcoming of linear PWAS analysis is that the model assumptions (slender piezo-electric wafer with $l_a \gg b_a \gg t_a$) are rarely met in practice. Usual PWAS transducers are either rectangular, square, or circular. The linear PWAS analysis does not approximate too well the first resonance of a square PWAS. A rectangular PWAS is approximated better provided the aspect ratio is 4 or higher (see the discussion in section 9.2.5 of Ref. [9]). For this reason, a 2D analysis needs to be used, as exemplified in the next section that discusses circular PWAS resonators.

## 6.7.2  Circular PWAS Resonators

Consider a circular PWAS (Figure 35) of radius $a$, and thickness $t$, excited by the thickness polarization electric field, $E_3$, produced by the application of a harmonic voltage $V(t) = \hat{V}\, e^{i\omega t}$ between the top and bottom surface electrodes.

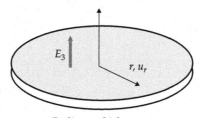

Radius, $a$; thickness, $t_a$

**FIGURE 35**    Circular PWAS resonator.

The analysis performed in section 9.3 of Ref. [9] shows that the admittance and impedance of a free circular PWAS resonator are given by

$$Y = i\omega C\left[1 - k_p^2\left(1 - \frac{(1+\nu)J_1(z)}{zJ_0(z) - (1-\nu)J_1(z)}\right)\right] \tag{138}$$

$$Z = \frac{1}{i\omega C}\left[1 - k_p^2\left(1 - \frac{(1+\nu)J_1(z)}{zJ_0(z) - (1-\nu)J_1(z)}\right)\right]^{-1} \tag{139}$$

where $J_0(z)$, $J_1(z)$ are Bessel functions, $z = \gamma a$, and $a$ is the PWAS radius, whereas

$$C = \varepsilon_{33}^T \frac{A}{t_a}, \quad A = \pi a^2 \quad \text{(electrical capacitance of a circular PWAS)} \tag{140}$$

$$k_p^2 = \frac{2}{(1 - \nu)} \frac{d_{31}^2}{s_{11}^E \varepsilon_{33}^T} \quad \text{(planar electromechanical coupling coefficient)} \tag{141}$$

### 6.7.2.1 Resonances and Anti-Resonances of Circular PWAS Resonators

The following conditions are considered:

- **Resonance**, when $Y \to \infty$, i.e., $Z = 0$
- **Anti-resonance**, when $Y = 0$, i.e., $Z \to \infty$

**Electrical resonance** is associated with the situation in which a device is drawing very large currents when excited harmonically with a constant voltage at a given frequency. At resonance, the admittance becomes very large, whereas the impedance goes to zero. As the admittance becomes very large, the current drawn under constant-voltage excitation also becomes very large because $I = Y V$. In piezoelectric devices, the mechanical response at electrical resonance also becomes very large. This happens because the electromechanical coupling of the piezoelectric material transfers energy from the electrical input into the mechanical response. For these reasons, the resonance of an electrically driven piezoelectric device must be seen as an **electromechanical resonance.** A piezoelectric wafer driven at resonance with high amplitudes may undergo mechanical deterioration and even break up.

**Electrical anti-resonance** is associated with the situation in which a device under constant-voltage excitation draws almost no current. At anti-resonance, the admittance goes to zero, whereas the impedance becomes very large. Under constant-voltage excitation, this condition results in very small current being drawn from the source. In a piezoelectric device, the mechanical response at electrical anti-resonance is also very small. A piezoelectric wafer driven at the electrical anti-resonance hardly moves at all. The resonance of an electrically driven piezoelectric device must also be seen as an **electromechanical anti-resonance.**

The **condition for electromechanical resonance** is obtained by studying the poles of $Y$, i.e., the values of $z$ that make $Y \to \infty$. The poles of $Y$ are roots of the denominator. These are obtained by solving the equation

$$z J_0(z) - (1 - \nu) J_1(z) = 0 \quad \text{(resonance)} \tag{142}$$

This equation is the same as the equation used to determine the mechanical resonances. This is not surprising, because, in our analysis of mechanical resonances, we only considered the axisymmetric modes, which couple well with a uniform electric field excitation. Hence, the frequencies of electromechanical resonance correspond identically to the frequencies for axisymmetric mechanical resonance.

The **condition for electromechanical anti-resonance** is obtained by studying the zeroes of $Y$, i.e., the values of $z$ which make $Y = 0$. Because electromechanical anti-resonances correspond to zeroes of the admittance (i.e., poles of the impedance), the current at anti-resonance is zero, $I = 0$. Equation (138) indicates that $Y = 0$ happens when

$$1 - k_p^2 \left( 1 - \frac{(1+\nu)J_1(z)}{zJ_0(z) - (1-\nu)J_1(z)} \right) = 0 \tag{143}$$

Upon rearrangement, one gets the anti-resonance condition as

$$\frac{zJ_0(z)}{J_1(z)} = \frac{1 - \nu - 2k_p^2}{\left(1 - k_p^2\right)} \quad \text{(anti-resonance)} \tag{144}$$

This equation is also transcendental and does not accept closed-form solutions. Its solutions are found numerically.

#### 6.7.2.2 Admittance and Impedance Formulae with Damping

The formulae of Eqs. (138), (139) predict that the admittance and the impedance become infinitely large at resonance and anti-resonance, respectively. In practice, this does not happen due to internal damping and electrical loss. Hence, Eqs. (138), (139) were modified to include internal damping and electrical loss effects, i.e.,

$$\overline{Y}(\omega) = i\omega\overline{C}\left[1 - \overline{k}_p^2 \left(1 - \frac{(1+\nu)J_1(\overline{z})}{\overline{z}J_0(\overline{z}) - (1-\nu)J_1(\overline{z})}\right)\right] \tag{145}$$

$$\overline{Z}(\omega) = \frac{1}{i\omega\overline{C}}\left[1 - \overline{k}_p^2 \left(1 - \frac{(1+\nu)J_1(\overline{z})}{\overline{z}J_0(\overline{z}) - (1-\nu)J_1(\overline{z})}\right)\right]^{-1} \tag{146}$$

where

$$\overline{k}_p^2 = \frac{2}{(1-\nu)} \frac{d_{31}^2}{\overline{s}_{11}^E \overline{\varepsilon}_{33}^T}, \qquad \overline{C} = \overline{\varepsilon}_{33}^T \frac{\pi a^2}{t_a}, \qquad \overline{c}_p = \sqrt{\frac{1}{\rho \overline{s}_{11}^E (1-\nu^2)}}, \qquad \overline{z} = \frac{\omega a}{\overline{c}_p} \tag{147}$$

$$\overline{s}_{11}^E = s_{11}^E(1 - i\eta), \quad \overline{\varepsilon}_{33} = \varepsilon_{33}^T(1 - i\delta) \tag{148}$$

The values of $\eta$ and $\delta$ vary with the piezoceramic formulation but are usually small ($\eta, \delta < 5\%$).

### 6.7.3 Constrained Linear PWAS Resonators

When affixed to a structure, the PWAS is constrained by the structure and its dynamic behavior is essentially modified. In this section, we will consider that the structure constraining the PWAS is represented by an unspecified dynamic structural stiffness, $k_{str}(\omega)$. Because this dynamic structural stiffness is frequency dependent, the way it interacts with the PWAS will also be frequency dependent and can significantly alter the PWAS resonances. As it will be shown in Section 8, the structural dynamics can overpower the inherent PWAS dynamics. In this case, the PWAS E/M impedance will closely follow the dynamics of the structure and the PWAS becomes a sensor of the dynamical modal behavior of the structure. Figure 36 shows a linear PWAS transducer constrained by a structural dynamic stiffness $k_{str}$. (The two springs with values $2k_{str}$ each acting in series produce a resultant spring $k_{total} = \left[(2k_{str})^{-1} + (2k_{str})^{-1}\right]^{-1} = k_{str}$.) Define the PWAS stiffness

$$k_{PWAS} = \frac{A_a}{s_{11}^E l_a} = \frac{b_a t_a}{s_{11}^E l_a} \quad \text{(PWAS stiffness)} \tag{149}$$

and the dynamic stiffness ratio

$$r = \frac{k_{str}(\omega)}{k_{PWAS}} \quad \text{(dynamic stiffness ratio)} \tag{150}$$

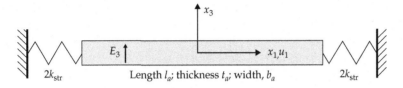

FIGURE 36    PWAS constrained by structural stiffness $k_{str}$.

The analysis performed in section 9.5.1 of Ref. [9] shows that the admittance and impedance of a constrained linear PWAS resonator are given by

$$
\begin{aligned}
\overline{Y} &= i\omega\,\overline{C}\left[1 - \overline{k}_{31}^2\left(1 - \frac{1}{\overline{\phi}\cot\overline{\phi} + \overline{r}}\right)\right] \\
\overline{Z} &= \frac{1}{i\omega\,\overline{C}}\left[1 - \overline{k}_{31}^2\left(1 - \frac{1}{\overline{\phi}\cot\overline{\phi} + \overline{r}}\right)\right]^{-1}
\end{aligned}
\tag{151}
$$

where $\overline{k}_{13}^2 = d_{31}^2/s_{11}^E \overline{\varepsilon}_{33}^T$ is the complex coupling factor, $\overline{C} = (1 - i\delta)C$, and $\overline{\phi} = \phi\sqrt{1 - i\eta}$. The values of $\eta$ and $\delta$ vary with the piezoceramic formulation but are usually small ($\eta$, $\delta$ <5%). The damping in the elastic constraint is similarly accounted for by assuming a complex stiffness expression, $\overline{k}_{str}$. As a result, the stiffness ratio will also take complex values, $\overline{r} = \overline{k}_{str}/\overline{k}_{PWAS}$.

## 6.7.4  Constrained Circular PWAS Resonators

When the circular PWAS is mounted on the structure, its circumference is elastically constrained by the dynamic structural stiffness (Figure 37). At the boundary $r = a$, we have the elastic constrain boundary condition $T_{rr}(r_a)\,t_a = -k_{str}(\omega)\,u_r(a)$.

Define the circular PWAS stiffness as

$$k_{PWAS} = \frac{t_a}{a\,s_{11}^E(1 - \nu)} \quad \text{(circular PWAS stiffness)} \tag{152}$$

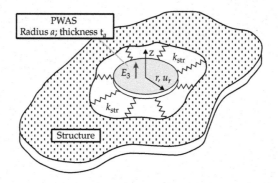

**FIGURE 37**   Circular PWAS constrained by structural stiffness, $k_{str}(\omega)$.

Also define the dynamic stiffness ratio for a circular PWAS as

$$\chi(\omega) = \frac{k_{str}(\omega)}{k_{PWAS}} \quad \text{(dynamic stiffness ratio)} \tag{153}$$

The analysis performed in section 9.5.2 of Ref. [9] shows that the admittance and impedance of a constrained circular PWAS resonator are given by

$$Y(\omega) = i\omega\overline{C}\left\{1 - \overline{k}_p^2\left[1 - \frac{(1+\nu_a)J_1(\overline{\phi})}{\overline{\phi}\,J_0(\overline{\phi}) - [(1-\nu_a) - (1+\nu_a)\overline{\chi}(\omega)]J_1(\overline{\phi})}\right]\right\} \tag{154}$$

$$Z(\omega) = Y^{-1}(\omega) = \frac{1}{i\omega\overline{C}}\left\{1 - \overline{k}_p^2\left[1 - \frac{(1+\nu_a)J_1(\overline{\phi})}{\overline{\phi}\,J_0(\overline{\phi}) - [(1-\nu_a) - (1+\nu_a)\overline{\chi}(\omega)]J_1(\overline{\phi})}\right]\right\}^{-1} \tag{155}$$

where $\overline{k}_p^2 = \dfrac{2}{(1-\nu_a)}\dfrac{d_{31}^2}{\overline{s}_{11}^E\overline{\varepsilon}_{33}^T}$, $\overline{C} = (1 - i\delta)C$, and $\overline{\phi} = \omega\dfrac{r_a}{\overline{c}_a}$.

## 6.8  HIGH-FREQUENCY VIBRATION SHM WITH PWAS MODAL SENSORS—THE ELECTROMECHANICAL (E/M) IMPEDANCE TECHNIQUE

The electromechanical (E/M) impedance technique is a high-frequency vibration method for damage detection with PWAS transducers. When a structure is excited with sustained harmonic excitation of a given frequency, the waves traveling in the structure undergo multiple boundary reflections and settle down in a standing wave pattern know as vibration. Structural vibration is characterized by resonance frequencies at which the structural response goes through peak values. The structural response measured over a frequency range including several resonance frequencies generates a vibration spectrum or frequency response function (FRP). When damage occurring in a structure induces changes in its dynamic properties, the vibration spectrum also changes. However, the conventional vibration analysis methods are not sensitive enough to detect small incipient damage; they can only measure structural dynamics up to several kHz, which is insufficient for the small wavelength needed to discover incipient damage. An alternative approach, which is able to measure structural spectrum into the hundreds of kHz and low MHz range, is offered by the E/M impedance method as described in chapter 10 of Ref. [9]. The E/M impedance method measures the electrical impedance, $Z(\omega)$, of a PWAS transducer using an impedance analyzer. The real part of the impedance Re(Z) reflects the mechanical behavior of the PWAS, i.e., its dynamic spectrum and its resonances. When the PWAS is attached to a structure, the real part of the impedance measured at the PWAS terminals reflects the dynamics of the structure on which the PWAS is attached, i.e., the structural dynamic spectrum and its resonances. Thus, a PWAS attached to a structure can be used as a structural identification sensor that measures directly the structural response at very high frequencies. Figure 7e illustrates the E/M impedance spectrum measured in the MHz range. An extensive treatment of this topic is given in chapter 10 of Ref. [9] to which the reader is referred to for a detailed analysis; here, we will just recall the main highlights.

## 6.8.1 Linear PWAS Modal Sensors

To facilitate understanding, we start the analysis with the consideration of simpler 1D problem that is easier to analyze while retaining all the important characteristics of the E/M impedance method and PWAS modal sensors. This simpler 1D problem permits closed-form solutions that can be verified against experiments conducted on simple-geometry specimens (here, metallic beams).

### 6.8.1.1 E/M Admittance of a PWAS Transducer Attached to a 1D Beam Structure

Consider a PWAS transducer attached to a 1D structure, e.g., a beam, undergoing axial–flexural vibrations (Figure 38). A detailed analysis of this situation, as given in section 10.2 of Ref. [9] reveals that the E/M admittance and impedance measured at the terminals of the structurally attached PWAS have the expression

$$\overline{Y}(\omega) = i\omega\,\overline{C}\left[1 - \overline{k}_{31}^2\left(1 - \frac{1}{\overline{\phi}(\omega)\cot\overline{\phi}(\omega) + \overline{r}(\omega)}\right)\right] \quad \text{(PWAS modal sensor admittance)} \quad (156)$$

$$\overline{Z}(\omega) = \frac{1}{i\omega\,\overline{C}}\left[1 - \overline{k}_{31}^2\left(1 - \frac{1}{\overline{\phi}(\omega)\cot\overline{\phi}(\omega) + \overline{r}(\omega)}\right)\right]^{-1} \quad \text{(modal sensor PWAS impedance)} \quad (157)$$

where

$$\overline{C} = \varepsilon_{33}^T\frac{b_a l_a}{t_a} \quad \text{(complex capacitance of the PWAS)} \quad (158)$$

$$\overline{k}_{13}^2 = \frac{d_{31}^2}{\overline{s}_{11}^E \overline{\varepsilon}_{33}^T} \quad \text{(complex coupling coefficient in PWAS material)} \quad (159)$$

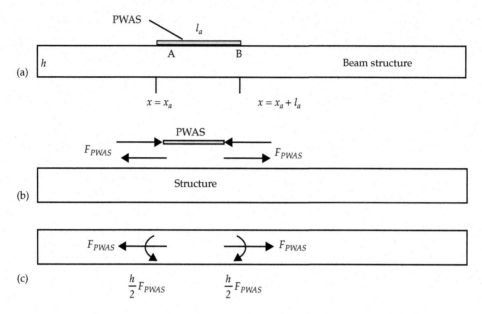

**FIGURE 38**   Interaction between a PWAS modal sensor a beam-like structural substrate: (a) geometry; (b) interaction forces at the PWAS ends; (c) excitation forces and moments resolved at the neutral axis.

$$\bar{\phi}(\omega) = \frac{1}{2}\frac{\omega}{\bar{c}_{PWAS}} \quad \text{(complex phase angle in PWAS material)} \tag{160}$$

$$\bar{c}_{PWAS} = \sqrt{\frac{1}{\rho_a \bar{s}^E_{11}}}, \quad \text{(complex wavespeed in PWAS)} \tag{161}$$

$$\bar{\varepsilon}_{33} = \varepsilon_{33}(1 - i\delta) \quad \text{(complex electric permittivity in the PWAS material)} \tag{162}$$

$$\bar{s}^E_{11} = s^E_{11}(1 - i\eta) \quad \text{(complex mechanical compliance in the PWAS material)} \tag{163}$$

The mechanical damping ratio $\eta$ and the electrical damping ratio $\delta$ vary with the formulation of the piezoceramic PWAS material, but are usually small ($\eta, \delta < 5\%$). The frequency-dependent complex stiffness ratio $\bar{r}(\omega)$ that appears in Eqs. (156), (157) is given by

$$r(\omega) = \frac{k_{str}(\omega)}{k_{PWAS}} \tag{164}$$

where $\bar{k}_{str}(\omega)$ is the structural stiffness given by Eq. (166) and $\bar{k}_{PWAS}$ is the PWAS stiffness given by

$$\bar{k}_{PWAS} = \frac{b_a t_a}{l_a \bar{s}^E_{11}}, \quad \text{(complex PWAS stiffness)} \tag{165}$$

with $t_a, b_a, l_a$ being the PWAS thickness, width, length, respectively.

### 6.8.1.2 Frequency-Dependent Structural Stiffness of 1D Beam Structure

The frequency-dependent structural stiffness $\bar{k}_{str}(\omega)$ of a 1D beam structure is calculated in section 10.2 of Ref. [9] by performing the dynamic analysis of a uniform undergoing forced axial–flexural vibration under PWAS excitation. Thus,

$$\bar{k}_{str}(\omega) = \rho A \left\{ \sum_{j_u=N_u^{low}}^{N_u^{high}} \frac{[U_{j_u}(x_a+l_a)-U_{j_u}(x_a)]^2}{\omega_{j_u}^2+2i\zeta_{j_u}\omega_{j_u}\omega-\omega^2} + \left(\frac{h}{2}\right)^2 \sum_{j_w=N_w^{low}}^{N_w^{high}} \frac{[W'_{j_w}(x_a+l_a)-W'_{j_w}(x_a)]^2}{\omega_{j_w}^2+2i\zeta_{j_w}\omega_{j_w}\omega-\omega^2} \right\}^{-1} \quad (166)$$

Equation (166) involves the use of the axial and flexural vibration modeshapes, $U_{j_u}(x)$ and $W_{j_w}(x)$, where $j_u$ and $j_w$ are mode indices that span the frequency band of interest. For free-free beams, the axial and flexural modeshape expressions used in Eq. (166) can be calculated with the following formulae:

$$U_{j_u}(x) = \sqrt{\frac{2}{l}}\cos\gamma_{j_u}x, \qquad \gamma_{j_u}=j_u\frac{\pi}{l}, \qquad \omega_{j_u}=j_u\frac{\pi}{l}\sqrt{\frac{E}{\rho}}, \qquad j_u = N_u^{low}...N_u^{high} \quad (167)$$

$$W'_{j_w}(x) = A_{j_w}\gamma_{j_w}\left[-\sin\gamma_{j_w}x - \beta_{j_w}\cos\gamma_{j_w}x + \frac{1}{2}\left(1-\beta_{j_w}\right)e^{\gamma_{j_w}x} - \frac{1}{2}\left(1+\beta_{j_w}\right)e^{-\gamma_{j_w}x}\right] \quad (168)$$

$$\gamma_{j_w} = \frac{z_{j_w}}{l}, \qquad \omega_{j_w}=\gamma_{j_w}^2\sqrt{\frac{EI}{\rho A}}, \qquad j_w = N_w^{low}...N_w^{high} \quad (169)$$

The eigenvalues $z_{j_w}$ are obtained as the solution of transcendental characteristic equation, i.e.,

$$\cos z \cosh z - 1 = 0 \quad (170)$$

The modeshape factors $\beta_{j_w}$ are calculated as

$$\beta_{j_w} = \frac{\cosh\gamma_{j_w}l - \cos\gamma_{j_w}l}{\sinh\gamma_{j_w}l - \sin\gamma_{j_w}l} = \frac{\sinh\gamma_{j_w}l + \sin\gamma_{j_w}l}{\cosh\gamma_{j_w}l - \cos\gamma_{j_w}l} \quad (171)$$

### 6.8.1.3 Detection of Structural Resonances from ReZ($\omega$) and ReY($\omega$) of 1D Structures

Section 10.2.5 of Ref. [9] shows that frequency spectrum of the E/M admittance real part, ReY($\omega$), and the frequency spectrum of the E/M impedance real part, ReZ($\omega$), resemble with fidelity the spectrum of the mechanical FRF imaginary part, ImFRF($\omega$), for the location at which the PWAS is attached to the structure. The peaks of ReY($\omega$) spectrum and ReZ($\omega$) spectrum resemble quite well the resonance peaks present in the ImFRF($\omega$) spectrum. This aspect is illustrated in Figure 39 which shows overlapped plots of ImFRF($\omega$), ReY, ReZ for a steel beam of $l = 100$mm, $b = 8$ mm, $h = 2.59$ mm with a 7-mm PWAS placed at $x_a = 40$mm from left end. The coincidence of the resonance peaks is quite apparent. The ReY($\omega$) and ReZ($\omega$) spectra reflect the structural dynamics, i.e., the peaks of the ReY($\omega$) and ReZ($\omega$) spectra coincide with the FRF peaks, which are the structural resonances. This observation stands at the foundation of using the E/M impedance method for high-frequency structural identification and SHM. Experimental confirmation of the veridicity of this theoretical prediction is given in section 10.2.6 of Ref. [9].

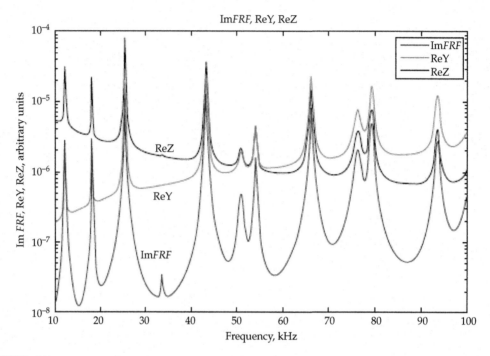

**FIGURE 39** Overlapped plots of Im*FRF*, Re*Y*, Re*Z* illustrating coincidence of the resonance peaks for a 1D structure (steel beam $l = 100$mm, $b_1 = 8$ mm, $h = 2.59$ mm with 7-mm PWAS placed at $x_a = 40$mm from left end).

## 6.8.2 Circular PWAS Modal Sensors

Circular PWAS modal sensors can be used on 2D structures, such as a plate or the thin-wall structure of an airplane. In order to develop a closed-form solution that can be compared with experimental data, section 10.3 of Ref. [9] considered the case of circular metallic plates instrumented with circular PWAS modal sensors (Figure 40).

### 6.8.2.1 E/M Admittance of a PWAS Transducer Attached to a 2D Circular-Plate Structure

The detailed analysis given in section 10.3 of Ref. [9] reveals that the E/M admittance and impedance measured at the terminals of the structurally attached PWAS have the following expressions:

$$\overline{Y}(\omega) = i\omega\overline{C}\left\{1 - \overline{k}_p^2\left[1 - \frac{(1 + \nu_a)J_1(\overline{\phi})}{\overline{\phi}\,J_0(\overline{\phi}) - [(1 - \nu_a) - (1 + \nu_a)\overline{\chi}(\omega)]J_1(\overline{\phi})}\right]\right\} \quad \text{(admittance)} \quad (172)$$

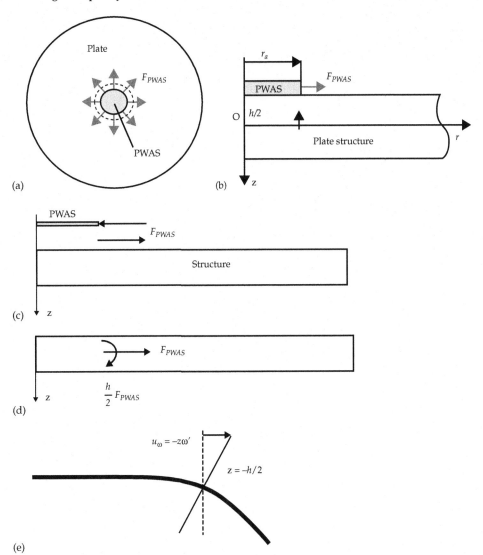

**FIGURE 40**   PWAS mounted on a circular plate: (a) geometry; (b) cross-section schematics; (c) interaction forces at the PWAS edge; (d) excitation force and moment resolved at the neutral plane; (e) kinematics of $u_w$ displacement due to flexural slope $w'$.

Inversion of the impedance $Z(\omega)$ yields the admittance $Y(\omega) = Z^{-1}(\omega)$, i.e.,

$$\overline{Z}(\omega) = \overline{Y}^{-1}(\omega)$$

$$= \frac{1}{i\omega\overline{C}} \left\{ 1 - \overline{k}_p^2 \left[ 1 - \frac{(1+\nu_a)J_1(\overline{\phi})}{\overline{\phi}\,J_0(\overline{\phi}) - [(1-\nu_a) - (1+\nu_a)\overline{\chi}(\omega)]J_1(\overline{\phi})} \right] \right\}^{-1} \quad \text{(impedance)} \quad (173)$$

where

$$\overline{C} = \overline{\varepsilon}_{33}^{T} \frac{\pi r_a^2}{t_a} \quad \text{(complex capacitance of the PWAS)} \tag{174}$$

$$\overline{k}_p^2 = \frac{2}{(1 - \nu_a)} \frac{d_{31}^2}{\overline{s}_{11}^{E} \overline{\varepsilon}_{33}^{T}} \quad \text{(complex planar coupling coefficient in PWAS material)} \tag{175}$$

$$\overline{\phi} = r_a \frac{\omega}{\overline{c}_a} \quad \text{(complex phase angle in PWAS material)} \tag{176}$$

$$\overline{c}_a = \sqrt{\frac{1}{\rho_a \overline{s}_{11}^{E}(1 - \nu_a^2)}}, \quad \text{(complex in-plane wavespeed in PWAS)} \tag{177}$$

$$\overline{\varepsilon}_{33} = \varepsilon_{33}(1 - i\delta) \quad \text{(complex electric permittivity in the PWAS material)} \tag{178}$$

$$\overline{s}_{11}^{E} = s_{11}^{E}(1 - i\eta) \quad \text{(complex mechanical compliance in the PWAS material)} \tag{179}$$

with $\nu_a$ being the Poisson ratio in the PWAS material. The mechanical damping ratio $\eta$ and the electrical damping ratio $\delta$ vary with the formulation of the piezoceramic PWAS material, but are usually small ($\eta, \delta < 5\%$). A brief derivation of this formula is given in section 10.3 of Ref. [9]. The stiffness ratio, $\overline{\chi}(\omega)$, is calculated as

$$\overline{\chi}(\omega) = \frac{\overline{k}_{str}(\omega)}{\overline{k}_{PWAS}} \tag{180}$$

where $\overline{k}_{str}(\omega)$ is the frequency-dependent structural stiffness and $\overline{k}_{PWAS}$ is the complex PWAS stiffness given by

$$\overline{k}_{PWAS} = \frac{t_a}{r_a \overline{s}_{11}^{E}(1 - \nu_a)}, \quad \text{(complex PWAS stiffness)} \tag{181}$$

with $t_a$ being the PWAS thickness.

### 6.8.2.2 Frequency-Dependent Structural Stiffness of 2D Circular-Plate Structure

The frequency-dependent structural stiffness $\overline{k}_{str}(\omega)$ of a 2D circular-plate structure is calculated in section 10.3 of Ref. [9] by performing the dynamic analysis of a circular plate undergoing forced axial–flexural vibration under PWAS excitation. Thus,

$$\overline{k}_{str}(\omega) = \frac{\rho h a^2}{2 r_a} \left[ \sum_{j_u = N_u^{low}}^{N_u^{high}} \frac{U_{j_u}^2(r_a)}{-\omega^2 + 2i\zeta_{j_u}\omega\omega_{j_u} + \omega_{j_u}^2} + \left(\frac{h}{2}\right)^2 \sum_{j_w = N_w^{low}}^{N_w^{high}} \frac{W_{j_w}'^2(r_a)}{-\omega^2 + 2i\zeta_{j_w}\omega\omega_{j_w} + \omega_{j_w}^2} \right]^{-1} \tag{182}$$

where

$$U_{j_u}(r) = A_{j_u} \, J_1\left(\gamma_{j_u} r\right) \quad \text{(axial modes of circular plate)} \tag{183}$$

$$z_{j_u} J_0(z_{j_u}) - (1 - \nu) J_1(z_{j_u}) = 0 \quad \text{(axial characteristic eq. of circular plate)} \tag{184}$$

$$\gamma_{j_u} = z_{j_u}/a \quad \text{(axial wavenumber)} \tag{185}$$

$$A_{j_u} = 1/\sqrt{J_1^2(z_{j_u}) - J_0(z_{j_u}) J_2(z_{j_u})} \quad \text{(axial amplitudes)} \tag{186}$$

$$W'_{j_w}(r) = A_{j_w} \gamma_{j_w} \left[ -J_1(\gamma_{j_w} r) + C_{j_w} \, I_1(\gamma_{j_w} r) \right] \quad \text{(slope of circular-plate flexural mode)} \tag{187}$$

$$\frac{z_{j_w} J_0(z_{j_w}) - (1 - \nu) J_1(z_{j_w})}{z_{j_w} I_0(z_{j_w}) - (1 - \nu) I_1(z_{j_w})} + \frac{J_1(z_{j_w})}{I_1(z_{j_w})} = 0 \quad \text{(flexural characteristic eq. of circular plate)} \tag{188}$$

$$\gamma_{j_w} = z_{j_w}/a \quad \text{(flexural wave number)} \tag{189}$$

$$C_{j_w} = -J_1(z_{j_w})/I_1(z_{j_w}) \quad \text{(flexural coefficients)} \tag{190}$$

$$A_{j_w} = \frac{a}{\sqrt{2}} \Bigg/ \sqrt{\int_0^a \left[ J_0(z_{j_w} r/a) + C_j I_0(z_{j_w} r/a) \right]^2 r \, dr} \quad \text{(flexural amplitudes)} \tag{191}$$

## 6.8.2.3 Detection of Structural Resonances from ReZ($\omega$) and ReY($\omega$) of 2D Structures

As already discussed in Section 8.1.3, the frequency spectrum of the E/M admittance real part, ReY($\omega$), and the frequency spectrum of the E/M impedance real part, ReZ($\omega$), resemble with fidelity the spectrum of the mechanical FRF imaginary part ImFRF($\omega$) for the location at which the PWAS is attached to the structure. The peaks of ReY($\omega$) spectrum and ReZ($\omega$) spectrum resemble quite well the resonance peaks present in the ImFRF($\omega$) spectrum. This aspect is illustrated in Figure 41 for a circular PWAS modal placed on 2D structure; Figure 41 shows overlapped plots of ImFRF($\omega$), ReY, ReZ for circular plate $a = 50$mm, $t = 0.8$mm with 7-mm PWAS ($r_a = 7$mm, $t_a = 0.2$mm) placed at its center. The peaks observed in Figure 41 correspond to the natural frequencies of axial and flexural vibration of the circular plate $f_{j_u} = 31.6$ kHz and $f_{j_w} = 3.05, 6.70, 12.44, 19.49, 28.1, 38.3$ kHz, respectively, as verified by the ImFRF($\omega$) plot. Note that the PWAS natural frequencies do not come into play here because the first PWAS resonance is way above the 40 kHz range (the first PWAS frequency is around 300 kHz). The coincidence of the ReY($\omega$) and ReZ($\omega$) spectra peaks with the ImFRF($\omega$) structural resonance peaks is quite apparent. Through electromechanical coupling, the ReY($\omega$) and ReZ($\omega$) spectra reflect the structural dynamics, i.e., the peaks of the ReY($\omega$) and ReZ($\omega$) spectra coincide with the FRF peaks, which are the structural resonances. This observation stands at the foundation of using the E/M impedance method for high-frequency structural identification and SHM. Experimental confirmation of the veridicity of this theoretical prediction is given in section 10.3.5 of Ref. [9].

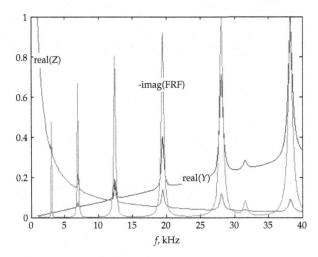

**FIGURE 41** Overlapped plots of Im$FRF$, Re$Y$, Re$Z$ illustrating coincidence of the resonance peaks for a 2D structure (circular plate $a = 50$ mm, $t = 0.8$ mm with 7-mm PWAS) ($r_a = 7$ mm, $t_a = 0.2$ mm) placed at its center.

## 6.8.3 Damage Detection with PWAS Modal Sensors and the E/M Impedance Technique

The advantage of using PWAS for damage detection resides in their very-high-frequency capability, which exceeds by orders of magnitudes the frequency capability of conventional modal analysis sensors. Since frequency and wavelength are inversely proportional, a very high frequency would ensure a very low wavelength which infers the capability of detecting small local changes in structural properties. Thus, PWAS are able to detect subtle changes in the high-frequency structural dynamics at local scales. Such local changes in the high-frequency structural dynamics are associated with the presence of incipient damage, which would not be detected by conventional modal analysis sensors operating at lower frequencies. Incipient damage in safety-critical hot-spot locations would be easily detected by the high-frequency PWAS modal sensors. (However, PWAS modal sensors would not be appropriate for the detection of low-frequency global vibration changes which, nonetheless, could be detected quite well with more conventional modal analysis equipment and techniques.)

Section 15.2 of Ref. [9] presents a case study of using the E/M impedance technique and PWAS modal sensors to detect simulated crack damage of increasing intensity in a set of circular plates. These experiments were performed on free circular plates because they have clean and reproducible resonance spectra and the effect of damage on the spectrum is easy to observe. Thus, a tractable development of the damage identification methods could be achieved. Experiments were performed on five statistical groups of free circular-plate specimens with increasing amounts of damage. Each group of five plates represented a different state of damage. Group 0 represented the pristine condition (Figure 42a). The specimens were 0.8-mm thick and 100-mm diameter circular aluminum plates. All plates were instrumented with a 7-mm PWAS placed in the plate center. To assess the statistical variation within each group, each group contained five nominally identical

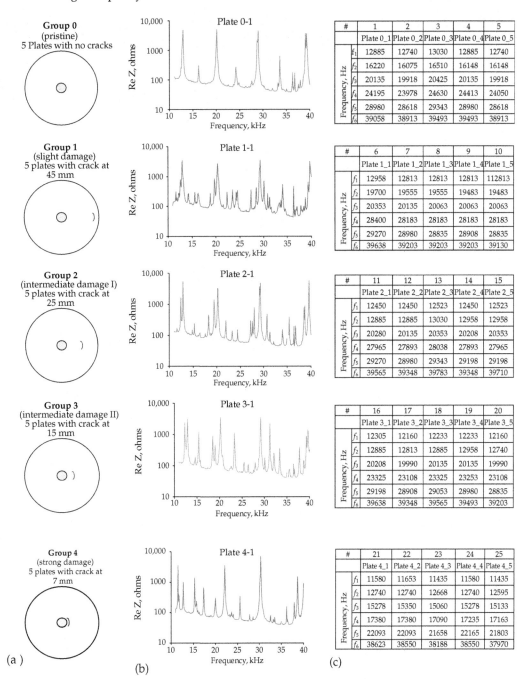

**FIGURE 42**    Damage detection experiments on circular plates with PWAS and the E/M impedance method: (a) schematic of damage; (b) E/M impedance spectra; (c) frequencies for each specimen.

specimens. A 10-mm circumferential through-the-thickness narrow slit was used to simulate an in-service crack. As shown in Figure 42a, the simulated crack was incrementally brought closer to the plate center $r = 45, 25, 15, 7$ mm, thus creating the five groups of specimens (Group 0 through Group 4).

Several damage identification methods were tried: (a) the damage index calculated with damage metric formulae applied to the raw spectrum; (b) a statistical method; (c) neural networks using spectral features, as detailed in section 15.2 of Ref. [9].

# References

[1] Chang, F. K. (1995) "Built-In Damage Diagnostics for Composite Structures", *10th International Congress of Composite Materials ICCM-10*, Canada, 1995, pp. 283–289.

[2] Chang, F.-K.; Lin, M. (2002) "Diagnostic Layer and Methods for Detecting Structural Integrity of Composite and Metallic Materials," US Patent 6,370,964 B1.

[3] Chang, F.-K. (1998) "SMART Layer—Built-in Diagnostics for Composite Structures", *4th European Conference on Smart Structures and Materials*, UK, 1998, pp. 777–781.

[4] Lin, M.; Quing, X.; Kumar, A.; Beard, S. J. (2001) "SMART Layer and SMART Suitcase for Structural Health Monitoring Applications", presented at the Smart Structures and Materials 2001 and Nondestructive Evaluation for Health Monitoring and Diagnostics, San Diego, CA, 2001.

[5] Lin, M.; Qing, X.; Kumar, A.; Beard, S. J. (2005) "SMART Layer and SMART Suitcase for Structural Health Monitoring Applications", DTIC.

[6] Lin, M.; Chang, F. K. (2002) "The manufacture of composite structures with a built-in network of piezoceramics", *Composites Science and Technology*, **62**(7–8), 919–939 , 2002, doi: Pii s0266-3538(02)00007-6.

[7] Lin, M.; Seydel, R. E.; Wang, C. S.; Ihn, J. B.; Wu, F.; Chang, F. K. (2000) "Built-in active sensing structural diagnostics", *Proceedings of US–Korea Workshop on New Frontier in Infrastructural/Seismic Engineering*, 2000, pp. 343–348.

[8] Lin, M.; Chang, F.-K. (1998) "Development of SMART Layer for Built-in Diagnostics for Composite Structures", *Proceedings of the American Society for Composites 13th Annual Technical Conference*, Baltimore, MD, 1998, pp. 55–63.

[9] Giurgiutiu, V. (2014) *Structural Health Monitoring with Piezoelectric Wafer Active Sensors*, 2nd Ed., Elsevier Academic Press, Oxford, UK.

[10] Giurgiutiu, V (2008) *Structural Health Monitoring with Piezoelectric Wafer Active Sensors*, 1st Ed, Elsevier Academic Press, Boston, MA.

[11] Ihn, J.-B. (2003) "Built-in Diagnostics for Monitoring Fatigue Crack Growth in Aircraft Structures", PhD Dissertation, Stanford University, Department of Aeronautics and Astronautics, 2003.

[12] Duke, J. C., Jr. (1988) *Acousto-Ultrasonics—Theory and Applications*, Plenum Press.

# Fiber-Optic Sensors

# 7.1 INTRODUCTION

Fiber-optic sensors offer several advantages for structural health monitoring (SHM): (i) immune to electromagnetic interference (EMI); (ii) corrosion resistance; (iii) possibility of multiplexing several sensors on the same optical fiber; (iv) the promise of direct embedment into the composite material along the reinforcing fibers, etc. Fiber-optic sensors also have several limitations that have impeded their widespread usage: (i) the need for considerable optoelectronic equipment to convert the optical changes into actual readings of the physical quantity being monitored (strain, or other material property); (ii) bandwidth limited by the bandwidth of the optoelectronic equipment that has to perform complicated processing of the optical signal; etc.

The promise of direct embedment of fiber-optic sensing into composite is very attractive; however, its full realization needs to be considered with caution: the overall diameter of a typical telecom optical fiber varies between 120 and 250 $\mu m$. When embedded in a composite, the optical fiber may create a discernible discontinuity (Figure 1); this may become an issue for composite structural integrity, and studies of this aspect are still under way. Development of small-diameter optical fibers (e.g., 52 $\mu m$) for composite SHM embedment has been reported [2], but these specialty fibers have a much larger cost than the run-of-the-mill telecom fibers.

Optical fiber at 0° to carbon fiber

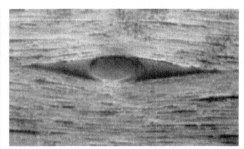

Optical fiber at 30° to carbon fiber

**FIGURE 1**  Micrographs of an optical fiber embedded in a carbon fiber-reinforced polymer (CFRP) composite at various inclinations with respect to the composite fibers: (a) 0°; (b) 30° showing resin pockets ahead and after the optical fiber [1].

A comprehensive review of fiber-optic sensors for SHM applications is provided in Refs. [1,3–5]. The fiber-optic sensing mechanisms for SHM applications can be one of four types [1]: (i) intensity modulation; (ii) phase modulation; (iii) spectral modulation; (iv) polarization modulation.

## 7.1.1  Intensity Modulation Fiber-Optic Sensors

Intensity modulation is one of the simplest to measure because it only requires a photodetector to measure the light intensity. The intensity of the light wave traveling through an optical fiber can be modified by microbending of the optical fiber, by a change in coupling of the fiber with the surrounding medium, or the fracture of the optical fiber. A photodetector is used to measure the intensity of the light transmitted through the fiber or reflected back to the input. One drawback of these simple sensors is that they cannot be multiplexed into sensor networks.

## 7.1.2  Polarization Modulation Fiber-Optic Sensors

Polarization modulation fiber-optic sensors rely on various polarization directions existing in an optical fiber. The cross-section of the optical fiber has a fast axis and a slow axis which are mutually perpendicular. Two light waves individually polarized about each of these axes will not usually interfere with each other and their intensity can be separately measured using polarizing filters. External stimuli such as pressure or twisting of the fiber may induce transfer between the two polarized modes, which may be quantified and used for measurement.

## 7.1.3  Phase Modulation Fiber-Optic Sensors

Phase modulation fiber-optic sensors rely on the difference in phase (arrival time) due to changes in the wave path. Mechanical strain or thermal expansion changes the optical path length through the fiber and hence affects the light wave phase. Interferometric methods are used to measure the phase shift between the light wave traveling through the sensor and a reference wave split from the same source. The reference fiber can be exposed to some of the external effects such as to decouple these effects from those to be measured by the sensing fibers. For example, if mechanical strain and temperature changes are both present, the reference fiber can be exposed to temperature only, whereas the measuring fiber is exposed to both strain and temperature, such that the resulting relative phase between the two fibers will be due only to the strain effects. Optoelectronic interferometers that convert the phase shift into some numerical readout are commercially available [6]. However, phase modulation measurements remain particular to a single fiber and multiplexing of several sensors on the same fiber is not generally possible.

## 7.1.4 Spectral Modulation Fiber-Optic Sensors

Spectral modulation fiber-optic sensors use the changes in signal wavelength produced by mechanical or temperature strain. This type of modulation is achieved through Fabry–Perot interferometers (FPI) and fiber Bragg gratings (FBG). Such sensors typically act as filters, transmitting certain wavelengths and radiating or reflecting others. Changes in mechanical strain, temperature, or other external parameters are converted into a shift of the characteristic wavelength of such a filter. The decoding of the wavelength-encoded signal is commonly done by one of three methods: (i) with an optical spectrum analyzer (OSA); (ii) with a tunable laser (TL) source that scans a band of wavelengths; or (iii) with a wavelength-tunable filter. Other, more complex methods based on three-way couplers or fast Fourier transforms have also been developed for high-rate data acquisition (DAQ) and sensor multiplexing. The cavity-based sensors are manufactured by inserting into a capillary tube the two ends of a cleaved fiber separated by a small distance (Figure 2). The cavity acts as wavelength filter; if strain is applied, the cavity size changes and its center frequency shifts. This wavelength shift is decoded.

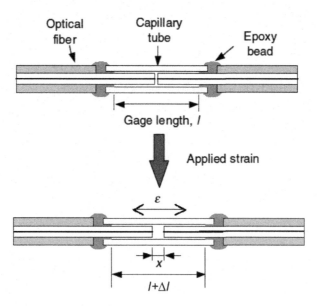

**FIGURE 2** Schematic of a cavity-based optical sensor obtained by splicing into a capillary tube the two parallel-cleaved ends of an optical fiber [7].

FBG sensors achieve performance similar to that of FP sensors but with a much simpler and more robust construction: instead of a physical cavity as used in the FP sensors, the FBG sensors rely on Bragg scattering from a grating of index of refraction discontinuities imprinted into a given small portion of the fiber. The filtering wavelength of the FBG sensor is dictated by the spacing in its Bragg grating. FBG sensors have gained considerable popularity because they are simple to manufacture with ultraviolet (UV) imprinting directly onto the fiber; several different FBG sensors imprinted on the same fiber can be individually addressed and multiplexed.

### 7.1.5 Scattering Modulation Fiber-Optic Sensors

Scattering modulation fiber-optic sensors rely on the intrinsic scattering properties of the material used in the construction of the optical fiber. Several backscatter waves are possible: Rayleigh, Brillouin, Raman, and Stokes. Optical backscatter reflectometry combined with currently available high-speed computational methods and equipment has enabled scattering the development of new modulation methods. Recently available scattering modulation-based equipment permits the continuous monitoring of strain along a simple telecom fiber without the need for any sensing portions (FP cavities, FBGs, etc.) being manufactured along the fiber.

## 7.2  GENERAL PRINCIPLES OF FIBER OPTIC SENSING

An optical fiber consists of a central core surrounded by an annular cladding with a protective coating; traditional optical fibers were made from high-quality silica; more recently, optical quality plastic materials have been extensively used in the manufacture of optical fibers. A typical overall diameter of an optical fiber varies between 120 and 250 $\mu m$, although some development on small-diameter optical fibers has been reported [2].

### 7.2.1 Total Internal Reflection

The light guiding properties of optical fibers are linked to the phenomenon of total internal reflection which is due to the difference between the cladding and core refractive indices $n_{cladding} < n_{medium} < n_{core}$. Total internal reflection takes place at the interface between the high-index core and lower-index cladding. Snell's law applied to these different refractive indices yields the result that no transmission in the cladding is possible and hence the light wave reflects back into core. Hence, the light traveling in an optical fiber is confined to the core. As a result, light can travel large distances through an optical fiber with very little loss (Figure 3).

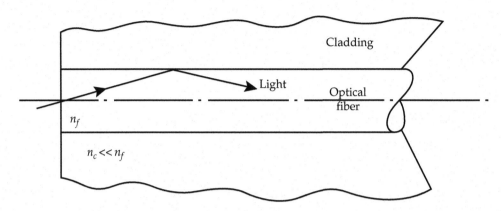

**FIGURE 3** Principle of total internal reflection in an optical fiber.

## 7.2.2 Single-Mode and Multimode Optical Fibers

Light propagates as an electromagnetic wave inside the optical fiber. This wave can propagate either in one mode, i.e., "single-mode fiber" or in multiple simultaneous modes, i.e., multimode fiber (Figure 4a). The number of modes that may propagate at a given wavelength depends on the diameter of the optical fiber core. The overall diameter of an optical fiber is made up of the diameter of the optical fiber core, the width of annular cladding, and the protective outer coating (Figure 4b). Optical fibers with a smaller core allow only a single mode; larger fibers allow multiple modes. When the core diameter is around 10 μm, the optical fiber may carry only the fundamental LP01 mode (Figure 4c, left); hence, it is called "single-mode fiber". If the core diameter is larger, say between 50 and 100 μm, the optical fiber allows more than one mode of the light wave to propagate (e.g., up to the LP21 mode, as shown in Figure 4c, right); hence, it is called "multimode fiber". A fiber-optic sensor can be constructed from either a single-mode or a multimode optical fiber depending on application. A single-mode optical fiber with a smaller core is much more sensitive than a multimode optical fiber; this may be a desirable feature in damage detection based on strain measurements [8]. A multimode fiber changes its properties more readily in response to microbends and other disturbances, hence it has been preferred in other damage detection sensors.

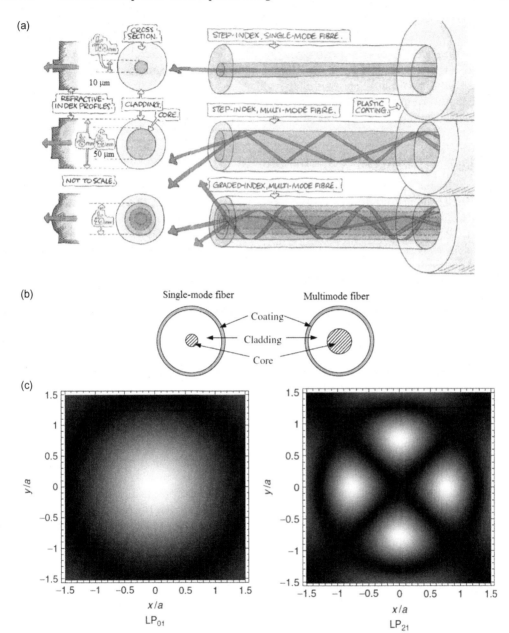

**FIGURE 4**  Single-mode and multimode optical fibers: (a) comparison between light distribution inside the optical fiber for modes LP01 (fundamental) and LP21 in a fiber [1]; (b) diagrammatical comparison between single-mode and multimode fiber cross sections [8]; (c) diagrammatical representation of single-mode and multimode propagation of light in a fiber [9].

## 7.3 INTERFEROMETRIC FIBER-OPTIC SENSORS

An interferometric sensor measures phase changes either in a single-mode optical fiber or developed between two single optical fibers, which is related to mechanical strain. It has high sensitivity, which is very useful in evaluating the effects of damage on the performance of the host structure [8]. Three types of interferometric sensors are commonly used, namely Mach–Zehnder, Michelson, and Fabry–Perot (FP).

### 7.3.1 Mach–Zehnder and Michelson Interferometers

A Mach–Zehnder interferometric sensor (Figure 5a) requires two single optical fibers that are used for both sensing and reference, respectively. However, the fact that the reference fiber is not embedded in the host composite makes the use of such a sensor impractical. In a Michelson interferometric sensor (Figure 5b), the sensing and reference fibers are both embedded in the host composite in close proximity to each other; their ends are mirrored to reflect the optical signal; they are of slightly different length, thus creating a sensing gage length inside the specimen. The sensing of specimen strain is based on the differential path length of the round-trip travel in the two fibers. Although this type of sensor is relatively easy to build with high sensitivity, it has three major drawbacks: phase preservation at the fiber–host interface, a greater degree of upsetting the host and being more vulnerable to noise interference.

### 7.3.2 Intrinsic Fabry–Perot Sensors

The disadvantages of the Mach–Zehnder and Michelson fiber-optic sensors stem mainly from the fact that these sensors require two optical fibers. An intrinsic Fabry–Perot interferometric (IFPI) sensor can be constructed from a single optical fiber and thus does not encounter such problems. A reflection region is created into the fiber by inserting through fusion splicing two reflective mirrors into the fiber, or by inserting one mirror into the fiber and putting the other mirror at the free end of the fiber as illustrated in Figure 5c. The fiber length contained between the two mirrors forms the sensing region which expands and compresses with the composite test specimen. The change in path length experienced by the light traveling through the sensing region is demodulated interferometrically as shown in Figure 5c.

Although IFPI sensors are less perturbative, a potential coupling with transverse strain may cause difficulty in strain interpretation in addition to the fact that fusion splicing the internal mirrors could significantly weaken the spliced cross sections of the optical fiber. Indeed, there are only few reported applications of IFPI sensors.

### 7.3.3 Extrinsic Fabry–Perot Interferometric Sensors

A more robust optical fiber sensor also based on the Fabry–Perot concept is the extrinsic Fabry–Perot (EFPI) sensor. Unlike IFPI sensors, an EFPI sensor (Figure 5d) has two optical fibers housed within a capillary glass tube by either fusion splicing or adhesive bonding [8].

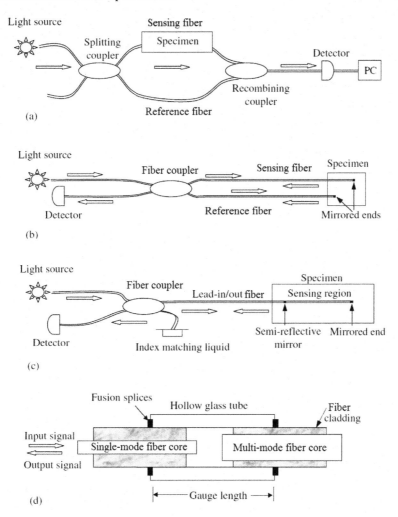

**FIGURE 5**    Schematic of interferometric fiber-optic sensors: (a) Mach–Zehnder; (b) Michelson; (c) intrinsic Fabry–Perot; (d) extrinsic Fabry–Perot [8].

A typical capillary tube has a length of 3–40 mm and an outer diameter of up to 0.3 mm, fractionally greater than that of the optical fibers. The mirrored ends of two optical fibers must be perpendicular to the fiber axes. The EFPI sensor mode of operation is based on the multi-reflection Fabry–Perot interference between the reflecting ends of the two cleaved optical fibers encapsulated in a capillary tube (Figure 2). One fiber acts as a lead-in/lead-out fiber, whereas the other fiber acts as a signal reflector. The gap between the two reflecting fiber ends forms a reflecting cavity and, for this reason, EFPI sensors are also known as cavity sensors.

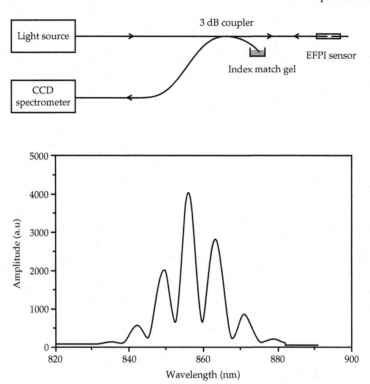

**FIGURE 6**   EFPI sensor: (a) CCD[1]-based conditioning system for the EFPI sensor; (b) optical spectrum of an EFPI sensor with a 50 μm cavity [10].

A single-mode optical fiber is used at the input and a multimode fiber is used at the output. The two fibers are inserted and fixed into a quartz capillary tube. The cavity comprises two reflectors that are parallel to each other and perpendicular to the axis of the optical fibers. The cavity length can be determined using a Fabry–Perot interferometer. The cavity length between the two surfaces of the optical fibers is changed when a strain is applied. A schematic illustration of an EFPI sensor system-based CCD spectrometer is shown in Figure 6a. The EFPI sensor is connected to the laser source and the CCD spectrometer through a 3-dB $2 \times 2$ coupler. One end of the coupler is immersed in an index matching gel to eliminate reflected light from the end faces. Figure 6c shows the optical spectrum of an EFPI sensor with a 50 μm cavity length. The absolute cavity length of the EFPI can be measured from the modulation in the reflection spectrum by counting the number of fringes over a specified wavelength range. The relationship between the cavity length and the wavelengths is

$$d = \frac{m\lambda_1 \lambda_2}{2(\lambda_2 - \lambda_1)} \tag{1}$$

where the phase difference at wavelengths $\lambda_1$ and $\lambda_2$ is $2m\pi$ and $m$ is an integer.

[1]CCD, charge-coupled device.

An EFPI sensor measures strain through a change in the cavity length, which is related to a phase change between the lead-in/lead-out and the reflection fibers. The two optical fibers in the sensor can be single mode, multimode, or a combination of them. Sensitivity of this type of sensor can be improved if metal or dielectric films are deposited at the mirrored ends. The major advantages of EFPI sensors are being well sensitive to longitudinal strain and quite insensitive to transverse strain. The major disadvantages of EFPI sensors are the stress concentrations around them when embedded. This is due to their relatively large size. Another disadvantage of EFPI sensors is the difficulty of deploying them in serial distributed sensor networks.

### 7.3.4 Transmission EFPI Fiber-Optic Sensors

The EFPI sensors can be based on the processing of either the reflected light or the transmitted light resulting from the sensor; in the latter case, they are abbreviated as TEFPI. The more recently developed transmission-type EFPI optical fiber sensor has both the advantages of reflection-type EFPI optical fiber sensors and a simpler and more effective function to distinguish strain direction than do reflection-type EFPI optical fiber sensors [11]. Therefore, this TEFPI optical fiber sensor could be easily applied to fatigue tests whereas the conventional reflection-type EFPI optical fiber sensor has difficulties in fatigue tests. The TEFPI sensor system needs only a laser diode, a photodiode, and a computer with an A/D board. The TEFPI sensor system is much faster and can sample data as you wish within the limit of the A/D board's sampling rate The disadvantage of the TEFPI sensor system is that the TEFPI sensor is composed with mechanically moving parts and is a handmade product [12].

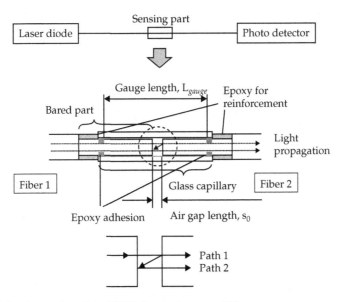

**FIGURE 7**    Principle of operation of the TEFPI fiber-optic sensor [12].

The structure of the TEFPI optical fiber sensor is shown in Figure 7. Conventional reflection-type EFPI optical fiber sensors have only one single-mode fiber for transmitting incident and reflected light. On the other hand, as shown in Figure 7, the TEFPI optical fiber sensor consists of two optical fibers (fibers 1 and 2) with an air gap between them. These two optical fibers are inserted into a glass capillary and used for transmitting light through the air gap. Fiber 1 is used for the incidence of light from the light source, and fiber 2 transmits the incident light from fiber 1 and the air gap to the optical receiver. Most of the incident light of fiber 1 is transmitted to fiber 2 through the air gap (path 1). However, 3.5% of the incident light from fiber 1 is reflected at the end of fiber 2. Of this light, 3.5% of this reflected light is reflected again at the end face of fiber 1 and then transmitted to fiber 2 (path 2). The phase difference of these two light paths causes interference in the optical receiver. Thus, the change in gauge length caused by the applied physical quantity in the longitudinal direction of the glass capillary results in the interferometric fringes as the output signal when the sensors are applied to the measuring objects. Because this TEFPI system differs from reflection-type EFPI optical fiber sensors, i.e., intensity loss from the spreading of light in the air gap occurs in both paths, the intensity of the transmitted light in the optical receiver clearly changes. Therefore, in this TEFPI optical fiber sensor system, the interferometric fringe counting indicates the measurement of the physical quantity and the degree of intensity loss from the spreading of light through the air gap reveals the distinction of the direction of strain. The measured strain can be calculated by the following equation:

$$\varepsilon = \frac{m\lambda}{4L_{gage}} \tag{2}$$

where $m$ is the number of half-periods of interferometric fringes, $\lambda$ is the wavelength of the light source, and $L_{gage}$ is the gage length.

### 7.3.5 In-line Fiber Etalon Sensors

Because of the physical discontinuity of EFPI sensors, an in-line fiber etalon (ILFE) sensor, a variant of EFPI, was developed [13] by fusion splicing a short segment of silica hollow-core fiber onto the short-stripped mirrored ends of the two optical fibers. The lateral dimension of an ILFE sensor is thus the same as the optic fiber diameter. This could be an advantage for its embedment in the host composite structures.

## 7.4 FBG OPTICAL SENSORS

The FBG sensor is an efficient embedded sensor for monitoring composite structures because it does not require mechanical splicing or coupling and is readily multiplexed. However, it requires elaborate instrumentation and data processing for achieving the strain detection, especially at high frequencies.

## 7.4.1 FBG Principles

FBG method is based on the action of a narrowband reflector consisting of a pattern of equally spaced gratings inscribed on the optical fiber at certain locations. The internal pitch of the grating determines the central wavelength of the light that is being reflected. One can get reflections of different wavelength from specific locations on the fiber by having gratings of different pitch values placed at those different locations. The strain sensing principle is based on the fact that a stretch of the fiber changes the central wavelength of the FBG reflector, i.e., generates a shift in the reflected light wavelength.

The periodic perturbations in the Bragg grating sensors essentially act as wavelength-specific mirrors. Although many wavelengths will be reflected at each perturbation, most are subsequently drowned out in the process of wave destruction as they undergo interference with other reflected signals. Only one particular wavelength will undergo constructive waveform addition. As a result, a narrowband spectrum is reflected with a central wavelength known as the Bragg wavelength (Figure 8). This condition is expressed as $\lambda_B = 2\eta_{eff}\Lambda$ where $\lambda_B$ is the Bragg wavelength or the peak reflected wavelength, $\eta_{eff}$ is the effective refractive index of the grating, and $\Lambda$ is the physical period of the grating.

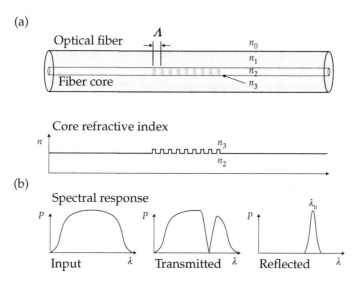

**FIGURE 8**    FBG optical sensors: (a) FBG principles; (b) details showing the notch in the transmission spectrum and the peak in the reflection spectrum at the Bragg wavelength $\lambda_B$ [6].

### 7.4.1.1 Strain Measuring Principle

The equation that governs the shift of the peak reflected wavelength under strain is

$$\Delta\lambda_B = \Delta\lambda_{BS} + \Delta\lambda_{BT} = \lambda_B(1 - p_e)\Delta\varepsilon + \lambda_B(\alpha_\Lambda + \alpha_n)\Delta T \tag{3}$$

where $p_e$ is the effective strain-optic coefficient, $\alpha_\Lambda$ is the thermal expansion coefficient, $\alpha_n$ is the thermo-optic coefficient, $\Delta\varepsilon$ is the strain increment, $\Delta T$ is the temperature variation,

$\Delta \lambda_{BS}$ represents the mechanical strain effects, and $\Delta \lambda_{BT}$ represents the thermal effects. If a constant temperature is maintained in the experiments, then Eq. (3) simplifies to

$$\Delta \lambda_B = \lambda_B(1 - p_e)\Delta \varepsilon \tag{4}$$

The effective strain-optic coefficient $p_e$ is a function of the Poisson ratio of the fiber $\nu$ and the Pockel's constants $\rho_{ij}$, i.e.,

$$\rho_\alpha = n_{eff}^2/2[\rho_{12} - \nu(\rho_{11} + \rho_{12})] \tag{5}$$

The relative wavelength shift is proportional to the strain, i.e.,

$$\frac{\Delta \lambda_B}{\lambda_B} = \kappa \, \Delta \varepsilon \tag{6}$$

where $\kappa$ is the FBG strain gauge factor. The theoretical value of $\kappa$ for pure silica glass has been calculated to be around $0.78/\varepsilon$ [19].

### 7.4.1.2 Refined FBG Patterns

The simple uniform FBG pattern produces considerable sidelobes in the reflection of the optical signal at the selected wavelength. This may be acceptable in simple applications; in more refined applications and in multiplexing, these sidelobes may impede the proper functionality of the FBG sensor. In order to reduce the sidelobes presence, apodizing of the FBG pattern is implemented. The apodizing method consists in modifying the intensity of each line in the grating such as to achieve a smoothed transition between the grated part of the fiber and the rest of the fiber. Other possible refinements of the FBG pattern are the chirped-FBG pattern and the $\pi$-phase-shifted FBG pattern [14].

## 7.4.2 Fabrication of FBG Sensors

Fabrication of the FBG gratings takes advantage of the photosensitivity of the optical fiber; it is done by the impression of UV light on the optical fiber. Low-reflectivity Bragg gratings can be produced in a Ge-doped single-mode glass fiber during the fiber drawing process and prior to coating. A pulsed excimer laser operating at 248nm with 450mJ pulses was used to photo induce Bragg gratings. The optical setup generates two converging UV beams that form an interference pattern within the core of the optical fiber (Figure 9a). This interference pattern has a discrete spacing that produced gratings with a single reflective wavelength of 1550nm. These gratings can be written into optical fibers at precise intervals along the length of the fiber ranging from every centimeter to many meters apart, simply by controlling the time interval between laser pulses. Each grating was 6mm in length and the reflectivity of individual gratings was measured to be approximately 100 parts per million. Fibers used in this research were coated with a polyimide resin and cured by a multistage in-line furnace [15].

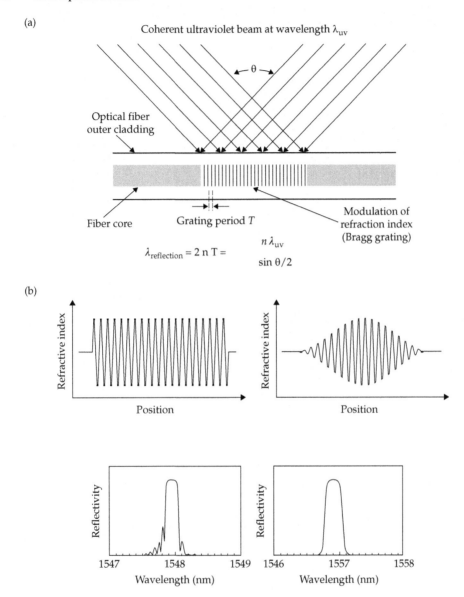

**FIGURE 9** (a) Fabrication of a Bragg grating on an optical fiber using two convergent UV beams that form an interference pattern on core of the optical fiber [16]; (b) improvement of the Bragg spectrum by apodizing the Bragg grating: left, uniform distribution with abrupt transition at the ends; right, apodized smooth in and out transition [17].

The Bragg reflection spectrum can be improved if the Bragg grating is apodized, i.e., by smoothing in and out the transition between normal fiber and the fiber imprinted with the Bragg grating. As shown in Figure 9b, apodizing results in suppression of spectrum sidelobes.

## 7.4.3 Conditioning Equipment for FBG Sensors

Reference [18] describes a number of strategies for determining the Bragg wavelength of signal reflected from a FBG sensor: scanning fiber filters; unbalanced fiber interferometers in combination with wavelength division multiplexers; wavelength swept lasers; and CCD arrays combined with a dispersive element. The latter method using a CCD array with a dispersive optical element is very promising because it can be implemented without moving parts. Its practicality has been hampered by the fact that the CCD technology has been mainly available at visible and near-infrared wavelengths while much of the Bragg grating technology and research effort has concentrated on the telecommunications wavelengths at 1300 and 1550 nm [18].

### 7.4.3.1 Optical Spectrum Analyzer Methods for Conditioning FBG Sensors

A simple FBG demodulation method relies on the use of an optical spectrum analyzer (e.g., ANDO AQ-6315B) which uses a tunable optical filter to measure the optical intensity at every wavelength over a scanned interval (Figure 10). Curve-fitting of the measured data determines a smooth profile and the identification of the wavelength of the Bragg reflected signal. Such a system has a scanning speed of about 1 Hz and thus is suited for quasi-static measurements only [19].

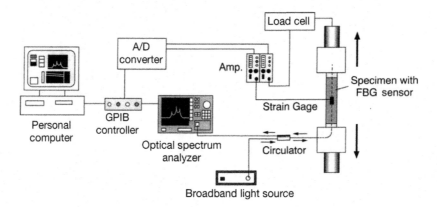

**FIGURE 10**    FBG system for measuring strain on an CFRP specimen [20].

### 7.4.3.2 Scanning FP Filter Methods for Conditioning FBG Sensors

The scanning FP filter method for FBG conditioning does not have the strain amplitude limitation of the interferometric systems. By scanning an optical FP filter with a fairly narrow transmission peak across the optical bandwidth of the source, signals from the FBG sensors are detected whenever the filter matches the wavelength of a particular FBG sensor (Figure 11a). Scan timing information allows us to find the filter transmission wavelength at the time of detection and thus determine the grating wavelength and corresponding strain. This method can in principle measure strains that move the Bragg peak across the full source spectrum, as long as the reflected wavelength bands from two FBG sensors do not overlap.

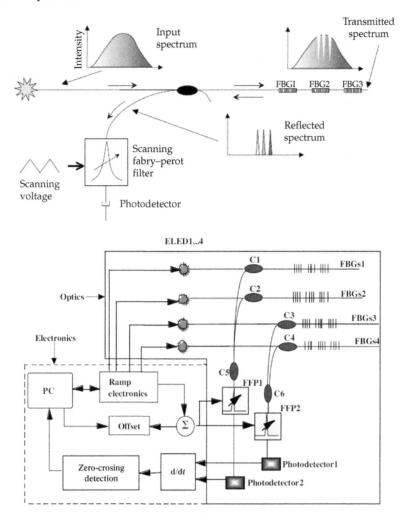

**FIGURE 11**    FBG conditioning with FP filters: (a) scanning FP filter method for conditioning the FBG sensor [18]. (b) FBG demodulation system using adjustable FP optical filter [16].

Reference [16] describes an FBG demodulating system that uses tunable FP filter to achieve demodulation of up to 64 channels, i.e., 64 FBG per fiber. The system uses four diodes (ELEDs[2]) as light sources, six couplers, two FP filters, two photodetectors, electronic parts, and a laptop computer (Figure 11b). The light emitted by the diodes and reflected back from the FBG sensor array is put through the tunable filters, which pass only a very narrowband wavelength. This passband is controlled by applying a rapidly stepped voltage to a piezoelectric actuator controlling the distance between the parallel mirrors of the tunable FP filter. The passed light signal is sent through a photodetector and differentiated. The zero crossings of the differentiated signal correspond to the peak

[2]ELED, entangled light-emitting diode.

wavelengths of the reflected light. The correlation between the ramp voltage level and the shifts in the zero crossings yields the strain for each sensor, since wavelength shifts are proportional to strain in the grating. The FP filters used in this work had a free spectral range of about 45 nm, thus allowing 16 individual sensors, spaced by approximately 2.7 nm, to be interrogated per filter scan. This spacing is sufficient to allow strains of about 1300 $\mu\varepsilon$ for each grating to be monitored. The resolution of the voltage ramp and the free spectral range of the filters determine the optimal strain resolution (minimum detectable strain); for the current system, a strain resolution of the order of $1-2\mu\varepsilon$ is typical. For the FP filters used, the ramp voltage can be applied at a maximum of 360 Hz, resulting in a Nyquist sampling rate of 180 Hz. Time averaging can be implemented in software for low-frequency applications where a 360 Hz sampling rate is not required; such averaging can result in improved resolutions to less than $1\mu\varepsilon$ [16].

Reference [11] describes optoelectronic FBG instrumentation developed by CEA-LETI[3] based on wavelength division multiplexing (WDM). Wavelength measurement is achieved in time domain with a scanning FP cavity. Specific software developed performs real-time DAQ, signal processing, results displaying, and storage (Figure 12).

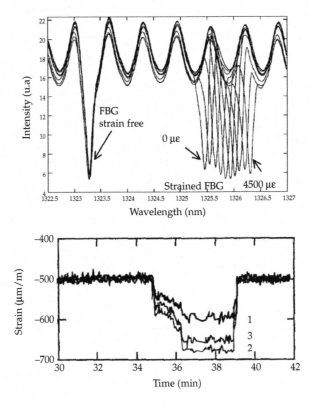

**FIGURE 12** FBG measurements with a CEA-LETI WDM system: (a) typical peak shifts due to strain loading up to 4500 $\mu\varepsilon$ in steps of 500 $\mu\varepsilon$ as measured by the FBG system; (b) loading and unloading strain measurements in a composite specimen instrumented with FBG sensors [11].

[3]CEA-Leti, a Grenoble, France-based research institute for electronics and information technologies.

### 7.4.3.3 **Interferometric Methods for Conditioning FBG Sensors**

Interferometer-based FBG conditioning systems exploit the fact that a change in wavelength at the input to the interferometer will manifest itself as a phase change at the output [18]. Techniques for measuring small phase changes with high accuracy have been developed for use with interferometric fiber-optic sensors, e.g., the phase-generated carrier technique. Signals from different Bragg gratings are routed to separate detectors and demodulators using fiber-optic WDM, which are optical bandpass filters (Figure 13). The interferometric technique has the advantage that it can be used to retrieve signals at frequencies of many kHz. However, the range of strain that can be detected is determined by the width of the passbands in the WDMs, typically in the range $\pm2000\mu\varepsilon$ for four-channel WDMs. This can be too little for some applications [18].

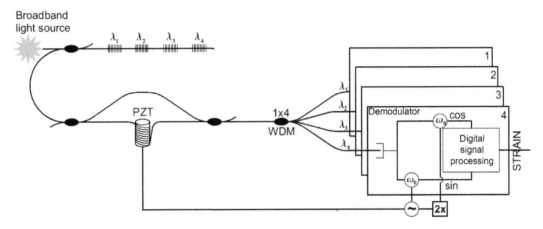

**FIGURE 13**    Interferometric FBG conditioning system using an unbalanced Mach–Zehnder interferometer for strain to phase conversion and phase-generated carrier demodulation for phase detection [18].

Reference [15] describes an FBG conditioning system using a TL source and a reference cavity consisting of a length of ordinary single-mode optical fiber between two air gap reflectors (Figure 14). Fiber-optic couplers act as beam splitters, sending laser energy both to the fiber instrumented with FBG sensors and to the reference cavity $L_{ref}$. The TL wavelength can be varied between 1510 nm and 1590 nm. The TL wavelength is varied in time for a particular "scan" of the sensing fiber. This scanning results in reflected fringes from the reference cavity that are detected and measured at specific wavelengths. This in turn triggers the recording of power coming back from the sensing fiber containing the gratings, resulting in a measure of reflected power as a function of wavenumber. The Fourier transform of this data set yields the reflected power as a function of the distance from a reference reflector located at the entry point into the sensing fiber. Thus the locations of the Bragg gratings relative to the reference reflector are accurately known. Reflected power as a function of distance information allows for the selection of, or "windowing" around, a particular FBG data set along the length of fiber. This data subset is Fourier transformed back into reflected power as a function of wavelength. The peak reflected power is used to determine the wavelength of the particular FBG at the time of measurement. Together with the initially measured wavelength, the current strain of any given grating can be determined.

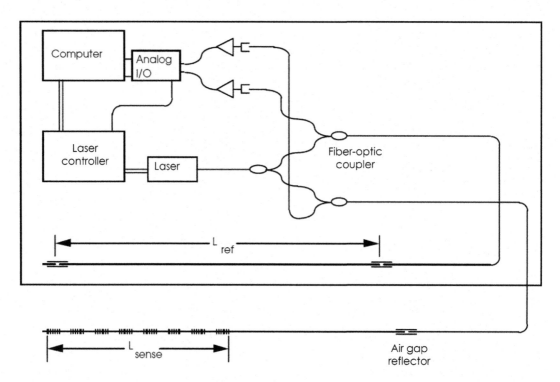

**FIGURE 14**   FBG demodulating system using a TL source and an interferometer [15].

Multipoint FBG systems use several fibers with several FBG systems on each fiber (Figure 15). Space division multiplexing (SDM) is used to individually address each fiber and wavelength division multiplexing (WDM) is used to individually address the FBG sensors on each fiber [21]. Signal processing electronics (Figure 15) identify the center of each FBG peak accurately using a slide and subtract technique: two versions of each pulse are generated and one is delayed with respect to the other. The delayed pulse is then subtracted from the other to produce a signal with a sharp zero crossing transition. The scan repetition rate is determined by the maximum desired refresh rate for each sensor and is fixed. Since the scan rate, Bragg grating width, and tunable filter passband width are all known, the pulse widths from the detectors are also known. These are constant because the scan rate of the system is fixed. By selecting the delay relative to the fixed pulse width, the gradient of the zero crossing transition can be maximized. The resulting signal has good noise immunity and can easily be converted into a transistor–transistor logic edge [21].

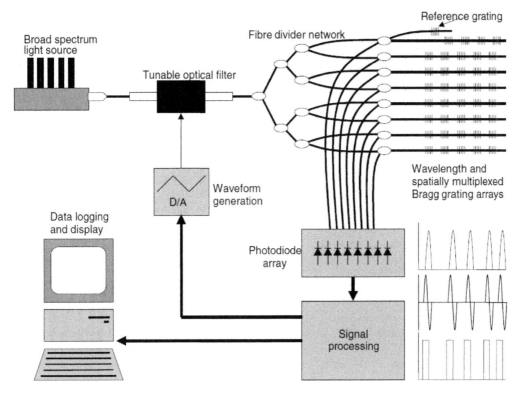

**FIGURE 15**    Multi-fiber multipoint FBG system [21].

### 7.4.4 FBG Demodulators for Ultrasonic Frequencies

The interpretation of FBG shifts at ultrasonic frequencies requires different approaches than at low frequencies or in quasi-static condition. The reason is that the optical spectrum analyzer (OSA) is a relatively slow device and hence OSA-based FBG conditioning is only appropriate for quasi-static measurements; the tunable optical filter is somehow faster and allows vibration measurements up to 180 Hz [16]. For ultrasonic frequencies, new principles of FBG conditioning must be used (Figure 16).

Reference [23] describes the use of a tunable matched fiber grating optical filter to demodulate the FBG signal generated by an acoustic emission (AE) event. A piezo transducer was used to generate ultrasonic vibration. A schematic of the experimental setup is shown in Figure 17.

Reference [24] describes a method of FBG conditioning that was used to collect ultrasonic strain data with 100-MHz sampling rate. The method is based on the overlap of two filters: (i) the FBG sensor and (ii) a fixed filter with a center wavelength just below the FBG filter. As shown in Figure 18b, when the FBG sensor is in tension and its center wavelength moves up, the quantity of light passing through the overlapped filters increases

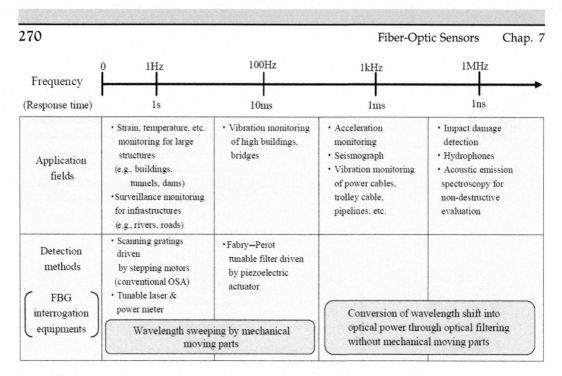

**FIGURE 16**   Various methods of FBG sensor demodulation as function of needed detection frequency and application type [22].

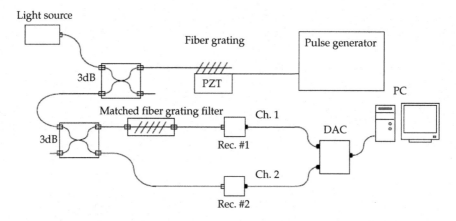

**FIGURE 17**   FBG sensor demodulator for ultrasonic frequencies using a tunable matched fiber grating optical filter [23].

(shaded area); when the FBG sensor is in compression and its center wavelength moves down, the passing light decreases. The variation in light intensity is captured by a photodetector and converted into strain. Since the photodetector is a very fast device, the speed of this process is only limited by the speed of the DAQ card.

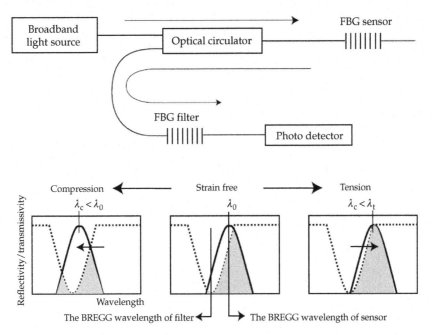

**FIGURE 18**    Method for dynamic FBG measurements based on the overlap of two filters [24].

### 7.4.4.1  Dual (Quasi-Static + Ultrasonic) FBG Sensing

Reference [25] describes a dual demodulator FBG sensor system. The dual demodulator is composed of (a) a demodulator with a tunable Fabry–Perot filter measuring the low-frequency signal with large magnitude such as strain and (b) another demodulator with a passive Mach–Zehnder interferometer measuring the high-frequency small-amplitude wave signal such as generated by impact or acoustic emission at the damage zone (Figure 19).

### 7.4.4.2  Fabry–Perot Tunable Filter for Ultrasonic FBG Demodulation

Reference [26] discusses the high-speed FBG interrogator (si920) provided by Micron Optics Inc. which uses a Fabry–Perot tunable filter (Figure 20). The interrogation method is based on matching the edge of the FP filter to the edge of the FBG primary peak. This interrogation method shares the inherent high-speed capability of edge filter methods; however, the FP transmission profile is devoid of spectral ripples, which would limit the sensitivity of grating of thin-film linear edge filters. Additionally, the interrogator has closed loop tracking of static peak wavelength changes such as due to thermal drifts through the tuning of the FP filter (active arm). The peak wavelength interrogation was performed at 625 kHz with one FBG sensor and at 295 kHz with three FBG sensors.

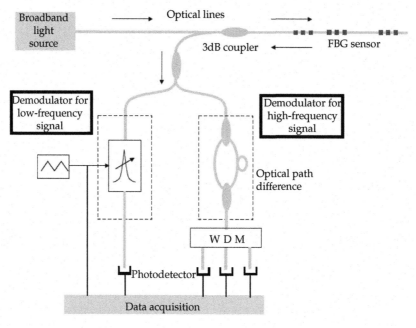

**FIGURE 19**   Dual demodulator FBG sensor system for simultaneous measurement of quasi-static strain and high-frequency strain wave components of the total strain [25].

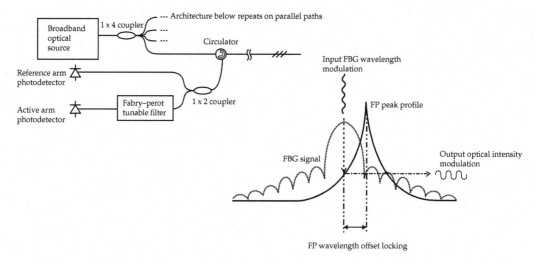

**FIGURE 20**   High-speed FBG sensor demodulation system using an adjustable FP filter tuned to the edge of the FBG primary peak [26].

### 7.4.4.3 Tunable Narrow-Linewidth Laser Source for Ultrasonic FBG Demodulation

Reference [27] discusses the principle of measuring small-amplitude ultrasonic strains with FBG sensors that are interrogated by a tunable narrow-linewidth laser source (e.g., a specially configured laser diode).

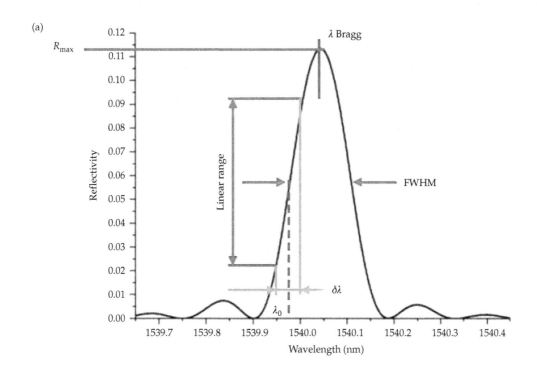

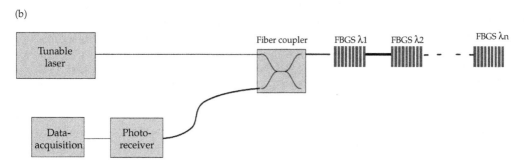

**FIGURE 21** The use of the full width at half-maximum (FWHM) principle for measuring ultrasonic strains with an FBG sensor and a TL source: (a) the FWHM principle; (b) TL-based equipment for demodulation FBG sensors to detect ultrasonic waves; (c) ultrasonic guided-wave signals captured with FBG sensors: "transducer" refers to the position of the transmitter transducer. Note that the FBG sensor is highly directional; the signal of the FBG sensor aligned with the wave propagation direction (top) being an order of magnitude stronger than the signal of the FBG sensor perpendicular to the wave propagation direction (bottom) [27].

(c)

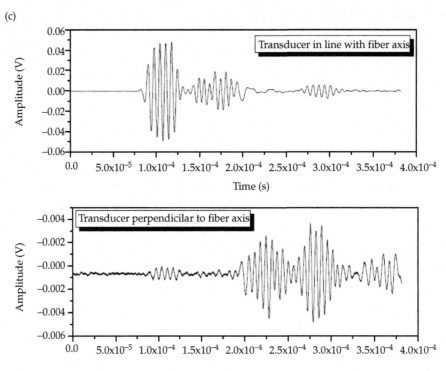

FIGURE 21    (Continued)

If the wavelength of the laser matches a certain part of the grating spectrum, any shift of the spectrum will as a consequence modulate the reflected optical power (Figure 21a). The use of a TL source allows the interrogation of several gratings within a single fiber line. In order to achieve the maximum range in both directions, the laser should be set to the wavelength corresponding to full width at half-maximum (FWHM). This results in a linear relation between the shift of the optical spectrum and the reflected power amplitude. With the laser being fixed at the wavelength $\lambda_0$ (Figure 21a), the time-varying reflection of the grating can be given as a function of the time-varying wavelength shift of the main peak due to the ultrasonic wave. A schematic of the equipment for demodulation the FBG sensor for ultrasonic frequencies is given in Figure 21b. Several FBG sensors are interrogated simultaneously through the WDM principle.

Experiments were performed on a perplex plate using both FBG and piezo receivers. The piezo transmitter was excited with a 150-kHz five-cycle tone burst. Good detection of ultrasonic waves was observed with both piezo and FBG sensors. Figure 21c shows the ultrasonic wave signals recorded with the FBG sensors. In this figure, the term "transducer" refers to the position of the transmitter transducer relative to the FBG fiber sensor. Note that the FBG sensor is highly directional, the signal of the FBG sensor aligned with the wave propagation direction (bottom) being an order of magnitude stronger than the signal of the FBG sensor perpendicular to the wave propagation direction (top). In contrast, the piezo sensors are multidirectional and can equally record waves coming from all directions.

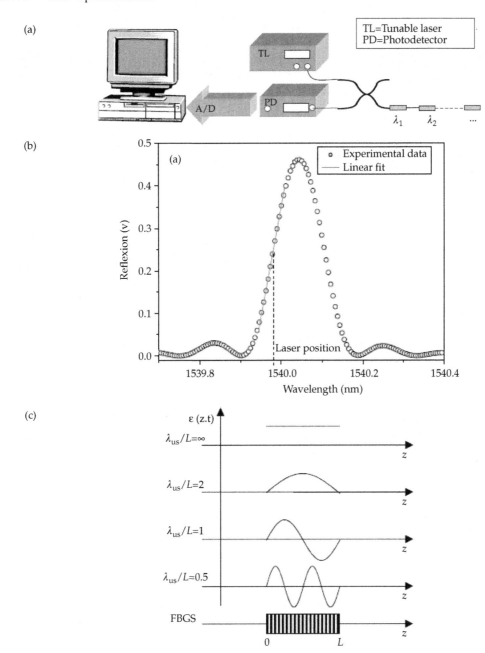

**FIGURE 22**   High-frequency interrogation of FBG sensors with a TL source: (a) principle of FBG demodulation with a TL source; (b) measured grating spectrum and linear fit between 20% and 80% of the peak value; (c) presentation of acoustic waves to the FBG sensor for various wavelength-to-grating length ratios [28].

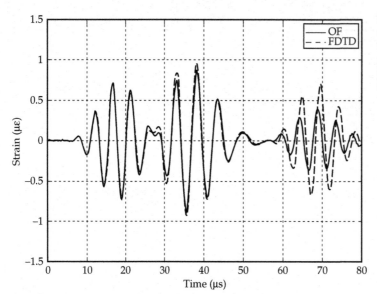

**FIGURE 23** Ultrasonic Lamb waves measured with the optical fiber (OF) AMAP sensor compared with calculations done with finite differences in time domain (FDTD) integration method [29].

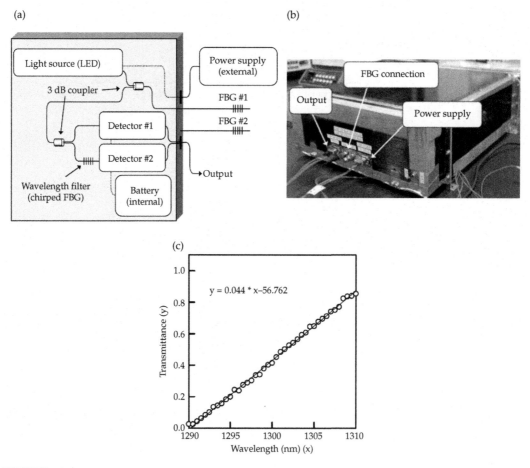

**FIGURE 24** On-board FBG sensor demodulation system tested on a reusable launch vehicle: (a) schematic; (b) actual space-qualified hardware weighing less than 2 kg; (c) characteristics of the chirped-FBG filter showing a linear relationship between transmissibility and wavelength of the incoming light [30].

Reference [28] further discusses the method for measuring ultrasonic strain with FBG sensors using a low noise narrow-linewidth TL diode and a high sensitivity photodetector (PD) as shown in (Figure 22a). The TL wavelength is set to the center of one of the slopes of the FBG spectrum; this results in a linear relation between the shift of the optical spectrum and the reflected power amplitude. The strain-induced shift of the FBG spectrum modulates the reflected optical power as detected at the PD receiver. The use of a widely-tunable laser source allows the interrogation of several FBG sensors placed on the same optical fiber be sequentially placing the operating points on the respective Bragg spectra.

This interrogation method uses the part of the spectrum where the grating spectrum can be assumed to be linear; for most gratings this happens at about 20–80% of the gratings' maximum reflectivity (Figure 22b). The TL wavelength is kept constant during the measurement of a specific FBG sensor; in order to achieve the maximum strain measurement range, the TL wavelength should be set to the FWHM position (Figure 22c).

Shift in the grating spectrum caused by slow varying temperature or static strain can be eliminated with high-pass filtering of the final signal because they are well below the frequency range of the ultrasonic signals. However, if such shifts are larger, the operating point may move in an undesirable range of the FBG pass spectrum; in this case, feedback control could be used to retune the TL to the FWHM operating point on another FBG sensor.

A similar approach is reported in Ref. [29] that also uses a TL source that is sequentially tuned to the FWHM operating point of the Bragg spectrum of the FBG sensor. In addition, Ref. [29] describes the clustering of the FBG sensors in an array of uniformly distributed FBG sensors that allows the modal separation of structural plate waves (Figure 23). The distributed array is termed the acoustic mode assessment photonic (AMAP) sensor. Good measurements of the ultrasonics waves transmitted from a piezoelectric transmitter were observed.

### 7.4.4.4 Chirped-FBG Filter for Ultrasonic FBG Demodulation

Reference [30] describes a compact FBG sensor demodulation system weighing less than 2 kg that was installed on board of a reusable launch vehicle (Figure 24). The system uses a chirped-FBG wavelength filter to achieve demodulation of the FBG sensor signal. The transmissibility of the chirped-FBG wavelength filter depends linearly on the wavelength of the passing light (Figure 24c). The light reflected from the FBG sensor is divided into two beams: one beam is the reference beam and is directly detected by photodetector #1. The other beam is transmitted through the wavelength filter and then detected by photodetector #2.

The intensity of the light passing through the wavelength filter is modulated in relation to the 1value of the Bragg wavelength of the FBG sensor. Refer to Figure 24a which shows two photodetectors named Detector #1 and Detector #2. The output of Detector #2 is divided by the output of Detector #1 in order to compensate the intensity drift of the reflected light. Consequently, this output ratio is proportional to the Bragg wavelength and allows the measurement of strain from the shift of the Bragg wavelength. Because this FBG sensor demodulation method has no moving parts, its frequency response is very high and is thus suitable for ultrasonic measurements. The upper frequency is only limited by the DAQ system used to digitize the output photodetectors.

### 7.4.4.5  Array Waveguide Grating Filter for Ultrasonic FBG Demodulation

References [22,31] describe the use of an arrayed waveguide grating (AWG) filter to detect ultrasonic strain waves. AWG filters consist of an array of many narrowband filters (Figure 25). The system was developed by the Hitachi Cable Ltd. [22]. When the reflected light from the FBG enters the AWG filter, the reflection spectrum passes over two adjacent peaks and the optical power is modulated to $P_1$ and $P_2$ by the filter 1 and filter 2, respectively.

The modulated optical powers are detected by the photodetectors, and the center wavelength shift, which is proportional to the strain applied to the FBG sensor, is obtained from interpolation between the powers $P_1$ and $P_2$. Since this system converts the wavelength into the optical powers directly without mechanical moving parts, the strain change can be measured sufficiently fast to detect ultrasonic strain waves.

### 7.4.4.6  Two-Wave Mixing Photorefractive Crystal for Ultrasonic FBG Demodulation

Reference [32] mentions a multiplexed approach that allows the monitoring of high-frequency dynamic strains ($f > 100$kHz). Two-wave mixing (TWM) photorefractive crystal (PRC) spectral demodulator is used to simultaneously acquire the ultrasonic waves from several FBG sensors. The schematic of the experimental setup is shown in Figure 26. The FBG sensor nominal wavelengths are separated by at least 4 nm with a 0.3-nm spectral line. A C-band (1530–1570 nm) amplified spontaneous emission source with 100 mW total CW power is used.

The light reflected from the FBG sensors went through a circulator to an erbium-doped fiber amplifier to be amplified to 500mW total power. Subsequently, it was split into "signal" and "pump" beams with a $1 \times 2$ coupler, and then spectrally demodulated through an adaptive InP:Fe PRC interferometer. The angle between the two beams was $6°$ at the input plane to the PRC. In the PRC, the signal and pump beams coherently interfere to form six independent index of refraction gratings through the photorefractive effect. Each grating corresponds to a particular wavelength, i.e., to a particular FBG sensor. The photorefractive gratings act as volume holograms. As such, in the direction of the transmitted signal beam, a diffracted pump is produced that has the same quasi-static information as the signal beams. These beams interfere after the PRC and produce an intensity change that is proportional to the phase change between the two beams. The phase change is linearly proportional to the optical path length difference between the signal and pump beams before the PRC, and the spectral shift experienced in the FBG sensors. A 6 kV/cm DC electric field was applied to the PRC to enhance the strength of the holograms and the two-beam coupling. With the application of the DC field, the static phase difference between the transmitted signal and diffracted pump beams is maintained at near quadrature. This allows for linear and sensitive conversion of the wavelength shifts of the FBG sensors into an intensity change. The two-TWM intensity gain for the interferometer was $0.2$cm$^{-1}$; this parameter quantifies the efficiency of optical beam diffraction in the PRC. After the PRC, the transmitted signal and diffracted pump beams were recoupled into an optical fiber and directed through a set of band-drop filters that split the light into six beams with different wavelengths corresponding to the respective FBG sensor signals. The

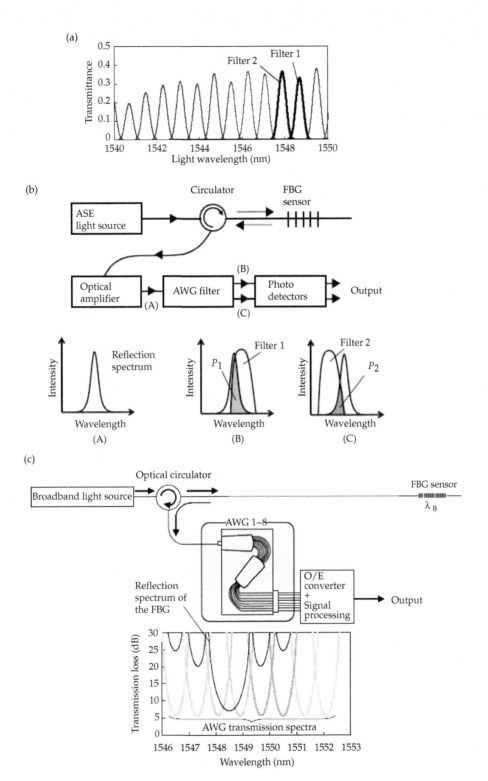

**FIGURE 25** FBG sensors demodulation to detect ultrasonic waves using an AWG filter: (a) multi-peak spectrum of the AWG filter; (b) schematic of the equipment principles [31]; (c) high-speed FBG sensor demodulator using AWG optical filter [22].

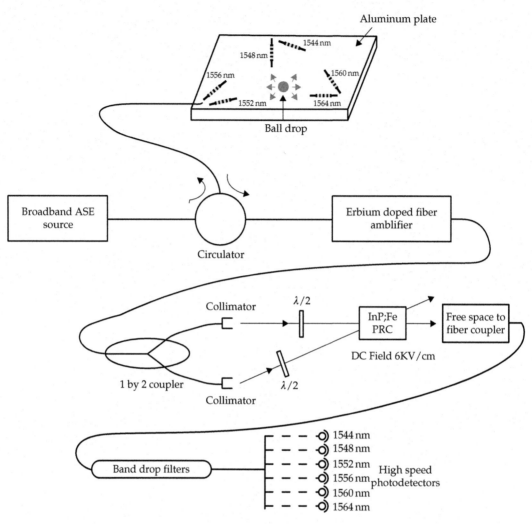

**FIGURE 26**   PRC-based system for FBG sensor demodulation at ultrasonic frequencies [32].

six beams were delivered to different high-speed photodetectors, and the output of the detectors was digitized in an oscilloscope and sent to a computer for data processing. TWM spectral interferometer is self-adaptive and only allows for the demodulation of wavelength shifts that occur on a time scale that is smaller than the grating formation time, which is about 1ms for the parameters used in this work. Changes that occur slower than this result in an adaptive reconfiguration of the photorefractive gratings, and the diffracted pump beams track the quasi-static changes in the signal beams. Therefore, the demodulator adaptively compensates for low-frequency drifts caused by large quasi-static strain and temperature drift and allows only high-frequency signals to pass through.

This makes the TWM spectral demodulator appropriate for monitoring dynamic strains on the FBG sensors such as those caused by vibration, acoustic emission, and stress waves due to impact. The low-frequency vibration of an aircraft wing caused by wind gusts and temperature changes will be automatically filtered by the TWM interferometer and high-frequency signals caused by fretting or crack propagation will be passed through the TWM interferometer for further analysis. A minimum separation of 4 nm is sufficient for WDM with negligible cross-talk [32].

## 7.4.5  FBG Rosettes

The directionality of the FBG sensors has been advantageously used in Ref. [33] to construct FBG sensor rosettes that can be processed to resolve the principal directions of strain wave propagation (Figure 27).

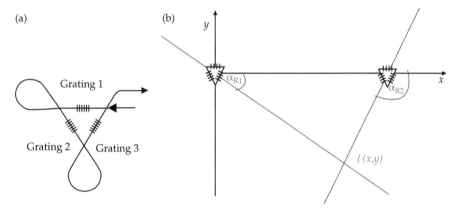

**FIGURE 27**    FBG sensor rosette: (a) rosette layout; (b) possible configuration for damage detection and localization [33].

The use of FBG rosettes is also hinted in Ref. [32] as illustrated in the corners of the plate specimen depicted in Figure 26.

## 7.4.6  Long-Gage FBG Sensors

Reference [34] describes the use of long-gage FBG sensors for distributed sensing. Long-gage FBG sensors are 10 times longer than ordinary FBG sensors (which are ~10 mm long). The long-gage FBGs can measure strain or temperature at an arbitrary position within the gage length with a high spatial resolution of submillimeter order when connected in an optical frequency domain reflectometer (OFDR). A low reflective index is required for the light to reach the whole grating (Figure 28). The OFDR measurement system consists of a TL source with adjustable wavelength (e.g., ANDO AQ4321), two photodiode detectors (D1, D2), three broadband reflectors (R1, R2, R3), three 3-dB couplers (C1, C2, C3), a long-gage FBG, and a PC with an A/D converter card (e.g., National Instrument PCI 6115). The part including D1, C2, R1, and R2 make up an in-fiber interferometer for an external clock. The part comprising D2, C3, R3 and the long-gage FBG is an interferometer for measurement. The interferometer for the external clock has an optical path length difference of $2nL_R$ where $n$ is the effective refractive index and $L_R$ (15.47 m) is the path difference of the two paths through the interferometer. $L_R$ has to be more than twice as long as $L_i$ to satisfy the sampling theorem.

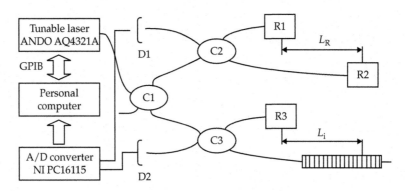

**FIGURE 28**   Long-gage FBG sensor and OFDR system for distributed strain sensing [34].

The laser light signal is separated at C1 and travels to C2 and C3. The light reflected between R3 and each grating is acquired by D2. The signal at D2 is given by

$$D2 = \sum_i R_i \cos(2nL_i k) \qquad (7)$$

Information about the sensor location is included in the signal at D2 as a frequency component. The spectrum at each grating position is derived by Fourier transform analysis. The external interferometer is required for Fourier transform analysis to allow the D2 signal to be acquired at a constant wavenumber interval, $\Delta k$. The light reflected between R1 and R2 is acquired by D1. The signal at D1 is given by

$$D1 = 2[1 + \cos(2nL_R k)] \qquad (8)$$

As the laser is tuned, the light intensity observed by D1 varies in a cycle depending on the wavenumber change, $\Delta k$, which is given by

$$\Delta k = \frac{\pi}{nL_R} \qquad (9)$$

The positive-going zero crossing of the signal at D1 is used to trigger the sampling of the signal at D2. The sensor signal which was acquired at D2 undergoes Fourier transform analysis with a sliding window along the wavenumber axis resulting in a spectrogram with the $x$- and $y$-axes indicating the wavelength and the distance, respectively. The central wavelength was defined from the center of the FWHM interval of the spectrogram which was acquired from the sensor signal at D2. Therefore, we could map the temperature or strain profiles along the gage length of FBGs. The readout resolution in the wavenumber domain is expressed as $(N_W - N_O)\Delta k$ where $N_W$ is the window width and $N_O$ is the number of overlaps. The following equation applies in the wavelength domain

$$k_2 - k_1 = -\frac{2\pi\Delta\lambda}{\lambda_1\lambda_2} = (N_W - N_O)\Delta k \tag{10}$$

$$\Delta\lambda = (N_W - N_O)\frac{\lambda_2 - \lambda_1}{\left(\frac{2\pi}{\lambda_1} - \frac{2\pi}{\lambda_2}\right)}\Delta k \tag{11}$$

In Ref. [34], $N_W = 4000$, $N_O = 3800$, $\Delta k = 0.140 \text{m}^{-1}$. The parameters $\lambda_1$ and $\lambda_2$ are the wavelengths for each window. The wavelength resolution $\Delta\lambda$ is approximately $\Delta\lambda \approx 0.011 \text{nm}$. Therefore, the readout strain resolution was approximately $\pm 5\mu\varepsilon$. The spatial resolution and the readout strain resolution depend on the window length and wavelength resolution, respectively.

## 7.4.7 Temperature Compensation in FBG Sensing

One issue with the use of FBG strain sensors is related to the separation of mechanical strains from thermal expansion strains. (This issue is common to the electrical wire strain gage too.) The strain applied to the FBG optical fiber inserted in a specimen may be mechanical or thermal in nature; hence, additional actions are taken to separate the two effects, because the basic FBG sensor cannot directly differentiate between mechanical strain and thermal expansion.

The effect of temperature expansion can be measured independently using a strain-isolated rosette, when feasible [21]; thus, the temperature effect can be subtracted from the total strain to obtain the structurally induced strain.

### 7.4.7.1 Temperature Compensation through Coefficients of Thermal Expansion Difference between Two Materials

Reference [17] describes a method of FBG temperature compensation based on the different coefficients of thermal expansion (CTE) of CFRP and glass-fiber-reinforced polymer (GFRP) composites. A hybrid CFRP/GFRP montage is used; the experimental setup of this method (Figure 29) works in combination with a thorough theoretical analysis [17].

### 7.4.7.2 FBG Temperature Sensor

Reference [35] describes a temperature sensor based on the FBG principles. The physical principle is based on the fact that temperature changes also contribute to a shift of the Bragg wavelength as indicated by Eq. (3). The FBG temperature sensor (Figure 30) consists of a 10-mm-long FBG encapsulated in a capillary steel tube (140/300 μm) to be insulated from external strain. The FBG is thus in a strain-free condition and only influenced by the temperature changes.

(a)

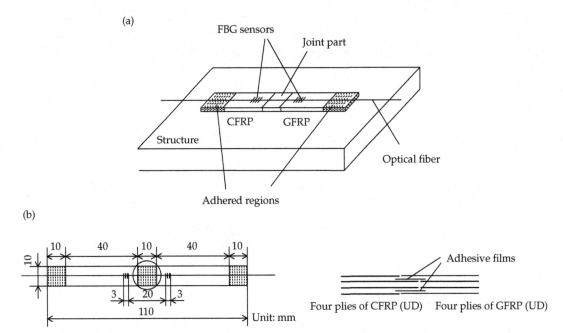

(b)

FIGURE 29  Hybrid CFRP/GFRP montage of FBG sensors for temperature compensation: (a) general layout; (b) cross-section view [17].

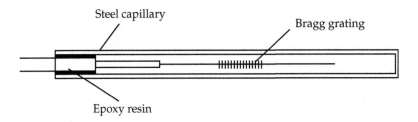

FIGURE 30    FBG temperature sensor [35].

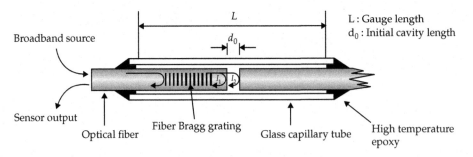

FIGURE 31    Hybrid FBG/EFPI sensor for separating the mechanical strain from the temperature strain [37].

### 7.4.7.3 Hybrid FBG/EFPI Sensors

References [36,37] report a hybrid FBG/EFPI that uses the FBG effect to measure temperature-induced strains and the extrinsic EFPI cavity to measure the total (temperature + mechanical) strain. Hence, the mechanical strain alone can be determined by subtracting the temperature strain from the total strain. A 10-mm-long FBG was encapsulated in a capillary silica tube (140/300 μm) to isolate it from the external strain (Figure 31). The EFPI cavity was formed between the fiber end face near to the FBG and a gold-coated fiber end surface inserted into the opposite end of the capillary tube.

## 7.5 INTENSITY-MODULATED FIBER-OPTIC SENSORS

Intensity-modulated sensors are perhaps the simplest fiber-optic sensor; it is based on a modulation of light intensity in a multimode optical fiber; the latter allows propagation of the light wave with greater intensity. It reflects losses of light intensity when any portion of the optical fiber is strained.

### 7.5.1 Typical Intensity-Modulated Fiber-Optic Sensors

Although intensity-modulated fiber-optic sensors are relatively easy to construct and do not require complex instrumentation and signal processing [3], they are susceptible to light fluctuations, or light loss associated with microbending of the optical fiber. They also suffer from the lead−fiber sensitivity and lack a clearly defined sensing region; hence they cannot determine the location of damage. Sensitizing of the optical fiber by selective removing of the cladding or microbending has been used to increase the response of this simple measurement method. The use of a networked grid of intersecting fibers has also been considered in order to provide a low-cost method for damage location identification [3].

### 7.5.2 Intensity-Based Optical Fiber Sensors

The intensity-based optical fiber (IBOF) sensor was first reported in 1995 by Ref. [38]; subsequently, it was further developed as described in Ref. [7]. The basic principle of measurement of the IBOF sensor is based on light transmission loss in an optical cavity as illustrated in Figure 32a,b.

Axial separation in the gap between two cleaved FO facing each other results in a change in the output light intensity. The change in output light intensity is related to the gap size. Fabrication of the IBOF sensors is basically the same as for extrinsic FP sensors (Figure 32b): two cleaved optical fibers were inserted into a capillary tube and then bonded with epoxy adhesive. An initial gap between the fiber ends ensures operation in both compression and extension strains. The gage length of the IBOF sensor is defined by

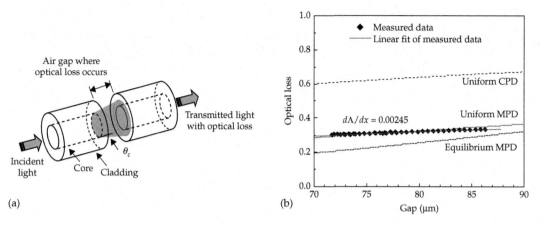

(a)                                                    (b)

**FIGURE 32**    IBOF sensor: (a) schematic of the IBOF sensing principle; (b) IBOF measurements versus various prediction methods [7].

the distance between the two bonding points. In Ref. [7], a quartz glass capillary tube with inner and outer diameters of 128 and 280 μm, and graded-index multimode fiber with a 50/125 μm core/cladding diameter and a numerical aperture of 0.201 were used. The light in the optical fiber can be generated by a tungsten halogen lamp which is launched through a microscope objective lens. The sensor output can be detected by a photodiode system, and the change in light intensity (optical loss, $\Lambda$) can be evaluated. Tensile experiments performed in Ref. [7] showed a good linear relationship between the optical loss $\Lambda$ and the gap in the cavity. Hence, the strain applied to the IBOF sensor can be determined as function of $\Lambda$ (Figure 32b).

## 7.6  DISTRIBUTED OPTICAL FIBER SENSING

The fused silica material used in the optical fiber has intrinsic scattering properties (Figure 33) that affect the light wave propagating through it. Several backscatter waves are possible: Rayleigh, Brillouin, Raman, and Stokes (Figure 33).

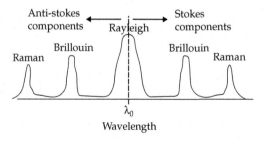

**FIGURE 33**    Spectrum of intrinsic scattering for fused silica optical fibers [3].

Brillouin scattering occurs due to interactions with acoustic waves known as phonons; its frequency shift can be related to applied strain or temperature. The weak Brillouin can be amplified by stimulating acoustic waves through the use of a second pulsed laser connected to the opposite end of the optical fiber. By scanning the frequency of the pulsed laser, the relative frequency shifts at which the stimulated Brillouin scattering occurs can be identified and converted into the applied strain or temperature. To determine the location of each Brillouin scattering event, the time of arrival of the scattered light is interrogated. In this manner, the strain or temperature distribution along an optical fiber could be measured. The strain and temperature resolution are around $5\mu\varepsilon$ and $0.25°C$, respectively; nominal spatial resolution with a single laser source is about 5 m. The use of advanced signal processing techniques, e.g., Brillouin optical correlation domain analysis has been shown to improve the spatial resolution significantly [3].

Various methods for distributed fiber-optic sensing using reflectometry principles exist, as described next.

## 7.6.1 Optical Time Domain Reflectometry

Optical time domain reflectometry (OTDR) is a method to detect changes in the structural strain from local reflection induced by an optical fiber sensitive to microbending. A laser diode launches very short pulses into the fiber. The fiber microbend acts as a reflector. The reflected light is detected by a photodetector coupled with a fast sampling electronics. Knowing the fiber properties, one can relate the round-trip travel time to the location of the reflector on the fiber (Figure 34). The OTDR sensor is suitable for distributed sensing but also requires relatively expensive instruments and a DAQ system with a high data sampling rate.

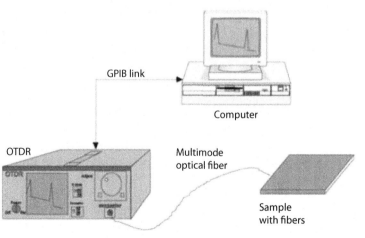

**FIGURE 34**   Schematic diagram of an OTDR measurement [11].

## 7.6.2 Brillouin Optical Time Domain Reflectometry

There are three important types of scattering in optical fibers that can be exploited in distributed sensing. They are Rayleigh, Raman, and Brillouin scattering. Among them, Brillouin scattering is a light that is inelastically scattered by interaction with acoustic phonons [39]. Brillouin scattering experiences a frequency shift with respect to the incident light proportional to the acoustic velocity within a fiber. The amount of the frequency shift is called Brillouin frequency shift $v_B$ and it is given by

$$v_B = \frac{2nV_a}{\lambda} \tag{12}$$

where $n$ is the fiber core refractive index, $V_a$ is the acoustic velocity, and $\lambda$ is the light wavelength. The shift is about 11 GHz at a wavelength of 1550 nm. It was reported that the Brillouin frequency shift in silica fibers varies greatly with strain and temperature [39]. The strain and temperature coefficients of $v_B$ change are given

$$\frac{dv_B}{d\varepsilon} = 49.3 \, \text{GHz}, \quad \frac{dv_B}{dT} = 1.00 \, \text{MHz}/^\circ\text{C} \tag{13}$$

This dependence of the Brillouin frequency shift on strain and temperature can be used for distributed sensing. Brillouin optical time domain reflectometry (BOTDR) measures local changes in the Brillouin shift along the length of a sensing fiber by using the OTDR method and spectral analysis. As with conventional OTDR, the spatial resolution, $\delta z$, of BOTDR is determined by

$$\delta z = \frac{vW}{2} \tag{14}$$

where $v$ is the light velocity in the fiber and $W$ is the incident width of the pulsed light [39]. Although this equation shows that the pulse narrowing improves the spatial resolution, the shorter pulse is detrimental for the frequency resolution of the measurements, namely, for the accuracy of strain and temperature measurements [39]. The spatial resolution of BOTDR is generally more than 1 m and the accuracy for strain measurement is about $\pm 30\mu\varepsilon$. Reference [40] describes a study of the possible use of the BOTDR method for strain measurements in civil engineering applications (Figure 35).

Reference [41] discusses the pre-pump pulse Brillouin optical time domain analysis (PPP-BOTDA) method, which measures the strain along an optical fiber by stimulated Brillouin scattering. The BOTDA system realizes a special resolution of 100 mm, a sampling interval of 50 mm, and a sensing range of more than 1 km. the PPP-BOTDA sensing system (Figure 36). It employs a stimulated Brillouin scattering technique and can measure axial strain at an arbitrary point along the optical fiber. Two laser beams, a pump pulse light having a unique wave profile, and a continuous wave (CW) probe light, are injected into both ends of an optical fiber. The interaction of these two laser beams (the pre-pump and the probe lights) excites acoustic waves, due to their different frequencies. The pulse part of the pump light is then backscattered by the phonons, and part of its energy is transferred to the CW. The power gain of the CW, which is called the Brillouin gain spectrum (BGS), is measured at the output end of the probe light while the frequency of the

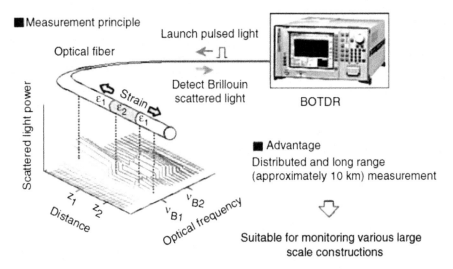

**FIGURE 35**    BOTDR principles [40].

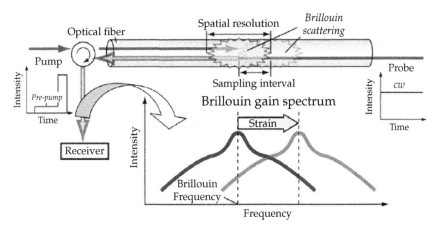

**FIGURE 36**    PPP-BOTDA system [41].

probe light is scanned as a function of the frequency difference between the two laser beams. The value of the axial strain can be estimated by measuring the peak frequency of the BGS, which is called a Brillouin frequency, while its position along the optical fiber is calculated from the light round-trip time. When the temperature is uniform, the Brillouin frequency $F(z)$ at a position $z$ is expressed as

$$F(z) = F_0(T) + C_s \varepsilon(z) \tag{15}$$

where $F_0(T)$ is the initial Brillouin frequency at temperature $T$, $\varepsilon(z)$ is the axial strain along the optical fiber, and $C_s$ is the strain proportionality constant.

### 7.6.3 Continuous Fiber Sensing Using Rayleigh Backscatter

The continuous fiber sensing technique uses inexpensive telecom-grade optical fiber to measure strain without the need of inserting FBG sensing sections into the fiber [42]. The physical principle is based on the Rayleigh backscatter principle. An external stimulus (like a strain or temperature change) causes temporal and spectral shifts in the local Rayleigh backscatter pattern. Optical backscatter reflectometry uses swept wavelength interferometry to measure the Rayleigh backscatter as a function of length in optical fiber with high spatial resolution. The changes in the local backscatter pattern produced by strain changes are measured and interpreted to give the actual strain values. The reflectometry principle is used to locate these readings at any desired position along an up to 20-m fiber with mm resolution. The strain measuring range is ±10,000 microstrain. A good combination of high-density sensing and dynamic acquisition can be achieved. The optical fiber can be meandered inside a composite layup to enable the measurement of strain value at many locations simultaneously at reasonable rates. Rates of up to 250 Hz can be achieved over a 2-m fiber and a 5-mm equivalent gage length with a ±5 microstrain repeatability. For a 20-m fiber, one can achieve 50-Hz acquisition rate with a 5-mm gage length and ±10 microstrain repeatability (Figure 37).

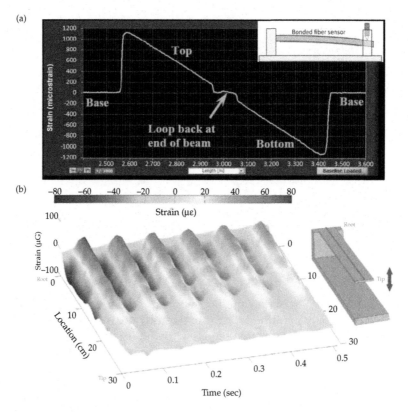

**FIGURE 37**    Rayleigh backscatter reflectometry strain measurement in a 30-cm cantilever aluminum beam using a plain fiber glue top and bottom on the beam surface: (a) static strain recording; (b) vibrational strain response as the beam oscillates at its natural frequency. The recording was done with Luna's Optical Distributed Sensor Interrogator OdiSI-B10 equipment [42].

## 7.6.4 Fiber-Optic Temperature Laser Radar

Raman scattering is a light that is inelastically scattered by interaction with optical phonons. It is known that the intensity ratio of the two components of Raman scattering, i.e., Stokes light and anti-Stokes light, is a function of temperature [39]. This ratio is given by Ref. [39] as

$$\frac{I_a}{I_s} = \frac{(\tilde{\nu}_0 + \tilde{\nu}_k)^4}{(\tilde{\nu}_0 - \tilde{\nu}_k)^4} \exp(-h\,\nu\,\tilde{\nu}_k/kT) \tag{16}$$

where $I_a$ and $I_s$ are intensity of anti-Stokes light and Stokes light, respectively, $\tilde{\nu}_0$ is the wavenumber of the incident light, $\nu_k$ is the wavenumber shift of the glass, $h$ is Planck's constant, $v$ is the speed of light in the optical fiber, $k$ is Boltzmann's constant, and $T$ is absolute temperature. By measuring Raman scattering distribution in an optical fiber using OTDR method, it is possible to measure the temperature distribution along the fiber. The fiber-optic temperature radar (FTR) is a Raman-distributed temperature sensor system of commercial instruments developed by Hitachi Cable, Ltd. The spatial resolution and the accuracy for temperature measurement by the FTR is about 3 m and $\pm 2°C$, respectively [39].

## 7.7 TRIBOLUMINESCENCE FIBER-OPTIC SENSORS

The term "triboluminescence" is now commonly used to refer to the light emission from certain materials when they are fractured. It is possible to procure polymeric optical fibers whose cores have been doped with a highly photoluminescent material. The concept of performing composite SHM with triboluminescent materials couples with a fiber-optic waveguide is described in Ref. [43]. Consider embedding a photoluminescent fiber and triboluminescent damage sensors into a composite. The light emitted by an embedded triboluminescent sensor upon fracture can be efficiently absorbed by the photoluminescent material, with the absorbed energy subsequently re-emitted as light which can match the waveguide modes of the polymeric fiber. To ensure efficient capture of the optical damage signal, the wavelength of light emitted by the triboluminescent sensor should be absorbed effectively by the photoluminescent fiber, i.e., there should be significant spectral overlap between the triboluminescent emission and the absorption profile of the photoluminescent material. To our knowledge, all currently available polymeric photoluminescent fibers are made from thermoplastics with relatively low softening temperatures ($\sim 100°C$), thus limiting the use of this light-capturing technique to low-temperature curing composites. Furthermore, currently available photoluminescent fibers possess an outside diameter no smaller than 250 μm, which may cause unacceptable degradation to the mechanical properties of the host structure. To extend the use of this light-capturing technique to higher temperature curing composites and those systems requiring thinner diameter fibers, hollow silica capillaries can be exploited [43]. These are available with outside diameters down to, at least, 100 μm, with outside diameters up to 150 μm generally being acceptable for most composite systems. After cure, the hollow silica capillaries can be filled with a photoluminescent material with higher refractive index than silica, thus generating photo-luminescent silica optical fibers. The use of polymeric and silica photoluminescent fibers in efficiently capturing and guiding the optical damage signal from an embedded tribolu-minescent sensor to a remote detector was demonstrated in Ref. [43].

# 7.8 POLARIMETRIC OPTICAL SENSORS

A polarimetric fiber-optic sensor is based on a single mode or an elliptical core two-mode birefringent polarization-preserving optical fiber. The birefringence is induced either by a residual strain field across the core or by an asymmetry in the core geometry [8]. The for-mer is achieved by introducing the "bow tie" element or two side holes in the cladding, whereas the latter is achieved by adopting an elliptical shape of the fiber core. A linearly polarized light launched into a birefringent fiber divides into two eigenmodes and travels along two orthogonal axes at different phase velocities. This difference in velocities defines a state of polarization (SOP). The SOP variations induced by the external load can then be related to mechanical strains applied to the fiber. Although the entire fiber can be used for sensing, the localized sensing region of a polarimetric sensor can also be formed by either two in-line splices (Figure 38a), or one in-line splice and a reflecting mirrored end (Figure 38b). The major disadvantages of polarimetric fiber-optic sensors are their high cost, complexity of the sensing system, low axial strain sensitivity, and three-dimensional nature in measured strains [8]. However, their high transverse sensitivity could be used for determining the location of impact; they also offer good potential for being developed into sensor networks.

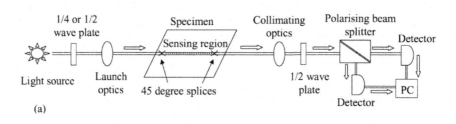

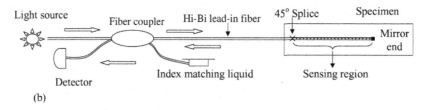

**FIGURE 38**   Schematic of polarimetric optical fiber sensors: (a) in-line splice; (b) single ended [8].

# 7.9 SUMMARY AND CONCLUSIONS

This chapter has offered a review of the various fiber-optic sensing methods that have been developed for SHM of composite structures and materials. This chapter started with a brief introduction in which the main fiber-optic sensing principles were enumerated: (i) intensity modulation; (ii) polarization modulation; (iii) phase modulation; (iv) spectral modulation; (v) scattering modulation. This was followed by a cursory presentation of the general principles that govern the fiber-optic technology such as total internal reflection and single-mode versus multimode propagation of the light wave inside the optical fiber.

Subsequently, this chapter has expanded on some of the sensing principles that have gained increased popularity and/or are more likely to transition to applications. Section 3 presented the interferometric fiber-optic sensors, whereas Section 4 was devoted to FBG sensors. The latter have gained extensive usage in various SHM composite applications; hence more discussion space was allocated to the FBG sensor than to other fiber-optic sensor types. The FBG section started with a presentation of the operational principles of the FBG sensor including the basic strain measuring concept and effect of refining the FBG patterns. Next, the fabrication of the FBG sensor was briefly reviewed. Attention was then focused on the conditioning equipment for the FBG sensors including OSAs, scanning FP filters, and interferometric equipment. The methods for the demodulation of the FBG signal at ultrasonic frequencies, which are of particular interest for active SHM, were discussed in some details next. Examples of actual implementation were given. The FBG strain rosettes, the long-gage FBG sensors, and the temperature compensation of the FBG sensors were discussed in the last part of the FBG section.

The next section of this chapter dealt with intensity-modulated fiber-optic sensors, whereas the following section dealt with distributed optical fiber sensing. The latter topic has gained extensive attention lately because technological advances have made possible the use of a simple telecom fiber as a distributed strain sensor. Based on the distinct intrinsic scattering spectrum of each fiber, this new development in fiber-optic sensing technology allows one to focus the sensing attention at specific locations along the fiber and to multiplex several such locations with rates of up to 100 Hz.

Final sections of this chapter describe the triboluminiscent fiber-optic sensors and the polarimetric optical sensors.

A general conclusion about the use of fiber-optic sensors in composite SHM applications is that, at present, there is no clear preferred technology, although FBG sensors seem to have been in use more than others in various SHM applications. The fact that so many fiber-optic methods and approaches have been attempted is an indication that the field has not yet converged and other concepts may also be developed and tried. The distributed optical fiber sensing, which has only recently become available at affordable costs, is an example of such new technological developments that may happen in the not too far future.

# References

[1] Peters, K. (2009) "Fiber-Optic Sensor Principles,". Chapter 59 in Encyclopedia of Structural Health Monitoring, C. Boller, F.-K. Chang, Y. Fujino (Eds.), Wiley, New York, NY.
[2] Mizutani, T.; Okabe, Y.; Takeda, N. (2003) "Quantitative Evaluation of Transverse Cracks in Carbon Fiber Reinforced Plastic Quasi-Isotropic Laminates with Embedded Small-Diameter Fiber Bragg Grating Sensors", *Smart Materials & Structures*, **12**(6), 898–903, Dec. 2003, http://dx.doi.org/10.1088/0964-1726/12/6/006.

[3] Peters, K. (2009) "Intensity-, Interferometric-, and Scattering-Based Optical-fiber Sensors,". Chapter 60 in Encyclopedia of Structural Health Monitoring, C. Boller, F.-K. Chang, Y. Fujino (Eds.), Wiley, New York, NY.

[4] Peters, K. (2009) "Fiber Bragg Grating Sensors,". Chapter 61 in Encyclopedia of Structural Health Monitoring, C. Boller, F.-K. Chang, Y. Fujino (Eds.), Wiley, New York, NY.

[5] Peters, K. (2009) "Novel Fiber-Optic Sensors,". Chapter 62 in Encyclopedia of Structural Health Monitoring, C. Boller, F.-K. Chang, Y. Fujino (Eds.), Wiley, New York, NY.

[6] Anon. (2014) "Fiber Bragg Grating", Wikipedia, accessed Aug. 2014, http://en.wikipedia.org/wiki/Fiber_bragg_grating. License: http://creativecommons.org/licenses/by-sa/3.0/.

[7] Lee, D. C.; Lee, J. J.; Kwon, I. B.; Seo, D. C. (2001) "Monitoring of Fatigue Damage of Composite Structures by Using Embedded Intensity-Based Optical Fiber Sensors", Smart Materials & Structures, 10(2), 285–292, Apr. 2001, http://dx.doi.org/10.1088/0964-1726/10/2/313.

[8] Zhou, G.; Sim, L. M. (2002) "Damage Detection and Assessment in Fibre-Reinforced Composite Structures with Embedded Fibre Optic Sensors—Review", Smart Materials & Structures, 11(6), 925–939, Dec. 2002, http://dx.doi.org/10.1088/0964-1726/11/6/314.

[9] Anon. (2002) UFO's Explained—Solving the Mystery of Unexplained Fibre Optics, Fiberoptics Industry Association http://www.fia-online.co.uk/pdf/Guide/L3814.pdf (accessed Oct. 2014).

[10] Leng, J. S.; Asundi, A. (2002) "Real-Time Cure Monitoring of Smart Composite Materials Using Extrinsic Fabry–Perot Interferometer and Fiber Bragg Grating Sensors", Smart Materials & Structures, 11(2), 249–255, Apr. 2002, http://dx.doi.org/10.1088/0964-1726/11/2/308.

[11] Bocherens, E.; Bourasseau, S.; Dewynter-Marty, V.; Py, S.; Dupont, M.; Ferdinand, P.; Berenger, H. (2000) "Damage Detection in a Radome Sandwich Material with Embedded Fiber Optic Sensors", Smart Materials & Structures, 9(3), 310–315, Jun. 2000, http://dx.doi.org/10.1088/0964-1726/9/3/310.

[12] Seo, D. C.; Lee, J. J.; Kwon, I. B. (2002) "Monitoring of Fatigue Crack Growth of Cracked Thick Aluminum Plate Repaired with a Bonded Composite Patch Using Transmission-Type Extrinsic Fabry-Perot Interferometric Optical Fiber Sensors", Smart Materials & Structures, 11(6), 917–924, Dec. 2002, http://dx.doi.org/10.1088/0964-1726/11/6/313.

[13] Sirkis, J. S.; Putman, M. A.; Berkoff, T. A.; Kersey, A. D.; Friebele, E. J. (1993) "In-Line Fiber Étalon for Strain Measurement", Optics Letters, 18(22), 1973–1975, (1993), http://dx.doi.org/10.1364/OL.18.001973.

[14] Liu, T. Q.; Han, M. (2012) "Analysis of π-Phase-Shifted Fiber Bragg Gratings for Ultrasonic Detection", IEEE Sensors Journal, 12, 2368–2373, (2012).

[15] Wood, K.; Brown, T.; Rogowski, R.; Jensen, B. (2000) "Fiber Optic Sensors for Health Monitoring of Morphing Airframes: I. Bragg Grating Strain and Temperature Sensor", Smart Materials & Structures, 9(2), 163–169, Apr. 2000, http://dx.doi.org/10.1088/0964-1726/9/2/306.

[16] Todd, M. D.; Johnson, G. A.; Vohra, S. T. (2001) "Depolyment of a Fiber Bragg Grating-Based Measurement System in a Structural Health Monitoring Application", Smart Materials & Structures, 10(3), 534–539, Jun. 2001, http://dx.doi.org/10.1088/0964-1726/10/3/316.

[17] Tanaka, N.; Okabe, Y.; Takeda, N. (2003) "Temperature-Compensated Strain Measurement Using Fiber Bragg Grating Sensors Embedded in Composite Laminates", Smart Materials & Structures, 12(6), 940–946, Dec. 2003, http://dx.doi.org/10.1088/0964-1726/12/6/011.

[18] Wang, G.; Pran, K.; Sagvolden, G.; Havsgard, G. B.; Jensen, A. E.; Johnson, G. A.; Vohra, S. T. (2001) "Ship Hull Structure Monitoring Using Fibre Optic Sensors", Smart Materials & Structures, 10(3), 472–478, Jun. 2001, http://dx.doi.org/10.1088/0964-1726/10/3/308.

[19] Kuang, K. S. C.; Kenny, R.; Whelan, M. P.; Cantwell, W. J.; Chalker, P. R. (2001) "Residual Strain Measurement and Impact Response of Optical Fibre Bragg Grating Sensors in Fibre Metal Laminates", Smart Materials & Structures, 10(2), 338–346, Apr. 2001, http://dx.doi.org/10.1088/0964-1726/10/2/321.

[20] Okabe, Y.; Yashiro, S.; Kosaka, T.; Takeda, N. (2000) "Detection of Transverse Cracks in CFRP Composites Using Embedded Fiber Bragg Grating Sensors", Smart Materials & Structures, 9(6), 832–838, Dec. 2000, http://dx.doi.org/10.1088/0964-1726/9/6/313.

[21] Read, I. J.; Foote, P. D. (2001) "Sea and Flight Trials of Optical Fibre Bragg Grating Strain Sensing Systems", Smart Materials & Structures, 10(5), 1085–1094, Oct. 2001, http://dx.doi.org/10.1088/0964-1726/10/5/325.

[22] Komatsuzaki, S. K., S.; Hongo, A.; Takeda, N.; Sakurai, T., "Development of High-Speed Optical Wavelength Interrogation System for Damage Detection in Composite Materials", SPIE Vol. 5758, paper #5758-7.

[23] Perez, I; Cui, H-L; Udd, E, "Acoustic Emission Detection Using Fiber Bragg Gratings", Smart Structures and Materials 2001: Sensory Phenomena and Measurement Instrumentation for Smart Structures and Materials, San Diego, CA, 2001, SPIE Vol. 4328, pp. 209–215.

[24] Tsuda, H.; Toyama, N.; Urabe, K.; Takatsubo, J. (2004) "Impact Damage Detection in CFRP Using Fiber Bragg Gratings", *Smart Materials & Structures*, **13**(4), 719–724, Aug. 2004, http://dx.doi.org/10.1088/0964-1726/13/4/009.

[25] Koh, J. I.; Bang, H. J.; Kim, C. G.; Hong, C. S. (2005) "Simultaneous Measurement of Strain and Damage Signal of Composite Structures Using a Fiber Bragg Grating Sensor", *Smart Materials & Structures*, **14**(4), 658–663, Aug. 2005, http://dx.doi.org/10.1088/0964-1726/14/4/024.

[26] Park, C.; Peters, K.; Zikry, M.; Haber, T.; Schultz, S.; Selfridge, R. (2010) "Peak Wavelength Interrogation of Fiber Bragg Grating Sensors During Impact Events", *Smart Materials & Structures*, **19**(4), Apr. 2010, http://dx.doi.org/10.1088/0964-1726/19/4/045015.

[27] Betz, D. C.; Thursby, G.; Culshaw, B.; Staszewski, W. J. (2003) "Acousto-Ultrasonic Sensing Using Fiber Bragg Gratings", *Smart Materials & Structures*, **12**(1), 122–128, Feb. 2003, http://dx.doi.org/10.1088/0964-1726/12/1/314.

[28] Betz, D. C.; Thursby, G.; Culshaw, B.; Staszewski, W. J. (2006) "Identification of Structural Damage Using Multifunctional Bragg Grating Sensors: I. Theory and Implementation", *Smart Materials & Structures*, **15**(5), 1305–1312, Oct. 2006, http://dx.doi.org/10.1088/0964-1726/15/5/020.

[29] Rajic, N.; Davis, C.; Thomson, A. (2009) "Acoustic-Wave-Mode Separation Using a Distributed Bragg Grating Sensor", *Smart Materials & Structures*, **18**(12), Dec. 2009, http://dx.doi.org/10.1088/0964-1726/18/12/125005.

[30] Mizutani, T.; Takeda, N.; Takeya, H. (2006) "On-Board Strain Measurement of a Cryogenic Composite Tank Mounted on a Reusable Rocket Using FBG Sensors", *Structural Health Monitoring-an International Journal*, **5**(3), 205–214, Sep. 2006, http://dx.doi.org/10.1177/1475921706058016.

[31] Okabe, Y.; Kuwahara, J.; Natori, K.; Takeda, N.; Ogisu, T.; Kojima, S.; Komatsuzaki, S. (2007) "Evaluation of Debonding Progress in Composite Bonded Structures Using Ultrasonic Waves Received in Fiber Bragg Grating Sensors", *Smart Materials & Structures*, **16**(4), 1370–1378, Aug. 2007, http://dx.doi.org/10.1088/0964-1726/16/4/051.

[32] Kirikera, G. R.; Balogun, O.; Krishnaswamy, S. (2011) "Adaptive Fiber Bragg Grating Sensor Network for Structural Health Monitoring: Applications to Impact Monitoring", *Structural Health Monitoring-an International Journal*, **10**(1), 5–16, Jan. 2011, http://dx.doi.org/10.1177/1475921710365437.

[33] Betz, D. C.; Thursby, G.; Culshaw, B.; Staszewski, W. J. (2007) "Structural Damage Location with Fiber Bragg Grating Rosettes and Lamb Waves", *Structural Health Monitoring-an International Journal*, **6**(4), 299–308, Dec. 2007, http://dx.doi.org/10.1177/1475921707081974.

[34] Eum, S. H.; Kageyama, K.; Murayama, H.; Uzawa, K.; Ohsawa, I.; Kanai, M.; Kobayashi, S.; Igawa, H.; Shirai, T. (2007) "Structural Health Monitoring Using Fiber Optic Distributed Sensors for Vacuum-Assisted Resin Transfer Molding", *Smart Materials & Structures*, **16**(6), 2627–2635, Dec. 2007, http://dx.doi.org/10.1088/0964-1726/16/6/067.

[35] Guo, Z. S. (2007) "Strain and Temperature Monitoring of Asymmetric Composite Laminate Using FBG Hybrid Sensors", *Structural Health Monitoring—an International Journal*, **6**(3), 191–197, Sep. 2007, http://dx.doi.org/10.1177/1475921707081108.

[36] Kang, H. K.; Kang, D. H.; Bang, H. J.; Hong, C. S.; Kim, C. G. (2002) "Cure Monitoring of Composite Laminates Using Fiber Optic Sensors", *Smart Materials & Structures*, **11**(2), 279–287, Apr. 2002, http://dx.doi.org/10.1088/0964-1726/11/2/311.

[37] Kang, H. K.; Kang, D. H.; Hong, C. S.; Kim, C. G. (2003) "Simultaneous Monitoring of Strain and Temperature During and After Cure of Unsymmetric Composite Laminate Using Fibre-Optic Sensors", *Smart Materials & Structures*, **12**(1), 29–35, Feb. 2003, http://dx.doi.org/10.1088/0964-1726/12/1/304.

[38] Badcock, R. A.; Fernando, G. F. (1995) "An Intensity-Based Optical Fibre Sensor for Fatigue Damage Detection in Advanced Fibre-Reinforced Composites", *Smart Materials & Structures*, **4**(4), 223–230, Dec. 1995, http://dx.doi.org/10.1088/0964-1726/4/4/001.

[39] Murayama, H.; Kageyama, K.; Naruse, H.; Shimada, A.; Uzawa, K. (2003) "Application of Fiber-Optic Distributed Sensors to Health Monitoring for Full-Scale Composite Structures", *Journal of Intelligent Material Systems and Structures*, **14**(1), 3–13, Jan. 2003, http://dx.doi.org/10.1177/104538903032738.

[40] Zhang, H.; Wu, Z. S. (2008) "Performance Evaluation of BOTDR-Based Distributed Fiber Optic Sensors for Crack Monitoring", *Structural Health Monitoring—an International Journal*, **7**(2), 143–156, Jun. 2008, http://dx.doi.org/10.1177/1475921708089745.

[41] Minakuchi, S.; Okabe, Y.; Mizutani, T.; Takeda, N. (2009) "Barely Visible Impact Damage Detection for Composite Sandwich Structures by Optical-Fiber-Based Distributed Strain Measurement", *Smart Materials & Structures*, **18**(8), Aug. 2009, http://dx.doi.org/10.1088/0964-1726/18/8/085018.

[42] Anon. (2014) "Optical Distributed Sensor Interrogator", Datasheets ODISI-A10 and ODISI-B10, Luna Inc., http://lunainc.com/odisi (accessed Oct. 2014).

[43] Sage, I.; Humberstone, L.; Oswald, I.; Lloyd, P.; Bourhill, G. (2001) "Getting Light Through Black Composites: Embedded Triboluminescent Structural Damage Sensors", *Smart Materials & Structures*, **10**(2), 332–337, Apr. 2001, http://dx.doi.org/10.1088/0964-1726/10/2/320.

*Structural Health Monitoring of Aerospace Composites*
DOI: http://dx.doi.org/10.1016/B978-0-12-409605-9.00008-8

# 8.1 INTRODUCTION

Structural health monitoring (SHM) relies on sensors that can be permanently placed on the structure and monitored over time either in a passive or in an active way. These sensors should be affordable, lightweight, and unobtrusive such as to not impose cost and weight penalty on the structure and to not interfere with the structural strength and airworthiness. Some of the sensor types that have been considered for SHM applications are:

- Conventional resistance strain gages
- Fiber-optic sensors, e.g., fiber Bragg gratings (FBG) strain sensors
- Piezoelectric wafer active sensors (PWAS)
- Electrical property sensors: resistance, impedance, dielectric, etc.

These sensors may operate in static and dynamic regimes, depending on the physical principle that is employed in monitoring the structure.

Other damage measuring methods based on large area measurements (ultrasonic C-scans, scanning Doppler laser velocimetry, thermography, etc.) have been used in SHM development for definition and confirmation of damage and/or for understating the proposed SHM approach; however, they do not seem appropriate for permanent installation onto the monitored structure and will not be discussed under the heading of "SHM sensors".

The PWAS transducers and the fiber-optic sensors have already been discussed in previous chapters. This chapter will discuss the remainder, i.e., the conventional resistance strain gages and the electrical property sensors.

# 8.2 CONVENTIONAL RESISTANCE STRAIN GAGES

## 8.2.1 Resistance Strain Gage Principles

The conventional resistance strain gages have been in operation since mid-1900s and have achieved wide recognition in the experimental stress analysis community. The strain gage operation principles are described well in standard references [1−4] and need not be discussed here in much detail. It is sufficient to say that their physical principle consists in converting a relative strain change into a relative resistance change that is read with a precision instrument (e.g., a Wheatstone bridge). The strain-induced resistance change may be due to purely geometric effects (e.g., foil strain gages) or may be enhanced by the piezo-resistive effect exhibited by some electronic materials. For a resistance strain gage (Figure 1a), the following general relation applies [1]:

$$\frac{\Delta R}{R} = S_g \Delta \varepsilon + S_T \Delta T + S_t \Delta t \tag{1}$$

where $S_g$ is the gage sensitivity to strain (a.k.a. gage factor, $S_g \approx 2$), $S_T$ is the gage sensitivity to temperature, and $S_t$ is the gage sensitivity to time (drift during in-service operation). Strain gage manufacturers thrive to minimize $S_T$ and $S_t$ by developing special strain gage materials/alloys. A good-quality strain gage is sensitive mostly to strain $\varepsilon$ with very little remnant sensitivity to temperature and little time drift. The foil technology has allowed the development of strain gages of various forms and shapes to serve specific application requirements (Figure 1b).

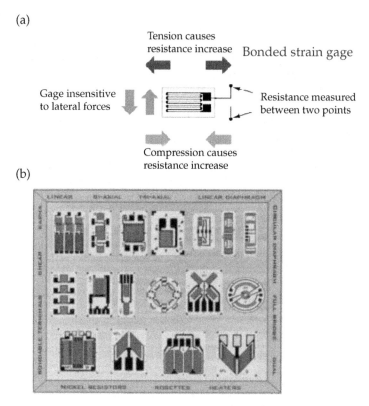

**FIGURE 1**    Strain gage fundamentals: (a) schematic of strain gage mode of operation; (b) various strain gage types [2].

Conventional resistance strain gages operate well in both static and dynamic regimes. The signal bandwidth is limited only by the bandwidth of the electronic conditioning equipment, because the strain gage coverts directly the strain change into a resistance change.

## 8.2.2  Strain Gage Instrumentation

The common instrument used in resistance strain gage measurements is the Wheatstone bridge developed in 1843 by Sir Charles Wheatstone for precise measurement of electrical resistances through comparison with known resistances. The Wheatstone bridge operation principles lie in balancing the bridge to zero voltage between opposing nodes. The balancing is done through a calibrated adjustable resistance. The resistance adjustments made to bring the bridge to zero represent how much the strain gage resistance has changed after a load was applied to the structure on which the strain gage is affixed. This approach is known as the *null method*; it is very precise, but elaborate.

An alternative approach is to use the bridge in *off-null mode*. In this case, the assumption is made that the voltage measured by the bridge is proportional to the resistance change

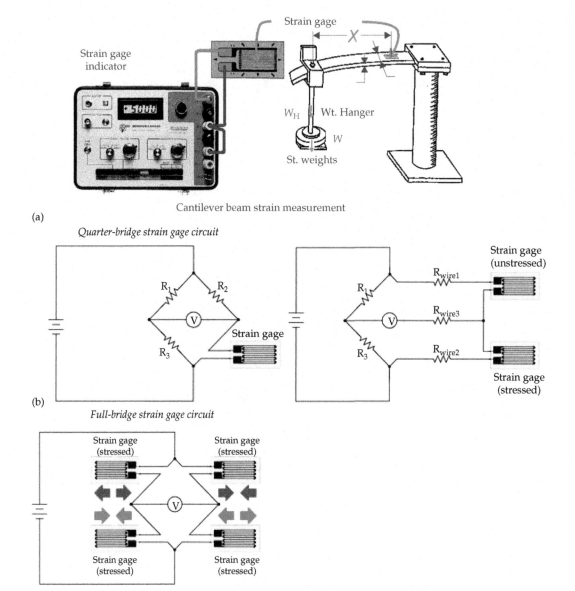

**FIGURE 2**    Strain gage instrumentation: (a) schematic of strain gage operation [5]; (b) various strain gage bridge configurations: quarter bridge, quarter bridge with compensation, full bridge [6]; (c) computer strain gage instrumentation [7].

(c)

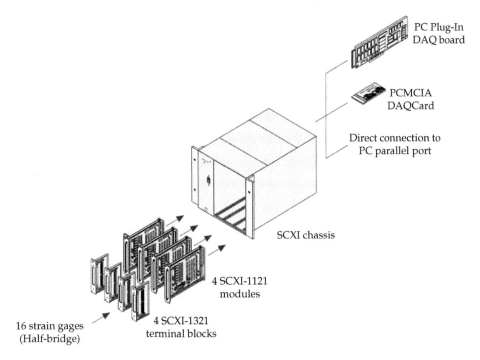

PC Plug-In
DAQ board

PCMCIA
DAQCard

Direct connection to
PC parallel port

SCXI chassis

4 SCXI-1121
modules

16 strain gages              4 SCXI-1321
(Half-bridge)                terminal blocks

(d)

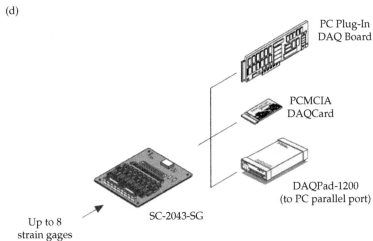

PC Plug-In
DAQ Board

PCMCIA
DAQCard

DAQPad-1200
(to PC parallel port)

Up to 8              SC-2043-SG
strain gages

**FIGURE 2**    (Continued)

in the measuring strain gage. This is true only for relative resistance change values, since for larger values the bridge output becomes nonlinear. Nonetheless, most automated strain gage bridge technology relies on the off-null approach because it is expeditious and convenient and because the strain gage resistance changes are relatively small for testing within the elastic range of the structural materials. However, this would not be appropriate for testing in the plastic range or for testing of polymers with large strain deformations.

The Wheatstone bridge can be energized in direct current (DC) mode or alternating current (AC) mode; low frequency AC operation at approximately 1 kHz has been shown to improve stability while producing similar results to DC operation since the spurious capacitive and inductive effects are negligible at this relatively low AC frequency.

Figure 2a shows a classroom schematic of strain gage operation [5]. Figure 2b illustrates the various configuration uses of the strain gage bridge; quarter bridge, quarter bridge with compensation, and full bridge are illustrated [6]. In the quarter bridge with compensation, one of the strain gages is loaded mechanically, whereas the other strain gage is only exposed to the temperature effects without being mechanically loaded. Thus, the temperature effects, which act on both gages, are compensated out and only the mechanical loading effects are left to be measured by the bridge. Figure 2c presents strain gage instrumentation interfacing with a computer for automatic data acquisition and multiplexing. Wireless strain gage instrumentation options are also available [8].

### 8.2.3 Aerospace Strain Gage Technology

The strain gage technology is well developed and relatively matured; known difficulties with using strain gage for long-term monitoring under various environmental conditions are usually related to the adhesion between the gage and the structure and with the adverse effect of the environment onto electrical wire connections that may result in strain drift or even signal loss.

The installation of strain gages on metallic aircraft structures requires the adhesion of the gages on the structural surface and extensive cabling. Figure 3a shows strain gage installation inside an aircraft wing, whereas Figure 3b shows strain gage installation on the outside of an aircraft fuselage.

### 8.2.4 Strain Gage Usage in Aerospace Composites

The use of strain gages in composite structures follows the general principles of use on metallic structures [11], but special attention has to be paid to the adhesion between the gage and the composite material and the extraction of stress information from the strain readings [12,13]. Figure 4 shows two recent examples of strain gage installation on composite structures. Figure 4a presents strain gages installed on the composite crew module pressure vessel Pathfinder development [14], whereas Figure 4b shows strain gages installed on a composite propeller [15].

An important feature of using strain gages on composite structures is the capability of inserting them inside the composite material, e.g., between the composite layers. However, the problem remains of the electrical wires that need to connect the strain gage

(a)

(b)

**FIGURE 3**   Examples of strain gage installation on metallic aircraft structures: (a) strain gages installed in the interior of AFTI/F-16 wing at NASA DFRC [9]; (b) strain gages installed on upper fuselage [10].

(a)                                          (b)

**FIGURE 4**   Examples of strain gage installation on composite aircraft structures: (a) strain gages installed on the composite crew module pressure vessel Pathfinder development [14]; (b) strain gages installed on a composite propeller [15].

to the measuring equipment. To resolve this issue, a special strain gage was developed [16] that can be inserted between the composite layers and has pins protruding through the composite to the outside where it can be trimmed and solder connected to the measuring equipment cable (Figure 5).

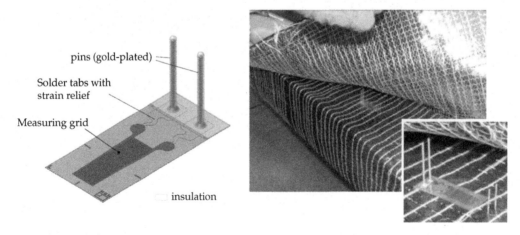

**FIGURE 5**   Insertion of strain gages in composite structures: (a) special HBM strain gage for insertion between the layers of a composite structure; the electrodes penetrate outside the composite and can be connected to the measuring cables after composite cure; (b) example of insertion of the special HBM strain gage between the layup plies [16].

## 8.3 ELECTRICAL PROPERTY SENSORS

The electrical SHM of composites relies on the material itself to act as sensor. Carbon fibers are electrically conductive; the epoxy resin is an insulator. The carbon fiber-reinforced polymer (CFRP) composite is somehow conductive because the densely packed carbon fibers may touch each other. As damage (e.g., cracks and delamination) takes place in the composite, the electric conductivity is expected to change.

The glass fiber-reinforced polymer (GFRP) composite is a nonconductive insulator with certain dielectric properties. Damage in GFRP composites creates microcracks and even sizeable delaminations, which may change the dielectric properties of the composite since the dielectric permittivity of air is smaller than that of GFRP. Ingress of water, which has a dielectric permittivity higher than GFRP, will change again the overall dielectric properties of GFRP composites.

These intuitive concepts stand at the foundation of the electrical SHM methods for composite materials. This approach is deemed "self-sensing" because it relies entirely on measuring a material property (i.e., electrical characteristic) and does not require an additional transduction sensor; the only instrumentation that needs to be installed on the composite structure consists of the electrodes. In the case of composite transport aircraft, the conductive screen skins currently used to mitigate lightning strike could potentially also serve as the measuring electrodes.

Electrical SHM methods range from the simple measurement of the electrical resistance measurements up to more sophisticated methods such as electrical potential mapping, dielectric measurement, electrochemical impedance, etc. Some of these methods will be discussed next. The focus will be on the instrumentation needed to achieve these measurements rather than on the actual results which will be presented in later chapters.

The methods for CFRP damage detection by electrical resistance monitoring fall into two large categories: (i) those that achieve monitoring of an average damage state of the whole structural component by performing overall resistance measurements and (ii) those that provide damage location capability through more elaborate measurements and data processing.

### 8.3.1 Electrical Resistance and Electrical Potential Methods for Composites SHM

The electrical resistance method involves the measurement of the electrical resistance between two points on the composite, or between an array of contact points. In the electrical potential method, the array of contacts is not used to measure the resistance, but to measure the electrical potential at each of them with respect to a reference point or the ground. The change in electrical resistance or in the electrical potential between a pristine specimen and a specimen with internal damage is used to infer the damage presence and its location. The latter is achieved through some mapping and interpolation approach.

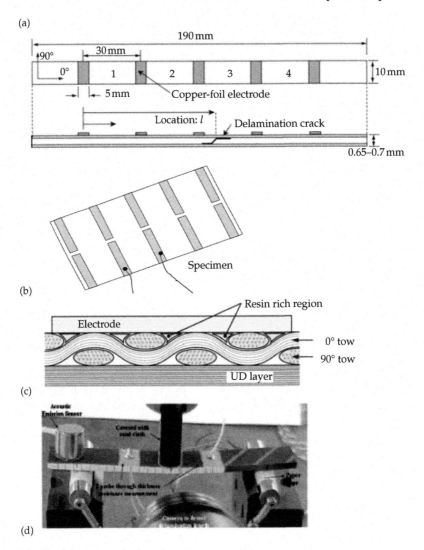

**FIGURE 6**   Single-sided surface-mounted copper-foil specimens co-cured with CFRP specimens: (a) schematic of electrodes installation of a CFRP beam [18]; (b) surface-mounted electrodes pattern on a CFRP plate [17]; (c) interface between the electrode and the woven CFRP ply [18]; (d) *in situ* manufactured silver paint electrodes on a CFRP specimen [19].

## 8.3.1.1 Electrodes Fabrication on Composite Materials for Electrical SHM Measurements

The reliable fabrication of the electrodes used for electrical measurements is essential for the success of the electrical monitoring approach. Hence, a major step using electrical methods for composites SHM is establishing a firm and reliable electrical contact with the composite material at various locations across the composite structure. Simply touching the composite material with an electrical probe would not do because the heterogeneous nature of the composite material which is made up of fibers (which may be more or less conductive) and a polymeric matrix (which is typically a good insulator). Using a point probe may lead one to get different electrical measurement values depending on whether the point probe is in contact with the matrix, or the fiber, or an inclusion, void, etc. To get a firm and reliable electrical contact, one has to cover an area in which the composite properties are evaluated in an averaged sense. This is achieved by creating finite-area contact electrodes in firm and reliable contact with the composite material.

References [17,18] report the integration of surface-mounted strip electrodes during the CFRP composite manufacturing process. Thin copper-foil strips were mounted on one side of CFRP specimens during the prepreg lamination and co-cured together with them (Figure 6a,b). However, as shown in Ref. [18], when a woven ply is used as the surface layer in a laminate of unidirectional plies, it may be expected that the surface woven ply may not have a good electrical contact to the electrodes due to resin-rich regions between the fibers (Figure 6c).

Reference [20] reports details about the fabrication of electrodes on an existing CFRP plate: each electrode site was first abraded with emery paper in order to reveal the surface fibers. A thin copper wire was then placed accurately on the electrode site and held in place with conducting silver paint and carbon cement. Reference [19] reports the *in situ* fabrication of electrical resistance measurement electrodes through the application of silver paint over specified areas on both sides of a CFRP specimen. Figure 6d shows the silver paint electrodes and the use of electrically conductive epoxy to connect to the respective wires in the instrumentation cable. The conductive epoxy was applied on top of soldered wire ends.

### 8.3.1.2 Measurement of the Electrical Resistance

Electrical resistance is usually measured with a multimeter, whereas the relative electrical resistance change $\Delta R/R$ may be measured directly with an electrical bridge (Figure 7a) similar to that used for reading the electrical resistance strain gages. DC or AC excitation may be used. The former avoids the possible effect of parasitic capacitances and inductances; the latter avoids inaccuracy due to polarization and may mitigate the nonlinear resistance effects thus being sometimes preferred in practical applications.

To increase accuracy, the four-point sensing method (a.k.a. 4-terminal, 4-wire, 4-probe, etc.) may be used (Figure 7b); this method may be more accurate because it uses a separate current-carrying path $(1 \rightarrow 4)$ and a voltage-sensing path $(2 \rightarrow 3)$ such that the contribution of the wiring and contact resistances may be eliminated.

### 8.3.1.3 Electrical Resistance and Electrical Potential Methods

8.3.1.3.1 Electrical Resistance Method

The electrical resistance method involves the measurement of the electrical resistance between two points on the composite. In order to find the damage location, some sort of surface interpolation method is applied to the electrical resistance values measured at these locations. Reference [22] describes a conductivity mapping approach that employs

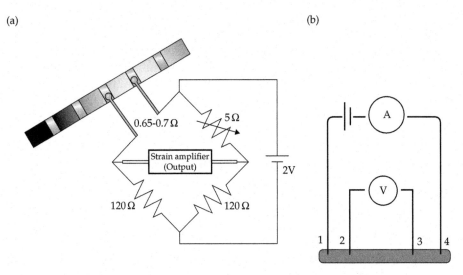

**FIGURE 7** Electrical resistance measurement methods: (a) strain gage bridge circuit for measuring directly the relative change of electrical resistance [18]; (b) four-point measurement of resistance between voltage sense connections 2 and 3. Current is supplied via connections 1 and 4 [21].

an array of contacts across the surface of the CFRP composite. Measurements could be taken between a contact and each of its neighbors in the array, to give conductivity values for each of these directions. By plotting the conductivity values over the specimen area and looking for anomalies in the conductivity surface, one could identify possible damage zones.

Figure 8 shows various electrode patterns to be used with the electrical resistance and electrical potential methods. For 1D specimens, the contacts are placed along the specimen (Figure 8a) [23]. For 2D specimens, the contacts can be distributed as arrays over the surface of the specimen, either in a rectangular grid covering the whole area (Figure 8b,c) or as a "circular" array confined closer to the boundary such as not to be damaged by the more severe impacts (Figure 8d,e).

### 8.3.1.3.2 Electrical Potential Method

The electrical potential method uses the array of contact not to measure the resistance, but to measure the electrical potential at each of them with respect to a reference point or the ground. The method is known as electrical potential mapping. Since the potential depends on the direction of electric current flow, the electrical potential would be done for all possible current flow configurations (electrical tomography).

### 8.3.1.4 Wireless Sensing for Remote Electrical Resistance Monitoring of Aerospace Composites

The electrical resistance method for monitoring composite materials can also be done wirelessly since it is similar in instrumentation to the electrical resistance strain gage method. Low-sampling-rate wireless strain and temperature sensors have become relatively wide

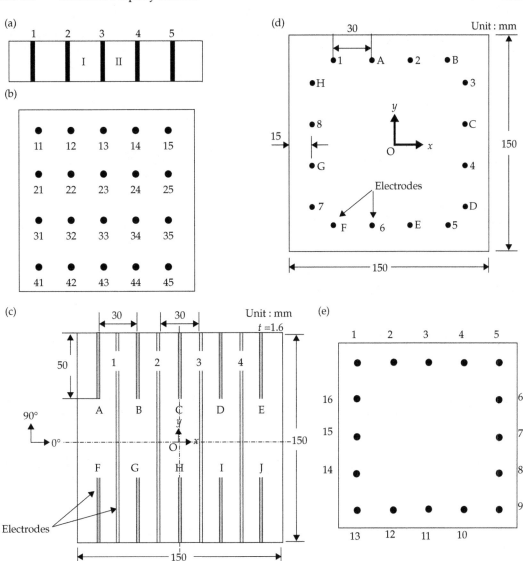

**FIGURE 8**  Possible distribution patterns for the electrical contacts on a composite specimen: (a) 1D specimen with contacts along it [23]; (b) 2D specimen with a grid array covering the whole specimen [23]; (c) matrix array electrodes [24]; (d) circular array electrodes [24]; (e) "circular" array confined to the specimen periphery to avoid damage by the testing impacts [23].

spread. Reference [25] describes the wireless detection of strain changes due to internal delaminations in CFRP composites using the electrical resistance change method and a wireless bridge that encodes the resistance change as a frequency shift (Figure 9).

This system was further refined in Ref. [26] to permit the measurement of the much smaller operational load strains (Figure 10).

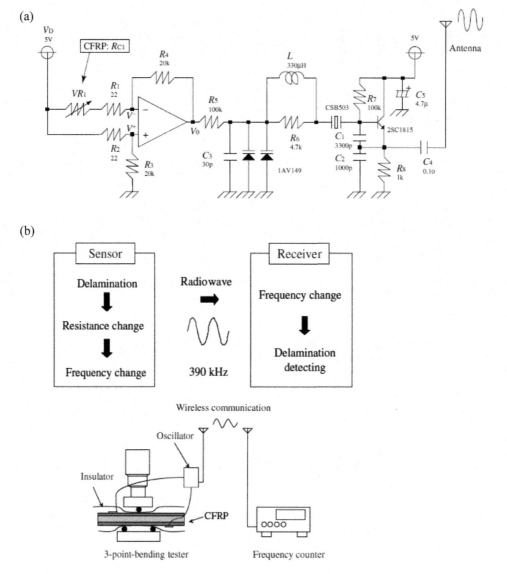

**FIGURE 9** Wireless monitoring of impact delamination in CFRP composites: (a) wireless circuit that encodes the resistance change as a frequency shift; (b) schematic of the complete system architecture [25].

Reference [26] also extended the work of Ref. [25] from a single sensor to an assembly of time-synchronized wireless sensors such that multiple regions on the composite structure can be simultaneously monitored remotely (Figure 10b). The wireless sensors were installed on CFRP specimens and loaded in a mechanical testing system. Normal loading and unloading of the specimen produces reversible strain and reversible change in the electrical resistance, which is encoded as frequency shifts in the wireless transmission.

(a)

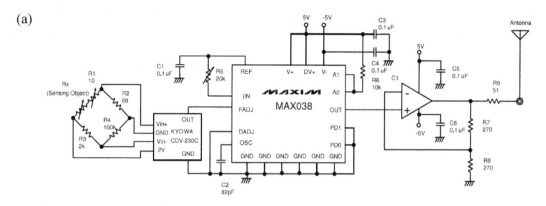

(b)

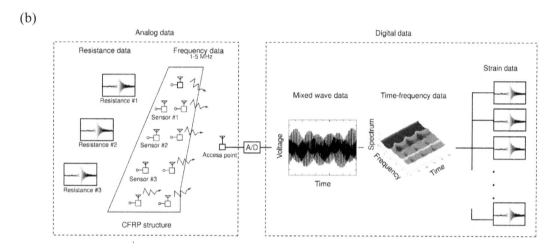

**FIGURE 10**    Time-synchronized wireless strain and damage monitoring of CFRP composites: (a) wireless circuit that encodes the resistance change as a frequency shift; (b) schematic of the complete system architecture permitting simultaneous monitoring of several locations on the composite structures [26].

### 8.3.1.5  Special-Built Composites for Resistance-Based Self-sensing Electric SHM

Reference [27] proposes a self-sensing resistance-based damage detection and location system involving the insertion of well-aligned conductors and glass fiber layers inside a CFRP composite laminate.

## 8.3.2  Frequency Domain Methods for Electrical SHM of Aerospace Composites

The electrical resistance and electrical potential methods are basically electrostatic methods. Even if AC excitation is used to improve the experimental procedure, that AC is of constant frequency (say 1kHz) and the measured parameter is still the resistance or voltage without any attention being paid to the phase information. However, more information about the composite damage may be extracted if the information contained in both the amplitude and phase is used. This can be achieved by using the electrochemical impedance spectroscopy (ECIS) method which determines the ratio in complex between the applied voltage and the resulting current in the form of the complex impedance $Z(\omega) = R + iX(\omega)$, where the real part of $Z(\omega)$ is the resistance $R$ whereas the imaginary part of $Z(\omega)$ is the reactance $X(\omega)$, which has inductive or capacitive components, i.e., $X(\omega) = i\omega L + \frac{1}{i\omega C}$. In an ideal circuit, the inductance $L$ and capacitance $C$ are assumed constant; however, when evaluating material state changes with ECIS method, we need to let the effective resistance, inductance, and capacitance vary with frequency, i.e., $R(\omega)$ and $X(\omega) = i\omega L(\omega) + \frac{1}{i\omega C(\omega)}$. In ECIS modeling of practical situation, equivalent circuit representations are more elaborate that this simplified three-parameter model may also be used.

### 8.3.2.1 Electrochemical Impedance Spectroscopy for Composites SHM

Reference [28] studied the use of electrochemical impedance spectroscopy (ECIS) for damage detection in composite materials. In Ref. [28], the ECIS testing was done inside an environmental chamber using a Gamry potentiostat controlled by a PC (Figure 11a). The Gamry potentiostat, which has a built-in signal generator and a frequency analyzer, acts like an impedance analyzer and produces the ECIS plots with the use of the PC software.

(a)

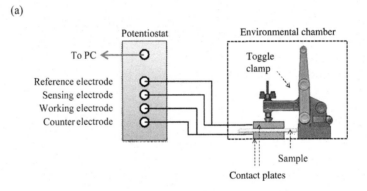

(b)

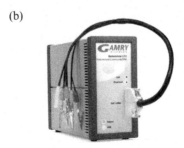

FIGURE 11   (a) Experimental setup for ECIS monitoring of fatigue damage in a GFRP composites [28]; (b) Gamry potentiostat.

### 8.3.2.2 Electromagnetic SHM of Composites

A hybrid electromagnetic method is also possible for CFRP composites in which a modulated high-frequency electric excitation is applied and the magnetic response of the material is recorded. Reference [29] describes such a hybrid method applied to a CFRP composite plate and compared in efficacy with the acousto-ultrasonic method. The instrumented CFRP specimen is depicted in Figure 12.

(a)

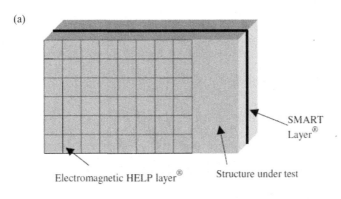

(b)

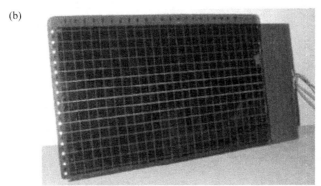

**FIGURE 12**   CFRP plate specimen instrumented with hybrid electromagnetic layer HELP© and with acousto-ultrasonic SMART Layer™: (a) schematic; (b) photograph [29].

The specimen was a 16-ply cross-ply CFRP plate. An acousto-ultrasonic SMART Layer™ was embedded in the midplane, and a hybrid electromagnetic layer HELP© with a 20mm × 20 mm mesh was applied to one side. A 700kHz high-frequency electric excitation with full 1 kHz modulation with was applied.

## References

[1] Sharpe, W. N. (2008) Springer Handbook of Experimental Solid Mechanics, Springer, New York, NY.

[2] Anon. (2014) "Strain Gauges Types", Micro-Flexitronics Ltd., http://www.mflstraingauges.com/index.htm (accessed Aug. 2014).

[3] Hoffmann, K. (1989) An Introduction to Measurements using Strain Gages, HBM Publication, Darmstadt, Germany.

[4] Anon. (2014) Precision Strain Gages, Accessories, and Instruments, http://www.vishaypg.com/micro-measurements/stress-analysis-strain-gages/ (accessed Aug. 2014).

[5] Kostic, M. (2014) "Calibration and Measurement with Strain Gages", Northern Illinois Univ., MEE 390 Experimental Methods in Mechanical Engineering, http://www.kostic.niu.edu/strain_gages.html (accessed Aug. 2014).

[6] Anon. (2014) "The Strain Gauge", Dokuz Eylul Univ., http://web.deu.edu.tr/mechatronics/TUR/strain_gauge.htm (accessed Aug. 2014).

[7] Anon. (1998) "Strain Gauge Measurement—A Tutorial", National Instruments Application Note 078, Aug. 1998.

[8] Anon. (2014) LORD MicroStrain Sensing Systems, http://www.microstrain.com/parameters/strain?gclid=CjwKEAjw4PCfBRCz966N9pvJ4GASJAAEdM_KVNvQRe1j0Bu2DgG2mpmIW2QC-Vh0U2fTOKOYISrSlhoC4Xvw_wcB.

[9] Anon. (1984) "Strain gage instrumentation installed on the interior of the AFTI/F-16 wing", (1/1/1984) NASA/Dryden Flight Research Center (NASA-DFRC), https://archive.org/details/NIX-ECN-28683 (accessed Aug.2014).

[10] Anon. (2006) High-Performance Instrumented Airborne Platform for Environmental Research—HIAPER, http://www.hiaper.ucar.edu/photo_gallery/img/092604/strain_gauges.jpg (accessed Aug. 2014).

[11] Anon. (2014) Stress Analysis Strain Gages—Application Notes, http://www.vishaypg.com/micro-measurements/stress-analysis-strain-gages/appnotes-list/ (accessed Aug. 2014).

[12] Anon. (2001) Strain Gage Measurements on Plastics and Composites, http://www.measurementsgroup.com (accessed Aug. 2014).

[13] Horoschenkoff, A.; Klein, S.; Haase, K.-H. (2006) Structural Integration of Strain Gages, HBM publication S2182-1.2en, http://www.hbm.com/ (accessed Aug. 2014).

[14] Anon. (2006) "NASA Composite Crew Module Pressure Vessel Pathfinder Development" http://www.nasa.gov/offices/nesc/home/Feature_CCM_prt.htm.

[15] Anon. (2009) "How Using Thermal Images Can Optimize Strain Gage Positioning for the Experimental Modal Analysis of Composite Rotors" http://www.hbm.com/pt/menu/tips-tricks/jornal-cientifico/artigos-tecnicos/datum/2009/10/14/how-using-thermal-images-can-optimize-strain-gage-positioning-for-the-experimental-modal-analysis-of-3/ (accessed Aug. 2014).

[16] Klein, S. (2008) "Integration of Strain Gages into Fibre Composite Structures", HBM PowerPoint Presentation, March 20, 2008 http://www.hbm.com/.

[17] Iwasaki, A.; Todoroki, A. (2005) "Statistical Evaluation of Modified Electrical Resistance Change Method for Delamination Monitoring of CFRP Plate", Structural Health Monitoring—an International Journal, 4(2), 119–136, Jun. 2005, http://dx.doi.org/10.1177/1475921705049757.

[18] Hirano, Y.; Todoroki, A. (2007) "Damage Identification of Woven Graphite/Epoxy Composite Beams Using the Electrical Resistance Change Method", Journal of Intelligent Material Systems and Structures, 18(3), 253–263, Mar. 2007 http://dx.doi.org/10.1177/1045389x06065467.

[19] Ngabonziza, Y.; Ergun, H.; Kuznetsova, R.; Li, J.; Liaw, B. M.; Delale, F.; Chung, J. H. (2010) "An Experimental Study of Self-diagnosis of Interlaminar Damage in Carbon-Fiber Composites", Journal of Intelligent Material Systems and Structures, 21(3), 233–242, 2010, http://dx.doi.org/10.1177/1045389x09347019.

[20] Angelidis, N.; Khemiri, N.; Irving, P. E. (2005) "Experimental and Finite Element Study of the Electrical Potential Technique for Damage Detection in CFRP Laminates", Smart Materials and Structures, 14(1), 147–154, 2005, http://dx.doi.org/10.1088/0964-1726/14/1/014.

[21] Various-authors. (2012). Wikipedia. Available: http://en.wikipedia.org/.

[22] Kemp, M. (1994) "Self-sensing Composites for Smart Damage Detection Using Electrical Properties," 2nd European Conference on Smart Structures and Materials, Glasgow, Scotland, 1994, pp. 136–139.

[23] Wang, D.; Chung, D. D. L. (2006) "Comparative Evaluation of the Electrical Configurations for the Two-Dimensional Electric Potential Method of Damage Monitoring in Carbon Fiber Polymer—Matrix Composite", Smart Materials & Structures, 15(5), 1332–1344 , Oct. 2006, http://dx.doi.org/10.1088/0964-1726/15/5/023.

[24] Todoroki, A. (2008) "Delamination Monitoring Analysis of CFRP Structures Using Multi-Probe Electrical Method", Journal of Intelligent Material Systems and Structures, 19(3), 291–298, Mar. 2008 http://dx.doi.org/10.1177/1045389x07084154.

[25] Matsuzaki, R.; Todoroki, A. (2006) "Wireless Detection of Internal Delamination Cracks in CFRP Laminates Using Oscillating Frequency Changes", Composites Science and Technology, **66**(3–4), 407–416, Mar. 2006, http://dx.doi.org/10.1016/j.compscitech.2005.07.016.

[26] Matsuzaki, R.; Todoroki, A.; Takahashi, K. (2008) "Time-Synchronized Wireless Strain and Damage Measurements at Multiple Locations in CFRP Laminate Using Oscillating Frequency Changes and Spectral Analysis", Smart Materials & Structures, **17**(5), Oct. 2008, http://dx.doi.org/10.1088/0964-1726/17/5/055001.

[27] Hou, L.; Hayes, S. A. (2002) "A Resistance-Based Damage Location Sensor for Carbon-Fibre Composites", Smart Materials & Structures, **11**(6), 966–969, Dec. 2002, http://dx.doi.org/10.1088/0964-1726/11/6/401.

[28] Reifsnider, K. L.; Fazzino, P.; Majumdar, P. K.; Xing, L. (2009) "Material State Changes as a Basis for Prognosis in Aeronautical Structures", Aeronautical Journal, **113**(1150), 789–798, Dec. 2009.

[29] Lemistre, M. B.; Balageas, D. L. (2003) "A Hybrid Electromagnetic Acousto-Ultrasonic Method for SHM of Carbon/Epoxy Structures", Structural Health Monitoring—an International Journal, **2**(2), 153–160, 2003.

# Impact and Acoustic Emission Monitoring for Aerospace Composites SHM

## 9.1 INTRODUCTION

Low-velocity impact of composite structures that produces barely visible impact damage (BVID) is one of the most researched areas of aerospace composites structural health monitoring (SHM) due to drastic effect that the presence of BVID could have on composite aircraft performance and safety. Various methods have been proposed and tried for capturing the impact event and monitoring the evolution of resulting BVID state inside the structure.

Both impacts and acoustic emission (AE) generate ultrasonic waves; hence, they benefit from commonality of sensors installation. However, the frequency bands in which the two events take place are different; the impacts are more strongly felt as relatively lower-frequency flexural waves (e.g., tens of kHz), whereas the AE events happen in a higher band (e.g., 150–300kHz). Reference [1] describes a proposed built-in damage diagnosis system for composite structures aimed at detecting damaging events, and monitoring the in-service structural integrity of the composites. The proposed system consisted of two major diagnostic processes: passive sensing diagnosis (PSD) and active sensing diagnosis (ASD). The PSD process utilizes the measurements done by the sensors to identify damage-inducing event [1]. The ASD process uses diagnostic signals sent by the actuators and received by the sensors to diagnose the change in structural integrity [2–5].

## 9.2 IMPACT MONITORING—PSD

Monitoring of damage-inducing impacts on a composite structure and determination of their location and amplitude has received considerable attention. Named "passive sensing diagnosis" (PSD) in Ref. [1], this process utilizes the measurements done by the sensors to identify damage-inducing event.

### 9.2.1 PSD for Impact Location and Force Identification

The PSD process was further developed in Refs. [6,7]: the impact force and location was determined from the processing of the sensor signals; a maximum likelihood estimator was used to solve the inverse nonlinear problem. Further improvements of the signal processing method and impact identification algorithm were reported in Refs. [8,9]. Figure 1 shows the PSD system schematic impact load history estimated from sensor signals using the PSD approach. Further developments of the PSD approach are described in Refs. [10–12]. Impact detection (ID) with piezoelectric wafer active sensors (PWAS) in composite materials was successfully demonstrated in Refs. [3,13], and others. The application of the PSD approach to an actual composite wing was described in Ref. [14].

Basically, the PSD method consists of using a sparse array of sensors (strain gages, piezo wafers, FBG sensors, etc.) that capture the guided waves generated by the impact. The use of fiber-optic FBG sensors is attractive because a single optical fiber can carry several FBG sensors which can be independently interrogated; hence, the installation cabling issues are greatly simplified if FBG sensors are used instead of the more conventional transducers. In addition, fiber-optic sensors are immune to electromagnetic interference, which can confer advantage in certain applications. The captured signals must be processed to determine the impact location using a triangulation algorithm. For illustration, we present a simple example as follows.

(a)

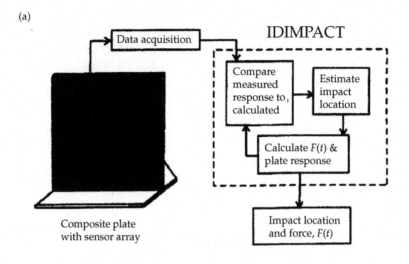

(b)

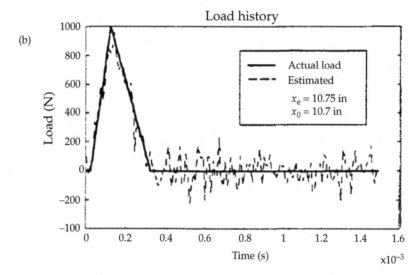

**FIGURE 1** PSD approach [1] for impact identification in a composite plate: (a) schematic [8]; (b) load history estimated from sensor signals [6].

## 9.2.2 Triangulation Example

Consider a network of PWAS transducers used to detect low-velocity foreign-object impact on a 1-mm aluminum plate. In our experiments, we used a 914-mm × 504-mm aluminum plate instrumented with a network of 11 PWAS transducers as shown in Figure 2a. A small steel ball (0.160 g) was dropped from a height of 50 mm. Signals were collected at PWAS #1, 5, 7, 9 (Figure 2a). The signals recorded at these PWAS #1, 5, 7, 9 are shown in Figure 2b. The high sensitivity of the PWAS transducers was very convenient for these

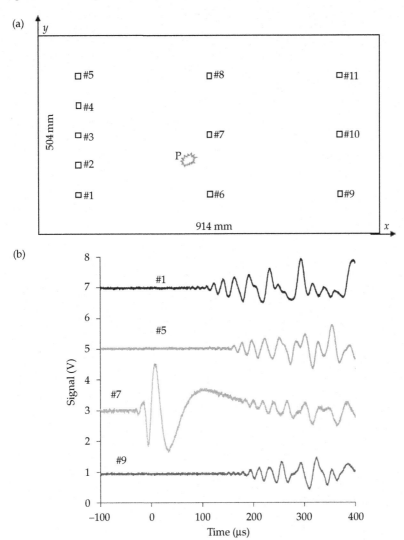

**FIGURE 2**    ID experiments: (a) location of impact and PWAS transducers #1, 5, 7, 9; (b) captured ID signals with arbitrary time origin due to oscilloscope trigger. The ID event was a 0.16-g steel ball dropped from a 50-mm height; the locations of the PWAS transducers and of the event are given in Table 1 (sensor #7 is closest to the impact) [15], page 698.

experiments because signals of up to $\pm 1.5$ V were directly recorded on a digital oscillo-scope without the need for any signal conditioning/preamplifiers. The corresponding signal time-of-flight (TOF) values relative to the oscilloscope trigger can be estimated as $t_1 = 126$ μs, $t_5 = 160$ μs, $t_7 = -27$ μs, $t_9 = 185$ μs (Table 1). Note the existence of a negative time $t_7 = -27$ μs; this phenomenon, which is due to oscilloscope trigger settings, indicates that the time origin cannot be specified from the recorded data and hence a time shift $t_0$ due to trigger delay must be considered in the calculations.

**TABLE 1**   Location of PWAS transducers by position $(x, y)$ and radial distance, $r$, from the AE and ID events

| Location | Coordinates (mm) | | TOF for ID ($\mu s$) | | Event Location Radius (mm) | | Resulting Travel Distance (mm) |
|---|---|---|---|---|---|---|---|
| | $x$ | $y$ | Captured | Adjusted | $r$ | Error | |
| PWAS #1 | 100 | 100 | 126 | 202.4 | | | 316 |
| PWAS #5 | 100 | 400 | 160 | 236.4 | | | 361 |
| PWAS #7 | 450 | 250 | −27 | 49.4 | | | 71 |
| PWAS #9 | 800 | 100 | 185 | 261.4 | | | 412 |
| Actual ID event | 400 | 200 | – | – | 447.2 | – | |
| Reconstructed ID event | 402.5 | 189.6 | – | – | 444.9 | − 0.5% | |

Note that the TOF was adjusted by 76.4 $\mu s$ to account for oscilloscope trigger.

The distance and TOF data given in Table 1 can be used to calculate the impact position using a version of the triangulation algorithm as follows: assume the unknown impact position is $(x, y)$ and write a set of simultaneous nonlinear equations represents the relation between travel distance, group velocity, and TOF.

$$(x_i - x)^2 + (y_i - y)^2 = [c(t_i + t_0)]^2, \quad i = 1, \ldots, 4 \tag{1}$$

These equations represent a set of four nonlinear equations with four unknowns $(x, y, t_0, c)$ which can be solved using error minimization routines. The unknowns are the impact location $(x, y)$, the wavespeed, $c$, and the trigger delay, $t_0$. In our studies, we tried two solution methods: (i) global error minimization and (ii) individual error minimization. It was found that individual error minimization gave marginally better results, whereas the global error minimization was more robust with respect to initial guess values. The impact location determined by these calculations was $x_{impact} = 402.5$ mm, $y_{impact} = 189.6$ mm. These values are within 0.6% and 5.2%, respectively, of the actual impact location (400 mm and 200 mm). The radial position of the actual ID event was 447.2 mm, whereas that of the reconstructed ID event was $r_{impact} = 444.9$, i.e., a 0.5% error. The trigger delay adjustment was found $t_0 = 76.4 \, \mu s$. The wavespeed was found $c = 1.560$ mm/$\mu s$, which corresponds to the tuning of the 7-mm PWAS into the quasi-flexural A0 mode at approximately 60 kHz.
A discussion of this method should address two important issues:

1. The correct estimation of the arrival time
2. The dispersive nature of the Lamb waves.

Parametric studies have revealed that the impact localization error is very susceptible to the arrival time estimates, $t_i$. This is mainly due to the difficulty of estimating the wave arrival times especially when a slow rise transition is present (cf PWAS 5 and #9 in Figure 2b). This aspect is due to the dispersive nature of the quasi-flexural A0 Lamb waves excited by the impact event. Because the impact is a rather broadband event (i.e., it excites a range of frequencies rather than a single frequency), the wave packet generated by the impact contains several frequency components, which travel at different wavespeeds. Hence, the wave packet generated by the impact disperses rather quickly (compare #7 with, say, #9). For example, the signal from PWAS #7 (which is close to the event) seems to form a compact packet, whereas the signals from PWAS #5 and #9 (which are farther from the impact) are much more dispersed.

To alleviate these difficulties, it seems intuitive that a different TOF criterion, e.g., the energy-peak arrival time, could be used. The application of energy-peak criterion would be quite easy on the compact wave packet signals such as PWAS #7. However, it would not be at all easy to apply this criterion on the dispersed signals such as PWAS #5 and #9. More research needs to be conducted to address the dispersion issues. Nonetheless, these experiments have proven that PWAS can act as passive transducers for detecting elastic wave signals generated by low-velocity impacts. It was also shown how data-processing algorithms could determine the impact location with reasonable accuracy.

The simple example discussed above has illustrated how triangulation method might work but has also highlighted the difficulties associated with the dispersive nature of the guided waves generated by the impact event into thin-wall structures. This aspect is even more difficult in composite structures which, due to their intrinsic anisotropy, have different wavespeeds and wave propagation characteristics in various directions. This makes the TOF determination through signal processing even more challenging when dealing with composite structures with complicated structural features.

The captured signals can also be processed to determine the impact amplitude. Two algorithmic classes have been explored for impact identification: (i) model based; (ii) data driven. The direct use of directional sensors has also been successfully used recently.

## 9.2.3 Model-Based Impact Monitoring

The PSD concept [1] uses a model-based approach, i.e., the experimental results are compared with model predictions and model parameter adjustments are performed through an optimization scheme until reasonable agreement is obtained. The model-based approach relies on a structural model that can simulate the signals to be received by the sensors for a given impact of assumed amplitude time history $F(t)$ and location $(x, y)$.

### 9.2.3.1 Structural Model Approach to Impact Identification

The structural model varies from simple composite plate flexure models [7] in which the effective stiffness and mass parameters were obtained through composite lamination theory (CLT) [8,10], through more-complicated models of stiffened plate [16], and finite element method (FEM) meshing of complicated composite structures [12]. The space dependence of the partial-differential equations was collapsed using either the space-domain Fourier transform [8] or projection on a set of shape functions [10]. The resulting time-domain differential equations were expanded into the state-space format and then discretized in time congruent with the experimental time series obtained through data acquisition.

### 9.2.3.2 System ID Approach to Impact Identification

An alternative approach is to perform a system ID process on the actual physical structure instead of modeling it. Reference [17] performed the system ID with the ARX technique (auto-regressive with exogenous inputs) and constructed a numerical model of the physical system based on a finite set of experiments. In the training stage, a set of experiments were conducted on the specimen; each experiment consisted in applying a known impact to a known location on the structure and recording the time series corresponding to the impact force and the signals received at the sensors (Figure 3).

The recorded time series data was used in ARX algorithm to determine the model parameters and implicitly the structural transfer functions between the impact location and the sensors. The simulated and the measured signal time series are compared and the model parameters are adjusted (impact location, amplitude, and time history). References [8,10] report the use of a two-stage optimization/fitting process to achieve this objective. In the validation stage, the ARX was used to synthesize sensor signals and compared them with the original experimental data; good agreement was observed [17]. Finally, the validated

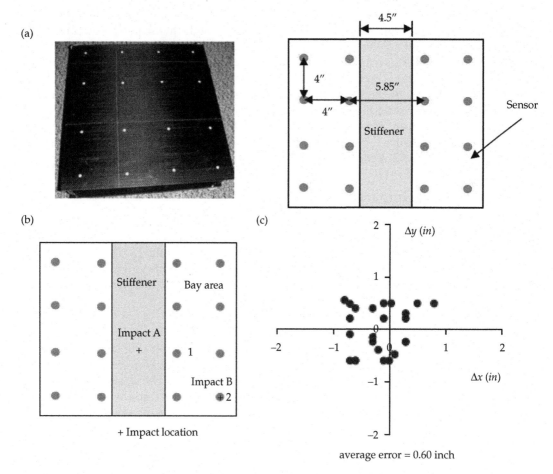

**FIGURE 3** Impact identification on stiffened composite panel using a system ID approach: (a) specimen description and sensor locations; (b) impact locations; (c) impact localization spread; (d) force reconstruction; (e) realistic specimen [17].

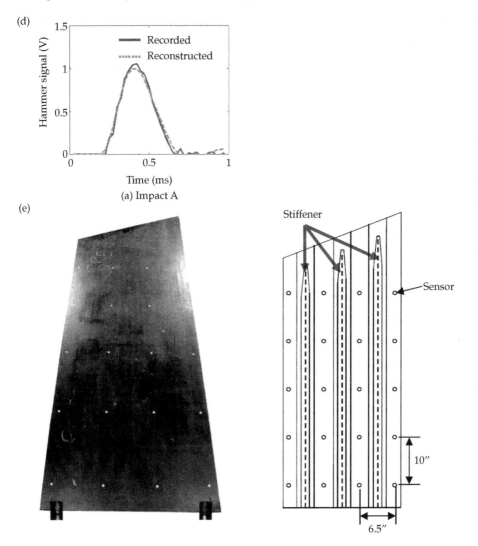

FIGURE 3    (Continued)

ARX model was used to detect impact at locations and time history different from those used in the training stage. The detection accuracy in terms of location and force time history was evaluated.

### 9.2.3.3 Hybrid System ID—Structural Model Approach to Impact Identification

More recent developments [12] have used an FEM simulation instead of actual hardware to construct the system ID model; again, the force of known time history was applied at given locations on the FEM structure and the response of the sensors was simulated. The simulated sensors response was used as input to the system ID algorithm. A two-step optimization using genetic algorithms (GA) was applied to fit the system ID model to the FEM-generated signals. This approach permitted the exploration of a wide parameter space in simulation in order to find the most appropriate sensor network configuration that would give acceptable probability of detection (POD) of the impact event with manageable sensor installation costs.

### 9.2.4 Data-Driven Impact Monitoring

The impact and damage identification methods described in previous sections used a structural model to interpret the sensor data. However, complicated structures may not have a structural model detailed enough to handle wave propagation simulation. Structural models for complicated composite structures may not be detailed enough to handle wave propagation simulation. Hence, techniques that do not require a structural model and rely entirely on the measured data are of interest. Such data-driven techniques have been investigated by several researchers.

Reference [18] describes impact monitoring on CFRP composite laminates using a sparse array of piezo wafer sensors and an artificial NN for signal processing.

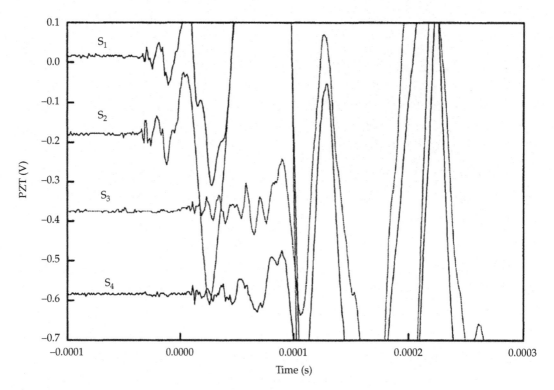

FIGURE 4   Impact-generated waves recorded at four piezo wafer sensors on an aluminum plate; the leading portions of the waves are used for neural network (NN) architecture [18].

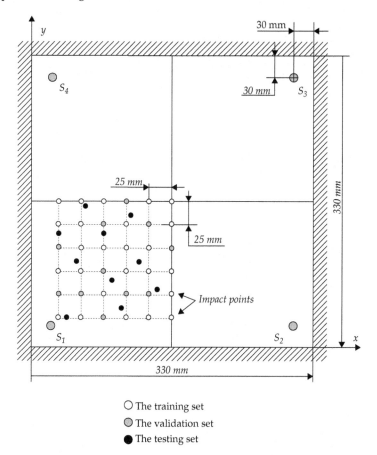

○ The training set
◎ The validation set
● The testing set

**FIGURE 5**    NN training, validation, and testing for ID on a rectangular plate instrumented to four piezo wafer sensors [18].

Only the beginning of the recorded waves was used in the NN because these small-amplitude leading waveforms have not yet been contaminated by the reflections from boundaries or scatter from structural features. For aluminum plate, these leading waves were found to have a frequency content in the range 20–80 kHz; in Figure 4, the leading waves are considered those up to 0.0001 s. For composite laminates, the leading waves were found to be in the 1–10 kHz range [18]. The input nodes in the NN architecture corresponded to the sampled values of the leading wave, whereas the output nodes were simply the $(x, y)$ coordinates of the impact. The NN was subjected to the usual developmental steps: (i) training; (ii) validation; (iii) testing (Figure 5).

Reference [19] used an artificial NN for ID in a carbon fiber-reinforced polymer (CFRP) composite plate instrumented with four piezo wafer sensors. NN training was done with an initial set of 100 low-energy impacts (0.3J) applied on the nodes of a 10 by 10 grid covering most of the plate. A second set of 30 low-energy impacts was performed at randomly selected sites on the same grid. The third set consisted of only one single severe energy impact (10J)

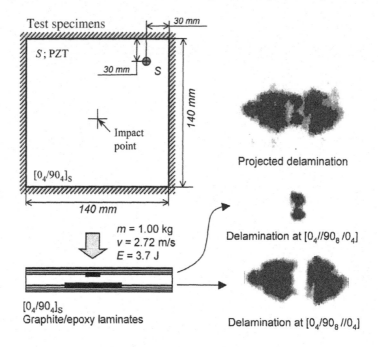

**FIGURE 6**  Delamination damage induced in a $[0_4/90_4]_S$ CFRP laminate by a 3.7-J impact as imaged by ultrasonic C-scan [18].

that produced an indentation on the front face of the panel, cracking on the back face, and internal delaminations between the plies. Finally, a fourth set of low-energy impacts was performed to teach the NN the "feel" of a damaged panel. The analysis considered several time and frequency domain features, namely: (i) time after impact of maximum response; (ii) magnitude of maximum response; (iii) peak-to-trough range of the response; and (iv) real and imaginary parts of the response spectrum, integrated over frequency. The input patterns to the network which proved most useful were (i) and (ii); hence, the networks required eight inputs, i.e., two values per sensor. The mean detection error was 5% (17 radial mm on a 340-mm × 340-mm plate). Features extracted from the signals were used to train the NN algorithm. The mean detection error was 5% (17 radial mm on a 340-mm × 340-mm plate). Further work by the same team, which includes a more-complicated NN, larger set of features, and the use of GA optimization, is reported in Refs. [20, 21]. Other artificial intelligence techniques, e.g., case-based reasoning (CBR), have also been explored [21].

Reference [22] discusses the use of a strain-amplitude algorithm for impact localization using fiber-optic FBG strain sensors. Bothe FEM simulations and actual experiments were performed (Figure 6).

## 9.2.5 Directional Sensors Approach to Impact Detection

The use of directional sensors can greatly simplify impact detection and localization. Directional sensors alleviate the difficulty created by the multimodal dispersive nature of the guided waves generated by the impact event. With directional transducers, the triangulation of the impact event is much simplified. Only two directional sensors are needed to locate an impact; each sensor generates a ray indicating the direction of the presumptive wave source and the impact is easily located at the intersection of these two rays (Figure 7a).

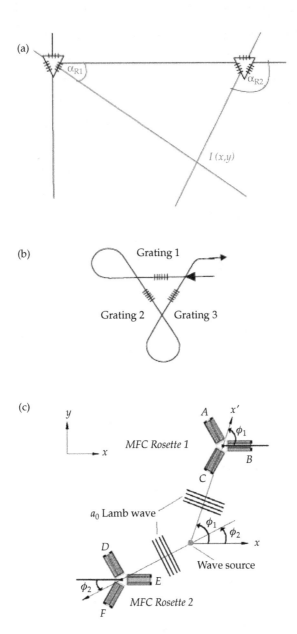

**FIGURE 7** Impact localization with directional sensors: (a) localization is obtained at the intersection of two directional sensor rays; (b) FBG rosette directional optical sensor [23]; (c) MFC rosette directional piezo sensor [24].

Two directional sensors types have been proposed, one based on fiber optics [23], the other based on piezo wafers [24]. Reference [23] constructed a rosette from three FBG optical sensors arranged in a triangular pattern (Figure 7b). Strain-gage rosette principles are used to resolve the principal directions and obtain the sensing ray direction. Reference [24] constructed a piezo rosette from three macro-fiber composite (MFC) sensors arranged in a star pattern (Figure 7c) and obtained very good impact localization without any structural model, neural net, or signal baseline (Figure 8). Piezo rosettes constructed from thin rectangular wafers of PMN crystals cut and poled in the $[011]_c$ direction were reported in Ref. [25].

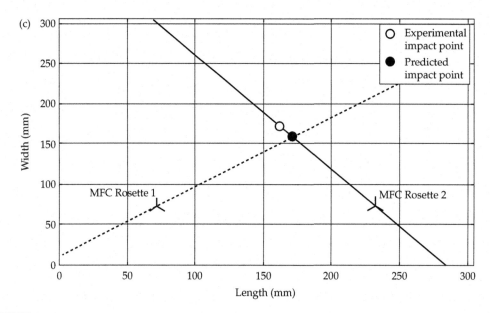

**FIGURE 8**   Model-free impact localization on quasi-isotropic CFRP woven composite panel using directional piezo sensors MFC rosettes [26].

A recent implementation of passive damage diagnostic approach using FBG optical strain sensors placed on an aircraft-like CFRP panel is discussed in Ref. [27]. The experiments were performed under realistic operational conditions consisting of vibration of the panel on a shaker (random spectrum between 10 and 2000 Hz). Both single FBG strain sensors and rosette FBG strain sensors were used. The results indicate that vibration environment does not constitute a major impediment in impact identification and location because vibration bandwidth and impact bandwidth are well separated.

## 9.2.6  AE Monitoring

AE can be monitored with both piezo and fiber-optic sensors. The former is relatively well established in conventional ultrasonic nondestructive evaluation (NDE); however, the conventional AE sensors are not quite appropriate for deploying in large numbers on a flight structure due to both their cost and size. The more compact SHM sensors (piezo wafers, fiber-optic sensors, etc.) have also been shown capable of AE monitoring. References [18,28,29] used piezo wafer sensors, whereas Refs. [30,31] used optical FBG sensors for AE emission monitoring. Existing AE monitoring signal capturing and interpretation methodology (noise filtering, AE events counting algorithms, etc. [32]) should also apply when SHM sensors are used.

### 9.2.7  Simultaneous Monitoring of Impact and AE Events

Since both impact events and AE events create guided waves in the monitored structures, it seems appropriate to monitor them both with the same sensor installation. First, the monitoring system would detect the impact event when it happens. Next, the same system would monitor the AE signals created by growth of the impact damage during operational loading.

In addition, the possibility exists to monitor the impact as well as the AE signal generated by the damage creation during the impact event. Reference [28] reports that the impact response recorded by a piezo wafer sensor mounted on a CFRP plate seems to show two separate wave types (Figure 9): (i) a stress wave due to the initial impact; (ii) a superposed high-frequency wave packed similar to an AE burst due to the composite laminate being damaged (delaminations and splits) by the impact.

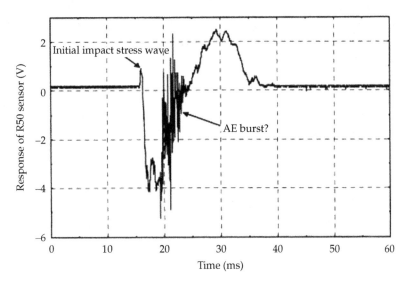

**FIGURE 9**    Impact response recorded by a piezo wafer sensor mounted on a CFRP plate: two separate wave types are apparent, an initial impact stress wave and a high-frequency wave packed similar to an AE burst [28].

A similar situation in which the signal received by the piezo wafer transducer during an impact event on a composite structure seems to contain two distinct components (impact-generated waves and AE waves) was reported in Ref. [18]. The wavelet transform (WT) analysis of the signals received by the sensors during the impact event showed higher-frequency components that could be associated with the damage being produced in the composite plate.

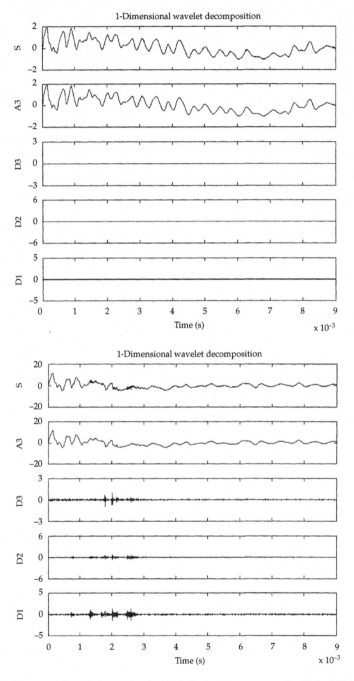

**FIGURE 10** WT decomposition of the received signal during impact on a $[0_4/90_4]_S$ CFRP laminate: (a) 0.1-J impact does not produce internal damage; (b) 3.7-J impact produces internal damage which generates AE waves that are picked up in the WT decomposition levels D1, D2, D3 [18].

Three impact energies were tested: low energy $E = 0.1J$ (non-damaging impact); intermediate energy $E = 3.7J$ (damaging impact producing delaminations); severe energy $E = 6.0J$ (damaging impact producing composite penetration). Figure 6 shows the $E = 3.7J$ delamination damage as imaged by ultrasonic C-scan. Figure 10 shows side by side the WT decomposition of the signals received by the piezo wafer sensor for the low energy $E = 0.1J$ event (Figure 10a) and the intermediate energy $E = 3.7J$ event (Figure 10b). It is apparent that the latter shows clear wave packets in the D1, D2, D3 decomposition levels whereas the former does not. These higher-frequency wave packets seem to be AE waves generated as the composite delamination was initiated and then propagated during the impact. Analysis of the sever-energy ($E = 6.0J$) impact signals showed similar higher-frequency packets but of a larger amplitude.

As illustrated in Figure 10, the higher-order WT components of the signal from a 3.7-J impact (Figure 10b) show AE bursts that were not present in the signal from a lower-energy impact(Figure 10a). The interpretation of these facts was that the 3.7-J impact produced internal damage (and hence AE signals) but the lower-energy impact did not.

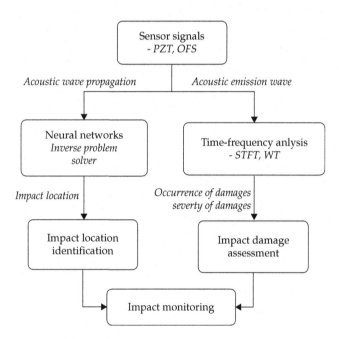

**FIGURE 11**  Block diagram of a dual-use impact monitoring system that would simultaneously perform impact location identification (left branch) and AE detection (right branch) to determine if the impact resulted in structural damage or not [18].

Reference [18] proposes a dual-use system for simultaneous monitoring of impact event and the accompanying AE event, if any. Figure 11 shows a block diagram of such dual-use impact monitoring system with the detection of impact location being performed on the left branch of the logic tree, and the assessment of whether or not damage resulting from the impact being done on the right branch of the logic tree. The event and location of an impact load can be identified using the propagating low-frequency acoustic waves using a data-driven neural net algorithm as discussed in Section 2.4. Simultaneously with the impact identification, the diagnosis of impact damage can be carried out to determine whether an incipient damage was initiated or during the impact; if such a damage event took place, then higher-frequency AE waves would also be present. This is achieved through time—frequency analysis of the captured signal.

Similar conclusions were previously reported in Ref. [13] which showed that "the apparition of high frequencies in the impact signal delivered by a sensor attached to the structure is a mean to discriminate between damaging and non-damaging impacts". Hence, Ref. [13] developed a high-frequency root-mean-square (HR-RMS) damage metrics that correlated well with the size of the delamination area.

$$\langle s \rangle = \sqrt{\frac{1}{T} \int_0^T s^2(t)dt} \qquad (2)$$

where $s(t)$ is the electrical signal filtered in a selected high-frequency band corresponding to creation of damage in the composite due to a high-enough impact energy. The HR-RMS parameter was used for full diagnosis: damage occurrence, localization, and severity. An almost linear correlation was established between the amplitude of the acoustic source at the impact location and the impact energy, with the clarification that a damage threshold exists in the impact energy (1—2 J) below which no damage was created in the composite plate [13].

## 9.3 IMPACT DAMAGE DETECTION—ASD AND ACOUSTO-ULTRASONICS

Active detection of impact damage consists of "interrogating" the structure with wave transmitters and picking up the structural response with wave receivers. The ASD process [1] uses piezo wafer transducers as both actuators and sensors of ultrasonic guided waves. This approach is also known as "acousto-ultrasonics" [33]. The "acousto" part of the name is associated with the reception of guided waves generated by the AE process at a propagating crack. The term "ultrasonics" infers that this is an active technique in which ultrasonic waves are generated by the transmitter; this is thus different from the AE technique that is just a passive technique. Acousto-ultrasonics requires two ultrasonic guided-wave probes, a transmitter and a receiver [34]. In SHM work, the transmitter has usually been a piezo wafer; the receiver has traditionally been also a piezo wafer, which imparts reciprocal transmitter—receiver capabilities to the setup and enables guided-wave tomography. The use of different receivers, e.g., fiber-optic FBG sensors, has also been reported [35]; this option is attractive because a single optical fiber can have several FBG sensors which can be independently interrogated; hence, the installation cabling issues are greatly simplified if FBG sensors are used instead of the more conventional piezo transducers. In addition, fiber-optic sensors are immune to electromagnetic interference, which can confer advantage in certain applications. However, FBG cannot usually act as transmitter and this imposes limitations on the methodology; reports of using optical fiber for guided-wave excitation also exist, but these attempts are still confined to the laboratory.

Another term used for the ASD method is "embedded pitch-catch" [36]. The pitch-catch NDE method is used to detect structural changes that take place between transmitter transducers and receiver transducers. In embedded pitch-catch NDE, diagnostic waves emitted by the transmitter piezo wafers are caught by the receiver piezo wafers. Guided waves traveling through a damaged region change their characteristic. The detection of damage is performed through the examination of the guided-wave amplitude, phase, dispersion, and TOF in comparison with a "pristine" situation. Guided-wave modes that are strongly influenced by small changes in the plate thickness (such as the antisymmetric quasi-flexural Lamb-wave modes) are well suited for this method. The piezo wafer transducers are either permanently attached to the structure (Figure 12) or inserted between the layers of composite layup. Typical applications include: (i) delamination detection in laminated composites, (ii) disbond detection in adhesive joints and composite patch repairs, etc.

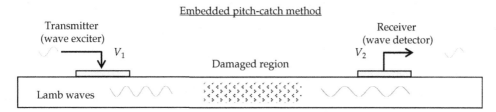

**FIGURE 12**    Pitch-catch method for embedded ultrasonics damage detection.

Two embodiments of the ASD approach have been proposed for detecting the damage in the composite: (a) a standing-wave approach (i.e., analysis of structural vibration) and (b) a propagating-wave approach.

In the *standing-wave approach*, the built-in actuators are used to excite structural vibration which is analyzed using a frequency domain method. Typical results are shown in Figure 13; the detection of damage was deduced from the differences in structural response magnitude over a bandwidth of up to 2 kHz. However, this approach was found to be sensitive to boundary conditions and requires that the damage be at least 10% of the structure length scale to be easily detectable [37,38].

In the *propagating-wave approach*, a suitable diagnostic signal is transmitted from the actuators through the composite structure and received at the sensors. Preliminary results of experiments performed on cross-ply CFRP composite specimens [2] showed significant signal changes due to structural damage caused by the impact (Figure 14).

## 9.3.1 ASD with Piezo Transmitters and Piezo Receivers

The ASD approach was developed [3,5] to identify impact damage location and size in composite specimens using the changes in the diagnostic signals due to wave scatter at the

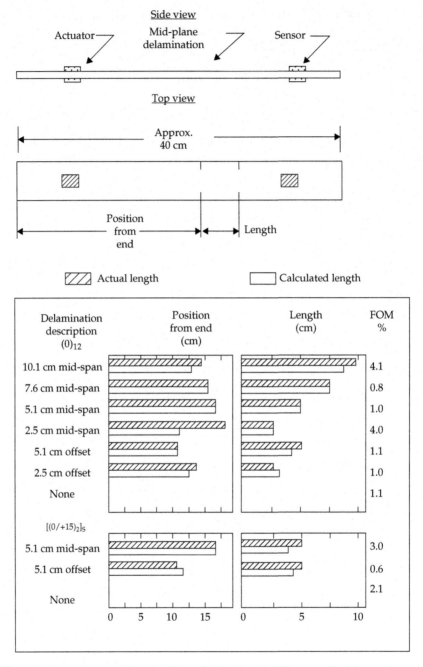

FIGURE 13　Delamination size estimated with a vibration-based ASD approach: (a) experimental schematic; (b) comparison of calculated and actual delamination length values [38].

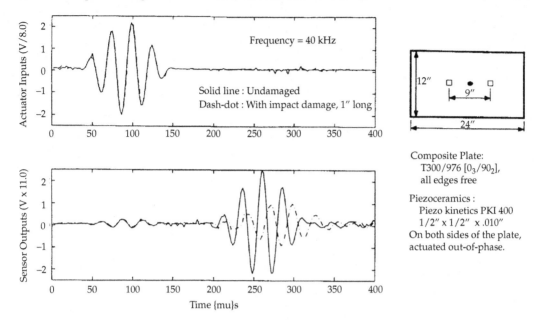

Composite Plate:
T300/976 [0₃/90₂],
all edges free

Piezoceramics :
Piezo kinetics PKI 400
1/2" x 1/2" x .010"
On both sides of the plate,
actuated out-of-phase.

**FIGURE 14** Impact-induced changes in a diagnostic signal traveling through a cross-ply CFRP composite (70-kHz 5-cycle smoothed tone burst traveling over a 9-in distance between a transmitter and a receiver piezo wafer active sensors) [2].

damage sites. A block diagram of the ASD process is presented in Figure 15. Kirchhoff plate model and effective stiffness and mass parameters obtained through CLT analysis were used. It is apparent that the PSD and ASD processes share the same structural model for simulating wave propagation phenomena. In addition, the ASD model also needs to model the interaction of the diagnostic waves with the damage site. A unit damage identification cell (UDIC), as defined by four actuator—sensor transducers, was considered [5]. A round-robin transmission—reception (pitch-catch) process between the four transducers takes place. The transducer acting in actuator mode generates a diagnostic wave that interacts with the composite structure and is modified by the damage. The wave is received at the other three transducers acting as receivers. The six possible transducer pairs define six path-dependent signals. For each path, a scatter signal is defined as the difference between the previously stored baseline signal (pristine specimen) and the current signal (damaged specimen).

Experiments were performed on CFRP (TS00/3900-2) composite plates with four PKI-400 piezo wafers mounted on one side of the plate using conductive epoxy; the whole composite plate was used as ground terminal. The piezo wafers were thin disks of 0.25-in (6.35 mm) diameter and 0.010-in (0.254) thickness. A schematic of the experimental setup is shown in Figure 16. An HP 33120A arbitrary waveform generator was used for transmitting stored diagnostic signals to the piezo wafer acting as actuator. The input signals were programmed with HP waveform generation software on a computer and downloaded into HP33120A through an IEEE488 bus. An industrial wideband power amplifier with a high slew rate was used to achieve piezo-transducer excitation of sufficient voltages

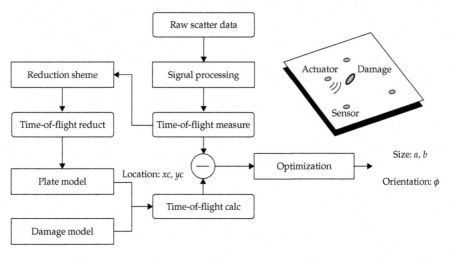

**FIGURE 15**   Block diagram of the pitch-catch damage identification procedure [3].

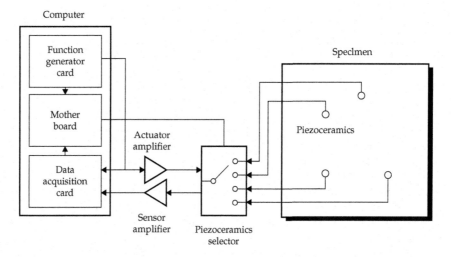

**FIGURE 16**   Schematic of the experimental setup for impact damage identification in composite structures using the pitch-catch method [3].

over the whole frequency range (40kHz through 150kHz). A purpose-built sensor amplifier using a general-purpose IC chip was used to boost the sensor signals. A high-speed GAGE A/D converter was used for data collection. The tone burst signals were sent from one of the piezo wafers and collected at all the other piezo wafers, in a round-robin fashion. The selection of the piezo transducers as actuators and sensors was done automatically with an electromechanical relay circuit controlled by the computer. Damage was introduced in the specimen with a quasi-static impact done by an MTS machine.

    The experimental signals were processed to yield the measured TOF estimates. The measured TOF is further processed to generate a reduced TOF which is used in a composite plate model to yield a calculated TOF for an assumed damage size and location. The measured and calculated TOF values are inserted in an optimization scheme which adjusts the damage size and location parameters to minimize the difference between the measured and assumed TOF values (Figure 15).

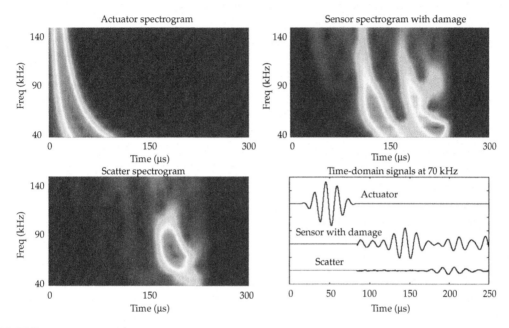

**FIGURE 17**    Spectrograms of actuator signal, sensor signal with damage, and scatter signal; the latter shows clearly a peak at around 180 μs, 70 kHz [5].

In order to correctly characterize composite damage from sensor measurements, diagnosis at a single frequency may not be sufficient because the interaction between the guided waves and the composite damage differs at the various wavelengths corresponding to various frequencies. Hence, a swept frequency method was adopted to generate sensor responses over a range of frequencies. To achieve this, the tone burst central frequency was swept from 40kHz through 150kHz [3,5]. The TOF determination was done through time–frequency analysis with the short-time Fourier transform to generate a spectrogram. Figure 17 illustrates a situation in which the maximum scatter occurred for the 70-kHz waves, as identified by a clear peak in the spectrogram around 180 μs arrival time and around 70 kHz frequency.

It is apparent that the spectrogram method offers advantages in finding TOF of frequency-swept narrowband wave packets. Recall that the scatter signal is the difference between two sensor signals recorded for two different specimen states, i.e., the reference state and the damaged state. Therefore, the scatter spectrogram can be obtained by subtracting the baseline sensor spectrogram from the current sensor spectrogram. The TOF of the scatter signal can thus be determined by scaling the scatter peak amplitude with the actuator peak amplitude. This TOF represents the time that the scatter signal takes to propagate from the actuator to the sensor via the damage site, which is the key information for the damage identification problem [3].

## X-ray image showing the damage

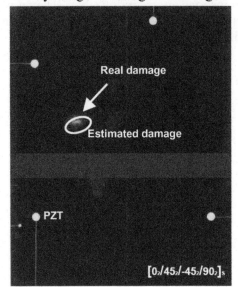

| [0₄/90₄]s | $a$ | $b$ | $\phi$ |
|---|---|---|---|
| Estimated | 1.4 | 0.6 | 2° |
| Real | 1.2 | 0.5 | 0° |

| [0₂/45₂/-45₂/90₂]s | $a$ | $b$ | $\phi$ |
|---|---|---|---|
| Estimated | 0.9 | 0.5 | 27° |
| Real | 0.8 | 0.4 | 32° |

$a$: Semi-major axis of the damage (in).
$b$: Semi-minor axis of the damage (in).
$\phi$: Angle between semi-major axis and 0° fiber direction.

**FIGURE 18**   Impact damage identification in a composite panel with the embedded pitch-catch method: the picture shows the "real damage" detected by X-rays and the "estimated damage" detected by the embedded pitch-catch method. The location of the four piezo wafer transducers (denoted "PZT" in the figure) is also shown [3].

In the damage identification procedure, it is assumed that the damage to be identified is located with the domain encompassed by the four piezo wafer transducers that form the UDIC domain. The UDIC contains six actuator–sensor (pitch-catch) diagnostic paths. Therefore, six scatter spectrograms can be generated and six TOF values can be obtained. These TOF values are function of the location, the size, and the orientation of the damage site. A theoretical model is used to estimate the scatter TOF values and compare them with the experimental measurements. In the model, certain assumptions were made regarding the description of the damage site: the damage was assumed elliptical; material degradation was assumed to be uniform over the damage area; the scattering was assumed to take place only at the damage site boundary. The parameters determining the damage location and size were as follows:

- Damage location: the coordinates $(x, y)$ of the ellipse center
- Damage size: the major and minor ellipse axis $(a, b)$ and the inclination angle $\theta$.

An additional unknown parameter that needed to be determined was the parameter $k$ indicating how much the guided wavespeed was changed when traversing the damage area.

In Refs. [3,5], axisymmetric flexure-like guided waves were assumed at relatively low frequency-thickness product values; these waves are highly dispersive at low frequency-thickness values and their speed tends to zero as frequency tends to zero; hence, a reduction in effective plate thickness due to damage affects a reduction in their speed. The wave velocity reduction factor was assumed to be constant within a damage area.

The damage location $(x, y)$ was determined through an optimization algorithm which minimized the error of the estimated distances among all the six pitch-catch paths. The damage size and orientation was determined by minimizing the differences between the measured arrival times and the calculated arrival times of the scattered waves for the six pitch-catch paths. Four parameters were used for minimization $(a, b)$, $\theta$, $k$.

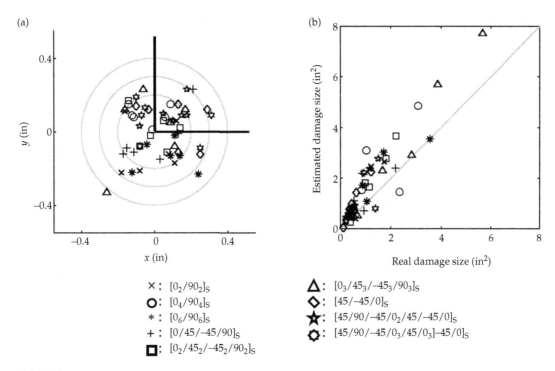

FIGURE 19    Study of damage identification accuracy: (a) damage location accuracy; (b) damage size accuracy [5].

The results of this damage identification process are illustrated in Figure 18 which shows identification results for two composite plates with different layup sequences, cross-ply and quasi-isotropic. An X-ray image of the quasi-isotropic plate is also shown. The actual delamination region embedded in the composite plate is shown as a lighter color spot in the X-ray marked "real damage". The elliptical contour drawn on the image represents the size and shape of the delamination as determined by the diagnosis process [3]. Further analysis was performed to determine the accuracy of this damage identification technique [5]. Extensive experiments were conducted on various composite plates; the results are presented in Figure 19 as accuracy plots, damage location accuracy on the right; damage size accuracy on the left. The center of the concentric circles in the damage location accuracy plot represents the actual damage center. It can be seen that 68.8% of the estimated damage centers are within 0.2 inches of actual and 90.1% are within 0.4 inches. Examining now the damage size accuracy plot, we note that that the size of damages was overestimated in most of the cases, and the prediction became worse as the damage size increased. The damage overestimation is a conservative estimate, though it might incur unwanted additional costs in practice.

The cause of this estimation may be due to the complexity of the damage which may differ substantially from the simple elliptical assumptions made in the analysis.

Similar pioneering work was done in about the same time frame at ONERA in France. References [39,40] report the use of a network of transmitter–receiver piezoelectric disks (5-mm diameter, 0.1-mm thick) to detect delamination in a composite plate. A 10-cycle 365-kHz tone burst transmitted from one of the piezoelectric disks generated S0 Lamb waves that were scattered by the damage into A0 and SH0 waves through diffraction and mode conversion. The received signals were analyzed with the discrete WT and compared with the pristine plate signals. The analysis was performed using simple TOF considerations. No analytical model of the damaged plate was needed.

More recent work on damage detection in composite structures has used more sophisticated data-processing methods; for example, Ref. [41] describes the use of time–frequency analysis to detect damage in a CFRP composite panel with stiffeners.

## 9.3.2 ASD with Piezo Transmitters and Fiber-Optic Receivers

Optical fiber sensors offer several advantages over piezo sensors such as resistance to electromagnetic interference and the capability of placing several sensors (e.g., FBG) on the same fiber. Culshaw and co-workers [42–45] have studied the use of piezoelectric transmitters and optical fiber receivers for damage detection in composite plates using ultrasonic Lamb waves. Conventional ultrasonic transducers, interdigital piezoelectric transducers, and simple PWAS were used. It was remarked that the PWAS gave good results in spite of their constructive simplicity.

Piezo-optical acousto-ultrasonics uses piezo transmitters and FBG sensor receivers. The use of fiber-optic FBG sensors is attractive because a single optical fiber can carry several FBG sensors which can be independently interrogated; hence, the installation cabling issues are greatly simplified if FBG sensors are used instead of the more conventional transducers. In addition, fiber-optic sensors are immune to electromagnetic interference, which can confer advantage in certain applications.

Reference [46] used a piezo-optical approach to detect impact delaminations in CFRP composites; a pitch-catch method based on the mode conversion between symmetric and antisymmetric guided waves at the delamination is used. A 3.4-mm-thick CFRP quasi-isotropic plate with an artificial midplane delamination was used. Piezo actuators were placed on both sides of the plate such that symmetric and antisymmetric modes could be independently excited. It was found that the intact 3.4mm quasi-isotropic CFRP specimen accommodates three guided-wave modes (quasi A0, S0, A1) at 300kHz. However, the delaminated region is composed of thinner branches, say 1.7 mm each. In these thinner plates, only two guided-wave modes (quasi A0 and S0) propagate at 300kHz. Hence, the A1 mode which is present in the pristine plate cannot propagate in the top and bottom branches of the delaminated plate. Thus, mode conversion takes place at the beginning of the delaminated region where the plate thickness changes from 3.4mm to 1.7mm: the A1 mode converts into S0 and A0 modes. Conversely, at the end of the delamination region, when the thickness changes from 1.7mm to 3.4mm, some of the energy of the S0 and A0 modes will be converted into the reemerging A1 mode. Therefore, if there is a delamination in the middle of the thickness of the laminate, it is expected that the frequency dispersion curves of the Lamb waves that pass through the delaminated area will change because of the mode conversions at both tips of the delamination.

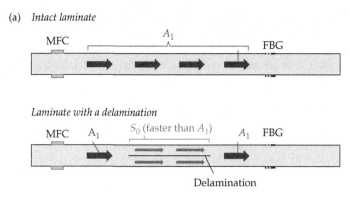

(a)  *Intact laminate*

*Laminate with a delamination*

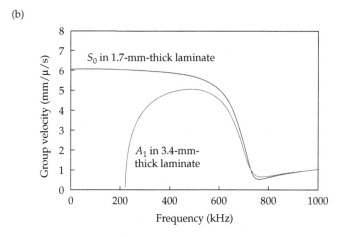

(b)

**FIGURE 20**  Conversion between A1 mode and S0 mode at the start and end of the delaminated region: (a) wave propagation schematic; (b) group velocity changes due to effective thickness change [46].

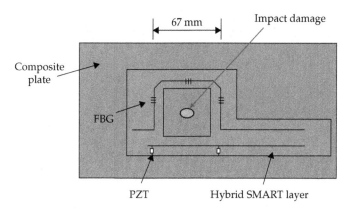

**FIGURE 21**  Hybrid PZT/FBG system for the detection of impact delamination damage in a CFRP composite plate [47].

When antisymmetric modes were excited (A0, A1) in the pristine plate, the A1 mode was converted into the S0 mode in the delaminated area (Figure 20a). In this case, the dispersion of the group velocity differs between the A1 mode in the intact region of 3.4-mm thickness and the S0 mode in the delaminated region of 1.7-mm thickness, as plotted in Figure 20b. Since the S0 mode is faster than the A1 mode, the arrival time of the A1 mode at the FBG sensors decreases with the increase of the delamination length. Furthermore, this difference in the velocity increases as the frequency decreases. Hence, the frequency dispersion of the A1 mode received in the FBGs is expected to change, depending on the delamination length. These phenomena were investigated numerically and experimentally and a damage index (DI) based on the A1 conversion was proposed for quantifying the delamination damage size. A similar mode conversion phenomenon was observed with scanning laser Doppler vibrometry (SDLV) by Ref. [48] who also reports the generation of standing waves that persist only in the delamination region.

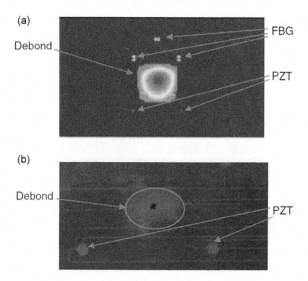

**FIGURE 22**   Detection of impact delamination damage in a CFRP composite plate using a hybrid PZT/FBG system: (a) reconstructed tomographic image; (b) X-ray image [47].

Reference [47] describes the use of a hybrid PZT/FBG system for active detection of impact delamination damage in a composite plate using acousto-ultrasonic tomography. Figure 21 shows a schematic of the sensors layout. The advantage of the FBG sensors is that several of them can be laid out on the same optical fiber, hence reducing the number of cable connections. As indicated in Figure 21, the FBG sensors surround the impact damage area and hence collect signals from several directions. Figure 22 shows the imaging of the damage through the guided-wave interrogation system and through conventional X-ray, respectively.

## 9.3.3 Guided-Wave Tomography and Data-Driven ASD

The guided-wave tomography has been extensively used in conventional ultrasonic NDE and some sophisticated data-driven imaging algorithms have been developed [49]. Similar approaches have tried for acousto-ultrasonics SHM imaging [50,51]. Reference [52] performs a comprehensive comparative study of various tomographic imaging algorithms: fan beam filtered back-projection (FBP), interpolated FBP, algebraic reconstruction technique (ART) with Bessel–Kaiser basis functions, the probabilistic reconstruction algorithm (PRA), etc. The authors of Ref. [52] recommends the use of their own RAPID[1] algorithm that accounts for wave scattering and reflection from damage using a probabilistic damage detection concept where the final tomogram is a superposition of ray ellipses.. The ray ellipses are initially constructed using simulated multimode guided-wave propagation in the composite; they can be modified based on experimental data for more accurate representation of defects (Figure 23).

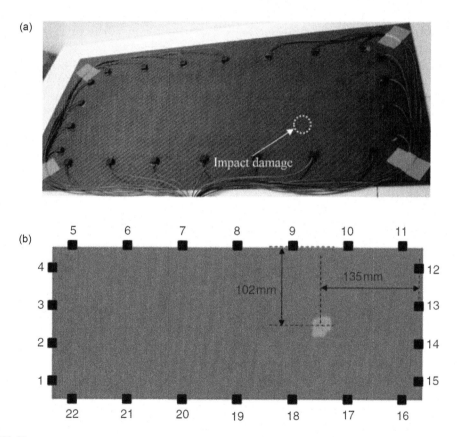

FIGURE 23    Imaging tomogram of impact damage in a composite plate: (a) photo showing damage location and sensors setup; (b) reconstructed tomogram showing predicted damage location and size [52].

[1]RAPID = reconstruction algorithm for probabilistic inspection of damage.

A considerable number of papers discuss imaging algorithms that are entirely data driven and do not require a structural model. Similar to the passive damage diagnostic approach, NNs have been proposed that would be trained on a number of damage scenarios and then used to identify an actual damage; however, this data-based approach seems to require a large number of tests and could be cost prohibitive.

### 9.3.4 PWAS Pulse-Echo Crack Detection in Composite Beam

In conventional NDE, the pulse-echo method has traditionally been used for across-the-thickness testing. For large area inspection, across-the-thickness testing requires manual or mechanical moving of the transducer over the area of interest, which is labor intensive and time consuming. It seems apparent that guided-wave pulse-echo would be more appropriate, because wide coverage could be achieved from a single location; hence, the use of guided Lamb waves for ultrasonic NDE pulse-echo inspection has been proposed [53]. The embedded pulse-echo method follows the general principles of conventional NDE pulse-echo, only that the transducer is permanently installed in the structure. A PWAS transducer attached to the structure acts as both transmitter and detector of guided Lamb waves traveling in the structure. The wave sent by the PWAS is partially reflected at the crack. The echo is captured at the same PWAS acting as receiver (Figure 24). For the method to be successful, it is important that a low-dispersion Lamb wave be used. The selection of such a wave, e.g., the S0 mode, is achieved through the Lamb-wave tuning methods [15].

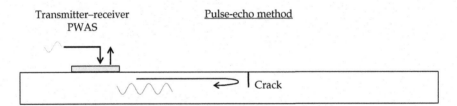

**FIGURE 24**    Pulse-echo method for embedded ultrasonics damage detection.

Reference [54] studied the detection of delaminations in a composite beam using the embedded pulse-echo method with low-frequency A0 Lamb waves. Figure 25a shows the experimental setup. Rectangular PWAS (20 mm × 5 mm) were used with the length oriented along the beam axis. This ensures that low-frequency Lamb waves were predominantly excited along the beam length. Two PWAS were used, one as transmitter (actuator) and the other as receiver (sensor). A 5.5-cycle 15-kHz Hanning-smoothed tone burst was applied to the transmitter PWAS. Figure 25b shows the signal recorded in the pristine beam. The initial bang and the reflection from the end of the beam are apparent. Then, a delamination was generated in the composite beam using a scalpel blade. The size of the delamination was progressively increased, as indicated in Figure 25c. The presence of the delamination crack produced an additional echo, as shown in Figure 25d. This work was extended to 2D composite plates by Refs. [34,55].

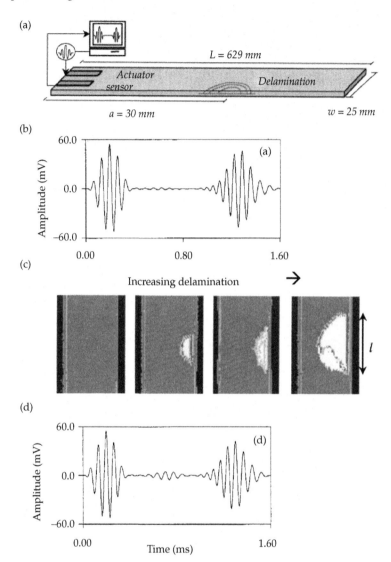

**FIGURE 25**   Detection of delaminations with the embedded pulse-echo: (a) experimental setup; (b) signals in the pristine specimen show only reflections from the beam end; (c) through-the-thickness C-scans of the specimen with conventional ultrasonics showing delamination increase; (d) signal in damaged specimen shows additional echo from the delamination crack [54].

## 9.3.5 Phased Arrays and Directional Transducers

Another model-free method for active damage detection is based on directional transducers. For passive sensing, Section 2.5 discussed the fact that directional sensors such as strain rosettes (both piezo and optical) can effectively detect an acoustic source without

(a)

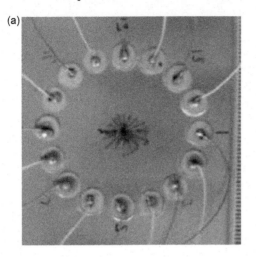

(b)

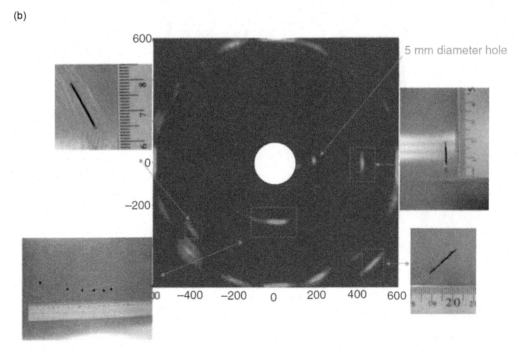

FIGURE 26  *In situ* phased array for monitoring multiple damages on an aluminum plate: (a) circular-fence array of piezo wafer active sensors placed at plate center; (b) simultaneous imaging of multiple crack and hole damage sites on the plate [52].

the need for a structural model. For active sensing, directional transmission as well as reception of guided wave can be achieved with phased-array transducers. Reference [52] reviews the main imaging techniques for guided-wave SHM and identifies the phased-array principles as the other major approach besides the tomography approach described in Section 3.3. Reference [56] constructed *in situ* phased array of piezo wafer active sensors tuned to low-dispersion S0 waves. They proposed the embedded ultrasonic structural radar (EUSR) phased-array SHM concept which performs the phased-array reconstruction in post-processing and thus does not need multichannel measuring capabilities. The EUSR phased-array SHM was used to *in situ* image the actual crack growth during fatigue testing of an aluminum plate [57]. Extension to 2D phased-array scanning was studied for different array geometries in Ref. [58] and demonstrated experimentally for a rectangular array [58] (Figure 26).

Reference [52] demonstrated the simultaneous imaging of multiple crack and hole damage sites using a single circular-fence phased-array placed at the center of an aluminum plate. Reference [59] studied the extension of phased-array principles to composite structures and developed appropriate signal processing algorithm; experimental results of actual damage imaging in a composite specimen have not yet been published.

Directional transducers, such as MFC, can be placed along directions in which an interrogating ultrasonic wave is desired. In addition, MFC transducers can be arranged in rosette configurations and then be excited with phase signals such that a beam steering effect is achieved. Reference [60] describes an extensive theoretical and experimental study of the capabilities of directional MFC transducers and MFC rosettes.

Another method of constructing transducers with intrinsic beam steering capabilities for large area SHM imaging is described in Ref. [61]. A steerable directional transducer is constructed with eight independently addressable sectors arranged around the circle at $45°$ pitch (Figure 27a). Each sector has an actuator outer part and a sensing inner part. An interdigitated electrode pattern was photolithography printed on a copper-clad Kapton film. The 0.2-mm-thick PZT-5A piezoceramic rings were diced into wedge-shaped fibers such that a fiber width of 0.36 mm was achieved at the inner radius. The fibers and electrodes were bonded together and cured in an autoclave and then poled (Figure 27b). The steering of the directional sensor is achieved by switching between the individual sector. Calibration experiments were performed on an aluminum plate; the SDLV instrument measured wavefield for various excitation directions agreed very well with the theoretical predictions.

A directional SHM transducer that does not require either phase-array delays or connection switching to achieve steering is described in Ref. [62]. This directional sensor achieves tuning into preferential direction at certain discrete frequencies which are the solution of a frequency-wavenumber equation. The transducer consists of a skew array of piezo wafer active sensors placed at pitch values $d_1, d_2$ and angles $\alpha, \beta$ about the 1 and 2 axes, respectively (Figure 28a). The transducer achieves directivity at certain frequencies; Figure 28b illustrates the directionality measured with SDLV on an isotropic aluminum plate: $45°@105kHz$, $120°@150kHz$, $−17°@200kHz$, $88°@280kHz$. Thus, the transducer directionality is controlled by the excitation frequency in discrete steps. Reference [62] also

(a)

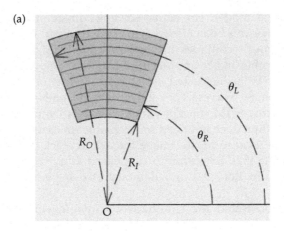

(b)

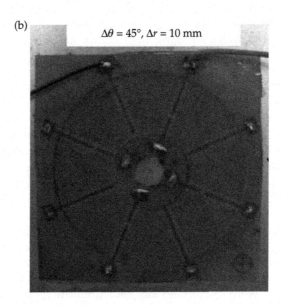

$\Delta\theta = 45°, \Delta r = 10$ mm

**FIGURE 27** Directionally steerable CLOVER transducer for 2D SHM scanning: (a) schematic illustrating the geometry of an individual sector; (b) actual implementation [61].

addresses directionality in a composite plate, which is more difficult to achieve because of the inherent anisotropy of the medium. Figure 28c shows such directional pattern obtained through guided-wave simulation in a unidirectional a GFRP plate. A further extension of this concept to achieve continuous steering through a double spiral shape is presented in Ref. [63].

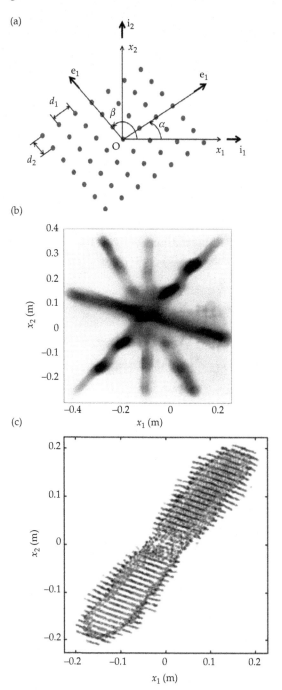

**FIGURE 28** Frequency-steerable direction SHM sensor: (a) grid pattern of the sensor elements; (b) preferential steering directions highlighted simultaneously through broadband excitation on an aluminum plate as measured with an SDLV; (c) directional wave pattern achieved on a composite plate at the first steering frequency: guided-wave simulation in a unidirectional glass fiber-reinforced polymer (GFRP) plate [62].

## 9.4 OTHER METHODS FOR IMPACT DAMAGE DETECTION

This section describes other methods for impact damage detection besides the ASD and acousto-ultrasonic approaches described in Section 3. This section will discuss direct methods for impact damage detection, strain monitoring methods, vibration SHM, frequency transfer function SHM, local-area active sensing with electromechanical (E/M) impedance spectroscopy (EMIS), etc.

### 9.4.1 Direct Methods for Impact Damage Detection

A number of publications describe attempts to directly detect the composite impact damage with damage-sensitive embedded sensors or through composite self-sensing.

Reference [64] uses optical intensity sensors consisting of small diameter (40 μm) multimode optical fibers with polyimide coating, which change the transmission characteristics when suffering microbending due to damage (Figure 29a).

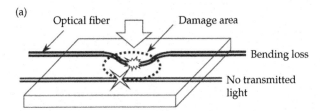

**FIGURE 29**   Measuring impact response in a composite with fiber optics through change of optical intensity due to fiber microbending and fiber fracture [64].

Reference [65] describes a method for the detection of damage in GFRP composites using conventional reinforcing E-glass fibers that were made to act as light guides and dubbed reinforcing fiber light guides (RFLG). These RFLG were used to detect damage induced in the composite by impact, indentation, and flexure. The E-glass fibers were converted into light guides by applying an appropriate cladding material. The coating resins were epoxy- and polyurethane-based resin systems. These self-sensing fibers or RFLG were surface mounted and also embedded at two specified locations within 16-ply glass-fiber-reinforced epoxy prepreg composites.

The data demonstrated that the self-sensing concept could be used to study *in situ* and in real time the failure processes in glass fiber-reinforced composites. The characterization of failure modes observed when the composites with the self-sensing light guides were subjected to impact, indentation, and flexural loading was attempted. The damaged areas in the GFRP composite were located by means of the "bleeding" light emanating from the broken self-sensing E-glass fibers (Figures 30 and 31). The RFLG approach was further refined by developing a sol-gel cladding fabrication technique [66].

Reference [67] describes the use of the triboluminescent effect to capture impact damage events in a CFRP panel. Hollow silica capillaries were incorporated into CFRP composite

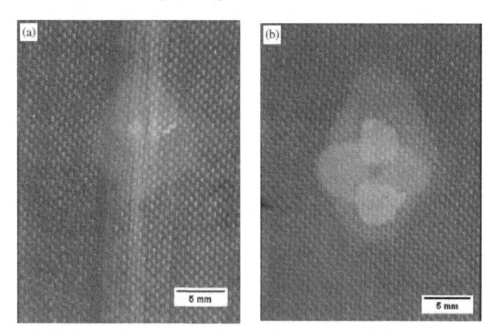

**FIGURE 30**   Detection of 2-J impact delamination in a GFRP composite with RFLG fibers placed on the top surface: (a) front face; (b) back face [65].

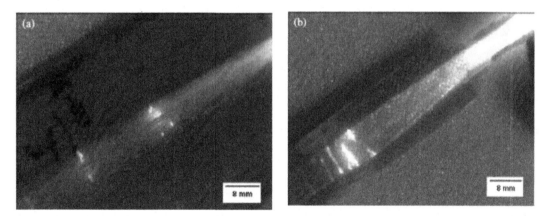

**FIGURE 31**   Light bleeding from broken E-glass optical guides after the GFRP composite was subjected to mechanical damage: (a) indentation; (b) three-point bending [65].

panels; the composite prepreg was doped with Terbium triboluminescent compound. After cure of the composite, the silica capillaries were filled with rhodamine 6G in benzyl alcohol and sealed. The panels were again impacted in the drop-weight tower, with the resulting light emission recorded (Figure 32).

(a)

(b)

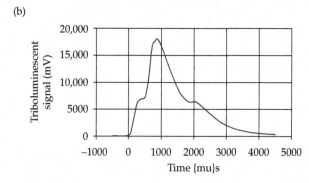

**FIGURE 32**   Use of triboluminescence to detect impact on a CFRP panel: (a) CFRP panel showing the place of impact and two photoluminescent hollow silica fibers used to capture the triboluminescence phenomenon; (b) triboluminescent signal captured during impact [67].

References [68,69] describe the use of optical time-domain reflectometry (OTDR) and FBG sensors to detect impact damage in GFRP sandwich specimens. The optical fibers were embedded in the sandwich core at various depths (Figure 33). The principle of detection consists in the fact that deformation and damage due impact leaves areas of permanent strain around the impact zone. The OTDR and FBG methods were employed to detect such changes in comparison with the pre-impact baseline (Figure 34). The small crosses in Figure 34a indicate the locations where OTDR indicates permanent strain changes, whereas the large cross indicates the actual impact location. The FBG detection is illustrated in Figures 35 and 36. It is apparent that both methods are able to detect the permanent strain changes produced by the impact provided the sensors are in the proximity of the impact zone (Figure 37).

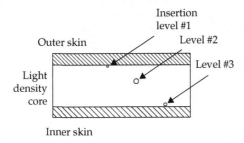

**FIGURE 33**    Various thickness-wise locations of the optical fibers in the sandwich core [68].

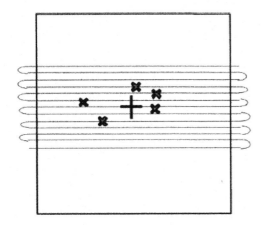

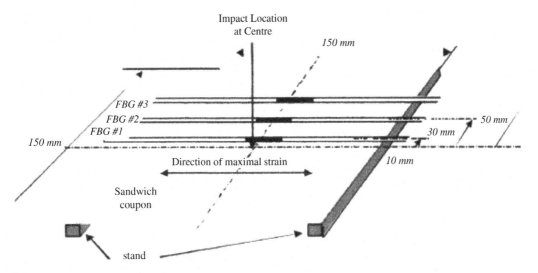

**FIGURE 34**    Arrangement of optical fibers in a GFRP sandwich for impact detection: (a) OTDR grid pattern over the specimen area; the small crosses indicate the locations where OTDR indicates permanent strain changes, whereas the large cross indicates the actual impact location; (b) parallel arrangements of fibers with FBG sensors [68].

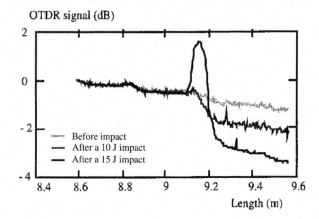

**FIGURE 35**   OTDR signal change with impact strength [69].

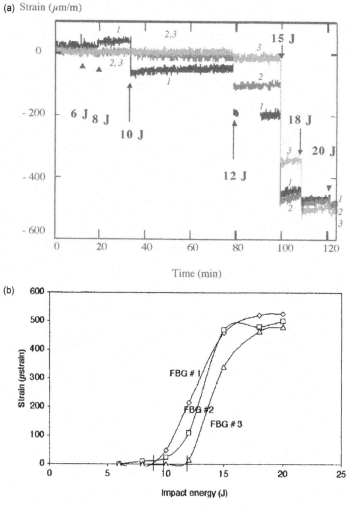

**FIGURE 36**   FBG sensor results: (a) permanent strain at various impact strengths; (b) permanent strain versus impact energy [68].

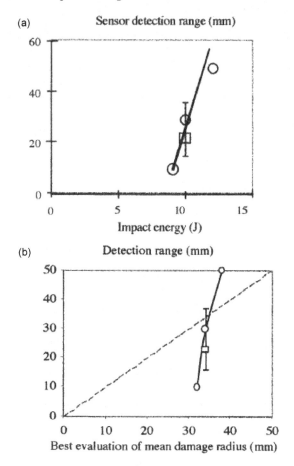

**FIGURE 37**   Comparison of OTDR and FBG detection of impact damage: (a) detection range versus impact energy; (b) evaluation of mean damage radius [68].

## 9.4.2 Strain-Mapping Methods for Damage Detection

Several investigators have tried to use the change of strain distribution inside a composite structure as an indicator of impact damage (e.g., delamination cracks) taking place into the structure. As discussed in Ref. [70], strain changes induced by a delamination caused by an impact are related to the residual strains built in the laminate during the curing and have only a local influence, limited to the delamination area. Such local cracks (although dangerous because they act as the failure initiation points during cyclic loading) will not change strongly the residual strain distribution and their presence in the strain field could be sensed only in their proximity (order of few centimeters or 2–3 inches). Hence, unless damage happens very close to the strain sensor, such delamination damage may easily go undetected. Reference [70] indicates that an optical fiber strain sensor bonded only 40 mm away from the impact damage produced in a CFRP plate did not record permanent strain changes. However, an impact that happened near the sensor (~10 mm) produced a change in the strain field that could be detected. For these reasons, a high-density strain sensor network is required. The distributed strain measurement technique utilizing Rayleigh backscatter and OTDR instrumentation allows the measurement of a large number of points on a single optical fiber. Reference [70] describes the use of this technique by applying a meandering optical fiber on a 200-mm by 200-mm CFRP panel and mapping the strain changes due to a delamination. A reasonable location of the delamination could be deduced.

Another approach of using strain measurements for detecting the effect of impact-induced damage is to subject the structure to progressive service loading and measure the strain distribution in the damaged structure in comparison with the strain distribution in the pristine structure before the damaging event. The premise of this approach is that the service loads will produce larger strains than the residual strains discussed in the previous paragraph and hence the changes due to structural damage may be easier to detect. Reference [71] describes the use of this approach on a 1.5-m CFRP wing section instrumented with four optical fibers each containing eight FBG strain sensors (32 FBG strain sensors total). A large amount of data was recorded first on the pristine specimen and then on the damaged specimen. The damage consisted of small delaminations induced with a razor blade. Bending-torsion loading of the specimen was progressively increased and then decreased. Strain measurements were taken at each of the 400 steps. The resulting large data set (32 sensors in 400 tests) was processed with the principal component analysis (PCA) method and the results between pristine and damage cases were compared. The results were encouraging and indicated that detection of damage is possible with the PCA method. However, the use of this approach in more complex structure may be more difficult due to the appearance of nonlinearities and other effects [71].

### 9.4.3 Vibration SHM of Composites

Structural vibration is characterized by resonance frequencies at which the structural response goes through peak values. At each resonance, the structure vibrates in salient modes call resonant modes. The structural response measured over a frequency range including several resonance frequencies generates a vibration spectrum or frequency response function (FRF). At a generic frequency that is not a resonance, the structure motion follows an operational deflection shape. If a structure suffers damage, then its dynamic properties (resonance frequencies and modeshapes, damping, operational deflection shapes, etc.) also change.

Finding damage from changes in structural vibration is one of the longest-researched SHM approaches [72]. A recent survey [73] classifies the vibration-based damage identification methods into four categories: (i) natural frequency-based methods; (ii) mode shape-based methods; (iii) curvature/strain mode shape-based methods; (iv) other methods based on modal parameters. Though most of these methods can use the vibration changes to tell that some damage is present, the finding of damage location and size proves to be very challenging. This fact is not surprising because vibration assessment is a global method whereas damage is a local event. Damage-induced local changes may have only limited influence at the global scale. In other words, loss of structural strength at a local scale may cause structural failure in virtue of the "weakest-link" principle but may not be detected at the global scale before failure actually occurs. An additional difficulty is that, in practical applications, the vibration changes due to damage may be confounded by those due to environmental vibrations or by temperature/humidity changes that affect the structural vibration but do not represent damage.

### 9.4.3.1 Model-Based Vibration SHM

Most of the vibration-based SHM work reported to date is model based; a common approach is to have a detailed structural model that is fitted to a pristine structure over the frequency range of interest. To detect damage, the model is manipulated by introducing simulated damage at various locations until the vibration measured on the damaged structure is being reproduced by the model. To achieve this data-fitting goal, a variety of optimization methods for model updating have been tried [73]. This approach was used by Ref. [74] to detect low-energy impact damage in quasi-isotropic 3.12-mm-thick CFRP beams.

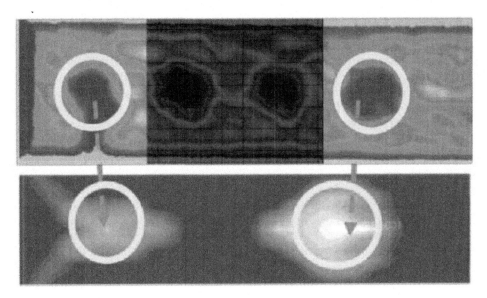

**FIGURE 38**   Vibration-based identification of impact damage in quasi-isotropic CFRP through model updating: top picture shows the ultrasonic C-scan of the specimen; bottom picture shows the damage areas fitted in the FEM model [74].

The beams were shaken with burst random excitation and FRF was measured at 33 points Doppler laser velocimeter (DLV) filtered in the $0 - 1.6$kHz band. A modal identification package was used to obtain the modal parameters; damage-induced decreases in natural frequencies and increases in modal damping values were observed. A detailed FEM model was used in a topological optimization loop to construct the damage scenario that best fits the measured vibration data. Figure 38 shows an FEM-generated image of the damaged specimen; the FEM fitted damaged zones compare well with the ultrasonic C-scan of the actual specimen.

### 9.4.3.2 Model-Free Vibration SHM

The methods using the curvature/strain modes hold the promise of detecting damage from the processing of experimental data without the need of a structural model. The premise of the method is that damage in thin-wall structures induces local discontinuities that affect strongly the curvature of the flexural modes of the structure [75]. The curvature of the modes (i.e., the curvature modes) can be determined through space-wise double differentiation of the measured displacement/velocity/acceleration modes. They can also be measured directly with surface-mounted strain gages.

If the modal displacement can be measured on a dense grid with an SDLV, then curvature modes method can be quite effective. Reference [76] demonstrates its successful application on a composite beam with a couple of saw-cut notches. The beam was excited with a surface-mounted piezo wafer. The curvature modeshapes are determined through space-wise double differentiation of the measured displacement and a strain energy ratio (SER) is calculated at each grid point. A method for self-baselining is also proposed. This method does not require measurements on a pristine structure; the baseline is simply determined from the damaged structure itself under the hypothesis that damage is a local phenomenon that sharply modifies the curvature modes of the structure in a local region. Hence, a baseline is obtained through low-order smoothing of the measured data such that higher-order variations are eliminated. If the data is very dense, this be achieved simply through undersampling [76]. Other possible approaches are low-order curve fitting, spline interpolation [76], WT, etc.

However, SHM application of the curvature-mode methods requires the use of *in situ* sensing instead of the SDLV approach. Reference [77] describes the use of surface-mounted flexible piezo sensors made of polyvinylidenefluoride (PVDF) placed at 16 locations on a cross-ply CFRP beam. Vibration excitation was done with a piezo wafer mounted at end of the beam energized by a sine signal swept up to 2 kHz. Various damages were applied on separate specimens (saw-cut notches; delaminations simulated by Teflon inserts; impact). A set of pristine specimens were used to construct a curvature modes baseline database. Subsequently, data was taken on the damaged beams. Further processing of these experimental results was done in Ref. [78]. A more recent development of this approach is reported in Ref. [79] which used continuous Bragg-grated fiber-optic distributed strain sensors coupled to a compact Fiber-Optic Sensing System (cFOSS) developed by the NASA Armstrong Flight Research Center (AFRC). This system allows individual axial strain measurements at 186 locations along the same fiber (location separation was 0.473 inches). This method shows promise although the experimental data still suffers from considerable scatter and signal-to-noise ratio (SNR) issues.

### 9.4.3.3 Statistics-Based Vibration SHM

Considerable effort has been dedicated to applying statistical signal processing methods to vibration signals [80]. Reference [81] describes shaker tests performed on a composite wing instrumented with FBG sensors and accelerometers. The wing had sandwich construction with composite faces and foam core. Consecutive low-velocity 7-J impacts were applied in a certain zone of the wing which was surrounded with FBG sensors. Vibration signals were recorded during stochastic Gaussian excitation in the 0–1500 Hz range. The FBG sensors were sampled at 3906 Hz. The vibration time series signals were processed with statistical method aimed at revealing the nonlinear response features associated with impact damage [81].

### 9.4.3.4 Nonlinear Vibration SHM

Nonlinear vibration effects have also been used to detect the presence of a delamination damage produced by an impact event. [82] showed that a 10-mm delamination in a CFRP beam produces clear side bands in the response spectrum under a 750-kHz high-frequency carrier excitation pumped with an intense 88-Hz vibration. Reference [83] applied this technique to detect impact damage in a CFRP skin-stiffener specimen. These nonlinear vibration methods are able to detect the presence of a delamination in the specimen but cannot yet detect its location.

### 9.4.3.5 Combined Vibration-Wave Propagation Methods for Impact Damage Detection

Reference [84] reports an attempt to use both the vibration spectrum approach and the wave propagation approach to develop damage metrics for quantifying the impact damage in a composite plate.

## 9.4.4 Frequency Transfer Function SHM of Composites

The frequency transfer function SHM method resembles the vibration SHM methods in the fact that it uses spectral representation of the data. However, its implementation is different; the frequency transfer function (FRF) between two structurally mounted piezo wafers is determined directly through sweep-sine or broadband random excitation. The complex quantity measured in this way is also known as transmittance. FRF SHM methods can be either model based or model free.

A model-based FRF SHM method is described in Ref. [85]. The analysis is performed in two steps: (a) direct problem; (b) inverse problem. The direct problem is stated as follows: given a composite structure with a known damage, find the FRF between a given pair of actuator–sensor wafer transducers. For solving the direct problem, a model of a composite structure with attached piezo wafer transducers is constructed, e.g., a composite beam with an actuator–sensor pair as depicted in Figure 39. The model is used to predict the FRF between the two transducers. Next, a delamination damage is introduced into the model and a new FRF is predicted (Figure 39). If the structure has several transducers, then all the possible combinations are used to determine a set of FRF data.

The inverse problem is more challenging: given a measured FRF between actuator–sensor pairs, find the location and size of the damage in the composite structure. For solving the inverse problem, one has to sift through all the possible combinations of damage size and location and then select that one which best fits the measured data. Reference [85] describes in quite detail this two-step process consisting of building the model, conducting experiments, and then adjusting iteratively the model to determine to

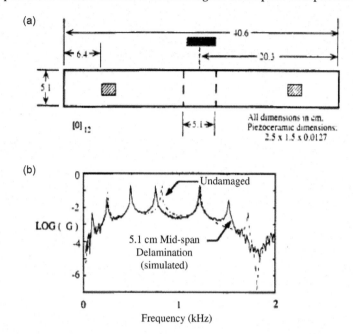

**FIGURE 39**    Model-based FRF SHM: (a) composite specimen geometry; (b) FRF up to 2 kHz showing changes due to damage [85].

the best possible approximation the size and location of the damage. It was found that the method was successful in determining delaminations that were greater than approximately 10% of the beam length but had difficulty with smaller delaminations. It was found that experimental noise can obscure the subtle changes in the amplitude response due to delaminations.

Reference [86] describes a model-free FRF SHM method named active damage interrogation (ADI). It also consists of multiple actuator–sensor pairs of piezo wafer transducers mounted on the structure. Broadband signals (white noise or chirp up to 100 kHz) are transmitted between all the possible pairs in a round-robin process and the associated FRF magnitude and phase spectra are calculated (Figure 40). The data is further processed extensively with statistical methods to determine a comprehensive DI for each actuator. The damage is then localized to a damage zone by identifying the actuator with the highest DI. Further processing of an array of DI values can be performed to further localize the damage within the damage zone. A revisiting of this approach through numerical modeling and improved placement of the transducers is reported in Ref. [87].

Reference [88] describes a frequency transfer method for detecting structural anomalies and damage in a composite structure using an array of FBG sensors and a simple vibrational excitation applied with an instrumented test hammer. The method has a broad area coverage and is relatively simple to implement because it does not require special excitation equipment. The method was validated on a composite panel specimen and on an actual composite rudder from a mine countermeasures ship [88].

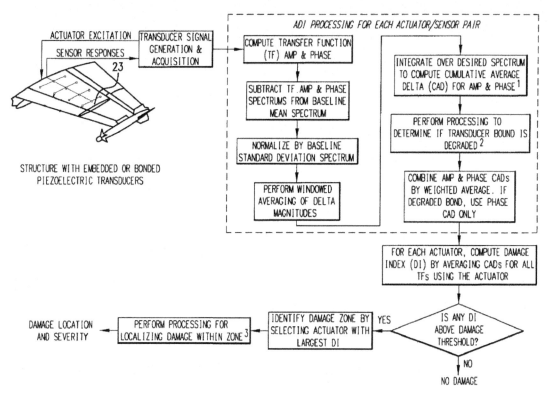

**FIGURE 40**    Model-free frequency transfer function SHM system based entirely on statistical signal processing [86].

## 9.4.5 Local-Area Active Sensing with EMIS Method

EMIS is a high-frequency standing waves method capable of detecting small local damage in the transducer vicinity. The essence of the EMIS method consists in using an electrical impedance analyzer instrument to measure the impedance, $Z(\omega)$, of a PWAS transducer firmly attached to the monitored structure. As proven through theoretical developments and experimental tests, the real part of the impedance Re(Z) reflects the mechanical behavior of the structure on which the PWAS is attached, i.e., its follows the dynamic spectrum and the resonances of the structure [15,89,90]. Thus, a PWAS attached to a structure can be used as a structural-identification sensor that measures directly the local structural spectrum at very high frequencies. EMIS has been shown to be very effective in the detection of proximal disbonds [91] and delaminations [92]. It has also been used to monitor material state change in composite specimens during fatigue testing [93,94].

Detection of delaminations has been tried by both traveling guided waves and standing ultrasonic guided waves. In the former case, mode conversion and scatter at the delamination boundary modifies the traveling wave pattern. In the latter case, disbonds and delamination manifest high-frequency "breathing" modes that were not present in the pristine structure.

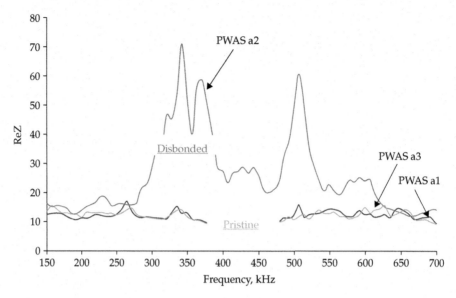

**FIGURE 41**  E/M impedance spectrum showing new resonant frequencies due to disbond (PWAS a2) that were not present in a well-bonded region on the same member [15], p. 551.

Conventional vibration analysis methods, which measure structural dynamics up to several kHz, are not sensitive enough to detect damage that is much smaller than the structural scale, such as local delaminations in a composite structure. These local resonances would manifest as new peaks in the high-frequency spectrum of the local structure. Local vibration measuring techniques that can sense the high-frequency breathing mode of the disbonded area and its effect on the dynamics of the surrounding structure are needed. LDV measurements have verified that piezoelectric wafer sensors attached to the structure can excite the high-frequency breathing modes of a disbonded or delaminated area [48,76,95]. A method to actually measure directly this high-frequency local vibration spectrum is offered by the E/M impedance method, which can measure the structural spectrum into the hundreds of kHz and low MHz range [15].

Figure 41 shows the application of EMIS for the detection of 20-mm-long delamination on a 2-mm-thick L-section stiffener bonded to an aluminum plate [91]; PWAS a2, which is placed in a zone proximal to the delamination finds the new resonances associated with disbond vibration at approximately 350kHz and 510kHz. In the same time, these new resonance are not present in the spectrum measured by the witness PWAS transducers a1 and a3 which were placed on the same stiffener but some 70 mm away on either side of the disbond. The fact that the a1 and a3 spectra are almost identical confirms the method consistency.

A comprehensive analysis of the use of EMIS for damage detection in laminated composites was presented in Ref. [92]. Figure 42 shows the experimental setup and the analytical model. An efficient transfer-matrix method approach was used to compute the high-frequency dynamics of the delaminated composite beam and then to predict the E/M

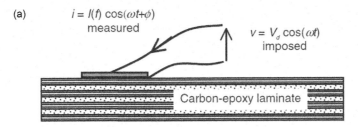

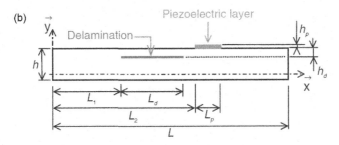

**FIGURE 42**  Modeling of EMIS method for delamination detection in CFRP composites: (a) experimental setup; (b) analytical model [92].

impedance at the piezoelectric sensor. An extensive parameter study of the effect of delamination length, position, and depth location on the E/M impedance spectrum was performed. The effect of the relative location of the PWAS transducer relative to the disbond was also studied. The prediction results were compared with experiments performed on quasi-isotropic CFRP beams.

EMIS has been used by several authors to measure disbonds and delamination: Ref. [96] used FEM simulation of EMIS disbond detection in bonded composite repairs; Ref. [97] measured EMIS for disbonding detection of CFRP composite strips used for strengthening and rehabilitation of concrete structures. EMIS was also been tried for monitoring the progressive changes in the composite material during cyclic fatigue loading: Ref. [98] applied EMIS to ceramic matrix composites during axial fatigue and compared the results with AE readings; Ref. [93] used EMIS to monitor woven GFRP specimens undergoing flexural fatigue and Ref. [94] performed comparison between measured results and FEM predictions at high ultrasonic frequencies (hundreds of kHz to tens of MHz).

## 9.5 ELECTRICAL AND ELECTROMAGNETIC FIELD METHODS FOR DELAMINATION DETECTION

### 9.5.1 Delamination Detection with the Electrical Resistance Method

Impact damage detection by electrical resistance monitoring of CFRP composites has been given a lot of attention. The methods for delamination detection fall into two large categories: (i) monitoring of an average damage state of the whole panel by performing overall resistance measurements; (ii) providing damage location capability through more elaborate measurements and data processing. Reference [99] describes preliminary work on a conductivity mapping approach that employs an array of surface-mounted wires on the CFRP composite. The material used was a 16-ply T800H-924C CFRP with a $[(0,90)_4]_s$ layup. A $6 \times 6$ array of sensing wires was used to generate a damage map. One face of the panel was lightly grit blasted to allow electrical contact to the carbon fibers. Sensing wires were attached on one face using silver paint and a protective coating of epoxy adhesive. Current input was through copper tags centrally positioned at opposite edges of the panel. Multiple averaging and interpolation was used to smooth out the data. The effect of an increasing level of impact damage produced a detectable change in the potential distribution. The damage sustained after the 6-J and 8-J impacts was fairly severe, comprising fiber fracture, splitting, and delamination, and this damage was readily detectable. Although it is not clear which component of the damage was responsible for the change, fiber fracture is the obvious candidate. At lower-energy levels where delaminations predominated, the potential distribution did not appear to correlate with the damage site, although it did cause perturbations. The cause of these perturbations was likely to be a reduction in the number of electrical conduction points between fibers due to ply separation (delamination), and it is unlikely that effects due to reactive components were responsible at the supply frequency used.

### 9.5.1.1 Electrical Resistance Change Method for Delamination Detection in Cross-Ply CFRP Laminates

Todoroki and coworkers [100–106] have studied extensively the use of electrical resistance change for damage detection in CFRP composites using surface-mounted electrode strips. Thin copper-foil electrodes were mounted on one side of CFRP specimens during the prepreg lamination and co-cured with them in order to get reliable electrical readings (Figure 43).

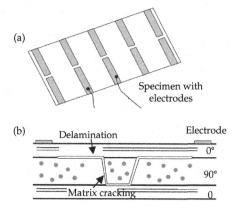

**FIGURE 43**    Electrical resistance monitoring with surface-mounted electrodes: (a) surface-mounted electrodes pattern; (b) schematic of delamination and matrix cracking phenomena in relation to the electrode placement [107].

Multiphysics FEM simulations were performed to understand the electric field and current distribution inside the specimen (Figure 44). These studies were initially done to understand the orthotropic electrical properties of the unidirectional CFRP layer [101] and were later extended to cross-ply CFRP laminates [102], and woven CFRP laminates [107].

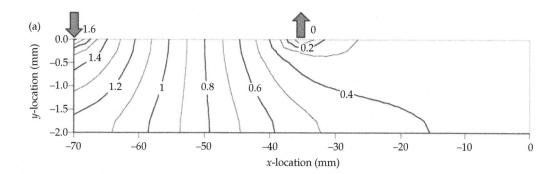

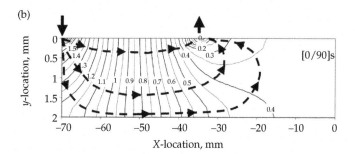

**FIGURE 44**    Multiphysics FEM simulation of electrical field distribution inside a CFRP composite specimen: (a) unidirectional specimen [101] cited in Ref. [108]; (b) cross-ply specimen [102].

In these studies [100–107], delaminations were created using a three-point bending technique. An indentation-type jig and cylindrical support was used. By changing the diameter of the cylindrical support jig from 10 to 50 mm, several sizes of delamination were created in the plate-like specimen. The indentation point is loaded from the opposite side surface where the electrodes are mounted. Since the specimen is a thin laminate, the loading creates a large delamination crack in the 0–90 interface near the electrodes. The damage size and location was verified with ultrasonic C-scan (Figure 45b). Electrical resistance change was monitored with strain-gage bridge instruments before and after indentation. Two-probe methods were usually employed; the four-probe method was also tried [105,109].

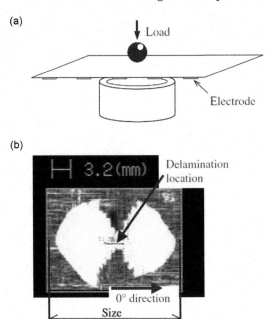

**FIGURE 45**   Indentation delamination damage for electrical resistance monitoring experiments: (a) damage is applied on the side of the plate opposite to the electrodes; (b) damage size and location was verified with ultrasonic C-scan [107].

References [102,107] processed the electrical resistance measurements with a "response surface" technique which comprises regression curve fitting to obtain approximate responses without the need for an actual model. In Ref. [107], a large number of cross-ply and quasi-isotropic specimens were tested such that statistical data processing could be applied. Copper-foil electrodes were mounted on one side of the CFRP specimens during prepreg layup and co-cured with the specimen (Figure 43). Impact-induced matrix cracking and delaminations were detected. Probability of location estimation and error bands were computed [107]. This represents a noteworthy statistical study of the use of surface-mounted electrodes and the response surface approach to detect matrix cracking and delamination with the electrical resistance method in composite materials.

### 9.5.1.2 Wireless Electrical Resistance Method for Delamination Detection

The electrical resistance method for monitoring composite materials can also be done wirelessly since it is similar in instrumentation to the electrical resistance strain-gage method. Low-sampling-rate wireless strain and temperature sensors have become relatively widespread. Reference [110] describes the wireless detection of strain changes due to internal delaminations in CFRP composites using the electrical resistance change method and a wireless bridge that encodes the resistance change as a frequency shift, as already described in Chapter 7. Cross-ply $[0_2/90_2]_S$ carbon/epoxy specimens (50-mm long, 20-mm wide, 1.8-mm thick) were fabricated. To create delamination, a short-beam three-point

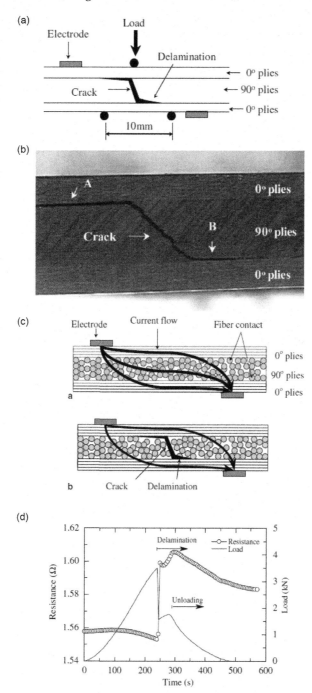

**FIGURE 46**   Delamination detection in a $0°/90°$ cross-ply CFRP composite with the electrical resistance method: (a) schematic of the experimental setup; (b) actual delamination and crack in the specimen; (c) delamination effects on the electrical path; (d) results showing changes of resistance due to delamination formation [110].

bending test was applied (Figure 46a). A micrograph of an actual specimen side view showing the delaminations and the crack path through the 90° ply is shown in Figure 46b. The delamination detection concept is described in Figure 46c. The test was conducted in displacement control; Figure 46d shows simultaneously the plot of load and of resistance versus testing time. The appearance of the delamination can be identified by a sudden drop in load accompanied by a sudden change in the electrical resistance. Prior to delamination failure, the specimen resistance displayed a slow downward drift. After the sudden increase at the moment the delamination occurred, the resistance continued to increase as the load recovered somehow. However, when the specimen started to experience profound failure and the load started to drop precipitously, the resistance started again a clear downward trend.

The experimentally measured resistance change was encoded into frequency change and transmitted wirelessly. The delamination size was also estimated and transmitted wirelessly using the resistance change magnitude and the corresponding frequency shift in the encoding equipment.

### 9.5.1.3 EIT Method for Delamination Detection with the Electrical Resistance Method

Conductivity mapping was also used in Ref. [111] but with edge contacts instead of a surface array. Named "electrical impedance tomography" (EIT), this method was used to reconstruct the damage state from a series of measurements between adjacent edge contacts. The EIT method gained wide recognition in the 1920s among geophysicists who placed arrays of electrodes into the ground. Oil-bearing rocks under the surface were identified by injecting current through a pair of electrodes and measuring the resulting voltage at the other electrodes. The same concept was used in Ref. [111] to determine the resistivity distribution in a CFRP laminate sheet. A schematic of the experimental setup is shown in Figure 47. Various electrodes are connected to the edges of the sample. An electrical current was injected via two electrodes and the potential drop between all other neighboring electrodes was measured. By taking various combinations of current · injecting electrodes and measuring the potential drops at the remaining electrodes, a set of measurements was obtained. This information dataset was utilized to extract resistivity distribution inside the specimen using a reconstruction algorithm based on fundamental electrodynamics theory [111].

This EIT approach (which requires a large number of measurements and involves calculations) proved capable of locating a hole drilled through a CFRP laminate and even determining the hole size [111].

Reference [108] describes the extension of the electrical resistance method to woven CFRP composites. In practical composite structures, a woven surface ply is usually applied to protect the carbon fiber unidirectional tows from surface damage. The electrical resistance change method developed for composite laminated from unidirectional plies utilizes the orthotropic electrical conductivity in the in-plane and thickness directions of CFRP laminae. The woven CFRP plies, however, have different electric properties: their in-plane electrical resistivity is rather isotropic, whereas the thickness direction electrical resistivity is rather high due to the undulating interaction of the carbon fiber bundles in the weave. These differences in electrical properties between the woven plies and the unidirectional

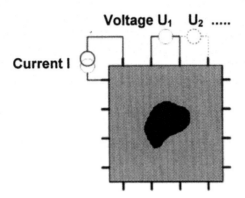

**FIGURE 47**    Schematic of the EIT experimental setup on a CFRP composite specimen [111].

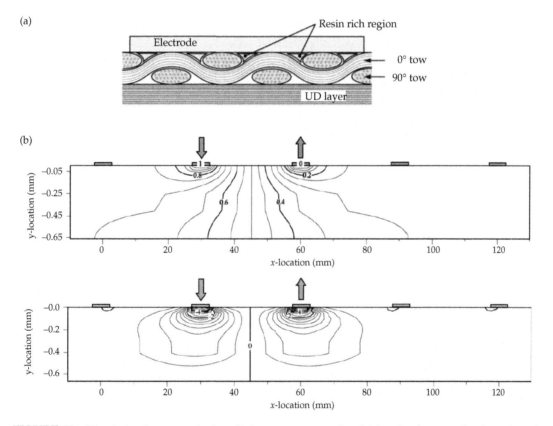

**FIGURE 48**    Electrical resistance method applied to woven composites: (a) interface between the electrode and the woven CFRP ply; multiphysics FEM simulation of the electrical resistance SHM method applied to woven CFRP composites; (b) electric field distribution; (c) electric current distribution [108].

plies may cause different responses when an electrical current is applied to a woven CFRP laminates in comparison to a laminate of unidirectional plies. In addition, when a woven ply is used as the surface layer in a laminate of unidirectional plies, it may be expected that the surface woven ply may not have a good electrical contact to the electrodes due to resin-rich regions between the fibers (Figure 48a).

Reference [108] describes a thorough investigation of this phenomenon including multi-physics FEM simulation of the electric field and current distributions (Figure 48b,c) and extensive experimental tests (Figure 49). It was concluded that, as expected, the composites containing woven plies pose more challenges for monitoring with the electrical resistance change method than composites made of unidirectional plies. Nonetheless, the differences in electrical properties between surface woven plies and unidirectional plies do not seem to affect the successful use of the electrical resistance change method to detect impact-induced delaminations and cracks.

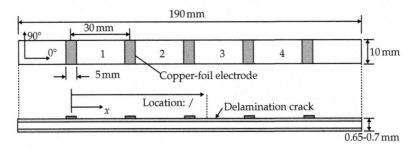

**FIGURE 49**    CFRP beam with surface-mounted electrodes on one side [108].

A number of good electrical contact regions between the woven surface ply and the electrodes were achieved at the top of the wavy tow. If the number of the electrical contacts is sufficient, a stable electrical contact between the surface woven layer and the electrodes is achieved. Although the surface woven ply does not ensure perfect electrical contact with the electrodes due to the crimp texture, the co-cured copper-foil electrode is sufficiently reliable for the application of the electrical resistance change method; the partial resin-rich regions created due to the crimp texture of surface woven ply do not seem to affect too much the transfer of electrical charge if a sufficient number of good electrical contacts with the electrode at the top of the wavy tows exist.

Reference [112] describes an EIT approach to detecting damage in CFRP composites using a multiwall carbon nanotube (MWCNT) and PVDF film. The method allows the detection of spatially distributed damage using spatial conductivity maps.

### 9.5.1.4 Delamination Detection in Filament-Wound CFRP Composite Cylinder with the Electrical Resistance Method

Reference [113] studied the use of electrical resistance measurements on a filament-wound CFRP cylinder using a number of electrical contacts that were connected in various patterns such that axial, radial, oblique, and circumferential resistances could be measured. The cylinder was impacted with progressively increasing impact energy. It was observed that resistance changes with damage (Figure 50).

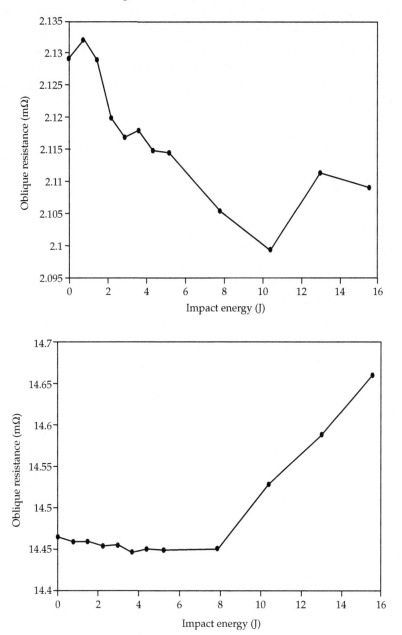

**FIGURE 50**   Resistance change with progressively increasing impact energy: (a) radial resistance; (b) oblique resistance [113].

The radial resistance decreased initially as the impact energy was increased up to 10 J, but subsequently increased. This behavior could be attributed to the fact that the damage at low impact energies caused the fiber layers in the wall of the cylinder to contact one another more directly, hence decreasing the electrical resistance; whereas the higher-energy impacts produce more profound damage that involved delamination and hence an increase in resistance. The oblique and axial resistances seemed to be rather insensitive to the lower-energy impacts but increase rapidly with higher-energy impacts that produced more profound damage. Hence, it was concluded that the radial resistance is more sensitive to minor damage than the oblique resistance, but the oblique resistance is more sensitive to major damage.

### 9.5.1.5 Multiphysics Simulation of the Electrical Resistance Change Method

Reference [114] performed multiphysics FEM simulation of two electrical conductivity techniques for delamination detection in CFRP composites: (a) the matrix-array method; (b) the circular-array method (Figure 51).

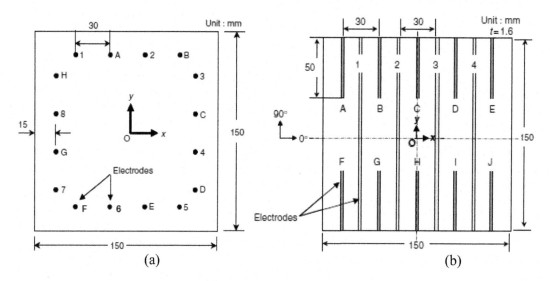

**FIGURE 51**   CFRP specimen dimensions and placement of electrodes: (a) circular-array method; (b) matrix-array method [114].

Rectangular plates of 150mm × 150mm × 1.6mm were simulated. The conductivity of each CFRP lamina was taken as $\sigma_0 = 32,000S$, $\sigma_{90} = 9.6S$, $\sigma_t = 8.3S$. The stacking sequence was $[0/45/-45/90]_s$. A rectangular delamination with dimensions varying from 12 to 48mm was simulated. The center of the delamination was located at various positions in the range $-50mm$ to $+50mm$ in both $x$ and $y$ directions. For delamination simulation, the FEM nodes at the inter lamina interface were doubly defined; the upper and lower nodes of the interface were normally joined; but this joint was released to simulate the presence of a delamination. An electric current of 0.1A was applied at the cathode and the voltage of the anode was set to 0V. The study generated plots of the estimation error for delamination location and delamination size with each of the two measuring methods. It was concluded that the matrix-array method was more effective and gave a smaller error in detecting delamination than the circular-array method [114]. However, the matrix-array method may be affected by severe impacts that may produce damage to the electrode strips mounted on the specimen surface. If this happens, the surviving electrodes which were more remote from the damage may be used to measure the large-scale electrical resistance changes as in the circular-array method. Such a hybrid monitoring method may be eventually the best practical compromise.

## 9.5.2 Delamination Detection with the Electrical Potential Method

We have seen that the electrical resistance method involves the measurement of the electrical resistance between two points on the composite. For 1D specimens, the contacts were placed along the specimen. For 2D specimens, the contacts were distributed as arrays over the surface of the specimen, either in a rectangular grid covering the whole area (Figure 52a) or as a "circular" array confined closer to the boundary such as not to be damaged by the more severe impacts (Figure 52b). In order to find the damage location, some sort of surface interpolation method was applied to the electrical resistance change values measured at these locations.

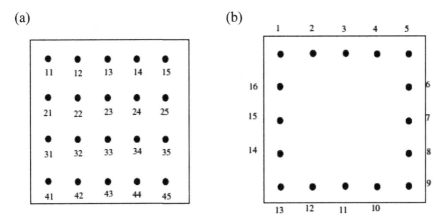

FIGURE 52    Distribution of electrical contacts on a 2D specimen: (a) grid array covering the whole specimen; (b) "circular" array at the specimen periphery [115].

The electrical potential method uses the array of contacts not to measure the resistance, but to measure the electrical potential at each of them with respect to a reference point or the ground. Since the potential depends on the direction of electric current flow, the electrical potential would be done for all possible current flow configurations (electrical tomography). Consider for example the array of contacts depicted in Figure 52b and assume that the current is applied from electrode contact #1 toward contact #9, while the electrical potential is measured at each of the other 14 contacts. After that, the current is applied, say, from #5 toward #13, while the potential is measured at each of the other 14 contacts. Since the current line and the potential gradient line (i.e., the line connecting the two points where potential is measured) do not overlap, this 2D method, called "electric potential method", does not correspond to resistance measurement, which involves overlapping of the current line and the potential gradient line.

Circular or grid arrays, as described in Chapter 7, may be used. Reference [116] describes a numerical and experimental study of using the electrical potential method to detect impact-induced delaminations in a CFRP quasi-isotropic laminate using a surface-mounted grid array consisting of 121 electrodes on a 25-mm pitch. The electrical current was applied between opposed electrodes and the electrical potential was measured at the other electrodes. Some of the probes were used to introduce current into the laminate. Additional probes, used for current introduction only, were placed on the top (instrumented) and bottom surface of the plate. A 100-mA current was introduced using a DC current calibrator. The potential values relative to a ground potential located on the plate center-line were monitored using a high impedance voltmeter. All data were recorded using a PC. A numerical study was also done with a multiphysics FEM package; particular attention was paid to properly modeling the interlayer interfacial conductance and the delamination zone. A rather large computer model resulted. Comparison of calculated and measured data revealed that the measured and calculated contours of equipotential change surrounding impact damage are superficially similar but differing in details (Figure 53). In both simulation and experiment, the potential changes are concentrated in a two-lobe region elongated in the surface fiber directions. There is a characteristic two-lobed appearance to the region in which there is a maximum and minimum point of potential change. The extent of the disturbed region extends between two and three times the maximum length of the delamination in the direction of the surface fibers, and little more than the width of the delamination perpendicular to the fibers.

Reference [115] presents an extensive study of the electrical potential method. Different methods of electric current injection were considered (use only one specimen surface; use both specimen surfaces and send current across thickness; use the specimen side edges). Relative orientation of electric current injection with respect to the direction of specimen fibers in the top layer was also investigated.

The direction of electric current flow was found to play an important role in the success of the electric potential method. The way that current flows is governed by the electrical contact configuration. Figure 54 shows several possible configurations:

(a) The current contacts A, D are on the surface of the laminate, and the current is confined to the specimen surface; the electrical potential measurement is done at contacts B, C.
(b) The current contacts A, D are on opposite surfaces of the laminate but not directly opposite one another, thereby providing an oblique current flow through the specimen thickness; the electrical potential measurement is done at contacts B, C.
(c) The current contacts A, D are at the edges of the specimen on the cross-sectional surfaces and the current flows in the plane of the laminate more or less uniform across the cross section of the specimen. The potential measurements are done at the surface contacts $B_1$, $C_1$ and at edge contacts $B_2$, $C_2$; in the latter case, the measurement is done in the location where the current is injected.

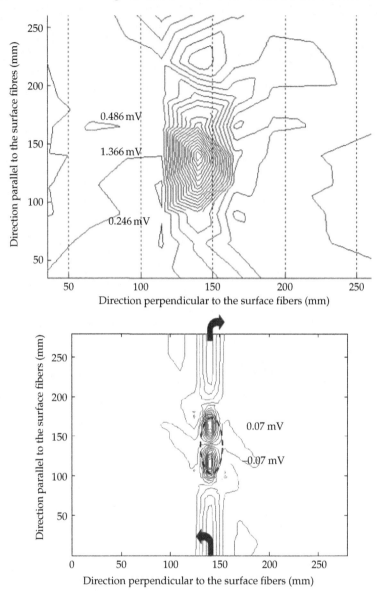

**FIGURE 53**   Contours of equipotential change on a quasi-isotropic CFRP composite specimen after impact: (a) measured; (b) calculated with multiphysics FEM [116].

(d) The current contacts A, D are again placed at the edges of the specimen on the cross-sectional surfaces and the electric current flows in the plane of the laminate more or less uniform across the cross section of the specimen. The potential measuring contacts are placed in through-the-thickness holes. This configuration is similar with (c) only that the electrical potential measurement is done at contacts B, C which are placed between the contacts A, D that inject the electric current.

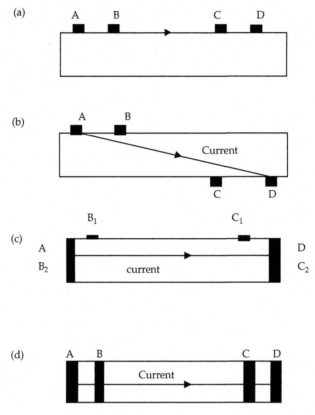

**FIGURE 54**  Possible directions of electric current flow in the electrical potential method. In each case, the current flows from A toward D: (a) surface flow; (b) diagonal over the thickness; (c) uniform over the thickness with potential measured along the surface ($B_1$, $C_1$) and uniform along the thickness ($B_2$, $C_2$); (c) uniform over the thickness with potential measured uniform over the thickness using contacts placed in through holes (B, C) [115].

Reference [115] performed experiments on quasi-isotropic $[0/45/90/-45]_{3S}$ CFRP specimens having the electrode pattern shown in Figure 55. The current was injected between electrodes 11 and 12 and electric potential measurements were done on the other 10 electrodes. Two sets of current contacts (Set A and Set B) were separately used as follows:

- Set A allowed surface (in the plane of the laminate) or oblique current application and consisted of contacts 11A and 12A, which were on the surface of the laminate. For surface current application (configuration Figure 54a), contacts 11A and 12A were on the specimen surface which was to receive impact. For oblique current application (configuration Figure 54b), the contact 11A was on the surface which was to receive impact, while contact 12A was on the opposite surface.
- Set B of contacts was for the current-injecting configuration described in Figure 54c and consisted of contacts 11B and 12B, which were conducting strips placed on the side edges of the specimen; each strip was 3-mm wide and covered the entire thickness of the specimen.

Three sets of specimens were considered with respect to the direction of the surface fibers:

- 0°, in which the fibers on the surface instrumented with electrodes were parallel to the long side of the rectangle in Figure 55
- 90°, in which the fibers on the surface instrumented with electrodes were parallel to the short side of the rectangle in Figure 55
- 45°, in which the fibers on the surface instrumented with electrodes were placed diagonally.

Since the direction of current flow is parallel to the short side of the rectangular specimens, these three configurations correspond to the current flowing perpendicular, parallel, and diagonal to the surface fibers, respectively.

Impacts of progressively increasing energy were applied successively at the locations P, Q, R as depicted in Figure 54. The impact energy was increased progressively up to approximately 16 J. The electrical potential was measured after each impact and compared with the baseline reading for the pristine specimen. An example of results obtained on the 0° for impacts on point P, electric current along 11A–12A, and surface-mounted electrodes is shown in Figure 56. Examination of these curves reveals that the sensitivity for the detection of major damage seems best for the potential gradient lines that are closest to the damage location; however, for the detection of minor damage, the potential gradient lines that are closest to the current line seem to be better suited.

The study of Ref. [115] covered all the other cases that result from combination of the four situation depicted in Figure 54 and the three surface fiber orientations 0°, 90°, 45°. The overall conclusion seemed to be that the effectiveness of the potential method depends on the direction of surface layer fibers. Surface voltage measurement was found effective for damage sensing in the specimens with 0° surface layer; in this case, the electric current applied perpendicular to fiber direction diffused in the direction of low resistivity, i.e., along the fibers. The surface potential method was ineffective for the 45° specimens. In the case of the 90° specimens (i.e., with the electric current spreading across the fibers, i.e., in the direction of high resistivity), this method is not recommended because it gave wild varying results.

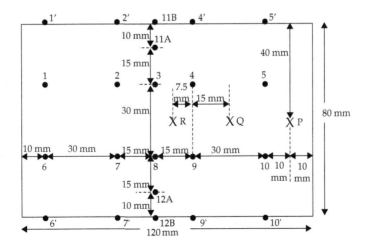

**FIGURE 55**   Specimen configuration and electrodes pattern used for electrical potential experiments [115].

Application of the electric current to all the laminae simultaneously by using current contacts on the sides of the specimen edges was found ineffective for the 0° specimens when in-plane surface voltage measurements were used, but was effective when voltage measurements were made on the edge electrodes. In this case, the potential gradient was found to decrease with increasing impact energy in the minor damage regime and to increase with increasing impact energy in the major damage regime. Application of current to all the laminae and the use of through-hole current and voltage contacts was also found effective and gave a similar variation of the potential gradient with increasing impact energy [115].

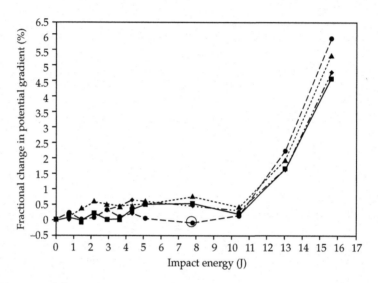

**FIGURE 56**  Variation of the potential gradient change with increasing impact energy applied at point P for various potential measuring electrodes placed on the surface of the 0° specimen; the circled data point may be a little off due to a data acquisition problem; ■: line 1−7; ♦: line 2−8; ▲: line 3−9; ●: line 4−10; [115].

A comparative study of the electrical resistance and electrical potential methods for damage detection in CFRP quasi-isotropic specimens of various thicknesses is described in Ref. [117]. CFRP quasi-isotropic laminate specimens of various thicknesses were studied:

• Thin 1.05-mm, 8-lamina specimens
• Intermediate 2.10-mm, 16-lamina specimens
• Thick 3.15-mm, 24-lamina specimens.

It was found that both the electrical resistance and the electrical potential methods were effective for damage sensing in the thin (8-lamina) composite specimen. In this case, the potential method was found superior to the resistance method, which showed some inconsistency in the variation of readings with increasing impact energy for the segments not containing the impact point. The superiority of the potential method over the resistance method for the 8-lamina composite was explained in terms of the current path distortion resulting from the impact damage.

For the 16-lamina and 24-lamina composites, the resistance method was effective, whereas the potential method was not. This means that the potential method does not work when the distance of separation between the applied current line and the potential gradient line is excessive. A distance of 2.10 mm (16 laminae) in the through-thickness direction was found excessive, whereas a distance of 1.05 mm (8 laminae) in the same direction was not.

### 9.5.3  Electromagnetic Damage Detection in Aerospace Composites

Electromagnetic methods for finding defects and damage in composite materials rely on the response of these materials to the excitation by electromagnetic fields. Dielectric composites, such as GFRP materials, can be interrogated with the use of an electric field, whereas conductive composites, such as CFRP materials, can be interrogated with a magnetic field. Figure 57 shows the magnetic imaging of a delamination in a CFRP plate. Both measurements were performed in transmission mode with excitation and detection fields being of the same type: electric→electric for GFRP and magnetic→magnetic for CFRP.

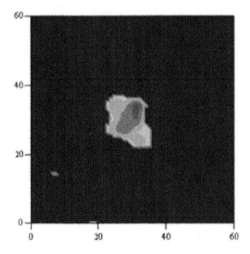

**FIGURE 57**  Electromagnetic methods for imaging of damage or defects in composite materials: magnetic imaging of a delamination in CFRP plate [118].

### 9.5.4  Hybrid Electromagnetic SHM of Aerospace Composites

A hybrid electromagnetic method is also possible for CFRP composites in which a modulated high-frequency electric excitation is applied and the magnetic response of the material is recorded. Reference [118] describes such a hybrid method applied to a CFRP composite plate and compared in efficacy with the acousto-ultrasonic method.

(a)

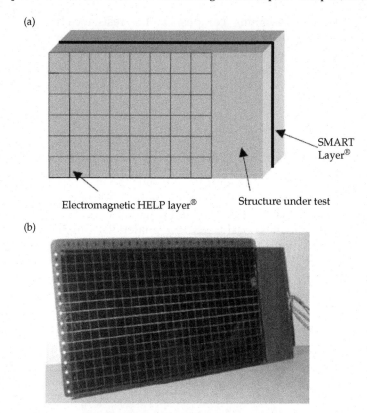

Electromagnetic HELP layer®        Structure under test

SMART Layer®

(b)

**FIGURE 58**  CFRP plate specimen instrumented with hybrid electromagnetic layer HELP© and with acousto-ultrasonic SMART Layer™: (a) schematic; (b) photograph [118].

The instrumented CFRP specimen is depicted in Figure 58. The specimen was a 16-ply cross-ply CFRP plate. An acousto-ultrasonic SMART Layer™ was embedded in the mid-plane, and a hybrid electromagnetic layer HELP© with a 20mm × 20 mm mesh was applied to one side. A 700kHzhigh-frequency electric excitation with full 1 kHz modulation was applied. Damage was applied in the form of impacts and burns. The impacts I1 and I2 were of 2 J and 4 J, respectively. The burns (simulating lightning strikes) were applied with high-energy sparks (30 V, 5 A) of various energies: 120 J, 40 J, 80 J, 400 J for B1, B2, B3, B4, respectively. The hybrid electromagnetic and acousto-ultrasonic imaging results are compared in Figure 59a.

The hybrid electromagnetic HELP© method could detect all damages but I1, whereas the Lamb-wave-based acousto-ultrasonic SMART Layer™ method could only detect damages I1, I2, B4. The fact that the HELP© method could not detect the low-energy damage I1 is attributed to the fact that this mild impact only produced delamination but no composite breakage; this indicates that the HELP© may not be able to detect mild delaminations. It is also noted that the lowest-energy burn B2 (40 J) is only lightly visible in Figure 59a, which may be indicative that this light burn damage is probably at the detection threshold of this method.

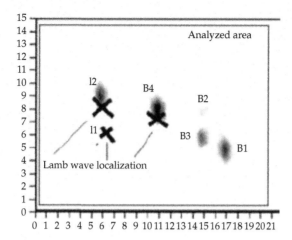

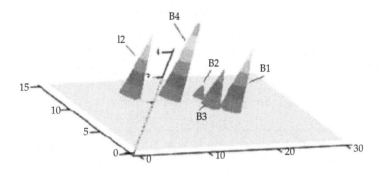

**FIGURE 59**   Comparative detection of impacts (I1, I2) and burns (B1, B2, B3, B4) on a CFRP plate instrumented with hybrid electromagnetic layer HELP© and with acousto-ultrasonic SMART Layer™: (a) image showing that the HELP© method could detect all damages but I1, whereas the Lamb-wave-based SMART Layer™ method could only detect I1, I2, and B4 damages; (b) image showing the detection levels of the electromagnetic HELP© method [118].

The fact that the SMART Layer™ method could not detect the B1, B2, B3 burns may be attributed to the fact that these damages were only superficial changes that did not affect sufficiently the guided-wave propagation used in the acousto-ultrasonic method. The severe damage B4 (400-J electric spark) was detectable with this method because it produced sufficient local changes in the material to scatter the guided waves used in the acousto-ultrasonic method.

## 9.5.5  Self-Sensing Electrical Resistance-Based Damage Detection and Localization

A self-sensing resistance-based damage detection and location system is proposed in Ref. [119]. Though the testing methodology seems simple and uses only a multimeter instrument, the construction of the specimen is rather elaborate involving the insertion of well-aligned conductors and glass-fiber layers inside a CFRP composite laminate.

## 9.6 PSD AND ASD OF SANDWICH COMPOSITE STRUCTURES

A typical aerospace sandwich structure is described in Ref. [120]: the sandwich is made of two orthotropic GFRP skins and an isotropic foam core. The skins are made of [0,90] plies. The stiffness (respectively mass density) of the core is 50 times (respectively 10 times) less than the stiffness (respectively mass density) of the skins. Therefore the sandwich is a very inhomogeneous structure. Moreover, the skins do not have the same thickness: one skin is three times thicker than the other. The core has the same thickness as the two skins taken as a whole. Low-velocity impacts induce specific damages in such sandwich plates (Figure 60). The damage appears to be a crush of the foam below the impact location, a debonding of the thin skin (zone 1), an oblique crack in the foam from the thin skin to the thick skin (zone la), and a large debonding of the thick skin (zone 2).

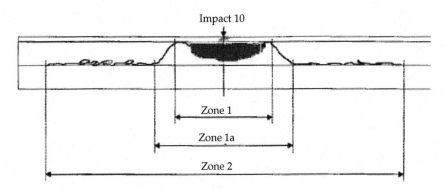

**FIGURE 60**   Schematic of the damage in an impacted composite sandwich structure [120].

For a damaged sandwich plate, the foam cannot, in the damaged area, resist to transverse shear stresses. Therefore, the quasi-flexural A0 waves are more sensitive to these defects than extensional waves; for this reason, one can use the interaction of flexural waves with these defects for damage detection analysis. The main modification in the behavior of the plate induced by the damage is the local decoupling of the two skins. In order to study a method to localize and estimate the size of such damage, Ref. [120] started with the study of sandwich plates with cylindrical holes in the foam core. Such holes make it possible to do numerical simulations [121] and allow a local decoupling of the two skins just like real damages.

Reference [120] used the embedded pulse-echo method to detect damage in the foam core of a sandwich plate with GFRP skins (Figure 61). The damage was simulated by a hole in the foam. Low-frequency A0 Lamb waves (10 and 20 kHz) were used. It was shown that damage location and intensity could be deduced from the echo analysis without an analytical model of the damaged plate. Reference [120] describes a system for health monitoring of sandwich structures based on the analysis of the interaction of Lamb waves with damage. The system consisted of a 20-mm diameter PWAS transmitter placed on the upper face of the sandwich and another 20-mm PWAS receiver placed on the lower face of the sandwich underneath the transmitter PWAS. The receiver PWAS is segmented into four quarters such as to distinguish the direction from which the received waves arrive. The same method was later used to detect actual impact damage in composite sandwich specimens [121].

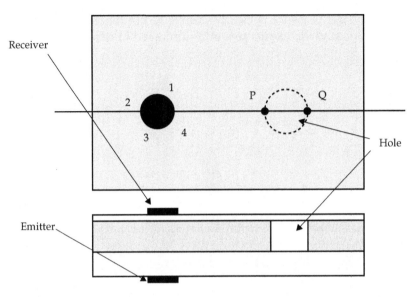

**FIGURE 61**    Placement of transmitter and received PWAS transducers on a composite sandwich structure with a simulated core damage (circular hole in the foam core) [120].

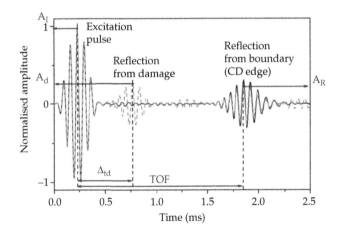

**FIGURE 62**    Comparative response of a pristine (solid line) and damaged (dashed line) CFRP sandwich beam after low-velocity impact [122].

Reference [122] studied *in situ* disbond detection in composite sandwich plates with PWAS transducers. In sandwich composites, the dispersion curves at low frequencies are similar to those for free plates with a greater attenuation due to the presence of the core. S0 mode at relatively high frequency was used for the detection of the debonding between the skin and the core and for low-velocity impact damage [122]. Figure 62 shows the comparative response of a pristine and impact-damaged CFRP sandwich beam. The additional echo due to the disbond presence is clearly visible. Electromagnetic methods for damage imaging in composite sandwich structures were discussed in Ref. [118].

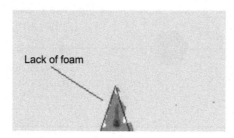

**FIGURE 63**   Electromagnetic methods for damage imaging in composite sandwich structures: electric imaging of a GFRP sandwich specimen showing the detection of a missing foam zone [118].

Electromagnetic methods for finding defects and damage in composite materials rely on the response of these materials to the excitation by electromagnetic fields. Dielectric composites, such as GFRP materials, can be interrogated with the use of an electric field, whereas conductive composites, such as CFRP materials, can be interrogated with a magnetic field. Figure 63 shows the detection of an internal defect (zone of missing foam) in a GFRP sandwich specimen by illumination with an electric field (e.g., with an electric antenna in near field) and by measuring the total electric component of the electric field crossing through the material.

Reference [123] describes the modeling of wave propagation and scatter from impact damage in sandwich structures with honeycomb core. Extensive FEM analysis was performed including multiphysics modeling of the piezo actuators/sensors. The FEM modeling results were compared with experimental measurements performed on actual sandwich specimens.

## 9.7  SUMMARY AND CONCLUSIONS

This chapter has covered the SHM methods for detection and quantification of impact damage in aerospace composites. Low-velocity impact of composite structures that produces BVID is one of the most researched areas of aerospace composites SHM due to drastic effect that the presence of BVID could have on composite aircraft performance and safety. Various methods have been proposed and tried for capturing the impact event and monitoring the evolution of resulting BVID state inside the structure.

The chapter started with a discussion of monitoring the impact event. This method is also known as PSD because it relies only on the signals generated by the impact event and captured by the sensors. Detection of impact location and impact intensity are the major aims of the PSD method. Several approaches using piezoelectric transducers and fiber-optic sensors were presented and discussed. The reconstruction of impact location and intensity, which is not an easy task, was found to be attempted with a variety of methods, some using a structural model (i.e., model-based approaches), others aiming to rely only on advanced data-processing techniques (data-driven approaches). The use of directional transducers for detection of the impact event was also discussed. It was also found that, besides the location and intensity of damage, the signals may also contain high-frequency information related to the material failure at the impact location, i.e., AE signals.

The chapter continued with the presentation of the methods aiming on finding the damage that was produced in the composite by the impact event. This method, known as ASD, relies on interrogating the structure with ultrasonic waves and recording the response of the structure to such interrogation. Piezoelectric transducers are usually used as ultrasonic wave transmitters, whereas either the same piezo transducers or other strain sensors, e.g., fiber-optic FBG sensors, may be used as receivers of the signal scattered by the damage. Various ASD methods such as pulse-echo, guided-wave tomography, data-driven ASD, phased arrays, etc. were presented and discussed.

Other methods for detecting the damage produced by an impact on a composite structure were also discussed. Direct methods for damage detection rely on changes that are produced by the impact on sensors located in the impact proximity (e.g., transmissibility of an optical fiber) or on the reinforcing glass fibers in the case of GFRP composites. Changes may also be produced in the composite appearance, such as in the case of triboluminescent composites or "bruising" composites. OTDR and fiber-optic FBG strain sensors were also used to detect the structural changes produced by the impact event in the composite material. Strain mapping of larger areas has now become possible with the advent of distributed strain sensing on a simple fiber using Rayleigh scatter OTDR.

Vibration SHM has also been tried for the detection of impact damage in composite structures. Model-based vibration SHM, model-free vibration SHM, statistics-based vibration SHM, and nonlinear vibration SHM, as well as the combined use of vibration and wave propagation SHM methods have been tried and are discussed in Section 4. The frequency transfer function SHM method was also presented in this section. The use of EMIS to perform local-area sensing of high-frequency vibration changes due to impact delamination was also discussed.

Section 5 covered electrical and electromagnetic field methods for delamination detection. Delamination detection in CFRP composites (which are mildly conductive) has received considerable attention and several examples were discussed in this section. Alternative methods using similar principles are the electrical potential method and the EIT. Electromagnetic damage detection, on the other hand, can be used in dielectric composites such as GFRP materials. A hybrid approach that uses both electromagnetic and ultrasonic interrogation of a CFRP plate was also discussed in Section 5.

Section 6 covered PSD and ASD method for impact damage detection in sandwich composite structures.

# References

[1] Chang, F. K. (1995) "Built-In Damage Diagnostics for Composite Structures," *ICCM-10: 10th International Congress of Composite Materials*, Canada, 1995, pp. 283–289.

[2] Roh, Y.; Chang, F. K., "Effect of Impact Damage on Lamb Wave Propagation in Laminated Composites," presented at the *ASME International Mechanical Engineering Congress and Exposition IMECE-1995*, San Francisco, CA, 1995.

[3] Wang, C. S.; Chang, F. K. (1999) "Built-in Diagnostics for Impact Damage Identification of Composite Structures," *2nd International Workshop on Structural Health Montoring*, 1999, pp. 612–621.

[4] Wang, C. S.; Chang, F. K. (2000) "Diagnosis of Impact Damage in Composite Structures with Built-in Piezoelectrics Network," *SPIE vol 3990*, San Diego, CA, 2000, pp. 13–19.

[5] Wang, C. S.; Chang, F.-K .(2000) "Diagnosis of Impact Damage in Composite Structures with Built-in Piezoelectric Network," *ASME IMECE-2000 International Mechanical Engineering Congress and Exposition*, 2000.

[6] Choi, K. Y.; Chang, F. K. (1994) "Identification of Foreign Object Impact in Structures using Distributed Sensors", *Journal of Intelligent Material Systems and Structures*, **5**(6), 864–869, Nov. 1994, http://dx.doi.org/10.1177/1045389x9400500620.

[7] Choi, K. Y.; Chang, F. K. (1996) "Identification of Impact Force and Location Using Distributed Sensors", *AIAA Journal*, **34**(1), 136–142, Jan. 1996, http://dx.doi.org/10.2514/3.13033.

[8] Tracy, M.; Chang, F. K. (1998) "Identifying Impacts in Composite Plates with Piezoelectric Strain Sensors, Part I: Theory", *Journal of Intelligent Material Systems and Structures*, **9**(11), 920–928, Nov. 1998.

[9] Tracy, M.; Chang, F. K. (1998) "Identifying Impacts in Composite Plates with Piezoelectric Strain Sensors, Part II: Experiment", *Journal of Intelligent Material Systems and Structures*, **9**(11), 929–937, Nov. 1998.

[10] Seydel, R.; Chang, F. K. (2001) "Impact Identification of Stiffened Composite Panels: I. System Development", *Smart Materials & Structures*, **10**(2), 354–369, Apr. 2001, http://dx.doi.org/10.1088/0964-1726/10/2/323.

[11] Park, J.; Ha, S.; Chang, F. K. (2009) "Monitoring Impact Events Using a System-Identification Method", *AIAA Journal*, **47**(9), 2011–2021, Sep. 2009, http://dx.doi.org/10.2514/1.34895.

[12] Markmiller, J. F. C.; Chang, F. K. (2010) "Sensor Network Optimization for a Passive Sensing Impact Detection Technique", *Structural Health Monitoring—an International Journal*, **9**(1), 25–39, Jan. 2010, http://dx.doi.org/10.1177/1475921709349673.

[13] Dupont, M.; Osmont, R.; Gouyon, R.; Balageas, D. L. (1999) "Permanent Monitoring of Damage Impacts by a Piezoelectric Sensor Based Integrated System," *International Workshop on Structural Health Monitoring 1999*, Stanford, CA, 1999, pp. 561–570.

[14] Park, C. Y.; Kim, I. G. (2008) "Prediction of Impact Forces on an Aircraft Composite Wing", *Journal of Intelligent Material Systems and Structures*, **19**(3), 319–324, Mar. 2008, http://dx.doi.org/10.1177/1045389x07083180.

[15] Giurgiutiu, V. (2014) *Structural Health Monitoring with Piezoelectric Wafer Active Sensors*, 2nd ed., Elsevier Academic Press.

[16] Seydel, R.; Chang, F. K. (2001) "Impact Identification of Stiffened Composite Panels: I. System Development; II. Implementation Studies", *Smart Materials and Structures*, **10**(2), 354–379, Apr. 2001, http://dx.doi.org/10.1088/0964-1726/10/2/323.

[17] Park, J.; Chang, F.-K. (2005) "System Identification Method for Monitoring Impact Events," *SPIE 2005 Smart Structures and NDE Symposium*, 2005, pp. 189–200.

[18] Sung, D. U.; Oh, J. H.; Kim, C. G.; Hong, C. S. (2000) "Impact Monitoring of Smart Composite Laminates Using Neural Network and Wavelet Analysis", *Journal of Intelligent Material Systems and Structures*, **11**(3), 180–190, Mar. 2000, http://dx.doi.org/10.1106/n5e7-m37y-3mar-2kfh.

[19] Maseras-Gutierrez, M. A.; Staszewski, W. J.; Found, M. S.; Worden, K. (1998) "Detection of Impacts in Composite Materials Using Piezoceramic Sensors and Neural Networks," *SPIE 1998 Smart Structures and Materials Symposium*, 1998, pp. 491–497.

[20] Haywood, J.; Coverley, P. T.; Staszewski, W. J.; Worden, K. (2005) "An automatic impact monitor for a composite panel employing smart sensor technology", *Smart Materials & Structures*, **14**(1), 265–271, Feb. 2005, http://dx.doi.org/10.1088/0964-1726/14/1/027.

[21] Mujica, L. E.; Vehi, J.; Staszewski, W.; Worden, K. (2008) "Impact Damage Detection in Aircraft Composites Using Knowledge-based Reasoning", *Structural Health Monitoring—an International Journal*, **7**(3), 215–230, Sep. 2008, http://dx.doi.org/10.1177/1475921708090560.

[22] Hiche, C.; Coelho, C. K.; Chattopadhyay, A. (2011) "A Strain Amplitude-Based Algorithm for Impact Localization on Composite Laminates", *Journal of Intelligent Material Systems and Structures*, **22**(17), 2061–2067, 2011, http://dx.doi.org/10.1177/1045389x11424214.

[23] Betz, D. C.; Thursby, G.; Culshaw, B.; Staszewski, W. J. (2007) "Structural Damage Location with Fiber Bragg Grating Rosettes and Lamb Waves", *Structural Health Monitoring - an International Journal*, **6**(4), 299–308, Dec. 2007, http://dx.doi.org/10.1177/1475921707081974.

[24] Matt, H. M.; Lanza di Scalea, F. (2007) "Macro-Fiber Composite Piezoelectric Rosettes for Acoustic Source Location in Complex Structures", *Smart Materials and Structures*, **16**(4), 1489–1499, Aug. 2007, http://dx.doi.org/10.1088/0964-1726/16/4/064.

[25] Zhao, P.; Pisani, D.; Lynch, C. S. (2011) "Piezoelectric Strain Sensor/Actuator Rosettes", *Smart Materials and Structures*, **20**(10), 102002, 2011, http://dx.doi.org/10.1088/0964-1726/20/10/102002.

[26] Salamone, S.; Bartoli, I.; Di Leo, P.; Lanza Di Scala, F.; Ajovalasit, A.; D'Acquisto, L.; Rhymer, J.; Kim, H. (2010) "High-Velocity Impact Location on Aircraft Panels Using Macro-fiber Composite Piezoelectric Rosettes", *Journal of Intelligent Material Systems and Structures*, **21**(9), 887–896, 2010, http://dx.doi.org/10.1177/1045389x10368450.

[27] Hirano, N.; Mamizu, H.; Kuraishi, A.; Itoh, T.; Takeda, N.; Enomoto, K. (2014) "Detectability Assessment of Optical Fiber Sensor Based Impact Damage Detection for Composite Airframe Structure," *EWSHM 2014—7th European Workshop on Structural Health Monitoring*, Nantes, France, 2014.

[28] Martin, T.; Hudd, J.; Wells, P.; Tunnicliffe, D.; Das-Gupta, D. (2001) "The Use of Low Profile Piezoelectric Sensors for Impact and Acoustic Emission (AE) Detection in CFRC Structures", *Journal of Intelligent Material Systems and Structures*, **12**(8), 537–544, Aug. 2001, http://dx.doi.org/10.1177/10453890122145339.

[29] Koh, Y. L.; Chiu, W. K.; Rajic, N.; Galea, S. C. (2003) "Detection of Disbond Growth in a Cyclically Loaded Bonded Composite Repair Patch Using Surface-mounted Piezoceramic Elements", *Structural Health Monitoring—an International Journal*, **2**(4), 327–339, 2003.

[30] Perez, I.; Cui, H.-L.; Udd, E., "Acoustic Emission Detection Using Fiber Bragg Gratings," presented at the *SPIE 2001 Smart Structures and NDE Symposium*, San Diego, CA, 2001.

[31] Koh, J. I.; Bang, H. J.; Kim, C. G.; Hong, C. S. (2005) "Simultaneous Measurement of Strain and Damage Signal of Composite Structures Using a Fiber Bragg Grating Sensor", *Smart Materials & Structures*, **14**(4), 658–663, Aug. 2005, http://dx.doi.org/10.1088/0964-1726/14/4/024.

[32] Pysical Acoustics Corp. (2012). *AEwin Acoustic Emission Software*. Available at: http://www.pacndt.com/index.aspx?go=products&focus=/software/aewin.htm.

[33] Duke, J. C. (1988) *Acousto-Ultrasonics—Theory and Applications*, Plenum Press.

[34] Diamanti, K.; Soutis, C. (2010) "Structural Health Monitoring Techniques for Aircraft Composite Structures", *Progress in Aerospace Sciences*, **46**(8), 342–352, Nov. 2010, http://dx.doi.org/10.1016/j.paerosci.2010.05.001.

[35] Kirikera, G. R.; Balogun, O.; Krishnaswamy, S. (2011) "Adaptive Fiber Bragg Grating Sensor Network for Structural Health Monitoring: Applications to Impact Monitoring", *Structural Health Monitoring—an International Journal*, **10**(1), 5–16, Jan. 2011, http://dx.doi.org/10.1177/1475921710365437.

[36] Giurgiutiu, V.; Cuc, A. (2005) "Embedded Nondestructive Evaluation for Structural Health Monitoring, Damage Detection, and Failure Prevention", *Shock and Vibration Digest*, **37**(2), 83–105, March 2005.

[37] Keilers, C. H.; Chang, F. K. (1995) "Identifying Delamination in Composite Beams using Built-in Piezoelectrics—Part I: Experiments and Analysis", *Journal of Intelligent Material Systems and Structures*, **6**(5), 649–663, Sep. 1995, http://dx.doi.org/10.1177/1045389x9500600506.

[38] Keilers, C. H.; Chang, F. K. (1995) "Identifying Delamination in Composite Beams using Built-in Piezoelectrics—Part II: An identification Method", *Journal of Intelligent Material Systems and Structures*, **6**(5), 664–672, Sep. 1995, http://dx.doi.org/10.1177/1045389x9500600507.

[39] Lemistre, M.; Gouyon, R.; Kaczmarek, H.; Balageas, D. (1999) "Damage Localization in Composite Plates Using Wavelet Transform Processing on Lamb Wave Signals," *IWSHM 1999*, Stanford, CA, 1999, pp. 861–870.

[40] Lemistre, M. B.; Balageas, D. L. (2001) "Structural Health Monitoring System based on Diffracted Lamb Wave Analysis by Multiresolution Processing", *Smart Materials and Structures*, **10**(3), 504–511, Jun. 2001, http://dx.doi.org/10.1088/0964-1726/10/3/312.

[41] Liu, Y.; Fard, M. Y.; Chattopadhyay, A.; Doyle, D. (2011) "Damage Assessment of CFRP Composites Using a Time–Frequency Approach", *Journal of Intelligent Material Systems and Structures*, **23**(4), 397–413, 2011.

[42] Pierce, S. G.; Philp, W. R.; Culshaw, B.; Gachagan, A.; McNab, A.; Hayward, G.; Lecuyer, F. (1996) "Surface-Bonded Optical Fibre Sensors for the Inspection of CFRP Plates Using Ultrasonic Lamb Waves", *Smart Materials and Structures*, **5**(6), 776–787, Dec. 1996, http://dx.doi.org/10.1088/0964-1726/5/6/007.

[43] Pierce, S.; Staszewski, W.; Gachagan, A.; James, I.; Philip, W.; Worden, K.; Culshaw, B.; McNab, A.; Tomlinson, G.; Hayward, G., "Ultrasonic Condition Monitoring of Composite Structures Using a Low Profile Acoustic Source and an Embedded Optical Fiber Sensor", presented at the *SPIE vol. 3041*, 1997.

[44] Pierce, S.; Culshaw, B.; Manson, G.; Worden, K.; Staszewski, W. (2000) "The Application of Ultrasonic Lamb Wave Techniques to the Evaluation of Advanced Composite Structures", *SPIE vol. 3986*, 2000, pp. 93–103.

[45] Gachagan, A.; Hayward, G.; McNab, A.; Reynolds, P.; Pierce, S.; Philip, W.; Culshaw, B. (1999) "Generation and Reception of Ultrasonic Guided Waves in Composite Plates Using Comformable Piezoelectric Transmitters and Optical-Fiber Detectors", *IEEE-TUFFC Transactions on Ultrasonics, Ferroelectrics, and Frequency Control*, **46**(1), 72–81, 1999.

[46] Okabe, Y.; Fujibayashi, K.; Shimazaki, M.; Soejima, H.; Ogisu, T. (2010) "Delamination Detection in Composite Laminates Using Dispersion Change Based on Mode Conversion of Lamb Waves", *Smart Materials & Structures*, **19**(11), Nov. 2010, http://dx.doi.org/10.1088/0964-1726/19/11/115013.

[47] Wu, Z.; Qing, X. P.; Chang, F.-K. (2009) "Damage Detection for Composite Laminate Plates with A Distributed Hybrid PZT/FBG Sensor Network", *Journal of Intelligent Material Systems and Structures*, **20**(9), 1069–1077, Jun. 2009, http://dx.doi.org/10.1177/1045389x08101632.

[48] Sohn, H.; Dutta, D.; Yang, J. Y.; DeSimio, M.; Olson, S.; Swenson, E. (2011) "Automated Detection of Delamination and Disbond from Wavefield Images Obtained Using a Scanning Laser Vibrometer", *Smart Materials and Structures*, **20**(4), 045017, 2011, http://dx.doi.org/10.1088/0964-1726/20/4/045017.

[49] Malyarenko, E. V.; Hinders, M. K. (2001) "Ultrasonic Lamb Wave Diffraction Tomography", *Ultrasonics*, **39**(4), 269–281, Jun. 2001, http://dx.doi.org/10.1016/s0041-624x(01)00055-5.

[50] Ihn, J. B.; Chang, F. K. (2008) "Pitch-Catch Active Sensing Methods in Structural Health Monitoring for Aircraft Structures", *Structural Health Monitoring—an International Journal*, **7**(1), 5–19, Mar. 2008, http://dx.doi.org/10.1177/1475921707081979.

[51] Su, Z.; Cheng, L.; Wang, X.; Yu, L.; Zhou, C. (2009) "Predicting Delamination of Composite Laminates Using an Imaging Approach", *Smart Materials and Structures*, **18**(7), 074002, 2009, http://dx.doi.org/10.1088/0964-1726/18/7/074002.

[52] Yan, F.; Royer, R. L.; Rose, J. L. (2010) "Ultrasonic Guided Wave Imaging Techniques in Structural Health Monitoring", *Journal of Intelligent Material Systems and Structures*, **21**(3), 377–384, Feb. 2010, http://dx.doi.org/10.1177/1045389x09356026.

[53] Rose, J. L. (2002) "A Baseline and Vision of Ultrasonic Guided Wave Inspection Potential", *Transactions of the ASME. Journal of Pressure Vessel Technology*, **124**(3), Aug. 2002, http://dx.doi.org/10.1115/1.1491272.

[54] Diaz Valdes, S. H.; Soutis, C. (2002) "Real-Time Nondestructive Evaluation of Fiber Composite Laminates Using Low-Frequency Lamb Waves", *Journal of the Acoustical Society of America*, **111**(5), 2026–2033, May 2002, http://dx.doi.org/10.1121/1.1466870.

[55] Diamanti, K.; Soutis, C.; Hodgkinson, J. M. (2007) "Piezoelectric Transducer Arrangement for the Inspection of Large Composite Structures", *Composites Part A—Applied Science and Manufacturing*, **38**(4), 1121–1130, 2007, http://dx.doi.org/10.1016/j.compositesa.2006.06.011.

[56] Giurgiutiu, V.; Bao, J. (2004) "Embedded-Ultrasonics Structural Radar for *In Situ* Structural Health Monitoring of Thin-Wall Structures", *Structural Health Monitoring-an International Journal*, **3**(2), 121–140, Jun. 2004, http://dx.doi.org/10.1177/1475921704042697.

[57] Giurgiutiu, V.; Yu, L. Y.; Kendall, J. R.; Jenkins, C. (2007) "*In Situ* Imaging of Crack Growth with Piezoelectric-Wafer Active Sensors", *AIAA Journal*, **45**(11), 2758–2769, Nov. 2007, http://dx.doi.org/10.2514/1-30798.

[58] Yu, L.; Giurgiutiu, V. (2008) "*In Situ* 2-D Piezoelectric Wafer Active Sensors Arrays for Guided Wave Damage Detection", *Ultrasonics*, **48**(2), 117–134, Apr. 2008, http://dx.doi.org/10.1016/j.ultras.2007.10.008.

[59] Yan, F.; Rose, J. L. (2007) "Guided Wave Phased Array Beam Steering in Composite Plates," *SPIE 2007 Smart Structures and NDE Symposium*, San Diego, CA, 2007, p. #15.

[60] Collet, M.; Ruzzene, M.; Cunefare, K. A. (2011) "Generation of Lamb Waves Through Surface Mounted Macro-Fiber Composite Transducers", *Smart Materials and Structures*, **20**, 025020, 2011, http://dx.doi.org/10.1088/0964-1726/20/2/025020.

[61] Salas, K. I.; Cesnik, C. E. S. (2009) "Guided Wave Excitation by a CLoVER Transducer for Structural Health Monitoring: Theory and Experiments", *Smart Materials and Structures*, **18**(7), 075005, 2009, http://dx.doi.org/10.1088/0964-1726/18/7/075005.

[62] Xu, B.; Senesi, M.; Ruzzene, M. (2011) "Frequency-Steered Acoustic Arrays: Application to Structural Health Monitoring of Composite Plates", *Journal of Engineering Materials and Technology-Transactions of ASME*, **133**(1), 011003, Jan. 2011, http://dx.doi.org/10.1115/1.4002638.

[63] Baravelli, E.; Senesi, M.; Ruzzene, M.; De Marchi, L.; Speciale, N. (2011) "Double-Channel, Frequency-Steered Acoustic Transducer With 2-D Imaging Capabilities", *IEEE Transactions on Ultrasonics Ferroelectrics and Frequency Control*, **58**(8), 1430–1706, Aug. 2011, http://dx.doi.org/10.1109/tuffc.2011.2000.

[64] Tsutsui, H.; Kawamata, A.; Sanda, T.; Takeda, N. (2004) "Detection of Impact Damage of Stiffened Composite Panels Using Embedded Small-Diameter Optical Fibers", *Smart Materials & Structures*, **13**(6), 1284–1290, Dec. 2004, http://dx.doi.org/10.1088/0964-1726/13/6/002.

[65] Kister, G.; Ralph, B.; Fernando, G. F. (2004) "Damage Detection in Glass Fibre-Reinforced Plastic Composites Using Self-Sensing E-Glass Fibres", *Smart Materials & Structures*, **13**(5), 1166–1175, Oct. 2004, http://dx.doi.org/10.1088/0964-1726/13/5/021.

[66] Wang, L.; Kister, G.; Ralph, B.; Talbot, J. D. R.; Fernando, G. F. (2004) "Conventional E-Glass Fibre Light Guides: Self-Sensing Composite Based on Cladding", *Smart Materials and Structures*, **13**(1), 73–81, Feb. 2004, http://dx.doi.org/10.1088/0964-1726/13/1/009.

[67] Sage, I.; Humberstone, L.; Oswald, I.; Lloyd, P.; Bourhill, G. (2001) "Getting Light Through Black Composites: Embedded Triboluminescent Structural Damage Sensors", *Smart Materials and Structures*, **10**(2), 332–337, Apr. 2001, http://dx.doi.org/10.1088/0964-1726/10/2/320.

[68] Balageas, D.; Bourasseau, S.; Dupont, M.; Bocherens, E.; Dewynter-Marty, V.; Ferdinand, P. (2000) "Comparison Between Non-Destructive Evaluation Techniques and Integrated Fiber Optic Health Monitoring Systems for Composite Sandwich Structures", *Journal of Intelligent Material Systems and Structures*, **11**(6), 426–437, 2000, http://dx.doi.org/10.1106/mfm1-c5ft-6bm4-afud.

[69] Bocherens, E.; Bourasseau, S.; Dewynter-Marty, V.; Py, S.; Dupont, M.; Ferdinand, P.; Berenger, H. (2000) "Damage Detection in a Radome Sandwich Material with Embedded Fiber Optic Sensors", *Smart Materials and Structures*, **9**(3), 310–315, Jun. 2000, http://dx.doi.org/10.1088/0964-1726/9/3/310.

[70] Guemes, A.; Fernandez-Lopez, A.; Fernandez, P. (2014) "Damage Detection in Composite Structures from Fibre Optic Distributed Strain Measurements," *EWSHM 2014—7th European Workshop on Structural Health Monitoring*, Nantes, France, 2014.

[71] Sierra-Perez, J.; Guemes, A.; Mujica, L. E. (2013) "Damage Detection by Using FBGs and Strain Field Pattern Recognition Techniques", *Smart Materials and Structures*, **22**(2), 2013, http://dx.doi.org/10.1088/0964-1726/22/2/025011/.

[72] Worden, K.; Inman, D. J. (2010) "Modal Vibration Methods in Structural Health Monitoring,". in *Encyclopedia of Aerospace Engineering*, R. Blockley, W. Shyy (Eds.), Wiley, pp. 1995–2004.

[73] Fan, W.; Qiao, P. (2011) "Vibration-Based Damage Identification Methods: A Review and Comparative Study", *Structural Health Monitoring—an International Journal*, **10**(1), 83–111, Jan. 2011, http://dx.doi.org/10.1177/1475921710365419.

[74] Shahdin, A.; Morlier, J.; Niemann, H.; Gourinat, Y. (2011) "Correlating Low Energy Impact Damage with Changes in Modal Parameters: Diagnosis Tools and FE Validation", *Structural Health Monitoring—an International Journal*, **10**(2), 199–217, 2011, http://dx.doi.org/10.1177/1475921710373297.

[75] Pandey, A. K.; Biswas, M.; Samman, M. M. (1991) "Damage Detection from Changes in Curvature Mode Shapes", *Journal of Sound and Vibration*, **145**(2), 321–332, 1991.

[76] Sharma, V. K.; Hanagud, S.; Ruzzene, M. (2006) "Damage Index Estimation in Beams and Plates Using Laser Vibrometry", *AIAA Journal*, **44**(4), 919–923, Apr. 2006, http://dx.doi.org/10.2514/1.19012.

[77] Hamey, C. S.; Lestari, W.; Qiao, P.; Song, G. (2004) "Experimental Damage Identification of Carbon/Epoxy Composite Beams Using Curvature Mode Shapes", *Structural Health Monitoring—an International Journal*, **3**(4), 333–353, 2004.

[78] Lestari, W.; Qiao, P.; Hanagud, S. (2007) "Curvature Mode Shape-Based Damage Assessment of Carbon/Epoxy Composite Beams", *Journal of Intelligent Material Systems and Structures*, **18**(3), 189–208, 2007, http://dx.doi.org/10.1177/1045389x06064355.

[79] Martins, B. L.; Kosmatka, J. B. (2015) "Detecting Damage in a UAV Composite Wing Spar Using Distributed Fiber Optic Strain Sensors," *AIAA SciTech 2015*, Kissimmee, FL, 2015, pp. paper AIAA-2015-0447.

[80] Staszewski, W. J.; Sohn, H. (2010) "Signal Processing for Structural Health Monitoring,". in *Encyclopedia of Aerospace Engineering*, R. Blockley, W. Shyy (Eds.), Wiley, pp. 2013–2025.

[81] Seaver, M.; Aktas, E.; Trickey, S. T. (2010) "Quantitative Detection of Low Energy Impact Damage in a Sandwich Composite Wing", *Journal of Intelligent Material Systems and Structures*, **21**(3), 297–308, Feb. 2010, http://dx.doi.org/10.1177/1045389x09347020.

[82] Chrysochoidis, N. A.; Barouni, A. K.; Saravanos, D. A. (2011) "Delamination Detection in Composites Using Wave Modulation Spectroscopy with a Novel Active Nonlinear Acousto-Ultrasonic Piezoelectric Sensor", *Journal of Intelligent Material Systems and Structures*, **22**(18), 2193–2206, Dec. 2011, http://dx.doi.org/10.1177/1045389x11428363.

[83] Ooijevaar, T. H.; Loendersloot, R.; Rogge, M. D.; Akkerman, R.; Tinga, T. (2014) "Vibro-Acoustic Modulation Based Damage Identification in a Composite Skin-Stiffener Structure," *EWSH 2014 7th European Workshop on Structural Health Monitoring*, Nantes, France, 2014.

[84] Mal, A.; Ricci, F.; Banerjee, S.; Shih, F. (2005) "A Conceptual Structural Health Monitoring System based on Vibration and Wave Propagation", *Structural Health Monitoring - an International Journal*, **4**(3), 283–293, Sep. 2005.

[85] Keilers, C. H.; Chang, F. K. (1995) "Identifying Delamination in Composite Beams Using Built-in Piezoelectrics—Part I: Experiments and Analysis; Part II: An Indentification Method", *Journal of Intelligent Material Systems and Structures*, **6**(5), 649–672, Sep. 1995, http://dx.doi.org/10.1177/1045389x9500600506.

[86] Lichtenwalner, P. F.; Dunne, J. P.; Becker, R. S.; Baumann, E. W., "Active Damage Interrogation Method for Structural Health Monitoring" US Patent US6006163-A, Dec. 21, 1999.

[87] Hu, N.; Fukunaga, H.; Kameyama, M. (2006) "Identification of Delaminations in Composite Laminates", *Journal of Intelligent Material Systems and Structures*, **17**(8–9), 671–683, 2006, http://dx.doi.org/10.1177/1045389x06055816.

[88] Davis, C. E.; Norman, P.; Ratcliffe, C.; Crane, R. (2012) "Broad Area Damage Detection in Composites Using Fibre Bragg Grating Arrays", *Structural Health Monitoring—an International Journal*, **11**(6), 724–732, 2012.

[89] Giurgiutiu, V.; Zagrai, A. N. (2002) "Embedded Self-Sensing Piezoelectric Active Sensors for On-Line Structural Identification", *Journal of Vibration and Acoustics-Transactions of ASME*, **124**(1), 116–125, Jan. 2002, http://dx.doi.org/10.1115/1.1421056.

[90] Zagrai, A. N.; Giurgiutiu, V. (2001) "Electro-Mechanical Impedance Method for Crack Detection in Thin Plates", *Journal of Intelligent Material Systems and Structures*, **12**(10), 709–718, Oct. 2001.

[91] Cuc, A.; Giurgiutiu, V.; Joshi, S.; Tidwell, Z. (2007) "Structural Health Monitoring with Piezoelectric Wafer Active Sensors for Space Applications", *AIAA Journal*, **45**(12), 2838–2850, Dec. 2007, http://dx.doi.org/10.2514/1.26141.

[92] Bois, C.; Hochard, C. (2004) "Monitoring of Laminated Composites Delamination Based on Electro-Mechanical Impedance Measurement", *Journal of Intelligent Materials Systems and Structures*, **15**(1), 59–67, 2004, http://dx.doi.org/10.1177/1045389x04039405.

[93] Pollock, P.; Yu, L.; Guo, S.; Sutton, M. (2011) "A Nondestructive Experimental Study on Damage Progression in Woven Glass-Epoxy Composites," *SPIE 2011 Smart Structures and NDE Symposium*, 2011, p. 127.

[94] Gresil, M.; Yu, L.; Giurgiutiu, V.; Sutton, M. (2012) "Predictive Modeling of Electromechanical Impedance Spectroscopy for Composite Materials", *Structural Health Monitoring—An International Journal*, **223**(8), 1681–1691, 2012..

[95] Lalande, F.; Chaudhry, Z.; Sun, F. P.; Rogers, C. A. (1996) "Debond Detection Using Broad-Band High-Frequency Excitation and Non-Contacting: Laser Vibrometer System", *Journal of Intelligent Material Systems and Structures*, **7**(2), 176–181, Mar. 1996.

[96] Chiu, W. K.; Galea, S. C.; Koss, L. L.; Rajic, N. (2000) "Damage Detection in Bonded Repairs Using Piezoceramics", *Smart Materials and Structures*, **9**(4), 466–475, Aug. 2000, http://dx.doi.org/10.1088/0964-1726/9/4/309.

[97] Giurgiutiu, V.; Harries, K.; Petrou, M.; Bost, J.; Quattlebaum, J. B. (2003) "Disbond Detection with Piezoelectric Wafer Active Sensors in RC Structures Strengthened with FRP Composite Overlays", *Earthquake Engineering and Engineering Vibration*, **2**(2), 213–224, Dec. 2003.

[98] Gyekenyesi, A. L.; Martin, R. E.; Morscher, G. N.; Owen, R. B. (2009) "Impedance-Based Structural Health Monitoring of a Ceramic Matrix Composite", *Journal of Intelligent Material Systems and Structures*, **20**(7), 875–882, 2009, http://dx.doi.org/10.1177/1045389x08099033.

[99] Kemp, M. (1994) "Self-Sensing Composites for Smart Damage Detection Using Electrical Properties," *SPIE vol 2361 2nd European Conference on Smart Structures and Materials*, Glasgow, Scotland, 1994, pp. 136–139.

[100] Todoroki, A. (2001) "The Effect of Number of Electrodes and Diagnostic Tool for Monitoring the Delamination of CFRP Laminates by Changes in Electrical Resistance", *Composites Science and Technology*, **61** (13), 1871–1880, 2001, http://dx.doi.org/10.1016/s0266-3538(01)00088-4.

[101] Todoroki, A.; Tanaka, M.; Shimamura, Y. (2002) "Measurement of Orthotropic Electric Conductance of CFRP Laminates and Analysis of the Effect on Delamination Monitoring with an Electric Resistance Change Method", *Composites Science and Technology*, **62**(5), 619–628, 2002, Pii s0266-3538(02)00019-2.

[102] Todoroki, A.; Tanaka, Y. (2002) "Delamination Identification of Cross-Ply Graphite/Epoxy Composite Beams Using Electric Resistance Change Method", *Composites Science and Technology*, **62**(5), 629–639, 2002, http://dx.doi.org/10.1016/s0266-3538(02)00013-1.

[103] Todoroki, A.; Tanaka, Y.; Shimamura, Y. (2002) "Delamination Monitoring of Graphite/Epoxy Laminated Composite Plate of Electric Resistance Change Method", *Composites Science and Technology*, **62**(9), 1151–1160, 2002, http://dx.doi.org/10.1016/s0266-3538(02)00053-2.

[104] Todoroki, A.; Tanaka, M.; Shimamura, Y. (2003) "High Performance Estimations of Delamination of Graphite/Epoxy Laminates with Electric Resistance Change Method", *Composites Science and Technology*, **63** (13), 1911–1920, Oct. 2003, http://dx.doi.org/10.1016/s0266-3538(03)00157-x.

[105] Todoroki, A.; Tanaka, Y.; Shimamura, Y. (2004) "Multi-Probe Electric Potential Change Method for Delamination Monitoring of Graphite/Epoxy Composite Plates Using Normalized Response Surfaces", *Composites Science and Technology*, **64**(5), 749–758, Apr. 2004, http://dx.doi.org/10.1016/j.compscitech.2003.08.004.

[106] Todoroki, A.; Tanaka, M.; Shimamura, Y. (2005) "Electrical Resistance Change Method for Monitoring Delaminations of CFRP Laminates: Effect of Spacing Between Electrodes", *Composites Science and Technology*, **65**(1), 37–46, Jan. 2005, http://dx.doi.org/10.1016/j.compscitech.2004.05.018.

[107] Iwasaki, A.; Todoroki, A. (2005) "Statistical Evaluation of Modified Electrical Resistance Change Method for Delamination Monitoring of CFRP Plate", *Structural Health Monitoring—an International Journal*, **4**(2), 119–136, Jun. 2005, http://dx.doi.org/10.1177/1475921705049757.

[108] Hirano, Y.; Todoroki, A. (2007) "Damage Identification of Woven Graphite/Epoxy Composite Beams Using the Electrical Resistance Change Method", *Journal of Intelligent Material Systems and Structures*, **18**(3), 253–263, Mar. 2007, http://dx.doi.org/10.1177/1045389x06065467.

[109] Ueda, M.; Todoroki, A. (2006) "Asymmetrical Dual Charge EPCM for Delamination Monitoring of CFRP Laminate", *Key Engineering Materials*, **321–323**, 1309–1315, 2006.

[110] Matsuzaki, R.; Todoroki, A. (2006) "Wireless Detection of Internal Delamination Cracks in CFRP Laminates Using Oscillating Frequency Changes", *Composites Science and Technology*, **66**(3–4), 407–416, Mar. 2006, http://dx.doi.org/10.1016/j.compscitech.2005.07.016.

[111] Schueler, R.; Joshi, S. P.; Schulte, K. (2001) "Damage Detection in CFRP by Electrical Conductivity Mapping", *Composites Science and Technology*, **61**(6), 921–930, 2001, http://dx.doi.org/10.1016/s0266-3538 (00)00178-0.

[112] Loyola, B. R.; Briggs, T. M.; Arronche, L.; Loh, K. J.; La Saponara, V.; O'Bryan, G.; Skinner, J. L. (2013) "Detection of Spatially Distributed Damage in Fiber-Reinforced Polymer Composites", *Structural Health Monitoring—an International Journal*, **12**(3), 225–239, 2013.

[113] Wang, S. K.; Chung, D. D. L.; Chung, J. H. (2006) "Self-Sensing of Damage in Carbon Fiber Polymer-Matrix Composite Cylinder by Electrical Resistance Measurement", *Journal of Intelligent Material Systems and Structures*, **17**(1), 57–62, Jan. 2006, http://dx.doi.org/10.1177/1045389x06056072.

[114] Todoroki, A. (2008) "Delamination Monitoring Analysis of CFRP Structures Using Multi-Probe Electrical Method", *Journal of Intelligent Material Systems and Structures*, **19**(3), 291–298, Mar. 2008, http://dx.doi.org/ 10.1177/1045389x07084154.

[115] Wang, D.; Chung, D. D. L. (2006) "Comparative Evaluation of the Electrical Configurations for the Two-Dimensional Electric Potential Method of Damage Monitoring in Carbon Fiber Polymer-Matrix Composite", *Smart Materials & Structures*, **15**(5), 1332–1344, Oct. 2006, http://dx.doi.org/10.1088/0964-1726/15/5/023.

[116] Angelidis, N.; Khemiri, N.; Irving, P. E. (2005) "Experimental and Finite Element Study of the Electrical Potential Technique for Damage Detection in CFRP Laminates", *Smart Materials & Structures*, **14**(1), 147–154, Feb. 2005, http://dx.doi.org/10.1088/0964-1726/14/1/014.

[117] Wang, D.; Wang, S.; Chung, D. D. L.; Chung, J. H. (2006) "Comparison of the Electrical Resistance and Potential Techniques for the Self-sensing of Damage in Carbon Fiber Polymer-Matrix Composites", *Journal of Intelligent Material Systems and Structures*, **17**(10), 853–861, 2006, http://dx.doi.org/10.1177/1045389x06060218.

[118] Lemistre, M. B.; Balageas, D. L. (2003) "A Hybrid Electromagnetic Acousto-Ultrasonic Method for SHM of Carbon/Epoxy Structures", *Structural Health Monitoring—an International Journal*, **2**(2), 153–160 , 2003.

[119] Hou, L.; Hayes, S. A. (2002) "A Resistance-Based Damage Location Sensor for Carbon-Fibre Composites", *Smart Materials & Structures*, **11**(6), 966–969, Dec. 2002, http://dx.doi.org/10.1088/0964-1726/11/6/401.

[120] Osmont, D.L.; Devillers, D.; Taillade, F. (2001) "Health monitoring of sandwich plates based on the analysis of the interaction of Lamb waves with damages", *Proc. SPIE 4327, Smart Structures and Materials 2001: Smart Structures and Integrated Systems* 290, (August 16, 2001), http://dx.doi.org/10.1117/12.436540.

[121] Barnoncel, D.; Osmont, D.; Dupont, M. (2003) "Health Monitoring of Sandwich Plates with Real Impact Damages Using PZT Devices," *IWSHM 2003*, 2003.

[122] Diamanti, K.; Soutis, C.; Hodgkinson, J. M. (2005) "Non-Destructive Inspection of Sandwich and Repaired Composite Laminated Structures", *Composites Science and Technology*, **65**(13), 2059–2067, Oct. 2005, http://dx.doi.org/10.1016/j.compscitech.2005.04.010.

[123] Luchinsky, D. G.; Hafiychuk, V.; Smelyanskiy, V. N.; Kessler, S.; Walker, J.; Miller, J.; Watson, M. (2013) "Modeling Wave Propagation and Scattering from Impact Damage for Structural Health Monitoring of Composite Sandwich Plates", *Structural Health Monitoring—an International Journal*, **12**(3), 296–308, 2013 http://dx.doi.org/10.1177/1475921713483351.

## 10.1 INTRODUCTION

This chapter deals with monitoring the damage of aerospace composites that appears during normal operational service. Such damage may be due to operational loads and environmental factors and may result in a gradual degradation of composite properties rather than the sudden changes that may appear due to accidental events, like the impact damage discussed in the previous chapter.

This chapter starts with a discussion of passive structural health monitoring (SHM) methods, e.g., monitoring strain, acoustic emission (AE), operational loads, etc. The aim of these passive methods is to record what is actually happening with the composite structure and to act accordingly. For example, the loads applied to the composite structure can be monitored during normal operation conditions and compared with the design loads (both limit loads and fatigue spectra). If the differences between design loads and actual loads are significant, then adjustments can be made to the operational life predictions and flight profiles. Another outcome of passive monitoring could be the identification of changes in the strain distribution pattern over the structure that might be indicative of partial local failures (e.g., microcracks and delaminations) that may affect the load paths in a fail-safe structural design. In addition, AE monitoring could detect when actual local failures happen by recording the waves created by the elastic energy released by a "popping" crack.

The next major section of this chapter deals with monitoring the actual fatigue damage induced in the composite by repeated application of service loads. As already detailed in Chapter 5, fatigue damage in composites is substantially different from fatigue damage in metallic structures. Most results reported in this section deal with monitoring of matrix microcracking and subsequent delamination that appears during cyclic loading tests of aerospace composites. Various passive SHM and active SHM methods are discussed, including fiber-optic measurements, pitch-catch piezo measurements, electrochemical impedance spectroscopy (ECIS), etc.

Another major section of this chapter is devoted to the use of electrical resistance method to monitor in-service degradation and fatigue of carbon fiber-reinforced polymer (CFRP) composites. This approach is specific to CFRP composites because their carbon fibers have electrical conductivity which is imparted to the overall composite through the fact that individual fibers embedded in the polymeric matrix occasionally make contact when bunched together in the composite system. Early tests have shown that as the CFRP material is loaded, its electrical resistance changes thus acting as a built-in indicator of microcracks, delaminations, and other fatigue damage types taking place in the CFRP composite. These initial concepts have been extended through a wide body of research including measurement and mapping of both the electrical resistance and the electrical potential of the composite using different electrode patterns and current injection methods.

The last section of this chapter covers various methods used in the monitoring of disbonds and delamination in composite patch repairs, composite adhesive joints, in nonconductive glass fiber-reinforced plastic (GFRP) composites, etc. Results obtained with guided-wave methods as well as with dielectric measurements are presented and briefly discussed.

This chapter ends with summary, conclusions, and suggestions for further work.

## 10.2  MONITORING OF STRAIN, ACOUSTIC EMISSION, AND OPERATIONAL LOADS

Monitoring of operational loads and strain on actual vehicles has been reported by several authors. Reference [1] reports multichannel fiber Bragg gratings (FBG) strain monitoring systems installed on a sailing yacht and a turboprop aircraft. Reference [2] reports fiber-optic multipoint strain measuring on two full-scale CFRP vehicles, an American's Cup class yacht and an experimental reentry vehicle. Reference [3] describes multipoint FBG strain measurements on a seagoing GFRP ship. The FBG sensors were installed at various locations to measure operational loads during sea trials. Both FBG strain rosettes and individual FBG sensors were used. Typical strain recordings are illustrated in Figure 1; strains due to wave-drive sagging and hogging of the whole ship and due to resonant whipping of the hull can be distinguished in these signals are shown in Figure 1a; equivalent stress resultants post-calculated from strain data and finite element method (FEM) modeling are shown in Figure 1b.

Reference [4] reports a space-qualified FBG system which uses FBG sensors to monitor the strains in a filament-wound CFRP tank during pressure testing. Reference [5] describes a space-qualified onboard FBG system used to monitor the strain on a CFRP composite LH2 tank installed on a reusable launch vehicle (RLV) test article. The FBG sensors were installed on the CFRP composite tank with ultraviolet-cured polyurethane adhesive that

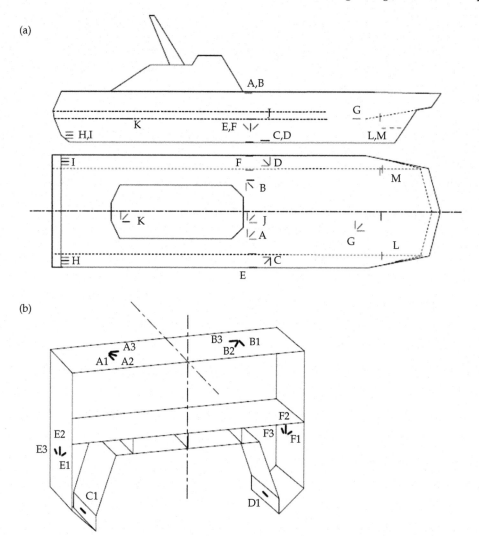

**FIGURE 1** Multipoint strain measurements on a seagoing GFRP ship: (a) strain-gage installation schematics; (b) installation details; (c) typical strain signals recorded with FBG sensors due to wave-drive sagging and hogging as well as resonant whipping; (d) stress resultants calculated from the strain data [3].

showed good performance at cryogenic temperatures. The schematic of the experimental setup for the cryogenic pressure test is shown in Figure 2a. The system (which weighs less than 2 kg) was installed, flown, and tested on a reusable rocket vehicle test (RTV). Data was acquired via telemetry (Figure 2b). Typical recorded data is shown in Figure 2c; the correlation between FBG measurements, conventional strain-gage measurements, and tank pressurization is apparent.

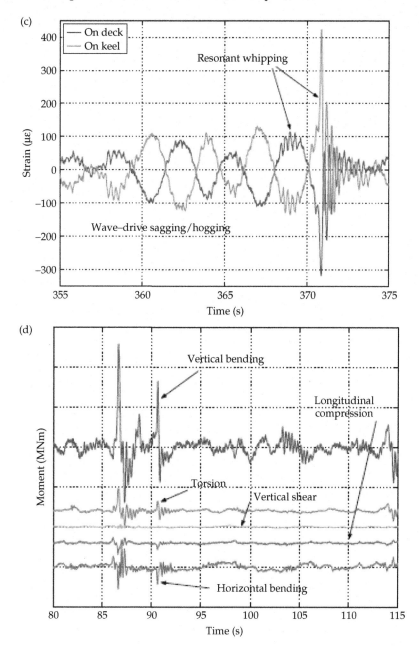

FIGURE 1    (Continued)

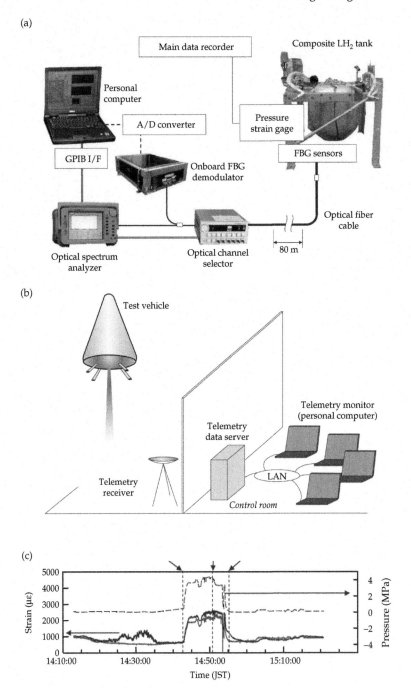

**FIGURE 2**  Space-qualified onboard FBG strain measurement on a composite LH2 tank: (a) schematic of the experimental setup for the cryogenic pressure test; (b) schematic of the telemetry measurement system for the RTV flight experiment; (c) typical recorded data [5].

## 10.2.1 Strain Distribution Monitoring

Passive sensing SHM is done by using sensors to just "listen" to structural signals and monitor them over time. The premise of strain SHM is that presence of damage will effect a redistribution of mechanical strain in the structure which can be monitored and interpreted. A sudden raise of strains in a safety-critical part of the structure should trigger an alarm of impending failure, whereas a sudden drop in strain may be indicative that, due to damage, the monitored part no longer carries its expected load share. Strain redistribution in the damage vicinity may be measured and interpreted as indicative of damage presence and extent. Strain monitoring can be done with conventional electrical resistance strain gages or with fiber optics. The latter offer the capability of having several measuring locations that can be individually addressed on the same fiber.

### 10.2.1.1 Strain Distribution Monitoring in a Composite Patch Repair

Reference [6] presented the use of embedded fiber optics to measure the strain distribution in a composite repair during static and fatigue testing. The sensing principle used in this work is the Rayleigh backscatter which allows one to pinpoint the strain values at various locations along the fiber. In this way, the strain distribution for two composite repair concepts could be compared. One composite repair concept used a stepped lap joint; the other used a scarf lap joint. The strain distribution measure during these experiments is shown in Figure 3: it is apparent that the scarf joint has a much lower stress concentration than the stepped joint. As a result of these severe stress concentrations, the lap joint repair specimen failed at only 66% of the load at which the scarf joint repair specimen failed. Based on these results, a full-scale specimen of a 6-m-long UAV wing spar repaired by the scarf joint concept was manufactured and successfully tested statically up to the ultimate design load, followed by fatigue and residual limit load testing.

### 10.2.1.2 Conventional Strain-Gage Monitoring for Delamination Detection

A delamination in a composite layup produces strain distribution in the surrounding area which is substantially different from that in a pristine zone. Reference [7] used conventional strain gages to monitor delamination in a composite T-joint (Figure 4). The strain gages were placed on the outside of the joint in zones deemed sensitive to strain redistribution due to delamination presence. Further investigations by the same authors focused on using artificial intelligence protocols for damage identification [8] and on comparing statistical outlier analysis and artificial neural net methods for damage detection and assessment [9].

### 10.2.1.3 Combined Fiber Optics and Conventional Strain-Gage Monitoring for Delamination Detection

Reference [10] reports the use of both conventional strain gages and fiber-optic FBG sensors to measure strain in an aircraft wing section having CFRP composite skins with both co-cured and co-bonded stiffeners. The boxes were fabricated with a glass-epoxy prepreg system; the stiffness of these boxes was lower than typically used in aircraft industry. Two testboxes with co-cured skins were fabricated. The dimension of both the boxes was 600mm × 500 mm × 150 mm. The spars and skins were made separately. The end spars were fastened to the bottom skin with bolts and bonded to the top skin with film adhesive

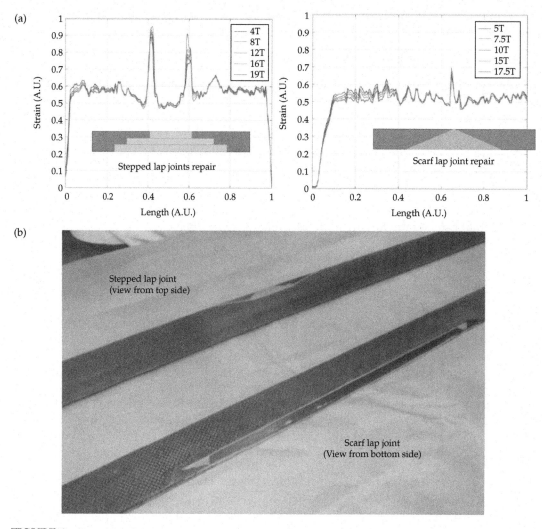

**FIGURE 3** Strain distribution measurement with optical fiber and Rayleigh backscatter method: (a) comparative distributions in stepped lap joint and scarf lap joint; (b) photo of the actual repairs specimens [6].

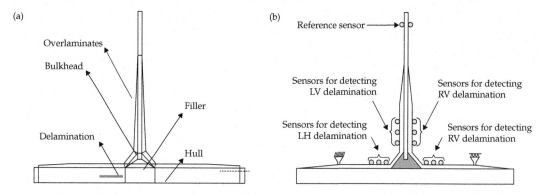

**FIGURE 4** Delamination monitoring with conventional strain gages: (a) composite T-joint susceptible to delamination failure; (b) placement of strain gages for detecting delaminations in the horizontal and vertical sections [7].

(Figure 5a). One testbox was kept "healthy" (BNDa); the other testbox (BNDb) was "damaged" by inserting a debonded region of 180-mm length at the center spar during the fabrication. Resistance strain gages and optical FBG sensors were installed on the testboxes at several locations. A photo of an actual testbox indicating the position of the end fixture for fastening during the static tests is shown in Figure 5b. The boxes were tested in cantilever configuration with the load applied at free end equally distributed over the three spars (left, right, and center spars). The panel dimensions were such that buckling was permitted. The debonds thus manifested themselves more easily due to the buckling of the skins.

(a)

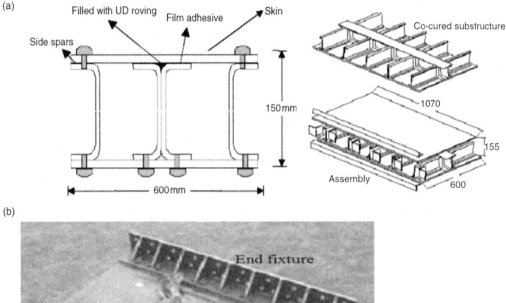

(b)

FIGURE 5    Co-cured composite wing section testbox: (a) construction schematic; (b) photo of the actual testbox indicating the position of the end fixture for fastening during the static tests [10].

The strain data from both the strain gages and the FBG sensors were recorded with the corresponding measurement systems at several intervals during loading. The strain data was processed in order to identify the difference in the strain pattern between the "healthy" specimen BNDa and the "unhealthy" specimen BNDb which had damage in the form of delamination between skin and stiffeners. For illustration, consider Figure 6 which compares the strain data recorded during loading at two locations: SG12 near the debond (Figure 6a) and SG18 further away from the debond (Figure 6b). It is apparent from both plots that, at the beginning, when the load is below the critical buckling load, the debond has a very small effect on the strain data. This means that this particular debond (delamination) does not change the strain distribution pattern when the loads are below the critical buckling load.

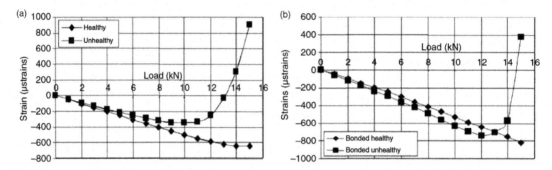

FIGURE 6   Comparative strain data in healthy and unhealthy bonded specimens: (a) strain measured at location SG12 near the debond; (b) strain measured at location SG18 further away from the debond [10].

However, as the load increased, the panel with debond started to buckle and the strains in the "unhealthy" specimen started to register deviations from those in the "healthy" specimen. These deviations become more and more intense as the loading progressed into the postbuckling behavior: they even changed sign from compression into tension. The effect of the sensor location can be estimated from comparing Figure 6a with Figure 6b: when the sensor is closer to the debond, the buckling due to the debond is sensed much earlier, e.g., at approximately 8 kN in Figure 6a. However, when the sensor is further away from the debond, the buckling due to debond is sensed only at a later stage, e.g., at approximately 12 kN in Figure 6b.

### 10.2.1.4 Delamination Occurrence Detection and Growth Monitoring with Specialty FBG Sensors

Reference [11] used specialty FBG sensors (chirped FBG) inserted in a cross-ply CFRP laminate to monitor the occurrence and growth of delamination under four-point bending. The changes in the optical spectrum of the chirped FBG were processed to yield the delamination occurrence and growth. It was observed that before the delamination was initiated from the crack tip, the spectrum had one broad peak. This shape of the spectrum was almost the same as that of the spectrum measured before the introduction of a crack because the influence of a transverse crack on the strain distributions of the FBG sensor was very small. As the delamination length increased, the long wavelength component shifted to the longer wavelength (Figure 7). These tendencies were investigated by both the measured and calculated spectra.

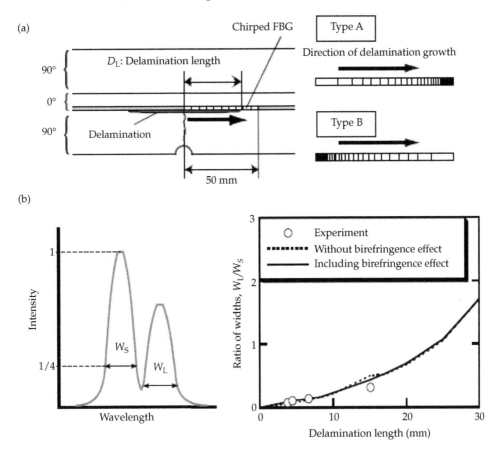

**FIGURE 7**    Chirped FBG sensor embedded in a cross-ply CFRP laminate to monitor occurrence and growth of a delamination under four-point bending: (a) location of sensor; (b) optical spectrum details used for quantifying delamination growth [11].

## 10.2.2 Composite Panel Buckling Monitoring

Reference [12] reports the monitoring of composite panel buckling with FBG sensors. The FBG sensors were used as strain sensors and buckling was detected by the change in the local strain measured by the FBG sensors. Three blade-stiffened CFRP panels with co-cured stiffener webs were tested under compressive load. Several FBG sensors were surface bonded on two of the stiffened panels and embedded into the stiffener webs of the third panel. The FBG sensors measure the strain distribution in the stiffener web and in the skin panels. The bucking onset was detected and the postbuckling behavior could be tracked.

## 10.3  ACOUSTIC EMISSION MONITORING

Reference [13] describes the use of the AE method to monitor the onset and propagation of delaminations in CFRP composites under quasi-static and fatigue loads.

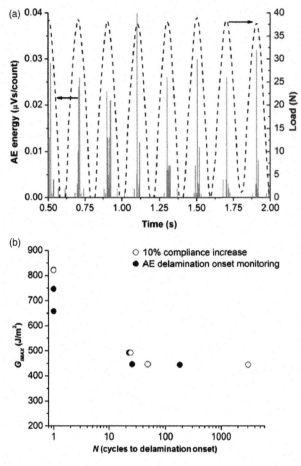

FIGURE 8   Composite delamination detection using the AE method during fatigue loading: (a) superposed plot of load variation during cyclic loading and energy of the AE hits; (b) AE detection of delamination onset that appears at various number of cycles depending on the corresponding strain energy release rate $G_{max}$ applied during the cyclic loading [13].

The experimental configurations used in specimen construction were (a) double cantilever beam (DCB); (b) mixed-mode bending (MMB); (c) end-notch flexure. AE data was collected with a MISTRAS-PAC Rk02 sensor using the PAC data acquisition and data processing systems. Fracture toughness testing was conducted in quasi-static regime, whereas fatigue delamination testing was conducted in cyclic loading of various intensities. It was found that delamination event can be correlated with peaks of AE "hits". Figure 8a presents a superposed plot of load variation during cyclic loading showing that maximum load points can be correlated with large AE energy values. Figure 8b plots the number of cycles where AE detection of delamination onset takes place for various values of the corresponding strain energy release rate $G_{max}$ applied during the cyclic loading. Also plotted in Figure 8b are the points where 10% compliance increase was observed during the cyclic testing. Note that compliance increase, aka stiffness loss, takes place when damage happens in a composite material.

# 10.4 SIMULTANEOUS MONITORING OF STRAIN AND ACOUSTIC EMISSION

Reference [14] describes the use of a dual-demodulator FBG system to measure both the strain and the AE signal emanating from the damage created in a cross-ply laminated composite under tensile loading. One FBG sensor, one electrical strain gage (ESG), and one piezo sensor (PZT in Figure 9) were mounted on a CFRP cross-ply composite specimen (Figure 9a,b). The dual demodulation was achieved with a tunable Fabry−Perot filter (TFPF) for quasi-static strain and a Mach−Zehnder interferometer (MZI) for strain waves.

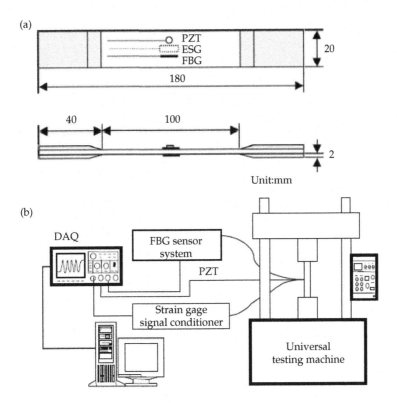

**FIGURE 9**   Simultaneous measurement of static strain and strain waves in a cross-ply composite using FBG sensors and dual-demodulator system: (a) specimen instrumented with FBG sensor, PZT sensor, and electrical strain gage (ESG); (b) experimental setup schematics; (c) quasi-static strain measured with the TFPF-demodulated FBG sensor signal; (d) strain wave signals measured with the MZI-demodulated FBG sensor compared with that measured by the PZT sensor [14].

The quasi-static strain and the strain wave signals measured by the FBG sensor showed that sudden strain shifts accompanied by high-frequency strain waves in the hundreds of kHz range were observed when transverse cracks propagate in the transverse layer of the cross-ply composite (Figure 9c,d). Note that the FBG signal in Figure 9d is proportional to the dynamic strain, whereas the PZT signal is proportional to the strain rate.

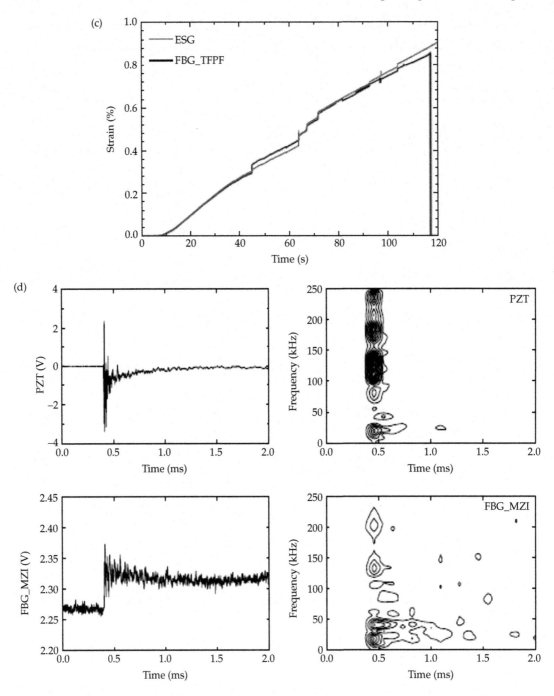

**FIGURE 9**   (Continued)

## 10.5 FATIGUE DAMAGE MONITORING

Composite structures are typically resistant to through-the-thickness cracks due to the inherent crack resistance of fiber reinforcement. However, in layered composite structures, cracks can easily propagate parallel to the wall surface, typically at the interface between layers. These cracks can be initiated by fabrication imperfections or low-velocity impact damage; subsequently, they propagated by cyclic fatigue loading. Even without a low-velocity impact initiation, cracks can initiate and propagate through the composite materials under the repeated loading occurring during normal operation. Matrix microcracks in the cross-ply locations are the first to appear, but interfacial cracks and delamination may also occur after exposure to a large number of fatigue cycles.

### 10.5.1 Fiber-Optic Monitoring of Transverse Cracks in Cross-ply Composites

In an early paper [15], Tanaka's group at the University of Tokyo, Japan, discussed the detection of transverse cracks in CFRP composites using embedded FBG sensors (Figure 10a). An uncoated optical fiber containing 10-mm-long FBG sensor was laid along the loading direction at the bottom of the longitudinal ($0°$) ply (Figure 10b). The sensing principle was based on the fact that, by removing the cladding and immersing the stripped region in the external medium represented by the matrix, it is possible to monitor the change in the matrix refractive index variation due to appearance of internal microcrack.

The strain was measured both by the FBG sensor and by a conventional resistance strain gage. The specimen was loaded in a tensile machine. Good agreement between the strain values measured by FBG sensors (using the highest peak in its spectrum) and conventional strain gage was observed (Figure 10c). Transverse cracks appeared in the $90°$ plies as the longitudinal strain was increased. During various loading stages of the experiment, the position and the number of the transverse cracks in the specimen were determined using an edge replica technique. It was found that the transverse crack density increases with longitudinal strain, as expected. As the crack density increased under increasing load, it was noted that the reflected FBG spectrum became distorted, besides being shifted. This phenomenon was explained through a detailed analysis of the modified strain distribution due to transverse cracks presence. It was concluded using simulation studies that the distortion of the spectrum can be correlated with the crack density, which could offer a baseline-free method for transverse cracks detection.

The spectrum width at the half-maximum was proposed as a quantitative indicator of transverse crack density. However, it was noted in practice that the residual thermal stresses resulting from the initial CFRP processing may pose a problem because the spectrum had already split into two peaks and became much broader before the tensile load was even applied. Hence, this proposed baseline-free method could be used only if the FBG sensor could be embedded without the non-axisymmetric residual thermal stresses being present. Nonetheless, the direct measurement of actual strain with the FBG sensor can be used as a baseline-based method of inferring the crack density from the strain

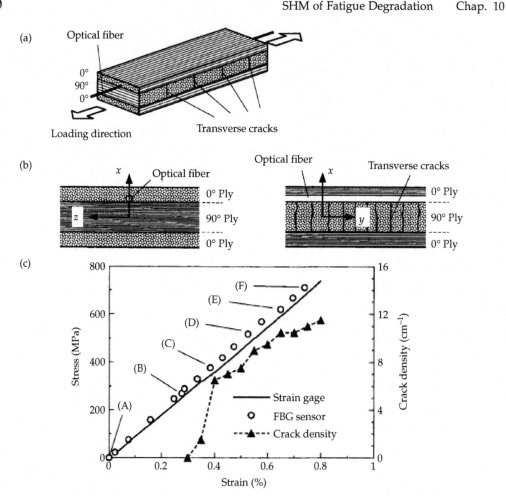

**FIGURE 10**  Detection of transverse cracks in a cross-ply CFRP specimen with fiber-optic sensors: (a) general view of the specimen; (b) placement of the optical fiber inside the specimen; (c) longitudinal strain measured by FBG sensor and conventional strain gage and density of transverse cracks in a typical cross-ply CFRP specimen [15].

values, although this would have to rely on an initial baseline measurement of the crack-free state before loading. This topic was further studied on by the same team to assess the effect of coating [16] and to test specialty small-diameter FBG sensors with 52 µm outside diameter [17].

## 10.5.2 Pitch-Catch Guided-Wave Detection of Fatigue Microcracking and Delamination

Larrosa et al. [18] studied detection and estimation of composite fatigue damage in the form of matrix microcracks and delamination. Dog-bone composite specimens of various layups were subjected to tension fatigue loading ($R \sim 0.14$, 5 Hz). The maximum load was approximately 85% of the static failure load. It was estimated that the strain in the specimen was 0.3–0.4%. Figure 11 shows the specimen geometry and instrumentation. The specimen had a waisted geometry to avoid failure in the gripping ends. A stress-concentration notch was machined on one side of the specimen midway between the grips. The pitch-catch instrumentation consisted of two Acellent Technologies SMART Layer™ strips placed toward the specimen ends.

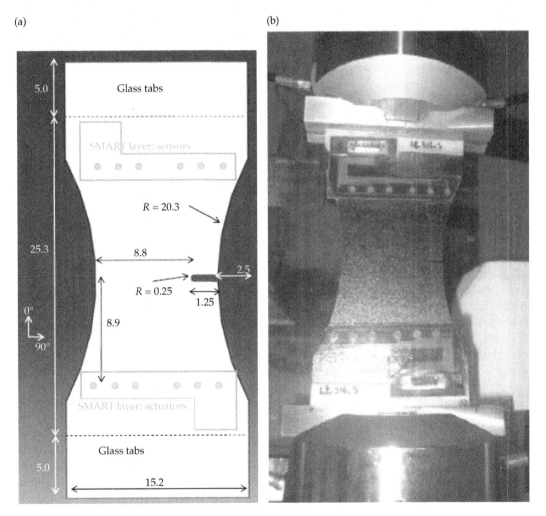

**FIGURE 11**  Composite specimen subjected to tension fatigue and monitored with pitch-catch guided waves: (a) specimen geometry (cm) and SMART Layer™ actuators and sensors locations; (b) actual photo of the specimen mounted in an MTS testing machine [18].

The fatigue loading was stopped periodically and the specimens were unloaded from the MTS machine in order to take measurements. The SMART Layer™ actuator sensors were used for guided-wave pitch-catch measurements. A Faxitron X-ray cabinet was also used to obtain X-ray images of the specimen. Dye penetrant was used to enhance the X-ray absorption. The X-ray images were analyzed: for each pitch-catch diagnosis path, the matrix microcracks were counted and delaminations were identified. A label of 0 was given to a diagnostic path that experienced only microcracks, whereas a label of 1 was given to a path that experienced delamination.

Typical experimental results are shown in Figure 12. An X-ray image after 400,000 cycles is shown in Figure 12a: the 23° path is showing matrix microcrack whereas the 0° path is showing delamination near the stress-concentration notch. A superpose plot of the diagnostic signals received after increasing levels of fatigue exposure is shown in Figure 12b. It can be observed that at the beginning of the fatigue testing when only matrix microcracks appeared, the changes in the received signal were not very large: slight shifts in the signal phase and slight decrease in the signal amplitude were observed. These shifts and amplitude changes were more pronounced as the number of microcracks grew. Later in the fatigue testing, when the specimen delaminated, the changes in the signal were much more pronounced, both in terms of the phase shift and in terms of the amplitude decrease.

The large amount of data collected during these experiments was processed as follows: first, all the data from the 11 composite specimens were preprocessed and merged into one data set. This data set contained 5141 data points with approximately 70% being matrix microcracking only and approximately 30% being delamination. Next, a Gaussian discriminant analysis (GDA) machine learning technique was combined with signal feature extraction methods to classify the recorded signals and identify the various composite fatigue damage types, i.e., matrix microcracking versus delamination. It was concluded that the use of the GDA algorithm in the tested sample of specimens yielded a 21% classification error, 71% precision, and 68% recall. Further studies are needed to assess how these error rates affect damage diagnostic algorithms in which per-path data provided by the classifier are used to estimate damage location and intensity.

A related study [19] was focused on developing numerical simulation models to predict the signal changes as function of various types of composite damage, e.g., matrix microcracks and delamination. The same specimen construction and instrumentation as presented in Figure 11 was used. ABAQUS multiphysics FEM analysis was performed. Convergence analysis and element tuning were applied to obtain a trustworthy simulation. The delamination was modeled by releasing the nodes of adjacent plies over a specified area. Matrix microcracks were modeled in as very thin (0.1 mm) slits. A sensitivity analysis of the received diagnostic signals with composite damage intensity was also attempted.

In a separate study, Ref. [20] compared the use of various signal analysis methods to monitor the damage initiation and progression in a composite wind turbine rotor blade during fatigue testing. A piezoelectric wafer active sensor (PWAS) sparse array was placed near the rotor blade root where the fatigue bending loads were deemed to be most intense. The pitch-catch method was applied with a centrally located PWAS acting as actuator and several other PWAS acting as receivers. Both ultrasonic guided waves (UGW) and diffused wave field (DWF) measurements, aka insonofication, were used.

(a)

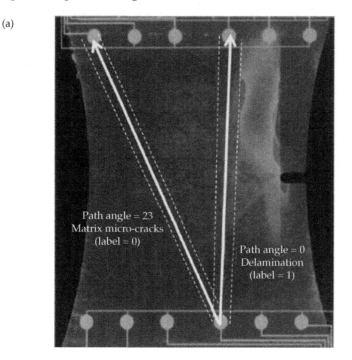

(b)

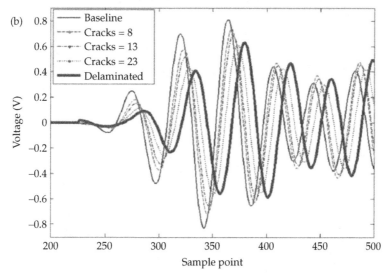

**FIGURE 12**  Experimental results for composite specimen subjected to tension fatigue and monitored with pitch-catch guided waves: (a) typical X-ray image showing matrix microcracks and delamination; (b) captured raw signals for various numbers of microcracks and after the apparition of the delamination [18].

### 10.5.3 ECIS Monitoring of Composites Fatigue Damage

Reference [21] used the ECIS technique to monitor woven GFRP composites during cyclic fatigue loading in flexure. Woven GFRP composite specimens were tested in cyclic flexure and the ECIS readings were taken at various stages during the fatigue life. The ECIS testing was done inside an environmental chamber (Figure 13).

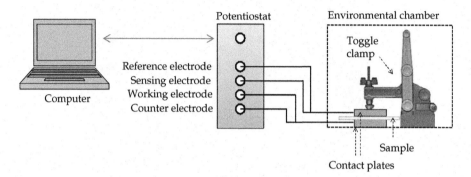

**FIGURE 13**   Experimental setup inside an environmental chamber for ECIS fatigue damage detection in GFRP composites [21].

The experimental setup for ECIS testing inside an environmental chamber using a potentiostat controlled by a PC is shown in Figure 14. The environmental chamber was used to precisely control the level of moisture in the ambient air surrounding the specimen since moisture permeates the damaged specimens to create conduction paths as damage develops. Several levels of ambient moisture and the presence of conductive ions (e.g., salt solutions) were studied and some multiphysics modeling of the moisture ingress was used to determine the best operating conditions. This is important for the practical implementation of this method since moisture condensation on the interior of airframes is a nearly ubiquitous condition which has served as one of the drivers for the adoption of composite materials, at least in transport aircraft. How to quantize the level of relative humidity, the conductivity of the moisture present, and the time for diffusion in a particular composite material is a question to be addressed by further research.

Damage began with matrix cracking, which had limited extent in the case of loading along fiber directions but was more extensive for loading along directions of 30° to 45° from the fiber direction. The pattern of matrix cracking is associated with tow geometry of the fiber weave pattern. Fracture of the specimen occurs when the damage propagates through the thickness of the specimens to create a continuous fracture path that crosses the fibers. The process of progressing damage and fracture path creation in this insulating heterogeneous material is related to a process of conduction paths creation due to moisture infiltration during damage development.

The measured data is presented in Figure 14. Each curve represents an $x-y$ plot of the Re–Im parts of the impedance for various frequency values in the 1kHz to 10kHz range. Low excitation frequencies (i.e., down to 1kHz) produce data to the right of Figure 14b,

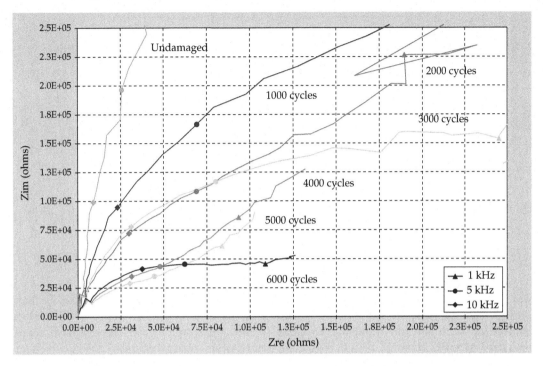

**FIGURE 14**    ECIS Nyquist plot of Re−Im impedance curves at various stages during low-cycle flexure fatigue testing of woven GFRP (specimen failure occurred at ∼7000 cycles) [21].

and high excitation frequencies (i.e., up to 10kHz) produce the data on the left-hand side of Figure 14b and toward the origin of the graph. It is apparent that the undamaged specimen is effectively a dielectric and its impedance is dominated by the capacitive response with the reactance $X_C(\omega) = 1/i\omega C$. In this case, the real part of the impedance represents the internal energy-dissipation losses during cyclic excitation.

As the frequency $\omega$ becomes higher, the capacitive reactance diminishes and the curve aims toward the axes origin. As damage progresses and conductive paths are created in the damage material, the relative proportion between the real and imaginary parts of the impedance changes. The greatest difference in Figure 14b data seems to be between the results for the undamaged specimen and the results for the specimen cycled to only 1000cycles, which represent the first 14% or so of the specimen life. As damage progresses, the lower frequency imaginary component of the impedance (to the right of Figure 14b) drops sharply in magnitude, indicating a sharp decrease in capacitive reactance.

It is apparent that the ECIS method is remarkably sensitive to the details of internal damage development in woven GFRP composites; in addition, the actual progress of the degradation in terms of life and remaining strength can be related to ECIS data. Figure 15 shows a comparison of different ways to quantify damage development in terms of loading cycles: (a) initial slope of the Nyquist plot; (b) impedance magnitude at a fixed frequency, say at 1kHz; (c) remaining strain to break in tension. Averaged multi-specimen

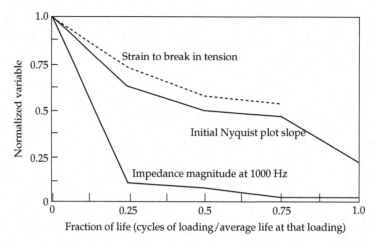

FIGURE 15   Comparison of normalized progressive damage accumulation as a function of the fraction of life of the tested specimens [21].

data recorded at 25%, 50%, 75%, and 100% of life are shown. The two selected ECIS characteristics (initial Nyquist slope and impedance magnitude at fixed frequency) show clear relationship with the fractional life variable; in fact, they seem good damage indicators, especially of early damage onset. However, many other characteristics could have been selected in the robust ECIS data set.

## 10.6  MONITORING OF IN-SERVICE DEGRADATION AND FATIGUE WITH THE ELECTRICAL RESISTANCE METHOD

Active sensing SHM consists of "interrogating" the structure with wave transmitters and picking up the structural response with wave receivers. The waves used in active SHM may be elastic or electromagnetic. The electrical SHM of composites relies on the material itself to act as sensor. Carbon fibers are electrically conductive; the epoxy resin is an insulator. The CFRP composite is somehow conductive because the densely packed carbon fibers may touch each other. For a CFRP lamina, representative values are $\sigma_0 = 32,000S$, $\sigma_{90} \sim 10S$, $\sigma_t \sim 8S$. As damage (e.g., cracks and delamination) takes place in the composite, the electric conductivity is expected to change.

The GFRP composite is a nonconductive insulator with certain dielectric properties. Damage in GFRP composites creates microcracks and even sizeable delaminations, which may change the dielectric properties of the composite since the dielectric permittivity of air ($\varepsilon_{air} \approx 1$) is smaller than that of GFRP ($\varepsilon_{GFRP} \approx 3$). Ingress of water, which has a dielectric permittivity higher than GFRP ($\varepsilon_{water} \approx 81$), will change again the overall dielectric properties of GFRP composites.

These intuitive concepts stand at the foundation of the electrical SHM methods for composite materials. This approach is deemed "self-sensing" because it relies entirely on measuring a material property (i.e., electrical characteristic) and does not require an additional transduction sensor; the only instrumentation that needs to be installed on the composite structure consists of the electrodes. In the case of composite transport aircraft, the conductive screen skins currently used to mitigate lightning strike could potentially also serve as the measuring electrodes. Electrical SHM methods range from the simple measurement of the electrical resistance measurements up to more sophisticated methods such as electrical potential mapping, dielectric measurement, electrochemical impedance, etc.

## 10.6.1 Fundamentals of the Electrical Resistance Method

The electrical resistance method may be the simplest and most intuitive technique among the electrical composite SHM method. Chung [22] gives an early review of how the DC electrical resistance method can be applied to the SHM of CFRP composites. Within a unidirectional carbon fiber lamina, the electrical conductivity is highest in the longitudinal direction (i.e., along the fiber tows); the conductivity in the transverse direction (i.e., across the fiber tows) is nonzero, even though the polymer matrix is insulating; the nonzero transverse conductivity is due to contacts between fibers of adjacent tows (Figure 16). In other words, a fraction of the fibers of one tow touches a fraction of the fibers of an adjacent tow here and there along the length of the fibers. These contacts result from the fact that fibers are not perfectly straight or parallel (even though the lamina is said to be unidirectional), and that the flow of the polymer matrix (or resin) during composite fabrication can cause a fiber to be not completely covered by the polymer or resin (even though, prior to composite fabrication, each fiber may be completely covered by the polymer or resin, as in the case of a prepreg, which is a sheet or tape of fibers impregnated with the polymer resin).

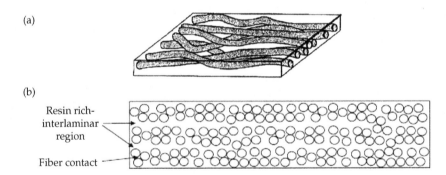

**FIGURE 16**    Explanation schematic of electrical resistance effects in CFRP composites: (a) fiber waviness; (b) nonuniform fiber distribution [23].

Thus, the transverse conductivity may depend on the average number of fiber–fiber contacts in the lamina. Similarly, the contacts between fibers of adjacent laminae cause the through-the-thickness conductivity to be also nonzero. Thus, the through-the-thickness conductivity may be related to the number of fiber–fiber contacts between adjacent laminae.

## 10.6.2 Electrical Resistance SHM of CFRP Composites

When a unidirectional CFRP composite specimen is stretched, its longitudinal resistance decreases. This phenomenon (which may be due to both a transverse Poisson effect and carbon fiber piezo resistivity) is reversible and the resistance returns to its initial value as the load is released (Figure 17).

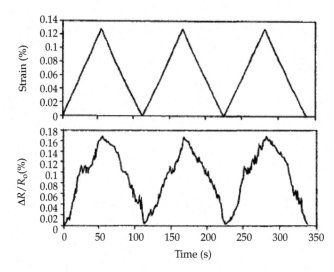

**FIGURE 17**   Variation of longitudinal electrical resistance with strain in a unidirectional CFRP composite specimen [22].

However, if damage occurs, irreversible resistance changes may occur. When transverse matrix cracking takes place, the transverse conductivity decreases; this may be due to decrease in the number of fiber–fiber contacts in the lamina. Similarly, matrix cracking between adjacent laminae (e.g., disbonds or delaminations) decreases the number of fiber–fiber contacts between adjacent laminae, thus decreasing the through-the-thickness conductivity. Thus transverse conductivity changes may be indicative of matrix cracking and disbonds or delaminations.

Fiber damage (i.e., small defects appearing in the fiber prior to actual fiber fracture) decreases somehow the fiber conductivity and hence the longitudinal conductivity of the lamina. When actual fiber fracture takes place, the decrease in longitudinal conductivity is more severe since, for that fiber, we now have an open circuit, i.e., zero conductivity. However, a lamina has a large number of fibers and adjacent fibers can make contact here and there. Therefore, the separate portions of a broken fiber still contribute to the longitudinal conductivity of the lamina. As a result, the decrease in conductivity due to fiber fracture is less than what it would be if a broken fiber did not contribute to the conductivity. Nevertheless, the effect of fiber fracture on the longitudinal conductivity is significant, so that the longitudinal conductivity changes can be indicative of fiber fracture damage. The through-the-thickness resistance of a laminate is the sum of the volume resistance of each of the laminae in the through-the-thickness direction and the contact resistance of each of the interfaces between adjacent laminae (i.e., the interlaminar interface). For example, a laminate with eight laminae has eight volume resistances and seven contact resistances, all in the through-the-thickness direction.

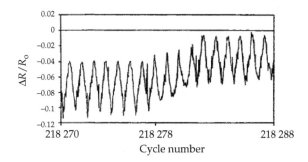

**FIGURE 18**   Variation of longitudinal electrical resistance during tension–tension fatigue testing of a unidirectional CFRP composite specimen. The cyclic change of resistance is due to reversible cyclic strain; the permanent increase in resistance around cycle #218,281 may be due to permanent damage, e.g., some fiber fracture [22].

Electrical resistance measurements have also been shown capable of detecting fatigue damage in CFRP composites. Figure 18 shows the variation of longitudinal electrical resistance during tension–tension fatigue testing of a unidirectional CFRP composite specimen. The reversible cyclic changes of resistance are due to the reversible cyclic strains. The permanent increase in resistance at cycle #218,281 may be due to permanent damage, e.g., matrix and some fiber fracture in the specimen. Reference [24] correlated the appearance of residual electrical resistance changes with optical microscope observation on the side of a cross-ply specimen. It was concluded that residual electrical resistance change became elevated just after the onset of matrix cracking at 0.25% strain.

Reference [25] describes the use of the electrical resistance technique to monitor crack growth in beam specimens under mode I (DCB) and mode II end-notch flexure (ENF) testing (Figure 19).

Unidirectional CFRP 12-ply beam specimens were fabricated with initial cracks (Teflon tape insertion) in the mid plane at one end. AE sensors were also used for comparison. Crack growth was monitored by observing the side of the specimen with an optical microscope. The electrical resistance method was found to be comparative in accuracy with the AE method for mode I DCB testing (Figure 20), but inferior to AE for mode II ENF testing [25].

Reference [26] proposes the concept of addressable conducting network (ACN) applied to the surface of CFRP composites for self-sensing with the electrical resistance method. To reduce the variation of the electrical network due to the laminate's configuration, a number of conducting lines are integrated on the top and bottom surfaces to make a grid arrangement (Figure 21). Monitoring the through-the-thickness resistance changes obtained from the pairs of conducting lines would locate the damage without the need of knowing the laminate's stacking sequences.

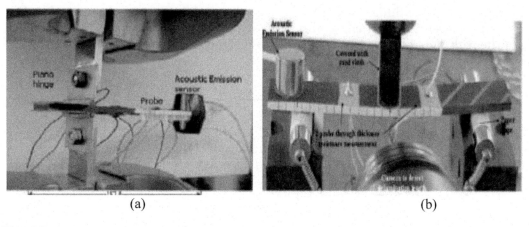

**FIGURE 19** Fracture crack propagation monitoring with electrical resistance and AE methods: (a) mode I fracture DCB setup; (b) mode II fracture end-notch flexing setup [25].

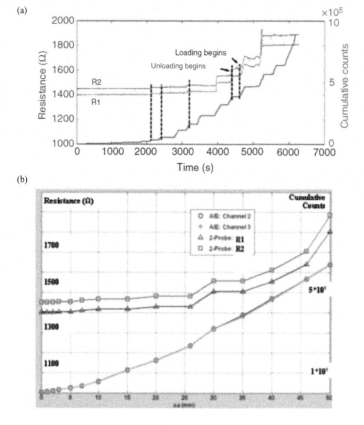

**FIGURE 20** Comparison between electrical resistance measurements (R1, R2) and AE cumulative counts during mode I testing of CFRP specimens: (a) resistance changes and AE cumulative counts versus testing time; (b) resistance changes and AE cumulative counts versus crack length [25].

**FIGURE 21**  Proposed ACN concept for monitoring the structural CFRP composite with the electrical resistance method [26].

Experiments performed on CFRP cross-ply specimens subjected to three-point bending revealed correlation between slight changes in the slope of the load–deflection curve and changes in the electrical resistance behavior indicative of damage taking place inside the specimen (Figure 22).

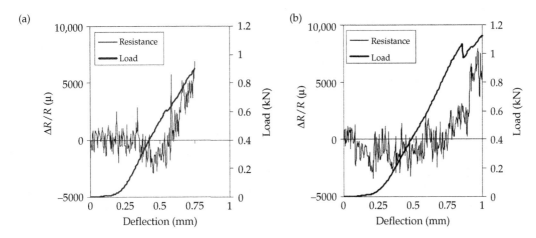

**FIGURE 22**  Correlation between electrical resistance behavior and changes in the load–displacement curve indicative of damage detection during three-point bending of CFRP cross-ply specimens: (a) loading up to 0.75mm deflection showing a possible internal damage event at around 0.60mm deflection; (b) reloading up to 1.00mm showing a new internal damage event at around 0.80mm accompanied by a loud noise [26].

During loading up to 0.75 mm, a change in the deflection slope indicated a possible internal damage event at around 0.60 mm deflection. This was accompanied by a sudden change in the behavior of the electrical resistance which started an upward trend (Figure 22a). After reaching 0.75 mm deflection, the specimen was unloaded and then reloaded monotonically up to 1.00 mm (Figure 22b); the electrical resistance did not change much up to the previously reached level of 0.75 mm because the creation of new damage requires an increase of the load beyond the previously reached level. When the mid-span deflection exceeded

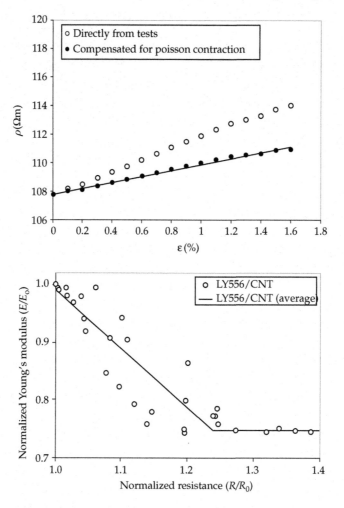

**FIGURE 23** Effect of 0.5% CNT doping: (a) change in resistivity of CNT doped resin specimens showing piezo resistivity after the elimination of the Poisson contraction effect; (b) correlation between damage (indicated by the decrease in Young modulus) and resistance change in a CNT doped cross-ply GFRP specimen [28].

0.8 mm, the load dropped sharply and a loud noise was heard from the specimen; it was presumed that a new internal damage event had taken place at around 0.80 mm as indicated by the sudden change in load–displacement curve and the loud noise. After the test, no visible damage or plastic deformation could be seen from outside, thus indicating the sensitivity of the electrical resistance method in detecting hidden internal structural damage [26]. The experimental tests were also compared with multiphysics FEM results.

## 10.6.3 Electrical Resistance SHM of CNT Doped GFRP Composites

Some recent studies [27,28] have addressed the improvement of the electrical resistance method by doping the composite matrix with conductive carbon nanotubes (CNT).

Polymers, which normally are insulators, become conductive even if very small amounts of nanotubes are added, i.e., the percolation threshold is attained for very low fractions of CNT ($\leq$1%). Reference [27] studied unidirectional CFRP composite with various fractions of CNT doping (0%, 0.1%, 1.0%). It was found that the transverse and through-the-thickness resistivity was clearly decreasing with CNT doping: it almost halved for 0.1% CNT doping and it further decreased with increased doping until reaching between 1/4 and 1/3 for 1.0% CNT doping. In the same time, the characteristic low longitudinal resistivity of the CFRP remained unaffected. It was concluded that addition of even very small amounts of CNT doping can significantly increase the sensitivity of the electrical resistance method, especially in detecting matrix cracking damage.

Reference [28] performed studies on neat epoxy resin doped with 0.5% CNT and found that resistivity changes with strain; besides the geometric Poisson effect, these changes also exhibited a piezo-resistive effect (Figure 23a). Further studies were done on GFRP cross-ply specimens fabricated with the 0.5% CNT doped resin. It was found that the CNT doping permits the use of the electrical resistance method to monitor damage evolution in GFRP composites, which was not possible in conventional GFRP composites because both their constituents (the glass fiber and the resin) are insulators. With the CNT doped specimen, it was possible to correlate damage (measured in terms of stiffness reduction) with electrical increase (Figure 23b). The interpretation of the results presented in Figure 23b is that the initial stiffness decrease is a consequence of transverse cracks damage. These cracks contribute to breaking up the electrical pathways within the resistive percolated network, thus resulting in increased electrical resistance. The electrical resistance increases with the increase of transverse cracks damage. Apart from transverse cracking, there is also another type of damage that occurs at higher load values, i.e., the delamination between the 90° and the 0° plies. This damage also reduces the electrical paths and increases the resistance, but does not affect the axial stiffness significantly. This latter damage type explains the last part of the curve in Figure 23b where the electrical resistance increases but no stiffness reduction occurs.

## 10.6.4 Wireless Sensing Using the Electrical Resistance Method

The electrical resistance method for monitoring composite materials can also be done wirelessly since it is similar in instrumentation to the electrical resistance strain-gage method. Low-sampling-rate wireless strain and temperature sensors have become relatively widespread. Reference [29] describes the wireless detection of strain changes due to internal delaminations in CFRP composites using the electrical resistance change method and a wireless bridge that encodes the resistance change as a frequency shift. This system was further refined in Ref. [30] to permit the measurement of the much smaller operational load strains. Reference [30] also extended the work of Ref. [29] from a single sensor to an assembly of time-synchronized wireless sensors such that multiple regions on the composite structure can be remotely monitored simultaneously, as already described in Chapter 7.

These wireless sensors were installed on CFRP specimens and loaded in a mechanical testing system. Normal loading and unloading of the specimen produces reversible strain and reversible change in the electrical resistance, which is encoded as frequency shifts in the wireless transmission. Figure 24a shows the reversible loading and unloading curves measured at three different positions on the specimen with three different wireless sensors, each with a different encoding frequency. These results clarify that the frequency change ratios due to the tensile strain are approximately the same for all sensors although the encoding basic frequencies are different for each sensor; this was possible because the relative shifts depend only on the relative electrical resistance change.

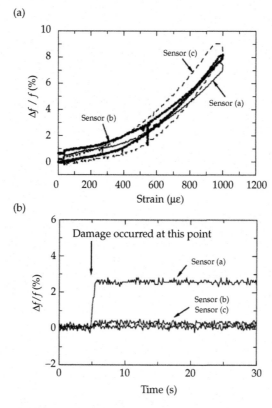

**FIGURE 24**   Use of wireless system to remotely detect damage in CFRP composite with the resistance change method. The sensor signal is encoded as frequency change: (a) reading curve during reversible loading and unloading of the specimen; (b) reading curve detecting the appearance of damage [30].

Figure 24b shows the relative frequency shifts measured using three sensors connected at different location of the same specimen subjected to tensile load. When a fiber breakage occurs in the section corresponding to the wireless electrical resistance sensor (a), a sudden jump is noticed in its relative frequency shift. The fiber breakage occurs at a time of around 5s. At the same time, the frequencies of the sensors (b) and (c) do not change because they are located outside the damage area. This shows that the fiber breakage can be successfully measured by the sensor at location (a) close to the breakage; it also showed that the other sensors located in areas outside the fiber breakage were not affected. It was concluded that the proposed system could successfully measure with time synchronization the applied strain and the damage at multiple locations on a CFRP specimen.

# 10.7 DISBONDS AND DELAMINATION DETECTION AND MONITORING

Bonded joints have a wide area of potential applications. However, the fact that the long-term durability and reliability of bonded joints is not fully understood slows down their implementation in critical applications. For this reason, the development of methods to perform reliably *in situ* SHM of bonded joints has received extensive attention.

In conventional nondestructive evaluation (NDE), composite cracks and delaminations are detected with ultrasonic probes that can sense additional echoes due to through-the-thickness P waves being reflected by the delamination. Area coverage is achieved with surface scanning (C scans) using manual means or mechanical gantries. The aim of composites SHM is to detect cracks and delaminations in composite structures using guided waves transmitted from one location and received at a different $(x, y)$ location. The disbond/delamination produces signal diffraction and mode conversion that can be analyzed and compared with the pristine signals. Analysis of the change in the guided-wave shape, phase, and amplitude should yield indications about the crack presence and its extension. In addition, the sensor network built into the structure can also be used for monitoring low-velocity impact events that may be the cause of composite damage.

## 10.7.1 Disbond and Delamination Detection with Conventional Ultrasonics Guided Waves

Ultrasonic testing of adhesively bonded joints using guided waves for both aerospace and automotive applications is gaining more and more attention. In the NDE of adhesively bonded joints of particular interest are the Lamb waves. Lamb-wave methods have considerable potential for the inspection of adherent assemblies for two reasons: they do not require direct access to the bond region, and they are much more amenable to rapid scanning than are pressure wave techniques. Lamb waves can be excited in one plate of a bonded assembly, propagated across the joint region, and received in the second plate of the assembly. Inspection of the joint would then be based on the differences between the signals received on one side of the assembly compared to those transmitted on the other side.

Rose et al. [31] used the UGW for NDE of adhesively bonded structures. They developed a double spring hopping probe (DHSP) to introduce and receive Lamb waves. This method was used to inspect a lap splice joint of a Boeing 737-222. Preliminary results showed the method's capability of disbond detection. Severe corrosion area was also pointed out using the DHSP handheld device.

## 10.7.2 Monitoring of Composite Patch Repairs

Repair patches are widely used in the aircraft industry for small repairs of the aircraft fuselage in order to extend the operational life of aging aircrafts. Lamb waves can be successfully used to detect disbonds of composite repair patches [32].

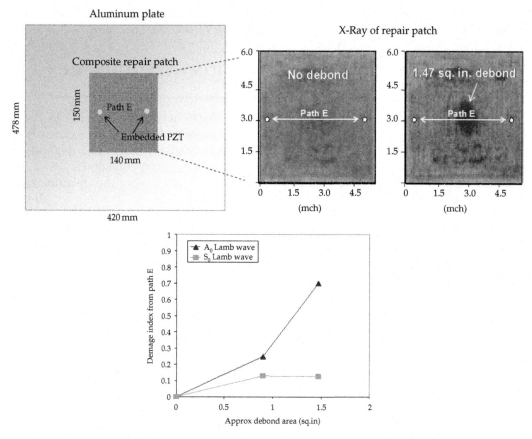

**FIGURE 25**   Embedded pitch-catch NDE for disbond detection of composite repairs of metallic structures [33].

References [33] and [34] report the use of the pitch-catch technique with SMART Layer™ instrumentation to detect disbonds in boron–epoxy composite patch repair of cracks in metallic structures (Figure 25). A SMART Layer™ was embedded between the composite repair layers. A0 Lamb waves, which are more sensitive to disbond and delaminations, were used. The A0 Lamb-wave frequency was 320 kHz. It was found that the damage index based on A0 scatter energy was able to correctly detect the disbond presence.

## 10.7.3  Monitoring of Composite Adhesive Joints

Reference [35] considered the example of an unmanned air vehicle constructed with extensive use of adhesively bonded joints. In particular, they studied the critical adhesive connection of the composite wing skin to the wing spar. Two approaches for guided-wave inspection of bonds were considered: (a) "across the bond" in which the waves are generated in the adherent on one side of the bond and received across the bond and (b) "within the bond" in which the waves are both generated and detected in the bond region (Figure 26a). The former approach was extensively discussed by Ref. [36] in relation to a lap joint of two metal plates (Figure 26b), in which the guided-wave leakage from one adherent into the other adherent through the bonded overlap is considered. The wave propagation is complicated by the mode conversions of the Lamb wave entering and leaving the bonded overlap. This happens because the dispersive behavior of a single plate is different from that of the multilayered overlap.

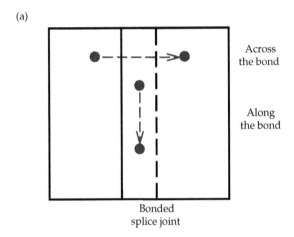

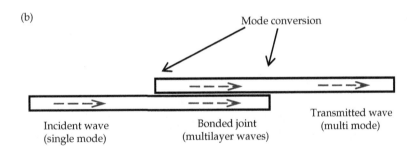

**FIGURE 26** Embedded pitch-catch method for disbond detection: (a) "across the bond" and "within the bond" directions [35]; (b) wave propagation schematics [36].

In the case of lap–shear joint, the dispersive behavior in the bonded overlap is modeled as a multilayer wave propagation problem for the three-bond states by using matrix-based methods. The mode conversion in the bonded overlap is a consequence of the different dispersive behavior of a single plate (the adherents) when compared with that of a multi-layer structure (the bonded overlap). Reference [36] investigated three different interfaces between two aluminum plates: a fully cured epoxy layer, a poorly cured epoxy layer, and a slip interface layer. The tests were based on the measurement of the acoustic energy transferred from one adherent to the other one through the adhesive layer. An efficient energy transfer through the bondline is an indication of a good bond. A disbond would cause a dramatic decrease in transferred energy, whereas a poorly cured adhesive will result in less severe attenuation. The authors demonstrated that the Lamb-wave modes with predominant in-plane displacements at the adherent outer edges are most sensitive to the presence of poorly cured adhesive, as the latter does not support shear-type wave propagation efficiently.

## 10.7.4 Dielectrical SHM of Delamination and Water Seapage in GFRP Composites

Since GFRP composites are non-conducting dielectrics, their electrical SHM is done with frequency domain methods. Reference [37] measured the changes in the dielectric constant of GFRP composites due to delamination and water seepage. The method rests on the different values of the relative permittivity of air ($\varepsilon_{air} \approx 1$), water ($\varepsilon_{water} \approx 81$) in comparison with that of GFRP composite ($\varepsilon_{GFRP} = 3.12$). GFRP plate specimens (12.5 mm thick) were fabricated with built-in delaminations (thin pockets) placed at different depths (2.0; 3.5; 4.0; 5.5; 6.0; 7.5 mm). The pockets could be filled with water and sealed. The measurements are done with a capacitance meter connected between two surface applied electrodes. Sensing of delamination depth was explored for both dry and wet pockets. It was found that the presence of delaminations (both dry and wet) could be detected well when they are close to the surface, but less well as their thickness position is more and more removed from the specimen surface.

Multiphysics FEM modeling was performed to optimize the surface electrodes design and compare measurement predictions with the experimental results. It was found that the surface measured capacitance increases with the depth of the delamination location for dry delaminations but decreases with depth for wet delaminations (Figure 27).

This result is to be expected: in the former situation (Figure 27a), the measured capacitance is reduced when the delamination is closer to the surface because the dielectric permittivity of the air is less than that of GFRP composite; this reduction diminishes as the delamination gets further away from the surface and the overall capacitance increases. In the latter situation, the measured capacitance is increased when the delamination is closer to the surface because the dielectric permittivity of the water is higher than that of GFRP composite; this increase diminishes as the delamination gets further away from the surface and thus the total capacitance decreases (Figure 27b).

The sensor sensitivity in detecting the presence of delamination was calculated from the relative capacitance change ratio (Figure 28).

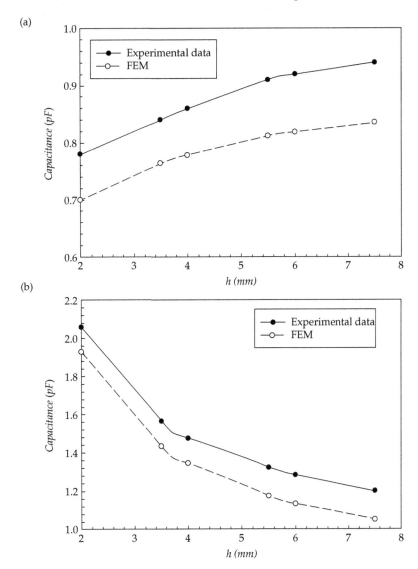

**FIGURE 27** Variation of measured capacitance with delamination depth: (a) dry delaminations; (b) wet delaminations, i.e., filled with water [37].

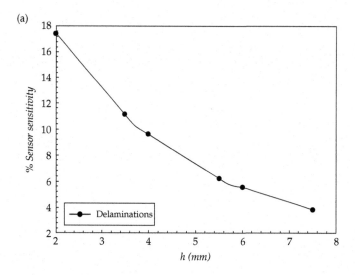

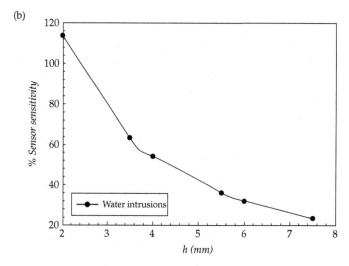

**FIGURE 28**  Variation of dielectric sensor sensitivity with delamination depth: (a) dry delaminations; (b) wet delaminations, i.e., filled with water [27].

## 10.8  SUMMARY, CONCLUSIONS, AND SUGGESTIONS FOR FURTHER WORK

This chapter has presented SHM methods and results related to the damage detection and monitoring that appear during normal operational service of aerospace composites. Such damage may be due to operational loads and environmental factors and may result in a gradual degradation of composite properties rather than the sudden changes that may appear due to accidental events, like the impact damage discussed in the previous chapter.

This chapter started with a discussion of passive SHM methods, e.g., monitoring strain, acoustic emission (AE), operational loads, etc. The aim of these passive methods is to record what is actually happening with the composite structure and act accordingly. For example, monitoring the loads applied to the composite structure during normal operation conditions allows one a comparison with the design loads (both limit loads and fatigue spectra) in order to make adjustments in the operational life predictions and flight profiles. Another outcome of passive monitoring could be the identification of changes in the strain distribution pattern over the structure that might be indicative of partial local failures (e.g., microcracks and delaminations) that may affect the load paths in a fail-safe structural design. In addition, AE monitoring could detect when actual local failures happen by recording the waves created by the elastic energy released when by a "popping" crack.

The next major section of this chapter dealt with monitoring the actual fatigue damage induced in the composite by repeated application of service loads. As already detailed in a previous chapter, fatigue damage in composites is substantially different from fatigue damage in metallic structures. Most results reported in this section are connected with the monitoring of matrix microcracking and of the subsequent delamination that appears during cyclic loading tests of aerospace composites. Various passive SHM and active SHM methods are discussed, including fiber-optic measurements, pitch-catch piezo measurements, electrochemical impedance spectroscopy (ECIS), etc.

Another major section of this chapter was devoted to the use of electrical resistance method to monitor in-service degradation and fatigue of CFRP composites. This approach is specific to CFRP composites because their carbon fibers have electrical conductivity which is imparted to the overall composite through the fact that individual fibers embedded in the polymeric matrix make occasional contacts when bunched together in the composite material. Early tests have shown that as the CFRP material is loaded, its electrical resistance changes thus acting as a built-in indicator of microcracks, delamination, and other fatigue damage taking place in the CFRP composite. These initial concepts have been extended through a wide body of research, including measurement and mapping of both the electrical resistance and the electrical potential of the composite using different electrode patterns and current injection methods.

The last section of this chapter covered various methods used in the monitoring of disbonds and delamination in composite patch repairs, composite adhesive joints, in nonconductive GFRP composites, etc. Results obtained with guided-wave methods as well as with dielectric measurements are presented and briefly discussed.

A major conclusion of this chapter is that monitoring of aerospace composites damage that may appear during normal operation is possible, but it has not received as much attention as the monitoring of impact events and subsequent damage that was reported in a previous chapter of this book. However, this situation may change in the future as major airlines acquiring and entering into service new composite-intensive aircraft such as Boeing 787 Dreamliner and Airbus A350 XWB. Hence, the need for fundamental and applied research into developing SHM methods for monitoring the in-service degradation and fatigue of aerospace composites is likely to increase significantly. It is apparent that sustained research programs for developing such methods and technologies should be put into place such that the fruits of discovery are made available before dramatic events happen into practice.

# References

[1] Read, I. J.; Foote, P. D. (2001) "Sea and Flight Trials of Optical Fibre Bragg Grating Strain Sensing Systems", *Smart Materials and Structures*, **10**(5), 1085–1094, Oct. 2001, http://dx.doi.org/10.1088/0964-1726/10/5/325.

[2] Murayama, H.; Kageyama, K.; Naruse, H.; Shimada, A.; Uzawa, K. (2003) "Application of Fiber-Optic Distributed Sensors to Health Monitoring for Full-Scale Composite Structures", *Journal of Intelligent Material Systems and Structures*, **14**(1), 3–13, Jan. 2003, http://dx.doi.org/10.1177/104538903032738.

[3] Wang, G.; Pran, K.; Sagvolden, G.; Havsgard, G. B.; Jensen, A. E.; Johnson, G. A.; Vohra, S. T. (2001) "Ship Hull Structure Monitoring Using Fibre Optic Sensors", *Smart Materials and Structures*, **10**(3), 472–478, Jun. 2001, http://dx.doi.org/10.1088/0964-1726/10/3/308.

[4] Kang, H. K.; Park, J. S.; Kang, D. H.; Kim, C. U.; Hong, C. S.; Kim, C. G. (2002) "Strain Monitoring of a Filament Wound Composite Tank Using Fiber Bragg Grating Sensors", *Smart Materials and Structures*, **11**(6), 848–853, Dec. 2002, http://dx.doi.org/10.1088/0964-1726/11/6/304.

[5] Mizutani, T.; Takeda, N.; Takeya, H. (2006) "On-Board Strain Measurement of a Cryogenic Composite Tank Mounted on a Reusable Rocket Using FBG Sensors", *Structural Health Monitoring-an International Journal*, **5**(3), 205–214, Sep. 2006, http://dx.doi.org/10.1177/1475921706058016.

[6] Bergman, A.; Ben-Simon, U.; Schwartzberg, A.; Shemesh, N.Y.; Glam, B.; Burvin, J.; Kressel, I.; Yehoshua, T.; Tur, M. (2014) "Evaluation of a UAV Composite Wing Spar Repair Using an Embedded Optical Fiber Rayleigh Back-Scattering Distributed Strain Sensing", *EWSH 2014 7th European Workshop on Structural Health Monitoring*, La Cite, Nantes, France, 2014, p. 0444.

[7] Kesavan, A.; John, S.; Herszberg, I. (2008) "Strain-Based Structural Health Monitoring of Complex Composite Structures", *Structural Health Monitoring—an International Journal*, **7**(3), 203–213, Sep. 2008, http://dx.doi.org/10.1177/1475921708090559.

[8] Kesavan, A.; John, S.; Herszberg, I. (2008) "Structural Health Monitoring of Composite Structures Using Artificial Intelligence Protocols", *Journal of Intelligent Material Systems and Structures*, **19**(1), 63–72, 2008, http://dx.doi.org/10.1177/1045389x06073688.

[9] Kesavan, A.; John, S.; Li, H.; Herszberg, I. (2009) "A Comparative Study of Statistical Outlier Analysis and ANN Methods for Damage Detection and Assessment in Composite Structures", *Journal of Intelligent Material Systems and Structures*, **21**(3), 337–347, 2009, http://dx.doi.org/10.1177/1045389x09343027.

[10] Kamath, G. M.; Sundaram, R.; Gupta, N.; Rao, M. S. (2010) "Damage Studies in Composite Structures for Structural Health Monitoring Using Strain Sensors", *Structural Health Monitoring—an International Journal*, **9**(6), 497–512, Nov. 2010, http://dx.doi.org/10.1177/1475921710365391.

[11] Takeda, S.; Okabe, Y.; Takeda, N. (2008) "Monitoring of Delamination Growth in CFRP Laminates Using Chirped FBG Sensors", *Journal of Intelligent Material Systems and Structures*, **19**(4), 437–444, Apr. 2008, http://dx.doi.org/10.1177/1045389x06074085.

[12] Guemes, J. A.; Menendez, J. M.; Frovel, M.; Fernandez, I.; Pintado, J. M. (2001) "Experimental Analysis of Buckling in Aircraft Skin Panels by Fibre Optic Sensors", *Smart Materials and Structures*, **10**(3), 490–496, Jun. 2001, http://dx.doi.org/10.1088/0964-1726/10/3/310.

[13] Sliversides, I.; Maslouhi, A.; LaPlante, G. (2013) "Acoustic Emission Monitoring of Interlaminar Delamination Onset in Carbon Fibre Composites", *Structural Health Monitoring—An International Journal*, **12**(2), 126–140, 2013, http://dx.doi.org/10.1177/1475921712469994.

[14] Koh, J. I.; Bang, H. J.; Kim, C. G.; Hong, C. S. (2005) "Simultaneous Measurement of Strain and Damage Signal of Composite Structures Using a Fiber Bragg Grating Sensor", *Smart Materials & Structures*, **14**(4), 658–663, Aug. 2005, http://dx.doi.org/10.1088/0964-1726/14/4/024.

[15] Okabe, Y.; Yashiro, S.; Kosaka, T.; Takeda, N. (2000) "Detection of Transverse Cracks in CFRP Composites Using Embedded Fiber Bragg Grating Sensors", *Smart Materials and Structures*, **9**(6), 832–838, Dec. 2000, http://dx.doi.org/10.1088/0964-1726/9/6/313.

[16] Okabe, Y.; Tanaka, N.; Takeda, N. (2002) "Effect of Fiber Coating on Crack Detection in Carbon Fiber Reinforced Plastic Composites Using Fiber Bragg Grating Sensors", *Smart Materials and Structures*, **11**(6), 892–898, Dec. 2002, http://dx.doi.org/10.1088/0964-1726/11/6/310.

[17] Mizutani, T.; Okabe, Y.; Takeda, N. (2003) "Quantitative Evaluation of Transverse Cracks in Carbon Fiber Reinforced Plastic Quasi-Isotropic Laminates with Embedded Small-Diameter Fiber Bragg Grating Sensors", *Smart Materials and Structures*, **12**(6), 898−903, Dec. 2003, http://dx.doi.org/10.1088/0964-1726/12/6/006.

[18] Larrosa, C.; Lonkar, K.; Chang, F.-K. (2014) "*In Situ* Damage Classification for Composite Laminates Using Gaussian Discriminant Analysis", *Structural Health Monitoring—An International Journal*, **13**(2), 190−204, 2014, http://dx.doi.org/10.1177/1475921713517288.

[19] Peng, T.; Saxena, A.; Goebel, K.; Xiang, Y.; Lu, Y. (2014) "Integrated Experimental and Numerical Investigation for Fatigue Damage Diagnosis in Composite Plates", *Structural Health Monitoring—An International Journal*, **13**(5), 537−547, 2014, http://dx.doi.org/10.1177/1475921714532992.

[20] Taylor, S. G.; Park, G.; Farinholt, K. M.; Todd, M. D. (2013) "Fatigue Crack Detection Performance Comparison in a Composite Wind Turbine Rotor Blade", *Structural Health Monitoring—An International Journal*, **12**(3), 252−262, 2013, http://dx.doi.org/10.1177/1475921712471414.

[21] Reifsnider, K. L.; Fazzino, P.; Majumdar, P. K.; Xing, L. (2009) "Material State Changes as a Basis for Prognosis in Aeronautical Structures", *Aeronautical Journal*, **113**(1150), 789−798, Dec. 2009.

[22] Chung, D. D. L. (2001) "Structural Health Monitoring by Electrical Resistance Measurement", *Smart Materials & Structures*, **10**(4), 624−636, Aug. 2001, http://dx.doi.org/10.1088/0964-1726/10/4/305.

[23] Todoroki, A. (2001) "The Effect of Number of Electrodes and Diagnostic Tool for Monitoring the Delamination of CFRP Laminates by Changes in Electrical Resistance", *Composites Science and Technology*, **61**(13), 1871−1880, 2001, http://dx.doi.org/10.1016/s0266-3538(01)00088-4.

[24] Todoroki, A.; Omagari, K.; Shimamura, Y.; Kobayashi, H. (2006) "Matrix Crack Detection of CFRP Using Electrical Resistance Change with Integrated Surface Probes", *Composites Science and Technology*, **66**(11−12), 1539−1545, Sep. 2006, http://dx.doi.org/10.1016/j.compscitech.2005.11.029.

[25] Ngabonziza, Y.; Ergun, H.; Kuznetsova, R.; Li, J.; Liaw, B. M.; Delale, F.; Chung, J. H. (2010) "An Experimental Study of Self-Diagnosis of Interlaminar Damage in Carbon-Fiber Composites", *Journal of Intelligent Material Systems and Structures*, **21**(3), 233−242, 2010, http://dx.doi.org/10.1177/1045389x09347019.

[26] Takahashi, K.; Park, J. S.; Hahn, H. T. (2010) "An Addressable Conducting Network for Autonomic Structural Health Management of Composite Structures", *Smart Materials & Structures*, **19**(10), Oct. 2010, http://dx.doi.org/10.1088/0964-1726/19/10/105023.

[27] Kostopoulos, V.; Vavouliotis, A.; Karapappas, P.; Tsotra, P.; Paipetis, A. (2009) "Damage Monitoring of Carbon Fiber Reinforced Laminates Using Resistance Measurements. Improving Sensitivity Using Carbon Nanotube Doped Epoxy Matrix System", *Journal of Intelligent Material Systems and Structures*, **20**(9), 1025−1034, 2009, http://dx.doi.org/10.1177/1045389x08099993.

[28] Fernberg, P.; Nilsson, G.; Joffe, R. (2009) "Piezoresistive Performance of Long-Fiber Composites with Carbon Nanotube Doped Matrix", *Journal of Intelligent Material Systems and Structures*, **20**(9), 1017−1023, Jun. 2009, http://dx.doi.org/10.1177/1045389x08097387.

[29] Matsuzaki, R.; Todoroki, A. (2006) "Wireless Detection of Internal Delamination Cracks in CFRP Laminates Using Oscillating Frequency Changes", *Composites Science and Technology*, **66**(3−4), 407−416, Mar. 2006, http://dx.doi.org/10.1016/j.compscitech.2005.07.016.

[30] Matsuzaki, R.; Todoroki, A.; Takahashi, K. (2008) "Time-Synchronized Wireless Strain and Damage Measurements at Multiple Locations in CFRP Laminate Using Oscillating Frequency Changes and Spectral Analysis", *Smart Materials & Structures*, **17**(5), Oct. 2008, http://dx.doi.org/10.1088/0964-1726/17/5/055001.

[31] Rose, J. L.; Rajana, K. M.; Hansch, K. T. (1995) "Ultrasonic Guided Waves for NDE of Adhesively Bonded Structures", *J. Adhesion*, **50**, 71−82, 1995.

[32] Rose, J.L.; Rajana, K.M.; Barnisher, J.N., "Guided Waves for Composite Patch Repair of Aging Aircraft", presented at the *QNDE 15* 1995.

[33] Ihn, J.B. "Built-in Diagnostics for Monitoring Fatigue Crack Growth in Aircraft Structures," PhD Dissertation, Dept. Aeronautics and Astronautics, Stanford Univ., 2003.

[34] Ihn, J.B.; Chang, F.-K. (2002) "Multi-crack Growth Monitoring at Riveted Lap Joints Using Piezoelectric Patches," *SPIE Vol. 4702*, San Diego, CA, 2002, pp. 29−40.

[35] Matt, H.; Bartoli, I.; Lanza di Scalea, F. (2005) "Ultrasonic Guided Wave Monitoring of Composite Wing Skin-To-Spar Bonded Joints in Aerospace Structures", *Journal of the Acoustical Society of America*, **118**(4), 2240, 2005, http://dx.doi.org/10.1121/1.2033574.

[36] Lanza di Scalea, F.; Rizzo, P.; Marzani, A. (2004) "Propagation of Ultrasonic Guided Waves in Lap-Shear Adhesive Joints: Case of Incident A0 Lamb Wave", *Journal of the Acoustical Society of America*, **115**(1), 146–156, 2004.

[37] Nassr, A. A.; El-Dakhakhni, W. W. (2009) "Non-Destructive Evaluation of Laminated Composite Plates Using Dielectrometry Sensors", *Smart Materials & Structures*, **18**(5), May 2009, http://dx.doi.org/10.1088/0964-1726/18/5/055014.

# Summary and Conclusions

## 11.1 OVERVIEW

This book has presented a review of the principal means and methods for structural health monitoring (SHM) of aerospace composite structures.

The first chapter of this book presented an overview of the problem. On the one hand, it recognized that composite materials have known an increasing acceptance into aerospace construction over an evolutionary period that spans more than four decades. Starting with initial use in secondary structures, composites have slowly penetrated into primary structures and gained more and more percentage participation into the airframe. Eventually, the aerospace community has come to see aerospace constructions dominated by composites, such as Boeing 787 Dreamliner and Airbus A350 XWB in which composites have 80% participation by volume (i.e., 50% participation by weight). On the other hand, the current widespread use of composite materials in commercial and military aircraft construction opens new avenues for study in terms of in-service performance, nondestructive evaluation (NDE), and structural health monitoring (SHM). This is indeed so because the damage and failure modes of composite structures are significantly more complicated and

**435**

diverse than those of metallic airframes. Hence, the rest of the book is dedicated to understanding these intricate phenomena and identifying sensors and methods by which they can be monitored in service through the SHM process.

## 11.2 COMPOSITES BEHAVIOR AND RESPONSE

The next three chapters dealt with the analysis of composite behavior and response. Chapter 2 was dedicated to the discussion of fundamental aspect of composite materials. Anisotropic elasticity due to the fibrous nature of the composite laminate was presented and the stiffness and compliance matrix, as well as the stress–strain relations, were discussed. The equations of motion in terms of stresses were introduced. This was followed by the equations of motion in terms of displacements The properties of the unidirectional lamina were discussed next and some formulae for the estimation of lamina elastic properties from the properties of the constituent fiber and matrix were presented in principal axes. Subsequently, expressions for the rotation of the stiffness and compliance properties of the lamina through an arbitrary angle about the $z$-axis were derived, for both 2D (plane-stress) and fully 3D formulations.

Chapter 3 was dedicated to the study of composites vibration, which may be used for SHM applications. It was shown that the axial and flexural vibrations of anisotropic composite plates are fully coupled, which is substantially different from the case of isotropic metallic plates in which the axial and flexural vibrations are decoupled and can be studied independently. The chapter started with the definition of displacements and stress resultants in terms of the motion and loading of a neutral mid surface of the composite plate. Subsequently, the equations of motion in terms of stress resultant were derived and the strains were expressed in terms of mid-surface strains and curvatures. The relation between stress resultants and mid-surface strains and curvatures was expressed through the use of the ABD matrices. Eventually, the vibration equations for an anisotropic laminated composite plate were derived in terms of displacements, mass distribution, and the ABD matrix components. Subsequently, the same approach was used for the case of an isotropic plate to allow direct comparison with the axial and flexural plate vibration formulations existing in literature.

Chapter 4 was dedicated to the study of wave propagation in thin-wall composite structures. Attention was focused on the propagation of ultrasonic guided waves in laminated composites which is an essential element of several SHM techniques discussed in subsequent chapters. The chapter started with an overview of the state of the art in modeling guided-wave propagation in laminated composites. It was found that several methods and methodologies exist for finding the dispersion curves and modeshapes of guided-wave propagation in composite materials: global matrix method (GMM), transfer matrix method (TMM), stiffness matrix method (SMM), etc. The fact that several methods coexist indicates that no unique methodology has yet been attained in solving this complicated wave-propagation problem. Subsequently, the chapter discusses the common foundation of these analysis methods and then presents each of them in details using the same common notations such that the reader easily identifies similarities and differences between these methods. The common foundation of all these methods lies in the Christoffel equations describing the three fundamental modes of wave propagation in a bulk composite material and the corresponding three polarization vectors. Each fundamental mode of wave propagation in a bulk composite is associated with a specific value of the squared wave-propagation speed, which means that the waves of that particular mode can propagate either forward (positive speed) or backward (negative speed) proving that, in fact, there are six modes of wave propagation in a bulk composite, but these six modes can be grouped in pairs sharing the same absolute value of the ± wave speed and a common polarization vector. Subsequently, the Christoffel equations were adapted to the case of a composite lamina through the imposition of the generalized Snell's law, which yielded the Christoffel equations for a lamina. Thus, in a lamina, the wave propagation is the superposition of six partial waves which can be grouped in three pairs, each with a common ± wavenumber and corresponding polarization vectors. Superposition of these six partial waves gave the solution for guided-wave propagation in a composite lamina. Next, the guided-wave modes and corresponding propagation speeds were found by imposing the traction-free boundary conditions at the upper and lower faces of the lamina. Evaluation of these parameters at various frequencies and/or wavenumbers allowed us to determine the dispersion curves for a composite lamina of arbitrary properties.

The analysis of guided-wave propagation in a composite laminate was performed by assembling the wave propagation in the constitutive laminae and imposing as boundary conditions displacement and traction continuity at the interface between two adjacent laminae, as well as traction-free boundary conditions at the upper and lower faces of the laminate. From here on, the solution process may proceed differently depending on adopted methodology.

In the GMM approach, the partial wave participation factors in each layer are kept as unknowns and assigned appropriate locations in a global unknown vector with size equal to six times the number of layers. The imposition of boundary conditions at the interfaces between adjacent laminae as well as at the upper and lower faces of the laminate yielded a $6N \times 6N$ set of homogenous algebraic equations which depended nonlinearly on wavenumber and wavespeed (or wavenumber and frequency, or wavespeed and frequency). Nontrivial solution of the $6N \times 6N$ set of equations only existed at specific wavenumber–wavespeed (or wavenumber–frequency, or wavespeed–frequency) combinations, i.e., at the system nonlinear eigenvalues. For each eigenvalues, the corresponding $6N$-long eigenvector gave at once all the partial wave participation factors in all the laminate layers. The partial wave participation factors were used to reconstruct the guided-wave modeshape at that particular eigen combination of wavenumber–wavespeed (or wavenumber–frequency, or wavespeed–frequency). The GMM approach is numerically stable, but the size of the corresponding matrix and associate determinant that has to be solved for nonlinear eigenvalues is quite large for practical laminates.

In the TMM approach, a transfer methodology was set up at the interface between adjacent layers using the state vector formulation that includes the particle displacements and the relevant stresses. Hence, the unknowns in a layer were expressed in terms of the unknowns in the previous layer and the problem size never exceeded a $6 \times 6$ formulation. After this transfer process was applied to all the layers in the laminate, a $6 \times 6$ system of algebraic equations was obtained to relate the state vector in the first layer of the laminate to the state vector in the last layer of the laminate. Imposition of the traction-free boundary conditions at the upper and lower faces of the laminate generated a $3 \times 3$ system of homogeneous algebraic equations in the unknown displacements at one of the free faces of the laminate as well as an additional $3 \times 3$ system of algebraic equations that related them to the displacements at the other free face of the laminate. The $3 \times 3$ homogeneous system depended nonlinearly on wavenumber and wavespeed (or wavenumber and frequency, or wavespeed and frequency). Nontrivial solution of this $3 \times 3$ set of equations only existed at specific wavenumber–wavespeed (or wavenumber–frequency, or wavespeed–frequency) combinations, i.e., at the system nonlinear eigenvalues. For each eigenvalues, the corresponding 3-long eigenvector gave the unknown displacements at the free surface at one end of the laminate. These displacements together with the stress-free conditions allowed us to assemble the state vector at one end of the laminate and then propagate it through the whole laminate to reconstruct the partial wave participation factors and hence the guided-wave modeshape at that particular eigen combination of wavenumber–wavespeed (or wavenumber–frequency, or wavespeed–frequency). The TMM approach was found to be computationally efficient because only small size ($6 \times 6$ or $3 \times 3$) matrices have to be dealt with. However, the TMM approach was found to be numerically unstable in certain wavenumber–wavespeed ranges (or wavenumber–frequency, or wavespeed–frequency) due to the "small differences of large numbers" phenomenon. This cause resides in the fact that the state vector combines displacements with stresses, and since these quantities have different units they may have hugely different numerical values. Hence, it was found that the TMM approach works well at low frequencies but breaks down at higher frequencies.

The SMM approach tried to alleviate the numerical problems of the TMM approach while maintaining its advantages such as the small size of the intervening matrices. The SMM approach rearranged the layer transfer across a layer from a state vector formulation to a stress–displacement formulation. While the state vector formulation related the displacements and stresses at one face of the layer to those at the other face, the stress–displacement formulation related the stresses at both faces of the layer to the corresponding displacements using the layer stiffness matrix expression. This approach eliminated inherent units issue identified in the TMM method but complicated the laminate matrix assembling process by replacing the simple matrix transfer across the interface used in the TMM approach by a recurrent stiffness buildup process. The SMM approach was found to be numerically stable at high frequencies; however, at low frequencies, it was found to be numerically unpredictable giving spurious results.

## 11.3 DAMAGE AND FAILURE OF AEROSPACE COMPOSITES

Chapter 5 was dedicated to a review of damage and failure in aerospace composites. The chapter started with a discussion of basic failure mechanisms inherent in stress–strain curves of the constituent fiber and matrix and then presented an overview of the basic failure modes of a unidirectional fiber-reinforced composite. Subsequently, various situations were treated in more detail. The tension damage and failure of a unidirectional composite ply was discussed in details starting with the strain-controlled tension failure due to fiber fracture, continuing with statistical effects on unidirectional composite strength and failure and culminating with the nonlinear aspects embedded in the shear-lag load sharing between broken fibers and the fiber pullout phenomenon.

Tension damage and failure in a cross-ply composite laminate was discussed next. The ply discount method was discussed. This method explains why and how the plies with unfavorable orientation relative to loading direction fail first, although their failure is not complete, but rather starts with matrix cracking and loss of transversal and shear stiffness. Discussion then progressed to the progressive failure of a cross-ply laminate that is due to more elaborate local failure phenomena. The effect of interfacial stresses at the laminate edges and around cracks was presented and the effect of matrix cracking on interfacial stresses was discussed. The characteristic damage state (CDS) was discussed next starting with its definition and the analysis of damage modes that modify local stress distribution. The evolution of stiffness and its decrease with damage accumulation was discussed here as well.

The next section in the discussion of damage and failure of aerospace composites was dedicated to the analysis of fatigue damage. The discussion started with a presentation of fatigue of unidirectional composites and then evolved into the presentation of cross-ply composite fatigue. The long-term behavior of aerospace composites was discussed next and three damage regions were identified: Region I in which damage accumulated until a widespread CDS condition was achieved without any apparent degradation of overall composite stiffness and strength; Region II in which crack coupling and delamination happened throughout the composite over a prolonged period of time in which overall composite damage accumulated and some degradation of overall composite stiffness and strength started to be observed; and Region III in which accelerated overall damage of the composite became apparent and overall failure becomes imminent finalizing in end of life termination of composite load-bearing capacity.

The next section of Chapter 5 dealt with compression fatigue and failure in aerospace composites. The quite different and particular compression failure mode found in fibrous composites as opposed to the failure modes found in isotropic metals was discussed and the fiber microbuckling phenomenon was presented. The dramatic effect of delaminations on accelerating compression damage accumulation and fatigue in aerospace composites was discussed. The compression fatigue damage under combined tension−compression loading was mentioned.

A separate section of Chapter 5 was reserved to the discussion of other composite damage types such as fastener hole damage, impact damage, and damage specific to sandwich composites and adhesive joints. It was found that fastener hole damage is a complex phenomenon that combines several damage types that can evolve during extended fatigue loading. The impact damage, which may occur accidentally but may go unnoticed, was found to be particularly disconcerting. Barely visible impact damage (BVID) may leave such a small surface footprint that it is not picked up by visual inspection; nonetheless, the internal effects of a BVID event on an aerospace composite structure, in the form of internal delaminations and/or fiber breakage, may be dramatic. In the case of a composite sandwich, such damage may manifest in the composite skins, in honeycomb or foam core, or in the interface skin/core interface.

The final section of Chapter 5 covered a discussion about what could and/or should be detected by a permanently installed SHM system and what should be expected to be detected by the nondestructive inspection (NDI), nondestructive testing (NDT), and nondestructive evaluation (NDE) processes during composite fabrication and during scheduled maintenance events.

## 11.4 SENSORS FOR SHM OF AEROSPACE COMPOSITES

Chapters 6–8 treated the subject of sensors that could be used for SHM of aerospace composites. Chapter 6 dealt with piezoelectric sensors, in particular the surface-mounted piezoelectric wafer active sensors, a.k.a. PWAS. The discussion started with an introductory coverage of current embodiments of these active sensors, such as SMART Layer™ product. The advantages of PWAS transducers over other SHM sensing options are briefly reviewed. Next, Chapter 6 covered the construction and operational principles of the PWAS transducers and coupling between the PWAS transducer and the guided ultrasonic waves traveling in the monitored structure. Shear-lag solutions for 1D and 2D guided-wave analyses (straight crested and circular crested, respectively) were reviewed. The principle of tuning between PWAS transducers and various modes of the structural guided waves was presented. Both the general solution and the simplified solution resulting from ideal-bonding assumption were given. Theoretically derived formulae were compared with experimental measurements. Subsequent discussion covered the wave propagation in a structure instrumented with PWAS transducers. The pitch-catch and pulse-echo methods were presented. The importance of high-frequency excitation was illustrated. The use of PWAS transducers as passive sensors for impact and acoustic emission (AE) detection was also discussed. Next, the use of PWAS arrays for directional beamforming using the phased-array principles was presented. The discussion of a simple linear PWAS array was followed by the presentation of the generic 2D PWAS array that allows 360° coverage of the monitored thin-wall structure. The concept of embedded ultrasonic structural radar (EUSR) was reviewed and discussed. PWAS resonators were covered next. The analysis and experimental validation of 1D (linear) and 2D (circular) PWAS resonator was briefly reviewed. The analysis was next expanded to cover the case of constrained PWAS resonator, such as those attached to the monitored structure during the SHM process. This analysis was naturally extended into the presentation of the electromechanical (E/M) impedance method which is a high-frequency technique for measuring the local dynamic spectrum of a structure using only electrical measurements, i.e., an impedance analyzer connected to a PWAS intimately attached to the monitored structure. The frequency-dependent structural stiffness as seen by the PWAS transducer was analyzed in both 1D and 2D geometries. The use of the E/M impedance technique for damage detection was illustrated.

Chapter 7 covered fiber-optic sensors for SHM of aerospace composites. The chapter started with a cursory review of the major fiber-optic sensing principles: intensity modulation; polarization modulation; phased modulation; spectral modulation; scattering modulation; etc. Then, the general principles of fiber-optic technology, such as total internal reflection and single mode versus multimode fiber options, were presented. A discussion of interferometric fiber-optic sensors followed. Mach−Zehnder and Michelson interferometers were presented. The intrinsic Fabry−Perot (FB) sensors were discussed. The extrinsic Fabry−Perot interferometer (EFPI) and the transmission extrinsic Fabry−Perot interferometer (TEFPI) were mentioned, as well as the in-line fiber etalon (ILFE). Attention subsequently focused on fiber Bragg gratings (FBG) optical sensors, which have become the dominant fiber-optic sensing technology in SHM applications. The ability of having several FBG sensors mounted on the same optical fiber and separately interrogated is of great importance for reducing the cost and complexity of SHM systems. The fabrication of FBG sensors, the conditioning equipment used with FBG sensors, and the FBG demodulation at ultrasonic frequencies were discussed. The latter issue presents great interest for composites SHM using ultrasonic guided waves and FBG optical sensors. Several high-frequency FBG demodulation methods were reviewed: tunable FB filters, tunable narrow-linewidth laser sources, chirped FB filters, array waveguide grating (AWG), two-wave mixing photorefractive crystal (PRC), etc. Other aspects related to FBG optical sensors, such as FBG rosettes, long-gage FBG sensors, temperature compensation of FBG sensors, were discussed next. Next, Chapter 7 covered other optical sensor types with good potential for application in composite SHM system. Intensity modulated fiber-optic sensors, including intensity-based optical fiber (IBOF) sensors, were discussed. The distributed optical fiber sensing principles based on optical time domain reflectometry (OTDR) were presented. Both Brillouin OTDR and Rayleigh backscatter OTDR were discussed. The possibility of measuring quasi-static and vibrational strain at any position along a plain optical fiber without the need for FBG engraving was emphasized. Other fiber-optic measurement methods discussed in Chapter 7 covered fiber-optic temperature laser radar (FTR), triboluminescence optical sensors, and polarimetric optical sensors.

Chapter 8 covered other sensors that could be used in aerospace composites SHM. Of particular interest were the conventional resistance strain gages and the electrical property sensors. The resistance-strain-gage sensing principles were briefly reviewed. The strain-gage instrumentation, both cabled and wireless, was discussed. The specifics of aerospace strain-gage technology were presented. Several examples of strain-gage usage in aerospace composites application were given. The electrical-property sensing principles used in conjunction with aerospace composites were presented. It was found that the electrical resistance measurements have found extensive applications in the SHM of carbon-fiber-reinforced polymer (CFRP) aerospace composites because the CFRP composites have a mild conductivity that changes significantly with the appearance of internal damage. The fabrication of surface-mounted and embedded electrodes for measuring the electrical properties of aerospace composites was discussed. The measuring of electrical resistance and that of electrical potential drop were comparatively discussed. Wireless sensing for remote electrical resistance monitoring of aerospace composites was briefly reviewed. The possibility of special-built composites for resistance-based self-sensing was also mentioned. Frequency domain methods for electrical SHM of aerospace composites were discussed next. The use of electrochemical impedance spectroscopy for detection of damage growth in aerospace composites was briefly reviewed. Electromagnetic methods that could be used with both conductive and nonconductive composite structures were also mentioned.

## 11.5  MONITORING OF IMPACT DAMAGE INITIATION AND GROWTH IN AEROSPACE COMPOSITES

Chapters 9 and 10 discussed methods for monitoring damage initiation and growth in aerospace composites. Chapter 9 focused on the detection of impact and acoustic emission (AE) and on monitoring of impact damage intensity and growth in aerospace composites. It was shown that this area of research has received extensive attention because of the drastic possible effects of undetected BVID events. For impact monitoring, it was found that the passive sensing diagnostics (PSD) approach can tell if an impact has taken place, locate the impact position, and even estimate the impact intensity. A simple triangulation example was presented to clarify the methodological approach. To obtain estimations of impact location and intensity, two major approaches were identified: (i) model-based approach and (ii) data-driven approach. In the model-based category, both the structural model approach and the system impact detection (ID) approach were found to have been used. A hybrid approach in which structural model and system ID approaches were combined was also found. An additional aspect in support of the PSD methodology is the use of directional sensors: both FBG rosettes and macro-fiber composite (MFC) rosettes were found to be reported. Another aspect of the PSD methodology is that AE events could also be detected with the same sensor network installation. It was also found that the simultaneous measurement of both AE and impact waves is possible, because the AE axial-wave signal, that travels at higher wavespeeds, arrives first, whereas the impact flexural-wave signal, which travels at lower wavespeeds, arrives later. Thus separation of AE and impact waves is possible in spite of the fact that the AE-related signals are much weaker than the impact-related signals.

The next part of Chapter 9 was dedicated to the discussion of active sensing diagnostics (ASD) methodology in which the transducers installed on the composite structure are used to send interrogative wave signals that interact with the damaged area and produce scatter waves that are picked up by same and/or other transducers. This methodology, which has many similarities to acousto-ultrasonics methodology, has been used with piezo transmitters and piezo receivers, as well as with piezo transmitters and fiber-optic receivers. In the latter case, the instrumentation complexity is greatly reduced by the fact that several FBG sensors can be mounted on a single optical fiber which thus requires only a single connection to the demodulation equipment. Data-driven ASD approaches and guided-wave tomography studies were also discussed in this section. PWAS pulse-echo crack detection in a composite beam was illustrated. Phased-array and directional transducer aspects of the ASD methodology were mentioned.

Other methods for impact detection discussed in Chapter 9 include direct methods for impact damage detection, strain mapping method for damage detection, vibration SHM methods, frequency transfer methods, and local area sensing with the E/M impedance spectroscopy method. The direct methods for impact damage detection are based on a network of optical fiber being embedded in the composite and being slightly damaged by the impact event. The strain mapping methods for damage detection rely on the possibility of redistribution of the strain under load and require strain measurement before and after an impact event under the same loading conditions. The vibration SHM methods rely on detecting the changes in the vibration characteristics (frequency, damping, modeshapes) of an aerospace composite structure after an impact event. Several vibration-based approaches were found to have been tried: model-based vibration SHM, model-free vibration SHM, statistics-based vibration SHM, nonlinear vibration SHM, combined vibration wave-propagation SHM, etc. However, it was generally concluded that the damage-related changes in vibration characteristics are not detectable in the low-frequency vibration modes and that one has to investigate high-frequency vibration modes local to the damage region in order to detect significant changes that could be related to the damage event. Such a high-frequency local vibration approach is offered by the E/M impedance spectroscopy method that produces the direct measurement of the local vibration spectrum in the tens and hundreds of kHz range through a permanently attached PWAS transducer connected to an impedance analyzer.

Also presented in Chapter 9 are electrical and electromagnetic methods used for impact damage detection in aerospace composites. The detection of delaminations with the electrical resistance method in CFRP composites was found to have received extensive attention with sequential improvements being reported year after year. Electrical resistance change method for delamination detection in cross-ply composites, wireless electrical resistance delamination detection, electrical impedance tomography (EIT) method for delamination detection with the electrical resistance method, multiphysics simulation of the electrical resistance change method, delamination detection with the electrical potential method, electromagnetic damage detection in aerospace composites, hybrid electromagnetic SHM of aerospace composites, and self-sensing electrical resistance-based damage detection and localization are presented in this section.

A separate part of Chapter 9 was reserved for impact damage detection in sandwich composite structures. Both PSD and ASD methodologies were illustrated and discussed.

## 11.6 MONITORING OF FATIGUE DAMAGE INITIATION AND GROWTH IN AEROSPACE COMPOSITES

Chapter 10 covered the monitoring of fatigue damage of aerospace composites that may appear during normal in-service operation of the composite aircraft or spacecraft. Such damage may be due to operational loads and environmental factors and may result in a gradual degradation of composite properties rather than the sudden changes that may appear due to accidental events, like the impact damage discussed in the previous chapter.

Chapter 10 started with a discussion of passive SHM methods, e.g., monitoring strain, AE, operational loads, etc. that were used in damage monitoring. These passive methods were used to record what is actually happening with the composite structure in order to compare the actual load to the design loads or/and detect unexpected developments. For example, monitoring the loads applied to the composite structure during normal operation conditions allows to compare design loads with measured loads (both limit loads and fatigue spectra) in order to make adjustments to the operational life predictions and flight profiles. Another outcome of passive monitoring could be the identification of changes in the strain distribution pattern over the structure that might be indicative of partial local failures (e.g., microcracks and delaminations) that may affect the load paths in a fail-safe structural design. In addition, AE monitoring could detect the moment when actual local failures happen by recording the waves created by the elastic energy released by a "popping" crack. Examples reported in Chapter 10 include strain distribution monitoring in a composite patch repair, conventional strain-gage monitoring for delamination detection, combined fiber optics and conventional strain-gage monitoring for delamination detection, and delamination occurrence and growth monitoring with specialty FBG sensors. AE monitoring and simultaneous monitoring of strain and AE are also discussed in this section of Chapter 10.

The next major section of Chapter 10 dealt with monitoring the actual fatigue damage induced in the composite by repeated application of service loads. As already detailed in Chapter 5, fatigue damage in composites is substantially different from fatigue damage in metallic structures. Most results reported in this section deal with monitoring of matrix microcracking and subsequent delamination that appears during cyclic loading tests of aerospace composites. Various passive SHM and active SHM methods are discussed including fiber-optic measurements, pitch-catch piezo measurements, electrochemical impedance spectroscopy, etc. Examples discussed in this section of Chapter 10 include fiber-optic monitoring of transverse cracks in a cross-ply composite, pitch-catch guided-wave detection of fatigue microcracking and delamination, electrochemical impedance spectroscopy monitoring of composite damage development and growth.

Another major section of Chapter 10 was devoted to the use of the electrical resistance method to monitor in-service degradation and fatigue of CFRP composites. This approach was found to be specific to CFRP composites because their carbon fibers have electrical conductivity which is imparted to the overall composite through the fact that individual fibers embedded in the polymeric matrix make occasional contact when bunched together in the composite system. Early tests have shown that as the CFRP material is loaded, its electrical resistance changes thus acting as a built-in indicator of microcracks, delamination, and other fatigue damage taking place in the CFRP composite. These initial concepts have been extended through a wide body of research including measurement and mapping of both the electrical resistance and the electrical potential of the composite using different electrode patterns and current injection methods. Examples presented in this section of Chapter 10 include electrical resistance SHM of CFRP composites, electrical resistance SHM of carbon nanotube (CNT) doped glass-fiber-reinforced polymer (GFRP) composites, and wireless sensing of the electrical resistance in a CFRP composite.

The last section of this chapter covered various methods used in the monitoring of disbonds and delamination in composite patch repairs, composite adhesive joints, in nonconductive GFRP composites, etc. Results obtained with guided-wave methods as well as with dielectric measurements are presented and briefly discussed. Examples include disbond and delamination detection with conventional ultrasonic guided waves, monitoring of composite patch repairs, monitoring of composite adhesive joints, and dielectric SHM of delamination and water seepage in GFRP composites.

A major conclusion of this chapter was that monitoring of aerospace composites damage that may appear during normal operation is possible, but it has not received as much attention as the monitoring of impact events and subsequent damage that was reported in Chapter 9. However, this situation may change in the future as major airlines are acquiring and entering into service new composite-intensive aircraft such as Boeing 787 Dreamliner and Airbus A350 XWB. Hence, the need for fundamental and applied research into developing SHM methods for monitoring the in-service degradation and fatigue of aerospace composites is likely to increase significantly. It is apparent that sustained research programs for developing such methods and technologies should be put into place such that the fruits of discovery are made available before dramatic events happen into practice.

## 11.7  SUMMARY AND CONCLUSIONS

Overall, it can be said that, whereas NDE technology is a rather mature technology, the SHM methodology and related technologies for aerospace composites have only just started to emerge. Considerable further research is needed to mature the development of aerospace composites SHM sensors and methods in order to achieve viable practical implementation of this promising new technology.

The introduction of SHM systems and SHM methodology should contribute to lower the maintenance costs of composite structures for which deterministic damage events, types, and limit sizes are difficult to predict. SHM would also facilitate the introduction of new composite materials and development of new composite structures by reducing the uncertainty component of the aircraft design cycle.

# Index

Printed in the United States
By Bookmasters